ANALOG AND DIGITAL CIRCUITS

THEORY AND EXPERIMENTATION

ANALOG AND DIGITAL CIRCUITS

THEORY AND EXPERIMENTATION

Volume 1

MANSOUR ESLAMI
State University of New York at Stony Brook

ROBERT E. KRIEGER PUBLISHING COMPANY
MALABAR, FLORIDA

Original Edition 1987

Printed and Published by
ROBERT E. KRIEGER PUBLISHING COMPANY, INC.
KRIEGER DRIVE
MALABAR, FLORIDA 32950

Printed in the United States of America

Library of Congress Cataloging-in-Publication Data

Eslami, Mansour.
   Analog and digital circuits theory and experimentation.

   Includes index.
   1. Electronic circuits—Laboratory manuals.   2. Electronic circuits.
I. Title.
TK7818.E48 1987   621.3815'3   86-15260
ISBN 0-89874-959-X (Vol. 1)
ISBN 0-89874-983-2 (Vol. 2)

10 9 8 7 6 5 4 3

TO
MY PROFESSORS
AT THE UNIVERSITY OF WISCONSIN-MADISON

AND

TO
MY TEACHERS
AT THE HADAF HIGH SCHOOL IN TEHERAN

CONTENTS

PART ZERO

PRE-LABORATORY INTRODUCTION

PART ONE

FAMILIARIZATION WITH BASIC ELECTRONIC INSTRUMENTATION

PART TWO

FAMILIARIZATION WITH SOME BASIC CIRCUIT COMPONENTS AND THEOREMS--MAGNETIC MATERIAL AND HYSTERESIS

PART THREE

ACTIVE TWO-PORT NETWORKS WITH TRANSISTORS

PART FOUR

OPERATIONAL AMPLIFIERS

PART SIX

ANALOG CIRCUITS DESIGN WITH APPLICATIONS

PART SEVEN

ELECTRONIC CIRCUITS DESIGN WITH INTEGRATED COMPONENTS

PREFACE

This textbook of experimentation on analog and digital circuits is prepared for the use of all electrical engineering students, both as a guideline for their sequence of basic laboratory courses and as a reference textbook for their core courses in this field. The material in this textbook aims at covering the basic experimental background required of an electrical engineering student at the sophomore-junior level in any four-year college program. Although a substantial amount of theory has been presented herein, it is not intended as a substitute for the textbook of the supporting lecture course (or courses) on the relevant subject matter. The theory has been included to help students understand their laboratory work better and to make this textbook self-contained, thereby saving students' time. Also, because this textbook is intended to be used for *several courses*, one should not be immediately concerned with every detail of Part Zero on pre-laboratory introduction. The approach in presenting the subject matter at hand has always been to minimize repetition in order to save space and ultimately production cost. Thus, entry-level readers who may see in the first few pages of Part Zero a section on such a topic as Laplace transform, should not be discouraged and feel that the level of the textbook is beyond their background.

A statement of why this textbook was written or what the difference is between this textbook and others, is now in order. Many outstanding textbooks on basic circuit theory, electronics, and especially, on electrical engineering laboratory experimentation are available and are being extensively used. Some of the problems with these textbooks are the following:

1. They are primarily written for one set of specific equipment. Thus, they can only be used by those institutions that have the proper equipment.

2. Most of these textbooks lack a good theoretical explanation for each experiment. Thus, the students are forced to consult outside references to understand the nature of the experiment. This search is, of course, time-consuming and sometimes, especially in the early experiments, not fully appreciated by students.

3. Above all, these textbooks lack step-by-step instructions, needed by students to complete their work uniformly. Thus, the faculty in charge of an experimental course is obliged to produce additional instructions, which is costly for both students and institutions.

The following features, in our opinion, form the specific advantages of the present textbook.

A. This textbook has a versatile but coherent set of experiments that form the core of an electrical engineering program, with sufficient explanations of the theory behind each experiment and an overall description of how to use the needed equipment. Thus, this textbook will enrich the overall capabilities and/or experience of the student and will result in a much better future engineer. We do, of course, realize that no textbook can produce all the necessary information about each piece of equipment that is being used in every institution. This is why we decided to present a sufficient *generic* explanation of the most commonly used equipment in everyday laboratory work. We do expect that each instructor will provide sufficient excerpts from manufacturers' manuals of the specific equipment that they are using in a given institution. The students must become familiar with this approach as early in their career as possible.

B. In this textbook a step-by-step and uniform explanation of what the students are supposed to be doing to complete their experimentation is provided in italics, which are missing entities in most available textbooks. Otherwise, to provide this additional instruction is a burden for the faculty in charge of the course.

C.   This textbook aims at providing the minimum or basic knowledge that should be gained by an electrical engineering student in perhaps four to five two-credit experimental courses. Accumulation of all these experiments in one place eliminates the possible overlapping that exists between various such courses taught by different instructors. In our opinion, a loose set of handouts or several such sets, no matter how neatly these are packaged, is a disservice to students in the long run. This is especially true if there is a lack of proper communication between various instructors.

D.   This textbook has a reasonable and balanced set of homework and pre-laboratory assignments that complements the actual instructions. The charts that are provided also help students to save their most important results for easy future review. These charts, of course, should be used when the experiment is completed.

E.   We believe that this textbook will not soon become obsolete and that the inevitable modifications required by a course instructor are minimal, to make it adaptable for the needs of a particular institution. This textbook is written for average students, in order that the instructor may delete some parts if it is felt that such an assignment is not feasible for a specific group of students. Therefore due to comprehensiveness of these experiments, it is not expected that the instructor will need to augment the present text of any experiment with additional assignments that require much effort. We do, however, realize that there may be an experiment in a new area designed at a particular institution that some instructor may wish to add to this set; that is beyond our control, although we welcome such suggestions for our future revisions. We also do not expect that all the experiments will be performed in a given institution. Therefore, the experiments are designed so that each instructor can easily make changes in the order as well as the number of experiments to be performed. There are cases in which the instructor may ask students to come up with their own designs. Note that some experiments require more than one laboratory session. Thus, there are options in using this textbook. If the size of a given class is manageable, then the instructor may delegate more freedom to students. But it is wrong, in our opinion, to assume that all students know what they are supposed to be doing in a laboratory and, therefore, to run the course format free. This is why we have constantly tried to keep the text of this book in a tight, well-structured framework.

F.   Finally, the instructions given at the beginning of each experiment regarding the needed equipment are extremely valuable for anyone using this textbook. If something is missing, it can be located before the experiment starts. Also, the course administrators can update the stock well in advance.

As an educator, this author is the first to acknowledge that there is still much room for improvement in this textbook. Nevertheless, this is perhaps the first textbook of its kind prepared solely from the educational point of view and for the main reason of helping students, teaching assistants, technical staff, and the faculty in charge of the basic laboratory courses. Comments and suggestions are most welcome.

MANSOUR ESLAMI

Stony Brook, New York

ACKNOWLEDGMENT

This textbook is the outcome of teaching various electrical engineering laboratory courses as both a teaching assistant and a faculty member supervising several teaching assistants. There is no way to mention the names of many people who have contributed in one form or another to the development of the material, although every effort has been made to give credit where credit is due. The major contribution of this writer is the arrangement of the material. Most theoretical aspects of these experiments have been already well developed and documented in the literature.

I am very grateful to Professor F.G. Stremler of the University of Wisconsin-Madison, who allowed me to use some parts of his laboratory notes (ECE 300). Without this initial condition I would never have dared to attempt this project. Our idea has been to help engineering students to learn from the best material available, and Professor Stremler has generously permitted his valuable work to be shared with them.

Also, special thanks are due to many of my other professors in the department of electrical and computer engineering at the University of Wisconsin-Madison, especially Professors Tiedemann, Marleau, Nordman, Bernstein, and Reitan, whose words are all through this textbook.

I am particularly inspired by many outstanding textbooks written by Professor M.E. Van Valkenburg of the University of Illinois. I have never formally met him, but I consider myself one of his students.

Many constructive suggestions provided by my dear friend Dr. S.N. Navid, now with Signetics, are fondly appreciated.

Special credit is due to many of my former students, both at the University of Wisconsin-Madison and at the State University of New York at Stony Brook, who have contributed to the development of this textbook.

This textbook is designed and produced by this writer in collaboration with several former graduate students: Haidar Chamas, who painstakingly developed the original computer program to print out this textbook and typed a major part of it; Kye Y. Lim, my former Ph.D. student, who drew all the figures, typed and proofread entire Part Five and made major contributions in the development of the last three experiments of this part; Hong T. Jeon, another former Ph.D. student, who typed and proofread Part Six; and Anand Y. Desai who typed the rest of the book. I also thank Mrs. Florence Trace for typing major parts of the first draft of this book. Special thanks are due to Charlie McKenna, Bob Martin, Yu Liu, Lois Koh, and Frank Kost for their assistance in this project.

The editorial assistance provided by the staff of R.E. Krieger Publishing Company is greatly appreciated. The careful reading and many constructive suggestions given by Dr. Edwin Strother of the Florida Institute of Technology are also most appreciated.

Finally, it is a great pleasure and distinct honor for this writer to acknowledge the true quality of the education he received while attending Hadaf High School in Teheran, Iran, from 1960 to 1966. That period was the best time of my life, thanks to the many outstanding educators who inspired me to be where I am now and taught me to care about the basic education of young people.

MANSOUR ESLAMI

PART ZERO

PRE-LABORATORY INTRODUCTION

PART ZERO

PRE-LABORATORY INTRODUCTION

0-0. INTRODUCTION

In this part we will briefly study basic concepts of circuits theory, and subsequently some basic aspects of error analysis. It is not, however, our intention to teach these materials to those who have not seen them before. There is also a section on circuit design which teaches the beginners how to construct a circuit using necessary parts and accessories. Meanwhile, we will be going over some important subjects essential for any laboratory operation, such as: safety rules and regulations. It is suggested that the students carefully read this part as soon as possible.

It is also recommended that the student bring this textbook to laboratory along with a calculator, a small ruler, a French curve, a carbon paper, and a few pages of white paper. Enjoy the laboratory and

GOOD LUCK!!

The table of contents for the remaining part of this review is as follows.

0-9.3. SOME BASIC NOTIONS OF PROBABILITY

0-9.4. REMARKS

0-1. REFERENCES

[0]   T. Bernstein, "Effects of electricity and lightning on man and animals," J. of Forensix Sciences, vol. 18, no. 1, pp. 3-11, January 1973.

[1]   A. Budak, Circuit Theory Fundamentals and Applications. Englewood Cliffs, NJ: Prentice-Hall, 1978.

[2]   C.F. Dalziel and W.R. Lee, "Reevaluation of lethal electric currents," IEEE Trans. Industry and General Applications, vol. IGA-4, no. 5, pp. 467-476, Sep./Oct. 1968.

[3]   W.B. Kouwenhoven, "Human safety and electrical shock," Electrical Safety Practices, Monograph 112, Instrument Society of America, Pittsburg, PA, pp. 91-97, 1969.

[4]   W.R. Lee, "Deaths from electric shock," Proc. IEE (London), vol. 113, pp. 144-148, January 1966.

[5]   F.G. Stremler, ECE 300 Laboratory Notes. Madison, WI: The University of Wisconsin-Madison, 1978.

[6]   A.T. Tiedemann, ECE 210/211 Experiment Notes. Madison, WI: The University of Wisconsin-Madison, 1976.

[7]   B.D. Wedlock and J.K. Roberge, Electric Component and Measurements. Englewood Cliffs, NJ: Prentice-Hall, 1969.

[8]   S. Wolf, Guide to Electronic Measurements and Laboratory Practice. Englewood Cliffs, NJ: Prentice-Hall, 1983.

0-2. BASIC CONCEPTS OF CIRCUITS THEORY

Although the basic concepts of electricity are well known to all readers, we present the following ideas to refresh their memory.

0-2A. Current

The concept of *current* in a conductor is considered to be the result of movement of electrons. In the MKS international system of units (SI) the unit of current is the *ampere* (1 mA$=10^{-3}$ A). One ampere is equal to one coulomb/second, where the *coulomb* is the unit of charge. If the current in a conductor remains constant for all time, then we call that a direct current or DC for short.

0-2B. Electric Field

The force exerted by one charge upon another represents the *electric field* of the first charge acting upon the second charge.

0-2C. Potential

The work done per unit of charge to move one charge in the field of another charge of the same sign is called *potential difference* or potential for short.

0-2D. Volt

The potential in the international unit of measurements (MKS) is measured in volts. One joule/coulomb equals one volt.

0-2E. Ground

The potential difference between two points is always the difference between two separate levels. Each of which is compared with a reference level called *ground.*

0-2F. Electrical Ground

The *electrical ground* means an electrical *sink* that accepts and/or supplies charge without changing its electrical characteristics. In most electrical testing we are content with the chassis *(local)* ground as opposed to the Earth *(global or universal)* ground. The potential of any ground is always considered to be zero. Further comments on electrical ground is given in Section 0-4.4.

0-2G. Resistance

Another relevant quantity is the *resistance* of a conductor which is a function of its geometry and substance. The resistance in the MKS international system of units is measured in *ohms*. The resistance R equals (resistivity ρ of the conductor multiplied by its length l) /(cross section A of the conductor). All relevant quantities are measured according to the international units of measurements.

0-2H. Power

Mainly *power* is used to indicate the measure of loss of energy as current flows in a conductor. Power in MKS is measured by *watt.* One watt equals (energy/second) or equivalently (joules/coulombs)$\times$(coulombs/second). This indeed is voltage times current.

The most fundamental laws of circuits theory are Ohm's law and Kirchhoff's laws. These laws are the most important *tools* of analyzing any network and must be completely understood by anyone who is trying to use the present textbook.

0-2I. Ohm's Law

The relationship between the voltage (V) applied across and the current (I) developed through any linear conductor (R) is as follows:
$$V = R\,I.$$
Ohm's Law is also valid for a linear capacitor or a linear inductor provided its complex resistance or impedance is used instead of R. In this case both V and I are also in the complex form.

0-2J. Generalized Ohm's Law

The relationship between the applied voltage (v) across any two-terminal device and the current (i) through this device is as follows. v=r(i). In general, this r is a nonlinear function of i and the expression is mostly written in the time domain.

0-2K. Kirchhoff's Voltage Law (KVL)

The algebraic sum of voltages of all elements of any mesh (or loop) is zero. By algebraic we mean the positive direction of the loop is chosen to be in agreement with at least one source. Sinks are always in the opposite direction.

0-2L. Kirchhoff's Current Law (KCL)

The algebraic sum of all currents into any node (or supernode) is zero. By algebraic we mean current coming into a node has a plus sign and current going out has a negative sign.

THE ABOVE TWO LAWS ARE TRUE IN BOTH TIME-DOMAIN AND FREQUENCY-DOMAIN MEASUREMENTS.

0-2M. Mesh and *Node Equations*

Generally speaking, circuits may have several nodes and meshes (or loops). Techniques to analyze any network are merely repetitive applications of the above three fundamental laws, which results in several linear equations in terms of currents and/or voltages of various elements. Setting up these equations *correctly* changes a network problem to an algebraic problem of solving several linear equations of several unknowns.

Perhaps we must point out that in general any circuit may contain all sorts of dependent or independent elements. (*Dependent* elements are those elements whose electrical characteristics are functions of electrical characteristics of other circuit elements.) Therefore to solve for various quantities of interest we may have to use all kinds of integrodifferential equations depending on the type of elements involved. One easy way around this difficulty is to use the Laplace transform of various currents and voltages of the elements. This transformation provides us a uniform set of algebraic equations in terms of currents and voltages of these various elements and their impedances. Also by substituting for the Laplace variable $s=\sigma+j\omega$, $(j=\sqrt{-1})$, we can easily develop the complex values of these quantities useful for phasor analysis. This subject of Laplace transforms may be new to some of our readers. Thus we briefly review this subject.

0-2N. Laplace Transforms

Given a signal $f(t)$ such that $|f(t)| < m\,e^{\gamma t}$ as $t\to\infty$, where $m > 0$ and γ is a real number (this means $f(t)$ is at most of order of an exponential), then for this function we introduce a Laplace transform $F(s)$, which is a function of complex-variable s, such that

$$\mathcal{L}\{f(t),\, t\to s\} \triangleq F(s) = \int_0^\infty f(t)\,e^{-st}\,dt \,, \ (t\to s: \text{time–domain to frequency–domain mapping}).$$

Here $s=\sigma+j\omega$, $j=\sqrt{-1}$, is also called the complex-frequency variable. We also require that $\mathrm{Re}(s)=\sigma > 0$ to be sufficiently large to guarantee the finiteness of the above integral. In the following we present some useful properties of the Laplace transform operator without any proof.

1. $\mathcal{L}\{a_1 f_1(t) + a_2 f_2(t),\, t\to s\} = a_1 F(s) + a_2 F(s),$

   here $F_i(s) \triangleq \mathcal{L}\{f_i(t),\, t\to s\}$, $i=1,2$, and a_1 , a_2 are constants.

2. $\mathcal{L}\{df(t)/dt,\, t\to s\} = sF(s) - f(0^+)$, provided $f(t)$ is differentiable and continuous at $t=0$. Otherwise it means derivative from the right. If $f(0^+) = 0$, then $\mathcal{L}\{df(t)/dt,\, t\to s\} = sF(s)$. *We call s a differential operator.*

3.   Under appropriate assumptions we have
   $\mathcal{L}\{d^n f(t)/dt^n ,\, t\to s\} = s^n F(s) - s^{n-1} f(0^+) - \cdots - s^0 f^{(n-1)}(0^+),$
   where $f^{(n-1)}(0^+) = d^{n-1}f(t) / dt^{n-1}$ evaluated at $t=0^+$.

4. $\mathcal{L}\{\int_0^t f(\tau)\,d\tau\} = F(s)/s.$

   We call $1/s$ an integral operator.

The above rules are sufficient for descriptions of most of our analysis. One question which may arise is how do we go from the frequency-domain function (i.e., F(s)) back to the time-domain function f(t). The answer is given by the following complex integration formula in which c is the abscissa of convergence. Here c is a real constant and is greater than the real parts of all singular points of F(s).

$$f(t) = \frac{1}{2\pi j} \int_{c-j\infty}^{c+j\infty} F(s)\, ds \ .$$

But we seldom use this expression. Instead we urge readers to avoid possible mistakes, by using the following table of Laplace transformations sufficient for most of our applications.

$$\mathcal{L}\{f(t),\ t\to s\} = F(s)$$

$$\mathcal{L}\{\text{unit impulse } \delta(t),\ t\to s\} = 1$$

$$\mathcal{L}\{\text{unit step } u(t),\ t\to s\} = 1/s$$

$$\mathcal{L}\{t^n,\ t\to s\} = n!/s^{n+1}$$

$$\mathcal{L}\{\exp(-at),\ t\to s\} = 1/(s+a)$$

$$\mathcal{L}\{t\exp(-at),\ t\to s\} = \mathcal{L}\{-d\exp(-at)/da,\ t\to s\} = 1/(s+a)^2$$

$$\mathcal{L}\{\sin\omega t,\ t\to s\} = \omega/(s^2+\omega^2)$$

$$\mathcal{L}\{\cos\omega t,\ t\to s\} = s/(s^2+\omega^2)$$

$$\mathcal{L}\{(at-1+e^{-at})/a^2,\ t\to s\} = 1/s^2(s+a)$$

$$\mathcal{L}\{\exp(-at)\,f(t),\ t\to s\} = F(s+a)$$

$$\mathcal{L}\{f(t-a)\,u(t-a),\ t\to s\} = \exp(-as)\,F(s)$$

$$\mathcal{L}\{t\,f(t),\ t\to s\} = -d\,F(s)/ds$$

$$\mathcal{L}\{f(t)/t,\ t\to s\} = \int_s^\infty F(s)ds$$

$$\mathcal{L}\{f(t/a),\ t\to s\} = a\,F(as) \ .$$

Theorem 1: If lim f(t) as t→∞ exists, then this limit equals lim sF(s) as s→0.
Theorem 2: If lim f(t) as t→0⁺ exists, then this limit equals lim sF(s) as s→∞.

0-20. The Laplace Transform of Inductors

In basic circuit theory we learned that the relation between current through and voltage across a linear inductor is $v_L(t) = Ldi_L/dt$. In Laplace transform representation we have

$$\mathcal{L}\{v_L(t),\ t\to s\} \triangleq V_L(s) = \mathcal{L}\{Ldi_L/dt,\ t\to s\}$$

$$= [Ls\, I_L(s) - Li(0^+)].$$

Letting $i(0^+) = 0$, we have

$$V_L(s) = (Ls)\, I_L(s).$$

The term (Ls) corresponds to the complex resistance or impedance of the inductor in the sense of the Ohm's law. If we let s=jω, we get the phasor diagram of Fig. 1, for the current and voltage relation of an inductor. We also recall that L is called *inductance* and is measured in *henrys* (H). L is a function of square of the number of turns of inductor L. In practice there is no ideal coil with only inductance L and zero resistance. In fact each practical coil must be modeled as a circuit with an ideal inductor L and a finite resistor r in series. Then a quantity, Q_L, defined as $Q_L = \omega L/r$ is a measure of the quality of the inductor for a given frequency $\omega = 2\pi f$. Given a frequency ω, higher Q_L means, of course, lesser r.

0-2P. Transformer

Another passive network element that is often used is the transformer. Suppose we put two coils L_1 and L_2 of turn ratios N_1 and N_2, respectively, next to each other as in Fig. 2. Then the expression for voltages of each coil becomes different than that of the previous ones. Here we no longer have just $v_1 = L_1 di_1/dt$ and/or $v_2 = L_2 di_2/dt$. Each of these voltage expressions now involves the contributions of current through the other coil. The reason is the fact that the magnetic flux ϕ_1 or ϕ_2 produced in each coil is a function of both currents i_1 and i_2. The new expressions for each voltage function becomes as follows:

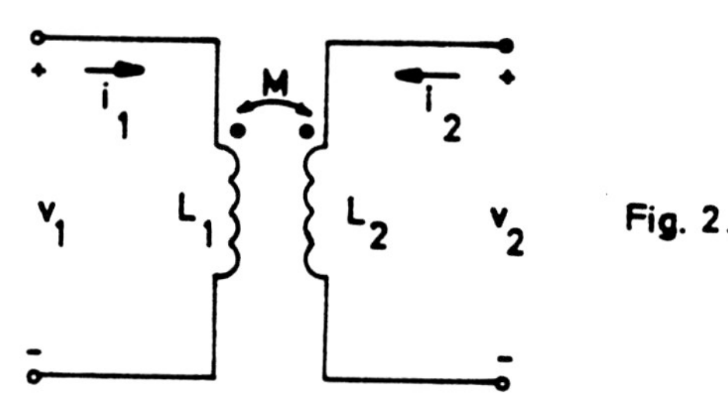

Fig. 2.

$$v_1 = L_1\, di_1/dt \pm M di_2/dt\,,$$

$$v_2 = L_2\, di_2/dt \pm M di_1/dt\,.$$

This phenomenon is called transformation of the flux from one coil to another and this device is known as a *transformer*. The coil on the left is the primary and that on the right is the secondary. The new parameter M (measured in Henrys) is called *mutual inductance* and is equal to $M = k\sqrt{L_1 L_2}$, where k is called the coupling coefficient. The plus and minus signs are chosen according to the following rule: If both currents enter (or leave from) the dotted end of the coils then we choose the plus sign. Otherwise if one enters and the other leaves the dotted end, then we choose the minus sign.

0-2P1. Equivalent Model of a Transformer

Suppose the set of equations that describe v_1 and v_2 in the above is given for the plus sign. Then the question arises as to what is the equivalent model of this set of equations. Although we will present a comprehensive discussion of two-port networks later, we mention that this set of equations represents a two-port network and one may easily derive following circuits that have exactly the same representation at each corresponding port.

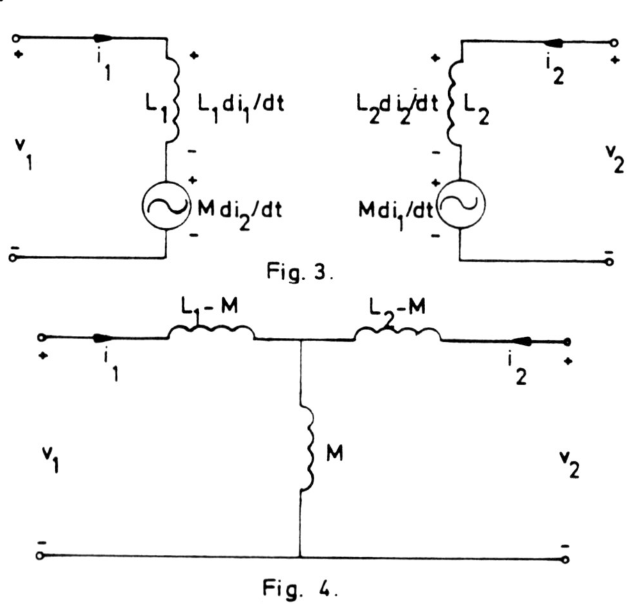

Fig. 3.

Fig. 4.

$$\begin{cases} v_1 = L_1 \dfrac{di_1}{dt} + M\dfrac{di_2}{dt}\,, \\[2mm] v_2 = L_2 \dfrac{di_2}{dt} + M\dfrac{di_1}{dt}\,. \end{cases}$$

Or

$$\begin{cases} v_1 = (L_1{-}M) \dfrac{di_1}{dt} + M\dfrac{d}{dt}(i_1 + i_2)\,, \\[2mm] v_2 = (L_2{-}M) \dfrac{di_2}{dt} + M\dfrac{d}{dt}(i_1 + i_2)\,. \end{cases}$$

Of course, in certain cases (L_1-M) or (L_2-M) may become negative which has no physical meaning. Nevertheless the above figure is an alternative to our transformer network for the purpose of analysis.

0-2P2. Ideal Transformer

This is the case in which we consider the transformer to be a voltage or current source transformer with no loss in either side. Here N_2/N_1 is written n, Fig. 5. Then we have $v_2 = nv_1$ and $i_2 = \frac{1}{n}i_1$. Thus $Z_2 = n^2 Z_1$, where Z_2 is the impedance of the secondary and equals $V_2(s)/I_2(s)$. When $Z_2(s)$ is brought to the primary it will be in parallel with the primary coil $Z_1(s) = V_1(s)/I_1(s)$. This case has many applications specially when we wish to analyze a circuit involving transformers. The approach is very simple. We study first, the effects of each element in either primary or secondary circuits on the electrical characteristics of secondary or primary, respectively. The rest is a simple circuit problem.

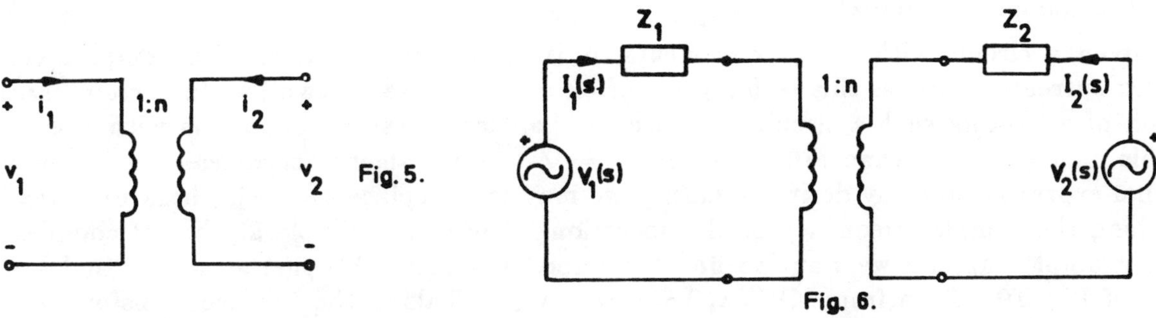

Fig. 5.

Fig. 6.

To clear this idea we present the circuit of Fig. 6, as an example, asking for $I_1(s)$ and $I_2(s)$. It is easy to show that the primary circuit is that of Fig. 7, with

$$I_1(s) = [V_1(s) - V_2(s)/n] / [Z_1(s) + Z_2(s)/n^2] .$$

Similarly, the secondary circuit becomes that of Fig. 8, where

$$I_2(s) = [V_2(s) - nV_1(s)] / [Z_2(s) + n^2 Z_1(s)] .$$

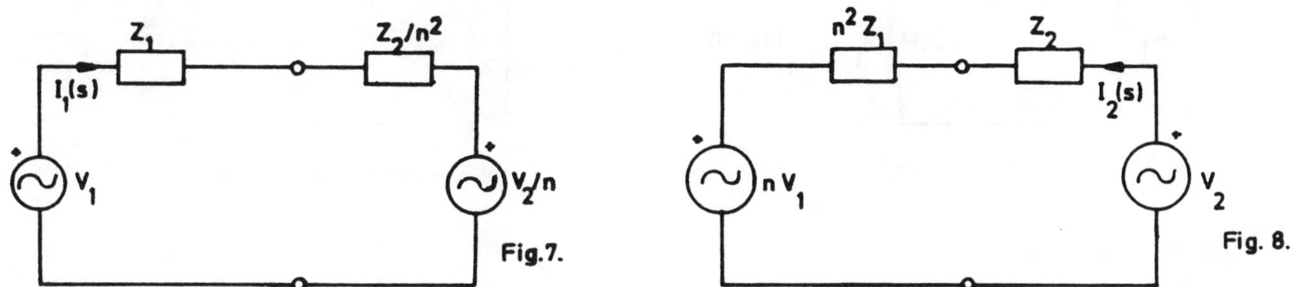

Fig. 7.

Fig. 8.

0-2Q. The Laplace Transform of Capacitors

Similarly, we have seen that for a linear capacitor the relation between current through and voltage across the capacitor is $i_c(t) = C\, dv_c(t)/dt$. Taking the Laplace transform of this equation results in $V_c(s) = I_c(s)/(sC)$. Here $1/(sC)$ is the impedance of the capacitor in the sense of the Ohm's law. C is called the *capacitance* and is measured in *farads (F)*. Here $C = \epsilon A/d$, where ϵ is called the dielectric constant of the medium, A is the cross section or common area of the plates and d is the distance between two plates whose capacitance is C. The phasor diagram of the above equation is shown in Fig. 9.

0-2R. *Resonant Frequency*

When a capacitor C and an inductor L are in series, the angular frequency $\omega_o = 1/\sqrt{LC}$ is called the *resonant* frequency of the LC network.

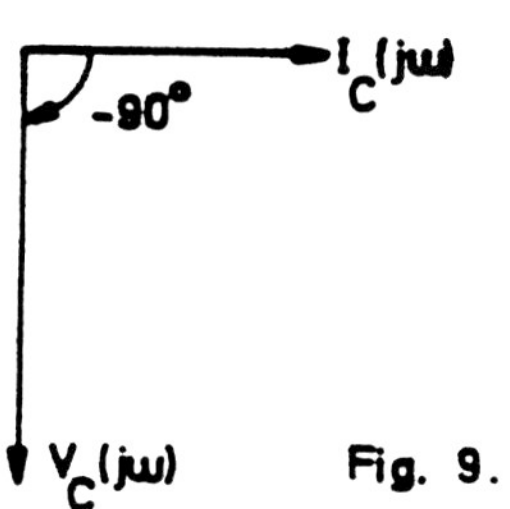

0-2S. *The Complex Analysis*

Given a circuit with complex elements, it is always easy to deal with complex voltages and complex currents *values* as this technique is often used in power networks. In *certain cases* one easy method of analyzing such a circuit is to use the Laplace transform equations with s as the Laplace variable. This way the chances that we miss $j = \sqrt{-1}$ in our algebra becomes less. When we develop the final expression for any desired quantity we may then replace $s = \sigma + j\omega$, by a numerical quantity describing the complex frequency of the operation. The rest is simple algebra of complex numbers. As an example suppose we want to find the impedance $Z(s) = V(s)/I(s)$ at $s=2+j$ and $s = j2$ of the circuit of Fig. 10. Then from KVL: $v(t) = v_R(t) + v_L(t)$. Taking the Laplace transform of this equation results in $V(s) = V_R(s) + V_L(s)$. From Ohm's law and the Laplace transform of $V_L(s)$ we get $V(s) = 2I(s) + sI(s) = (2+s)I(s)$, $i(0)=0$. Here $Z(s) = V(s)/I(s) = 2+s$. For $s=2+j$ and $s=j2$, we get $Z(2+j) = 2+2+j = 4+j$, and $Z(j2) = 2+j2$. The phasor diagrams of any of these impedances can be drawn similarly to that of Fig. 11. The real part is always drawn on the x-axis and the imaginary part on the y-axis, provided that the sign of each component is observed.

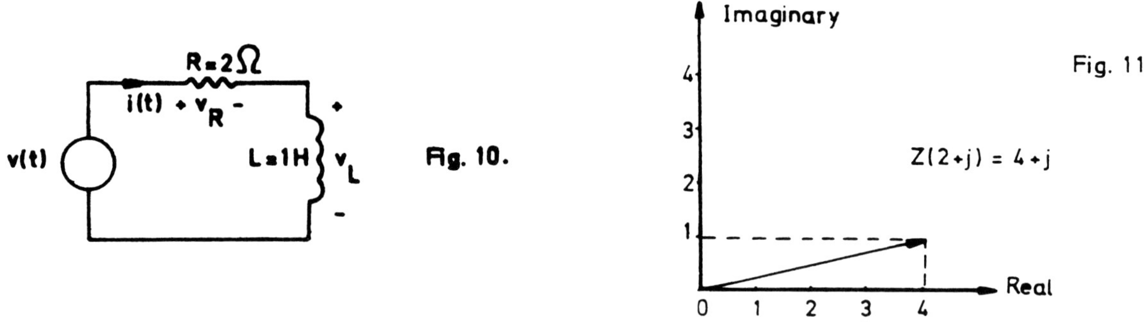

0-2T. *Bode Diagram*

Any physically realizable transfer function $T(s) = N(s)/D(s)$ represents a ratio of two arbitrary but desired signals of a given system or circuit. The numerator $N(s)$ and the denominator $D(s)$ of this $T(s)$ are rational polynomials of s with real constant coefficients and $\deg N(s) \leq \deg D(s)$. The numerator and denominator correspond, respectively, to the output and the input of this particular transfer function. By substituting $s = \sigma + j\omega$ in $T(s)$ we have a complex number that can be expressed as $T(\sigma + j\omega) = |T(\sigma+j\omega)| \angle T(\sigma+j\omega)$. Bode diagrams refer to the drawing of $20 \log|T(\sigma+j\omega)|$ and the phase function versus ω (or versus $\log\omega$). These graphs become very handy when $T(s)$ is factorized in terms of the first-or the second-order polynomials in s. To give a better picture of these particular graphs we present the following three basic Bode diagrams for $s=j\omega$ that are extensively used for a more complicated T(s).

1. $T_1(s) = \dfrac{1}{1+sT}$.

     $A_1(j\omega) = 20 \log\left|\dfrac{1}{\sqrt{1 + \omega^2 T^2}}\right| = -10 \log (1 + \omega^2 T^2)$.

     for $\omega << \dfrac{1}{T}$, we have $A_1(j\omega) = -10 \log 1 = 0$.

     for $\omega >> \dfrac{1}{T}$, we have $A_1(j\omega) = -20 \log \omega T$.

     for $\omega = \dfrac{1}{T}$, we have $A_1(j\omega) = -10 \log 2 = -3.01$.

Here the Bode diagrams become the lower half of Fig. 12 and Fig. 13.

2. $T_2(s) = 1 + sT$.

     For this case we get the upper graphs in Fig. 12, and Fig. 13.

3. $T(s) = \omega_n^2 / (s^2 + 2\xi\omega_n s + \omega_n^2)$.

     For this case we will have the standard Bode diagrams of Fig. 14.

4.   T=K is a constant.

     The Bode diagram for this case is a straight line parallel to the frequency axis at the distance of 20 log K.

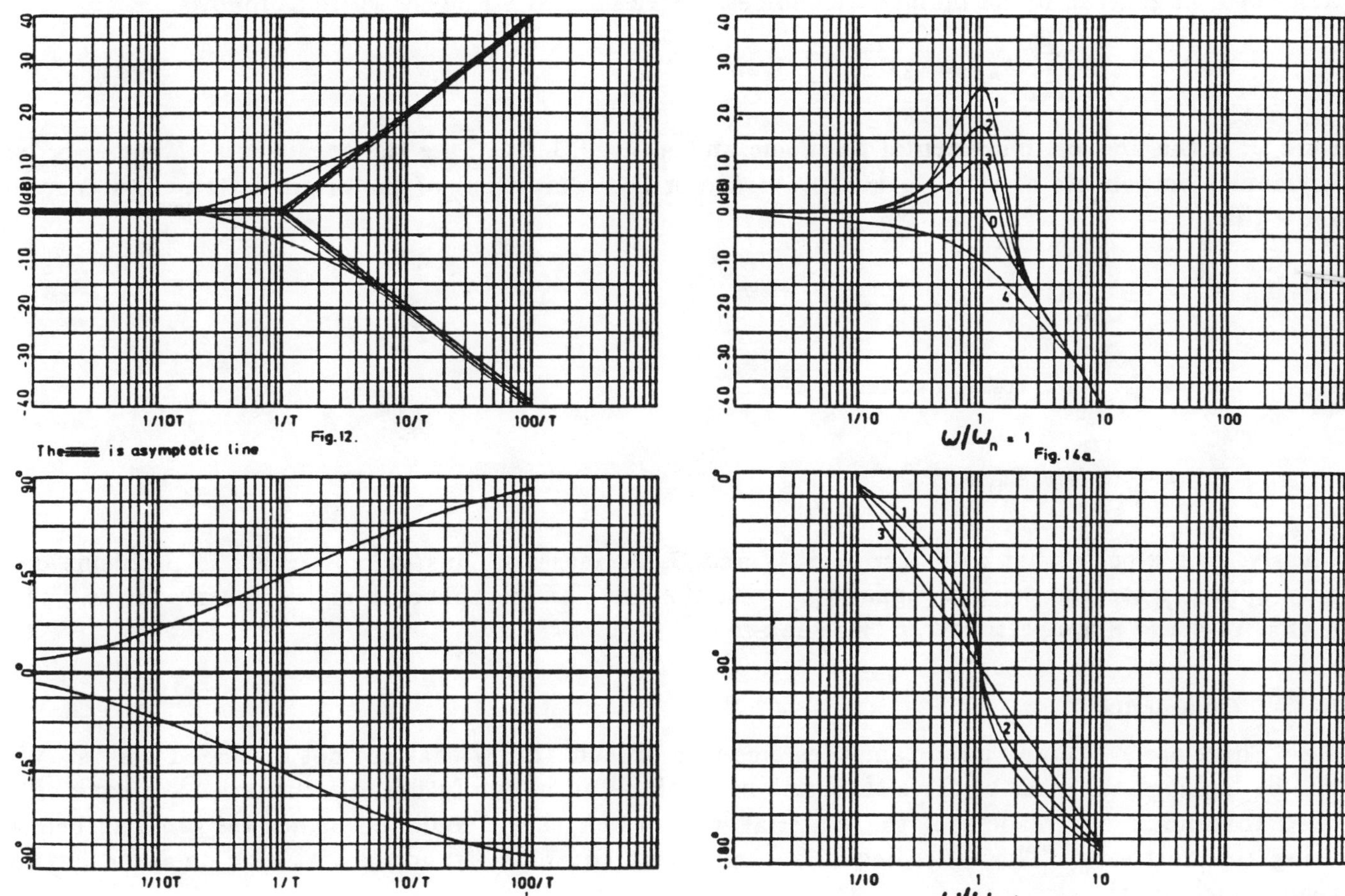

Remark(Fig. 14): Line #0 is asympt. & curves #1 to #4 correspond to increasing ζ.

0-2U. Decibels (dB)

This is a unit of measuring the ratio of two power levels. If P_r is the reference power level then dB is defined as

$$(\text{dB}) \triangleq 10 \log (P / P_r) .$$

If the reference power level P_r is one watt, then the power P is expressed in "decibels above one watt". Similarly if P_r is one milliwatt, P is expressed in "decibels above one milliwatt (dBm)", [5]. Substituting for P and P_r in terms of V and V_r, we get

$$(\text{dB}) = 10 \log (V^2 / R) / (V_r^2 / R_r),$$
$$= 20 \log (V / V_r) - 10 \log (R / R_r) .$$

Letting $R = R_r$ and $V_r = 1$ volt, we have

$$(\text{dB}) = 20 \log V .$$

Sometimes the above equation is taken as a definition of the decibel. This is valied only as long as the proper normalizing factors are considered whenever any comparison is made to the power calculations using $(\text{dB}) = \log (P / P_r)$, [5].

0-2V. Fourier Series

It is a well known fact that any well-behaved periodic function $v(t)$ of period T can be described, in general, by an infinite trigonometric series called a Fourier series as follows:

$$v(t) = A_0 + \sum_{n=1}^{\infty} A_n \cos n\omega_o t + \sum_{n=1}^{\infty} B_n \sin n\omega_o t .$$

Here ω_o is the first or fundamental harmonic and $A_1, A_2, \ldots, B_1, B_2, \ldots,$ are the coefficients of higher-order harmonic components of $v(t)$. A_0 is the average or DC-component of the original $v(t)$. It is easy to show that

$$A_0 = \frac{1}{T} \int_{-T/2}^{T/2} v(t) \, dt ,$$

$$A_n = \frac{2}{T} \int_{-T/2}^{T/2} v(t) \cos n\omega_o \, dt , \text{ and}$$

$$B_n = \frac{2}{T} \int_{-T/2}^{T/2} v(t) \sin n\omega_o \, dt .$$

Clearly, one expects that as n increases A_n and B_n decreases. Thus in most realistic problems one may approximate the original signal $v(t)$ with the first few components of its Fourier series. The EXPs. VI-4 and 8 are pertinent to this subject.

0-2W. Comments

There are still many important theorems from circuit theory that are not discussed herein. We will be looking into some of these theorems in much detail in our future experiments. We have only tried to remind our readers of the material with which they should be *somewhat* familiar before using this textbook. Those needing more help must consult a standard textbook on basic circuit theory such as [1].

0-3.   ACCESSORIES FOR CIRCUIT DESIGN
0-3.1.   INTRODUCTION

So far, we have only studied the mathematical symbols of a circuit that are being used in the analysis. In actuality to construct and to test any circuit we need usual network elements and certain accessories. One must obtain the following brief list of accessories and must become acquainted with them prior to the work in the laboratory. Even though the purpose of this textbook, and perhaps the supporting course, is to teach laboratory skills (and this can best be done in the laboratory), we believe the following introduction to the circuit testing accessories will become very useful. Please read carefully.

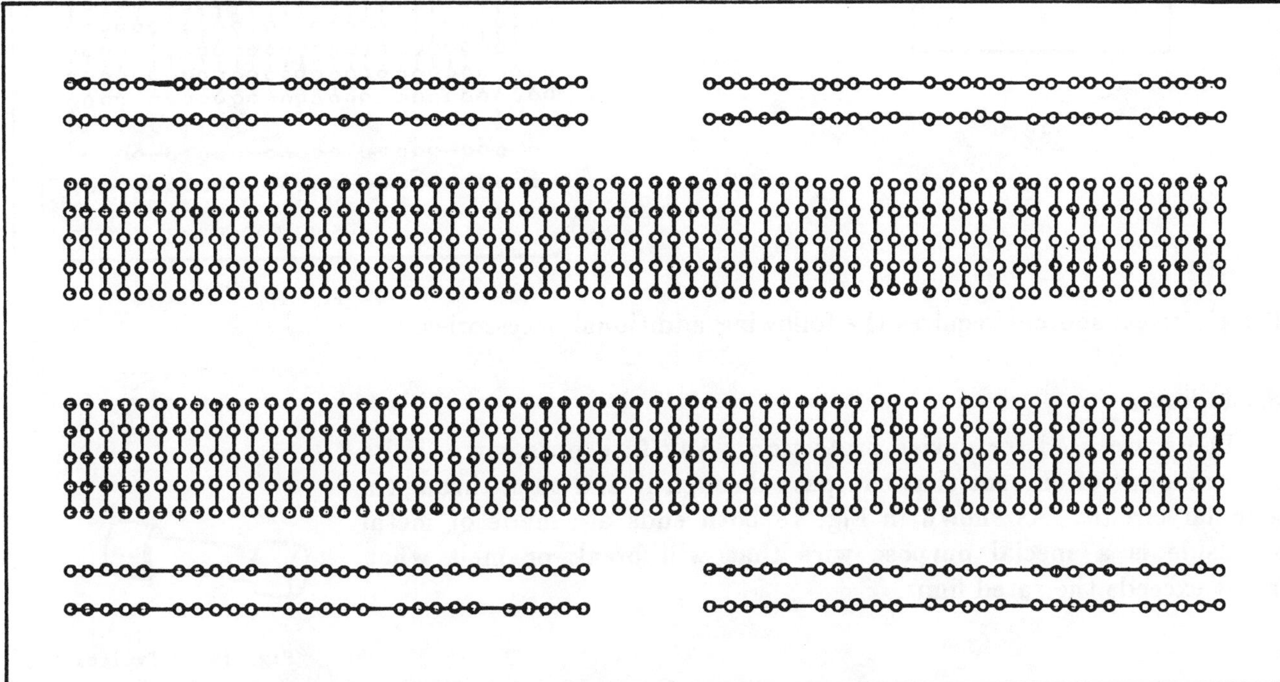

Fig. 15.   Schematic Diagram of a Typical Multiple Connection Bread Board.

This drawing is not in the same scale as any manufacturer's design.
It is, however, similar to what mostly are in the market.

Any circuit consisting of several elements can be assembled in several different ways. In the earlier days circuits were mostly built by *soldering* even for simple testing purposes. Sometimes, depending upon the application, *welding* was also used. But the easiest way, as far as we are concerned, is using a multiple connection bread board of a type that is built by several vendors. A schematic diagram of a typical such bread board is given in Fig. 15. The vertical and horizontal holes are internally connected as depicted in the picture. These boards make laying out a circuit component according to a particular diagram very easy. Certain manufacturers make several of these bread boards with the binding posts on one frame which makes it easier to lay out more complicated circuits.

As an example suppose we want to set up the circuit shown in Fig. 16 for testing. Fig. 17 shows how we lay out this circuit. The *art* is to use the least or the shortest amount of wiring between connections. *These circuit boards should always be neat and multi-colored to make debugging easy.*

Now, this is only the first phase of circuit testing. What will we do with the typical circuit layout of Fig. 17? The answer is we definitely need more accessories than just a multiple connection bread board or simple wiring. How to interface this typical circuit with measuring devices and/or

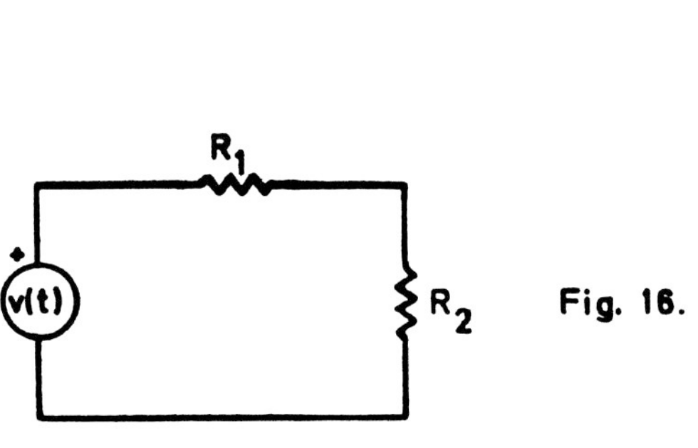

Fig. 16.

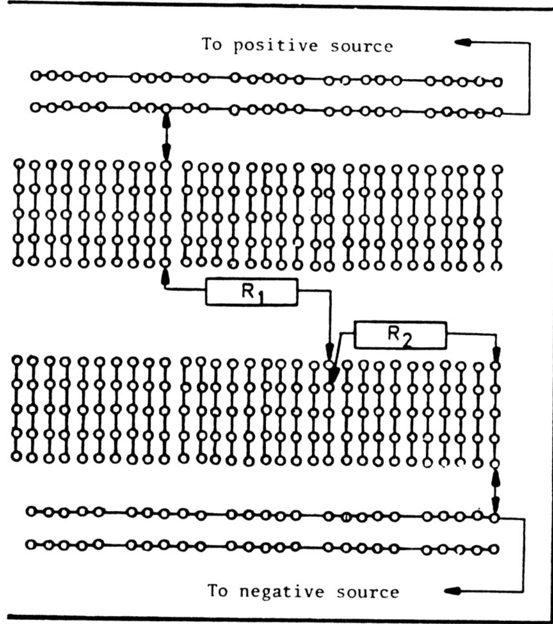

Fig. 17.

other electrical sources requires the following additional accessories.

0-3.1A. Fuses

These are used for device protection. Any fuse puts some limits on the amount of current that can pass through it and then through a particular circuit. As shown in Fig. 18 both ends are made of metal and inside is a special purpose wire that will break or melt when current exceeds the rated limit.

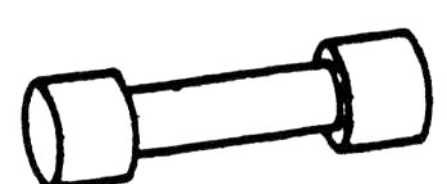

Fig. 18. A Typical Fuse.

0-3.1B. Wires and Cables

Wires are the bread and butter of any circuit designer and everyone has seen one kind of it or another, specially the *hookup wires* and *test probe wires* that are used in ordinary applications. Generally speaking, the latter has higher flexibility than the former. *Cables* are combinations of several shielded-conductors or wires in one place and under one outer shield. Some major types of cables are *shielded cables, multiple conductor* cables, and *coaxial cables.*

0-3.1C. Connectors

(a) General Purpose Test Clips and Binding Posts

Some general configurations of these test clips and binding posts are shown in Fig. 19. Usually, the binding posts are mounted next to the bread boards and these are either single or dual binding posts depending on their usage.

(b) Jumpers, Patch Cords, Test Leads

These are typical connectors used to connect the binding posts of the circuit test board to those of the measurement devices, *etc.* These are simply wires of various sizes that on either end have various types of terminals, plugs, or clips as described above.

0-3.1D. *Switches*

A switch is a device that is used to turn equipment on or off. There are several different switches in use and readers may consult any textbook on laboratory devices such as [8] for further information. A special class of switches is called relay switches. The main function of this type of switch is the same as any switch, but its mechanism is based on the electromagnetic field produced by a current carrying wire around a coil.

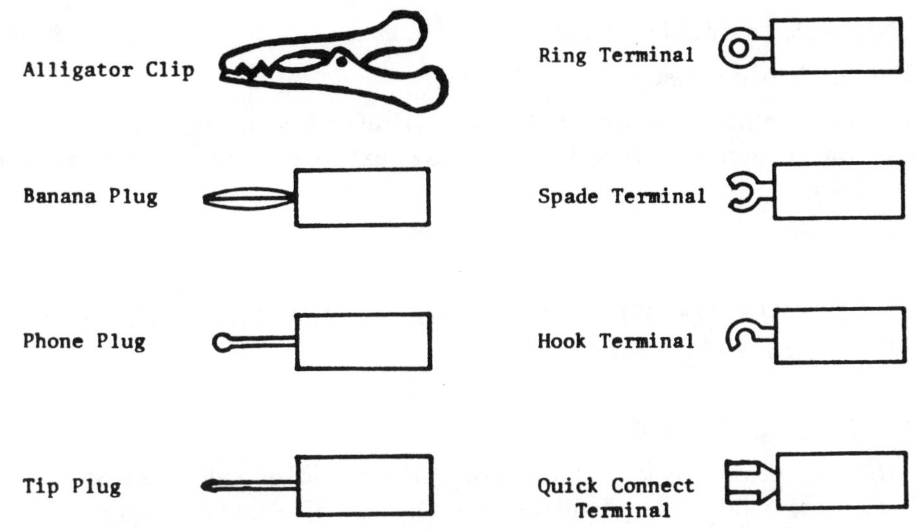

Fig. 19a. General Purpose Pluges, Clips, *etc*.

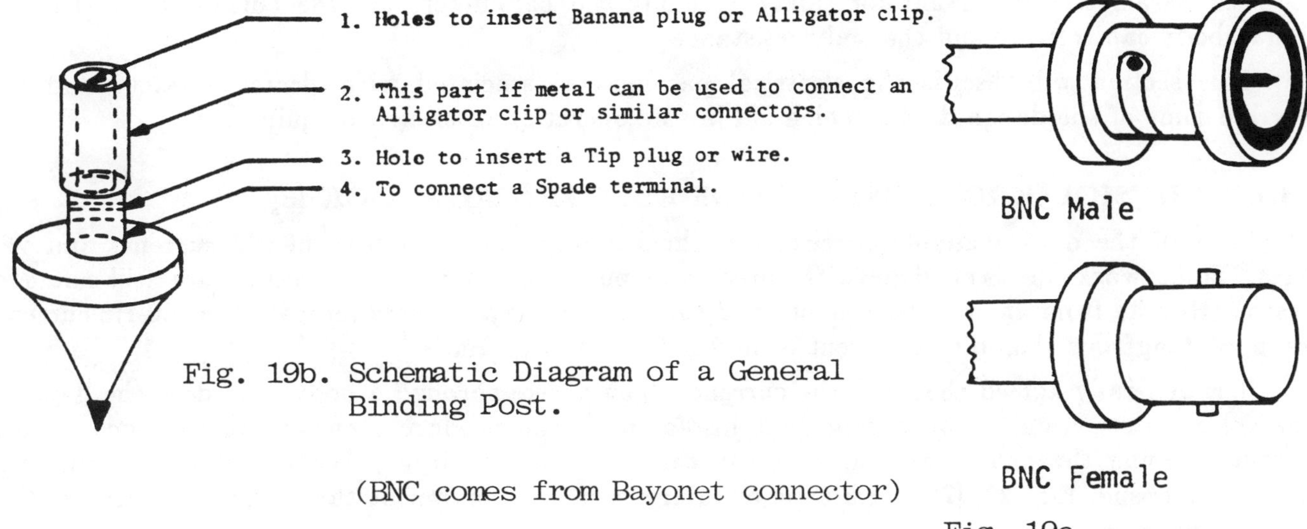

Fig. 19b. Schematic Diagram of a General Binding Post.

(BNC comes from Bayonet connector)

Fig. 19c.

0-3.2. SUGGESTED LIST OF ACCESSORIES

Before we close this section we make the assumption that any educational laboratory using this textbook is at least equipped with the following set of units and accessories for each group of two or three students. There is no biasing toward any particular vendor.

1. A dual power supply of range ± 20 volts.
2. A standard digital multimeter.
3. A standard multi-function or wave generator.
4. A dual oscilloscope.
5. One BNC male plug to alligator clips.
6. One BNC male plug both ends.
7. Four BNC male to double plug in.
8. Two BNC male (dual instrument adapter).
9. One multiple connection bread board. (Preferably more than one board which are mounted on a metallic board with easy external binding connection posts already mounted on.)
10. Six patch cords.

FOR THE SAKE OF TIME WE WILL CALL THE ABOVE SET THE ELECTRICAL EQUIPMENT PACKAGE OR EEP FOR SHORT.

0-4. ELECTRICAL SAFETY

Preliminary to the actual work in laboratory, the topic of safety must be addressed. Since much of the activity in this laboratory will involve the use of electrical equipment, consideration of electrical shock hazards is essential.

Strange as it may seem, most fatal electric shocks happen to people who should know better.

In designing electrical equipment or in studying electrical accidents, it is necessary to know the electrical parameters important for safety. Among the parameters are the current level that the human body can tolerate and the body resistance.

This section will discuss the electrical parameters associated with electrical safety and will describe some of the design details which aid in safe operation of electrical equipment.

0-4.1. PHYSIOLOGICAL EFFECTS OF ELECTRICAL SHOCK [5]

Most of the data available on electrical shock parameters deal with 60 Hz currents and voltages. Some work has been done with direct and pulse currents but the results are still not conclusive. Results from animal experiments and some human experiments indicate that 60 Hz currents are more dangerous than direct current or higher frequency currents.

It is generally agreed that it is the current which passes through a body that does the damage. The voltage in a circuit is only important insofar as it can produce a current in the body. Large currents passing through or around a person can cause serious injury because of the heating and burning of tissue, Fig. 20, [7]. This type of injury is more common at the higher voltages used by power companies.

Death at lower voltages, such as 120 to 240 V found around homes, can usually be attributed to one of three causes [5]:

0-4.1A. VENTRICULAR FIBRILLATION

0-4.1B. RESPIRATORY ARREST

0-4.1C. ASPHYXIA

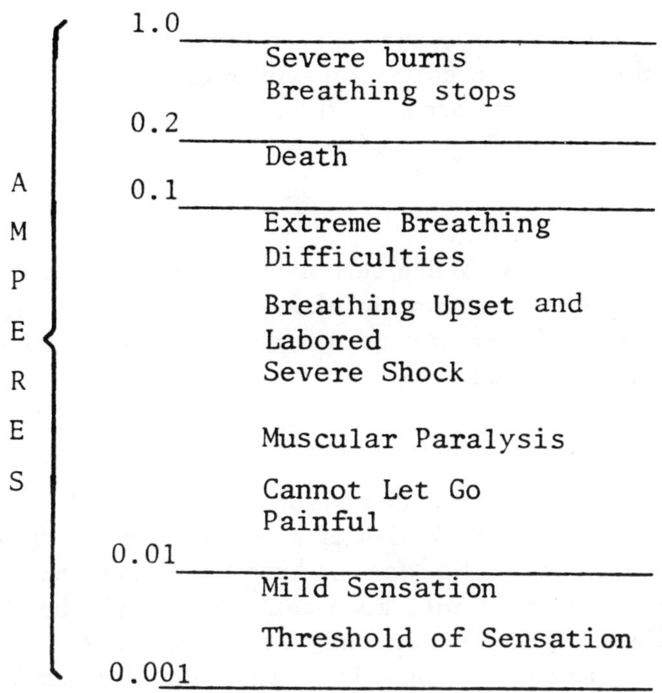

Fig. 20. The Physiological Effects of Electric Currents [7]. Reprinted by Permission from B.D. Wedlock and J.K. Roberge, Electronic Components and Measurements. Englewood Cliffs, NJ: Prentice-Hall,1969.

0-4.1A. *VENTRICULAR FIBRILLATION* [5]

Ventricular fibrillation is an uncoordinated asynchronous contraction of the ventricular muscle fibers of the heart in contrast to their normal coordinated and rhythmic contraction. The heart seems to quiver rather than to beat. This condition is caused by an electrical shock where the path of current is through the chest, such as between two arms or between an arm and a leg. Once a person goes into ventricular fibrillation, the only way to stop the fibrillation is to use a defibrillator which applies a pulse shock to the chest to restore the heart rhythm. Closed chest massage and artificial respiration may help until the victim can be defibrillated.

A typical electrocardiogram is shown in Fig. 21. The "P" wave is the beginning of the heart cycle, and indicates atrial contraction. The wave of excitation triggers the "QRS" complex which occurs as the ventricles contract as a unit and pump blood. When contraction is completed, the repolarization of the heart takes place during the "T" phase. Shocks during the "T" phase can cause the heart to go into ventricular fibrillation. To defibrillate the heart, it is necessary to employ a counter-shock which is large enough to depolarize all the muscle fibers of the heart causing all of them to contract simultaneously. When the heart relaxes after this shock, the next "P" wave will usually restart the heart cycle unless ventricular fibrillation has continued too long.

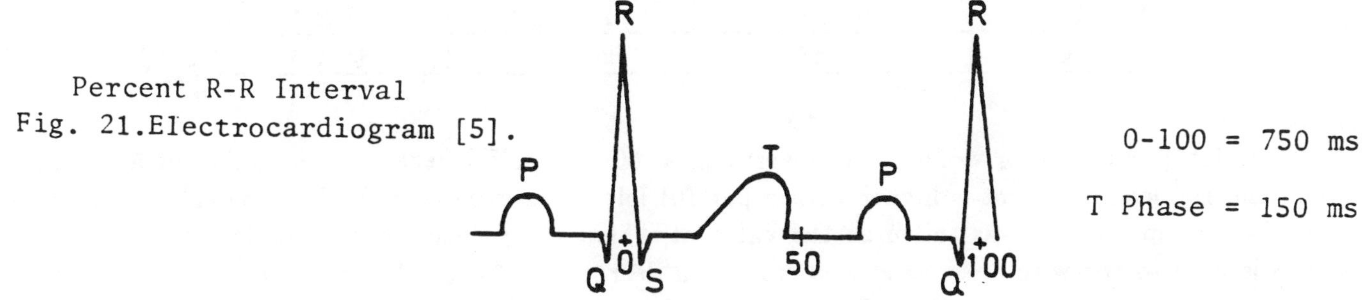

Percent R-R Interval
Fig. 21.Electrocardiogram [5].

0-4.1B. RESPIRATORY ARREST [5]

Shocks with a current path through the respiratory center can cause respiratory arrest. The respiratory center is at the base of the skull slightly above a horizontal line from the back of the throat. Thus shocks from the head to a limb or between two arms could lead to respiratory arrest. Artificial respiration can help in this case.

0-4.1C. ASPHYXIA [5]

Asphyxia is caused by contraction of the chest muscles. When current is above a certain level, a person cannot let go of an electrically hot wire. Currents somewhat above this level may not be sufficient to cause ventricular fibrillation but may be sufficient to cause the contraction of the chest muscles and asphyxia since the victim cannot let go of the wire.

0-4.2. ELECTRICAL PARAMETERS [5]

The electrical resistance between the limbs of an individual is highly variable. It depends on the contact conditions such as dry skin versus moist skin, the tough skin of a laborer versus a baby's tender skin, and so on. Tests indicate that a good approximation for the lowest resistance between any two limbs is 500 ohms. This is the estimated resistance with good contact through the skin. Using this figure, a person across a 120 V line, touching the line with any two limbs, might have a current of 240 mA through his body. Across 240 V, the current might be on the order of 480 mA.

The threshold for perception at 60 Hz is 0.2 mA. A current of 0.36 mA can be perceived by 50% of a group of men while 50% of a group of women can perceive 0.24 mA. This is an important parameter since a shock of such low level in itself is not dangerous but it might startle an individual so that he falls from a ladder or has some other involuntary action which could be hazardous to the individual or an associate.

At a somewhat higher current level than the threshold of perception, a person cannot let go of an energized conductor. Dalziel has run tests on 134 men and 28 women to determine at what current level the individual cannot let go [2]. Since this let go current level varies among individuals, he has used statistical methods to provide data for the following table:

LET GO CURRENT

Frequency	1/2% of Men (One Man in 200) Cannot Let Go	50% of Men Cannot Let Go	99.5% of Men Cannot let Go
5 Hz	15 mA	25 mA	36 mA
60 Hz	9 mA	16 mA	22 mA
1 kHz	13 mA	24 mA	35 mA

The let go current for women was lower; i.e., at 60 Hz 1/2% of the women could not let go at 6 mA. The DC let go current is higher with a painful jolt being experienced when the circuit was broken. DC let go current was taken as the value at which the subject refused to let go because the victim is held to the wire. His resistance may then decrease so that lethal currents are 40 to 60 mA. The victim should be pulled off the line or the line should be deenergized to allow the chest muscles to relax and permit breathing.

Respiratory arrest is not as common as asphyxia and ventricular fibrillation since the current path in accidents is usually not through the respiratory center. There is no figure readily available as to current level causing this condition but it is probably of the order of 0.1 A between the head and limb, artificial respiration will certainly help.

Ventricular fibrillation is very serious since the only real relief involves the use of a defibrillator available at a good hospital. From experiments with animals extrapolated to possibly human application, the 60 Hz current range which will produce ventricular fibrillation in one out of 200 people is given by the expression

$$I = \frac{116 \text{ to } 185}{\sqrt{T}} \text{ RMS mA,}$$

where T is in seconds and the limits show that the tests were valid for a range of $T = 8.3$ ms to $T = 5$ seconds. Tests seem to indicate that from 5 to 20 or 30 seconds the threshold is essentially the same. The RMS stands for root mean square (cf., EXP. I-3).

From this information it should be evident that currents as low as 0.2 mA can be dangerous. A current of 9 mA might prevent a person from letting go of a conductor while a current of approximately 150 mA for one second could cause ventricular fibrillation.

0-4.3. DANGER - LOW VOLTAGES

The following excerpt is gathered from [7] and some lecture notes prepared by the earlier instructors at the State University of New York at Stony Brook and is brought here due to its extreme importance.

It is a well known fact that the real measure of shock's intensity lies in the amount of current (amperes) forced through the body, and not the voltage. Any electrical device used on a house wiring circuit can, under certain conditions, transmit a fatal current.

It is common knowledge that victims of high-voltage shock usually respond to artificial respiration more readily than the victims of low voltage shock. The reason may be the merciful clamping of the heart, owing to the high current densities associated with high voltages. However, lest these details be misinterpreted, the only reasonable conclusion that can be drawn is that 75 volts are just as lethal as 750 volts.

The actual resistance of the body varies depending upon the points of contact and the skin condition (moist or dry). Between the ears, for example, the internal resistance (less the skin resistance) is only 100 ohms, while from hand to foot it is close to 500 ohms. The skin resistance may vary from 1000 ohms for wet skin to over 500,000 ohms for dry skin.

When working around electrical equipment, move slowly. Make sure your feet are firmly placed for good balance. Don't lunge after falling tools. Kill all power, and ground all high-voltage points before touching wiring. Make sure that power cannot be accidentally restored. Do not work on underground equipment.

Don't examine live equipment when mentally or physically fatigued. *Keep one hand in pocket while investigating live electrical equipment.*

Above all, do not touch electrical equipment while standing on metal floors, damp concrete or other well grounded surfaces. Do not handle electrical equipment while wearing damp clothing (particularly wet shoes) or while skin surfaces are damp.

Do not work alone! Remember the more you know about electrical equipment the more heedless you are apt to become. Do not take unnecessary risks.

0-4.4. GROUNDING AND DESIGN FEATURES FOR SAFE ELECTRICAL EQUIPMENT

Electrical equipment should be designed so that no electrically energized ("Hot") part of it becomes accessible. Further, to avoid possible accident in the case of overheating and/or melting of various insulators used to cover all the hot spots, the equipment must be well grounded. Generally speaking, the electrical ground is a point that accepts or supplies charges without changing its electrical characteristics. There is a so-called universal ground which is the same as earth and is considered as the absolute reference. The symbol for this ground is ⏚ . Under certain conditions, such as high voltage applications it is necessary to use this type of ground. Then there is the so-called local ground which is essentially the same as chassis ground or floating ground. Its symbol is ⏚ . This is the same as the common point of the equipment or circuit. Its voltage with respect to the earth is a nonzero value, but for practical reasons it is considered as zero reference.

Since our only concern is using safely various types of electrical equipment and/or appliances, we notice that this can only be done by a proper grounding procedure. To make this point clear consider the following situation:

Suppose an appliance has the wiring diagram of Fig. 22. In case of an accident one of the supportive insulators such as (a) or (b) may be changed and the appliance wiring may become loose and touch the body of the appliance as shown in Fig. 23.

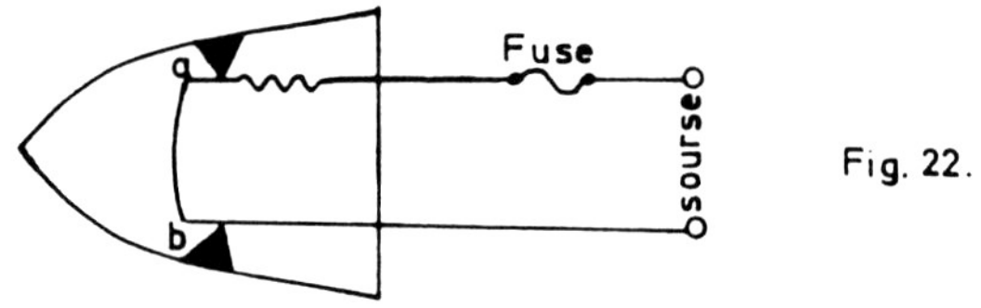

Fig. 22.

Now, since the chassis is not grounded, the appliance body becomes hot and acts like a conductor, an extremely dangerous situation. Therefore, for proper protection of the appliance and the safety of the user we make the arrangement given by Fig. 24.

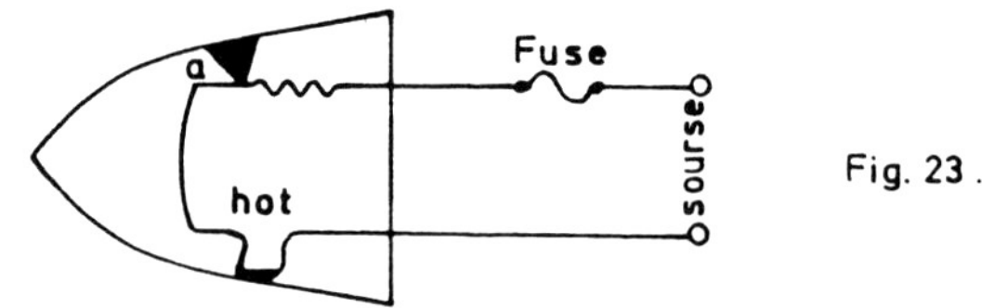

Fig. 23.

Here the returning current will go to ground through a much lower resistance path (grounding wire) than the actual appliance wiring. Thus more current, which is also called the unbalanced current, will be drawn and the protecting fuse will open immediately.

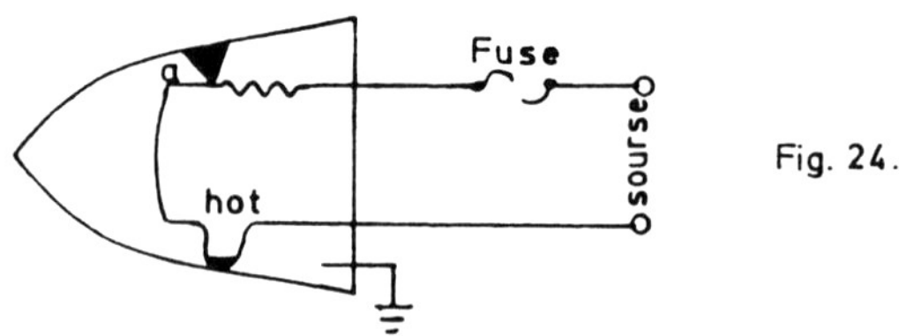

Fig. 24.

There is also a so-called ground fault circuit interrupter (GFCI) that detects the unbalanced current of the "hot" and ground wires going to electrical equipment. Thus a GFCI protects against the hazard of electric shock by de-energizing the faulted circuit. With this device fault or unbalanced currents of only 5mA can be detected even though the actual load on the circuit is 15 or more amperes [5].

0-4.5. WHAT TO DO FOR VICTIMS OF SERIOUS ELECTRICAL SHOCK

Remove the victim by means of a non-conductor such as dry wood, rope, blanket, *etc.*, from the contact area immediately. BUT MAKE SURE YOU DO NOT TOUCH THE VICTIM WITH YOUR BARE HANDS. If you have, however, access to the power switch you may try to cut the voltage instead of removing the victim. After disconnecting the victim from the electricity then start artificial respiration and wait for medical help to arrive. It is generally considered to be unwise to work alone in the laboratory for the obvious reason that in case of emergency at least someone else can make the above elementary steps to save a life.

0-5. GENERAL LABORATORY GUIDELINES

PLEASE MAKE SURE THAT THE FOLLOWING COMMENTS ARE OBSERVED BY EACH AND EVERYONE OF YOU.

1. Make sure you have received everything you need to perform your experiments - the list is given at the beginning of each experiment. If something is missing immediately ask for it. It is your time that will be wasted if you do not have the proper facilities.

2. Before making any testing make sure all energy sources in the circuit have been turned off; all multirange meters are at the highest scales; and all control switches are properly located.

3. Bring defective units to the attention of your teaching assistant and do not change these units with those on other laboratory tables.

0-6. LABORATORY CONDUCT

It is generally believed among engineering educators that what makes an engineering laboratory course different than other courses is the discipline imposed behind any such course. This kind of course is the only chance the students have to feel the pressure of a real engineering environment. They are taught how to interact with other people and how to interface with the equipment which possibly is not in optimal operating condition. Obviously when students leave their institutions they will be working with different sets of equipment and different kinds of pressure. Believe it or not many of them might wish they were still in school! Thus our main emphasis and the spirit of our teaching is to help the students to develop a sense of discipline and responsibility in themselves that would help them in their future jobs. We want the students to be able to adapt themselves to new situations as they arise.

In particular we believe that the following rules are useful and should be observed by the students.

1. The students must be present in laboratory within 5 minutes from the start of the laboratory and they must stop 10 minutes before the end of class to give teaching assistants time to close the laboratory.

2.   Students must make extra efforts to return all the accessories that they have borrowed to do the experiment to their teaching assistant and leave behind a clean work station.

3.   The students *shall not* tamper with equipment calibration, nor attempt to make any repairs or move any equipment from one station to another. In the event of equipment failure, the students must inform their instructor. Normal wear and tear of equipment can be expected, but students who make careless mistakes should be prepared to defray the costs of such actions.

4.   Normally no one is allowed to leave the laboratory when the class is in session. But if need arises to leave the room, students must inform their teaching assistants. Visitors should not attend the laboratory.

5.   Each group submits a copy of the raw data sheet at the end of each laboratory, to the instructor. This sheet includes each group members name, ID number, station number, and the experiment name and date. Teaching assistants will compare this with the original copy that is part of the laboratory report handed in the following session. The teaching assistants will make sure that each time a different student in the group takes the data. Therefore it will be necessary for the students to have a carbon copy handy and to make arrangements to rotate this task among themselves.

6.   The completed laboratory report for each session is always due the following session *at the beginning* of the laboratory period and upon arrival to the laboratory. Neatness is a must and the report should not exceed *6* pages. This report will be graded and will be returned to the student at the following session.

0-7. GENERAL GUIDELINES FOR COLLECTING DATA IN LABORATORY

The laboratory notebook is a complete log or record of all work done by the experimenter. While there is no universal set of rules for keeping this notebook the following guidelines which have been suggested by earlier instructors at the State University of New York at Stony Brook should be helpful.

1.   All information should be recorded directly into the notebook as soon as it is obtained. Memory or scraps of paper should not be used for a primary recording medium. The risk of loss or error is too great. The experimenter should be in possession of her/his notebook whenever she/he is working in the laboratory. Rough calculations and initial conclusions, should be recorded directly in the notebook.

2.   Do not crowd entries. Notebook paper is undoubtedly the least expensive item in any experimental investigation. Leave ample margins at the top, bottom and sides of each page for reference notes which may be added subsequently. Entries should be in ink or ball point pen. Never erase or obliterate. Cross through an entry if you must, but leave it legible. Only immediate errors of entry should be crossed through. Observations which are subsequently found to be in error should never be crossed out; the fact that they are in error will be recorded on a subsequent page and a margin reference to this page can be

inserted. Finally be sure that your writing and numerals are legible.

3.   Entries in the notebook should be in chronological order. Pages should not be skipped in the body of the notebook to be used at some later date. Each page of the notebook should be numbered. The experimenter's name, date, section, and the names of any co-workers, should appear on each page. The notebooks to be used in the course must have the pages prenumbered.

4.   There are some records which cannot conveniently be put into the laboratory notebook, such as long strips of recorder paper or a stack of Polaroid prints. Such records should be immediately marked with identification numbers and logged, together with information on their source, into the laboratory notebook. Identification systems need not be elaborate, but they must permit unequivocal access in both directions between the laboratory notebook and such separate records. The identification system must satisfy any labeling requirements for specimens or supplies. A beaker, bottle or box used in the experiment should either be empty or contain a label identifying its contents and recorded in the notebook.

In most cases a Laboratory Research Notebook, which is available at the Campus bookstore may be used, although in many cases this can be avoided to save money. At the end of each laboratory session you are *required* to submit a raw data sheet of your work to your laboratory instructor. Your formal report to be prepared and given to the instructor at the assigned date will then be compared with this actual raw work to determine your laboratory grade.

0-8. THE FORMAL REPORT

The particular style to employ in the writing of a report will vary with the subject matter to be covered; but it is believed that the following suggested order of arrangement will help the student to organize her/his material and will help someone with a similar background to duplicate the experiment in its entirety [6]. But more importantly learning how to differentiate between the following 12 sections is the most rewarding education that a student may gain while she/he is still in school. The following order may change according to someone's opinion and certainly we cannot please everyone. But in our opinion if a student knows what actually constitutes the result of an experiment and what does the objective, then writing these two sections, *per se*, in any order as one pleases is only a matter of seconds.

1.   *Title Sheet.* (On the title sheet are given the name and number of the experiment, date performed, name of experimenter, and names of the co-workers. Also give your station number and name of your instructor.)

2.   *Objective.* (Concise statement of the purpose or objective of the experiment.)

3.   *Equipment Required.* Complete nameplate rating of machines or apparatus under test. (Ratings of appurtenant manufactured apparatus such as shunts, instrument transformers, and so forth, together with the instrument data, are recorded on the data sheet. This is important in that if a certain unit of device fails, you will not be penalized. Also give a

list of the parts that you have used, specially when you design your own circuit.)

4. *Theory and Procedure.* (This portion usually consists of a concise discussion of the basic theory of the experiment together with the general procedure followed in the performing of the experiment.)

5. *Circuit Diagrams.* (Diagrams of the test circuits should show the exact relative positions of all instruments and auxiliary equipment. Conventional symbols should be used throughout.)

6. *Observations-Corrected Data Sheet.* (The data "as taken" are corrected by multiplying the actual scale divisions by the various factors as outlined previously and listed as "corrected" data.)

7. *Sample Calculations.* (The necessary formulas in general form and one detailed calculation of each different type should be shown.)

8. *Curve Sheets.* (All curves and vector diagrams should be drawn in ink. All test points should be shown with small circles, and if no discontinuities are anticipated a smooth curve is drawn through the experimentally determined points. Each sheet should have a complete title. If several curves are drawn on the same sheet, each should be properly labeled. Only related curves should be drawn on the same sheet. The independent variable should be plotted as abscissa and the dependent variable as ordinate with clear scale. In general, curves should be plotted with zero origins, although there are certain notable exceptions to this procedure. Do not put a table of observational data on the curve sheet. Also depending on the nature of the experiment one may use the regular Cartesian graph paper or one of the following graph paper: semilog; log-log, *etc.* Semilog paper has one logarithmic axis and one linear axis. This paper is very useful for plotting data having an exponential dependence. The log-log paper has logarithmic scales on both axes.)

9. *Discussions and Conclusions of Results.* (A complete discussion of the results obtained must be given. This should be the essence of the report and should tell how much of the theory was verified and how much apparently contradicted, including probable reasons for the results obtained. *Any departure of the results from the theoretical expectations must be carefully explained.* Perhaps an error analysis as appropriate should be included.)

10. *Safety Precautions and Your Comments.* (Described below.)

11. *Appendix* (If you wish to discuss something which is not closely related to your experiment, but you have written it for your own curiosity and information, then place it in an Appendix. Your instructor will not read this part, unless requested.)

12. *References.* (If you are using a handbook or your report is taken from certain sources other than your textbook, then you must give the names of these sources. Use the IEEE standard for your references, as we have used in several occasions in this textbook.)

WARNING: ALWAYS SAVE *A* COPY OF YOUR REPORT (PERHAPS A SCRATCH COPY) THAT IF IN CASE OF AN ACCIDENT YOUR TEACHING ASSISTANT LOSES YOUR REPORT YOU WOULD HAVE SOMETHING TO PREPARE FOR QUIZZES, ETC.

0-8.1. SAMPLE REPORT

In the following we present a sample laboratory report for your information. Please follow the same procedure for your other reports.

1.   *Title Sheet:*

State University of New York at
Stony Brook
Dept. of Electrical Engineering
ESG 211 - EXP. I-3
AC-Measurements
April 20, 1983
Prepared by: Diana Isaza
Co-Workers: Thomas Semetsis and Joanne Miele
Station: 14
Instructor: Mr. David S. Bayard

2.   *Objective:* To study the current voltage characteristics of a diode and compare it with Ohm's law.

3.   *Equipment Required:* Variable Power Supply-PS503A, Voltmeter-DM501A, Ammeter-VOM-Heath, p-n diode and 1 kΩ resistor. (These names are just for the sake of example.)

4.   *Theory:* For a given voltage V and current I Ohm's law states:
$$V = R\,I,$$
where R is resistance of the material. If the value of R is constant, then V is proportioned to I. Therefore voltage and current will exhibit a linear relationship.

The resistance of a p-n junction diode is not a fixed quantity; it depends upon the mobility of carriers. Therefore, the V-I characteristic of a p-n junction diode will not exhibit a linear relationship.

4.1.   *Procedure:* Connect the circuit as shown in Fig. 25. Make sure the diode is forward biased and a resistor of 1 kΩ is placed in the circuit to limit the current. For various values of the applied voltage the corresponding values of current are recorded. (Notice: We broke Section 4 into two parts.)

5.   *Circuit Diagram:*

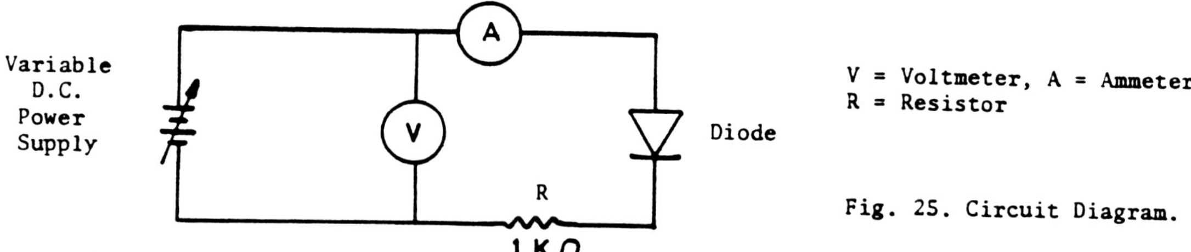

Fig. 25. Circuit Diagram.

6.   *Observations:*

voltage (volts)	1	2	3	4	5	6
currents $(mA=10^{-3}A)$	2	4	8	16	25	36

7.   *Sample Calculations:* None

8.   *Curve Sheets:* Please refer to fig. 26. If you have more than one graph, you may draw all of them on one page to save paper.

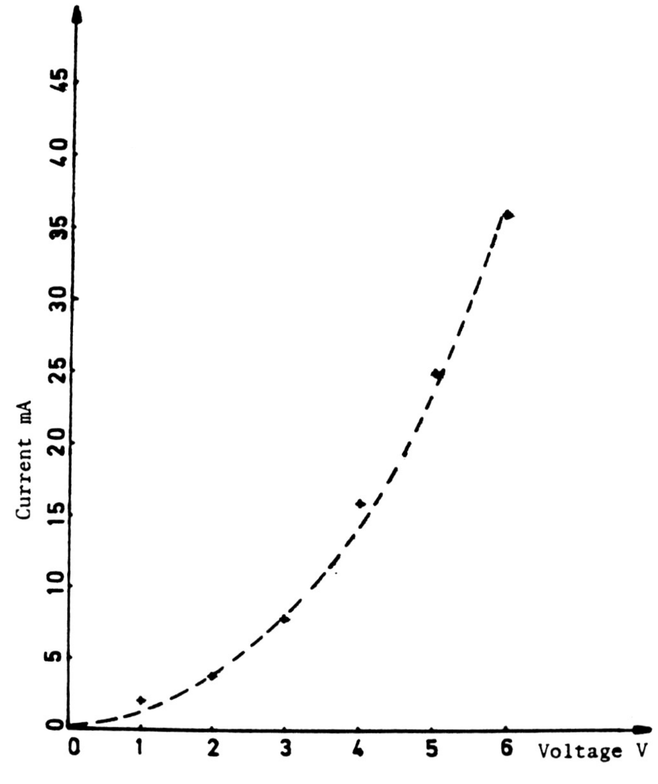

Fig. 26.   (V-I) Characteristics of a Diode.

9.   *Discussions and Conclusions of Results:* The voltage current characteristics of the diode is plotted on a graph paper. Fig. 26 shows that the V-I characteristic is not linear. For each set of voltage and current values, the diode has a different value of resistance. Therefore Ohm's law, $V = RI$, for a constant value of R, does not apply to the given diode.

10.   *Safety Precautions and Comments:*

    1.   Applied voltage and the current in the circuit should not exceed the voltage and current rating of the diode.

    2.   Make sure ammeter is not connected in parallel with the power supply. That will burn the ammeter.

    3.   Do not touch the live circuit with your bare hands.

11.   *Appendix:* None

12.   *References:*

(Sample for a book)
> [1] B.D.Wedlock and J.K. Roberge, Electric Component and Measurements. Englewood Cliffs, NJ (place of publisher): Prentice-Hall (name of publisher), 1969.

(Sample for a paper)
> [2] J.L. Moll, "The evolution of the theory for voltage - current characteristic of p-n junctions, "Proc. IRE (name of journal-if well known use abbreviation), vol. 46, pp. 1076-1082, June 1958.

0-9. BASIC ERROR ANALYSIS

0-9.1. INTRODUCTION

The word error is considered to be *a* generalized difference (or difference function if appropriate) between two quantities of interest. One of these two quantities is called the reference value. In a positive manner, we may replace the word error with *accuracy.* Accuracy is always associated with a physical measurement and represents closeness with compared to the actual or reference values of the readings of a particular measuring device. Error, on the other hand is an abstract word referring mainly to the mathematical differences between various quantities. To be specific about an error we must be specific about method of generating that error. Especially we must know how accurately the reference value is developed. It is simply wrong to compare blindly a test result with a reference so long as no indication of degree of *confidence* of this reference has been presented. In other words there is always a degree of *uncertainty* associated with even those values that seem to be absolutely *true.*

In general to analyze or to describe or merely to utilize a phenomenon we may use one of the three following routes. We do understand, however, that no matter how we try these routes there is simply no way to achieve a *true* value or a *true* result for a given experiment. These three approaches are as follows.

0-9.1A. Completely Theoretical Method

In this approach a general mathematical model based on physical laws governing the system operation is developed. It is well known that always there is some missing information that is neglected in this model for various reasons. Often this mathematical model is further simplified to what we call the *local* model, for the sake of practicality and the ease of computations. This local model is also called *reduced-order* model. Using reduced-order model instead of *global* mathematical model may result in erroneous conclusions. To be specific, suppose a system has a linear tenth-order global model and we are using its third-order *local* model instead. If this system over a certain frequency range, becomes unstable, then we may not see this situation from the *reduced-order* model and we may conclude that the system is always stable for all frequency ranges. We can produce many such examples. The irony is that we are always using a *local* model and pretending that this is the *global* model! Thus no matter what we get, this sole mathematical result always becomes an approximation to the *true* result which generally is unknown and even unattainable.

0-9.1B. *Completely Experimental Method*

This approach refers to cases in which the system is so complicated that we cannot use mathematical models to describe its performance. Here we rely on our experiment to describe the system behavior. But we bear in mind that we have some information about what we should be getting. Thus inherently some information about the system behavior must be available and used to achieve further knowledge about the system. It is perhaps extremely naive to wish to study a totally *unknown* system. The very moment that we use a particular measuring device to work on a system we are saying that we know something about the experimental result. At any rate this is an approach that helps us by trial and error and step by step research to learn about a system's operation. In our set of experiments are some cases where the goal is to identify the parameters of a given network. One such experiment is the Thevenin equivalent network for a given black box. The fact that we chose a particular method of experimenting emphasizes that we already know something about this black box. Experimental methods inherently are local due to limited experimental trials. Also these methods involve considerable time and many specific apparatus. Nevertheless in many cases it is the only feasible approach. We will, of course, always use our experimental results and will study these results by techniques from statistical theory to predict a mathematical model for the future operation of the system. We will emphasize this idea further in the next approach.

0-9.1C. *Partially Experimental and Partially Theoretical Method*

In this approach the experimenter must make a judgment how to choose either one of the above two methods and/or what balance must be made between the applications of these two methodologies. The following thoughts give a tentative idea on how we may consider breaking an experiment between these two methods. Perhaps it is wise to take advantage of approach *0-9.1A*, and simply extract as much information as possible about the global model of the phenomenon. If not quite possible then we may test experimentally the accuracy of the simplified or reduced-order model to see its validity. If this is not possible at all, then we may use experimental results to predict a model - though local--for understanding the phenomenon. Or perhaps we use our knowledge to improve this model or in some cases to learn more about this system to develop a relevant theory. At any rate, *simultaneously* employing both *0-9.1A* and *0-9.1B* and using a keen judgment on how to balance these two methods is the key issue of this approach. This is as much an *art* as a science.

0-9.2 VARIOUS MEASUREMENT ERRORS

Recall that by *error* we mean the difference between what we read from our experiment and what we chose as the *true* value or reference for this particular experiment. Since most of our experiments are performed according to that of *0-9.1C* above, the nature of error is both analytical and experimental. In summary we consider following as possible sources of errors. The order in which these are presented is by no means an indication of their relative importance.

1. The analytical errors due mainly to *imperfect* mathematical modeling and further approximation of the model. The experimenter must expect this fact and must try to compensate for such errors as much as possible. One method of reducing this type of error is applications of theory of sensitivity.

2. The experimenter's reading errors due to fatigue or other human errors such as simple miscalculation of the scale factor.

3. Improper choice of measuring devices or any other related device malfunctions that are not corrected by the experimenter prior to its usage.

4. Various equipment errors due to a lack of proper maintenance or calibration, aging, or which arise from using the equipment at the extreme limits (very high or very low) of its range of operation.

5. Various errors caused by changing the working conditions such as extreme heat or cold, humidity, or existence of various parasitic signals acting on measurement devices, *etc*.

6. Finally, all other and unexpected errors caused by unknown events such as the sudden change in line voltage prior or during the course of experiment.

The above list is by no means complete. The main purpose of the above categorization is to remind the readers that no matter how and under what circumstances an experiment can be performed, the *true* (mostly unknown) values of data are absolutely unattainable. Thus the main issue in any experiment is the level of confidence the experimenter can express about the collected data. Since inherently the collected data must be compared with the appropriate reference in each category, we must be concerned with two quantities, both of which are nondeterministic, in general. But to make the analysis simpler we assume these two quantities are developed in such a manner that the reference quantity is always very close to the *true* value and thus we only are concerned with the actual measurement data. Fig. 27 presents our thoughts in this regard.

Referring to Fig. 27, we see that no matter how an experiment is performed there are always some errors and based on these errors we make several important decisions. In path one we conclude that both theoretical and experimental results are valid. This occurs, in general, within some degree of tolerances, nevertheless we are content with our decision. In paths two and three we have also made definitive conclusions as described in the figure. Finally, in path four we conclude that the results could have been better or more accurate than what we have but due to our imperfect experimentation we are confronted with a set of data that do not represent the *true* values. Here we may have a second question. How much deviation do these data have compared with the *true* values? To answer this question and provide some measure of certainty about our experimental results we must appeal to the theory of probability. This theory is the main tool of handling these kinds of problems.

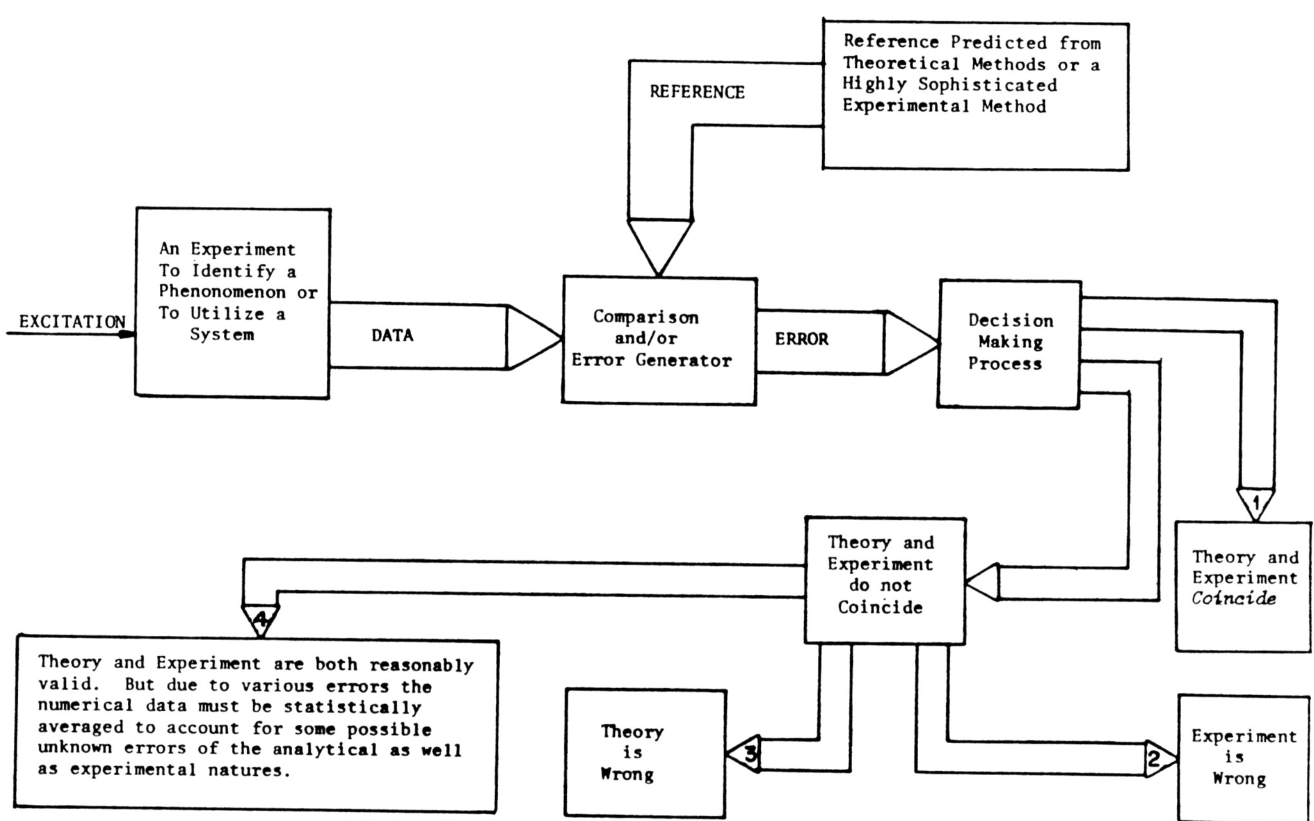

Fig. 27. Error Evaluation in an Experiment.

0-9.3. SOME BASIC NOTIONS OF PROBABILITY

Suppose a physical experiment that is performed 30 times repeatedly results in the set of readings shown in Fig. 28.

Trial number	Readings	Trial Number	Readings
1	2.6012	2	2.8117
3	2.3117	4	2.4667
5	2.7018	6	2.2198
7	2.5220	8	2.9532
9	2.7088	10	2.5897
11	2.1587	12	2.3288
13	2.4566	14	2.5755
15	2.6277	16	2.6337
17	2.9012	18	2.2266
19	3.1414	20	2.6211
21	2.5332	22	2.6422
23	2.1717	24	2.9312
25	2.3244	26	2.4827
27	2.5867	28	2.3781
29	2.7122	30	2.7301

Fig. 28.

How do we make some sense out of these readings? Which one is correct, if any, and why? To answer this question we make the following observations and we introduce a slightly different approach to handle this problem.

Suppose we make the assumption that the actual result of this experiment is somewhere between the lowest reading of 2.1587 and the highest reading of 3.1414. This is an important decision. It is, of course, obvious that any physical system that is excited gives reading that in nature are finite. In other words,

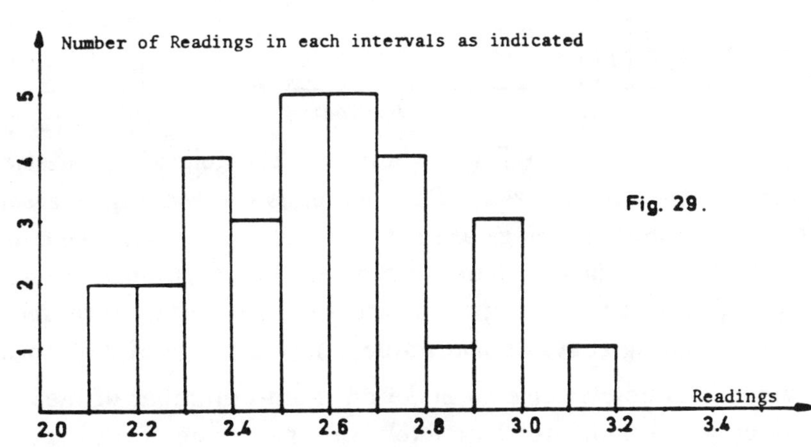

Fig. 29.

even though these experiments may not be perfect we do not expect that the readings be in the range of $\pm\infty$. Then we ask in which of the intervals (2.1 to 2.2, or ..., or 3.1 to 3.2) is the correct reading? To answer this question we find it convenient to draw a simple diagram that maps our readings into a set of rectangles whose heights are equal to the number of readings in each of the above intervals, as in Fig. 29. We may even wish to draw the above diagram for cases in which the above intervals are divided by two, as in Fig. 30.

Since we are dealing with a physical system and most likely we know how to perform this experiment, therefore we may assume that if we increase the number of readings substantially and further *refine* Fig. 30, for smaller intervals we will be getting a figure that is chopped from both the highest and the lowest

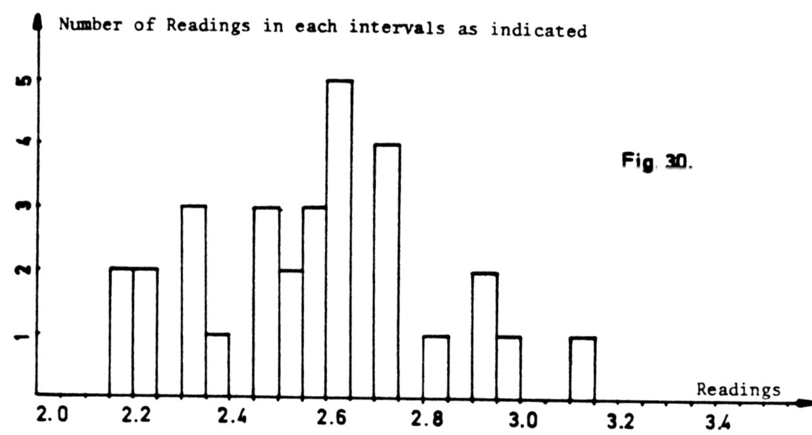

ends (due to physical constraints mentioned before) and mostly concentrated in the area which is the closest to the actual or *true* reading of the experiment. We understand that we have no control over an interval; in other words, these numbers are experimentally determined. It is a fact of statistical theory that as the number of trials increases eventually Fig. 30, will approach that of Fig. 31, which forms nearly continuous readings between the intervals.

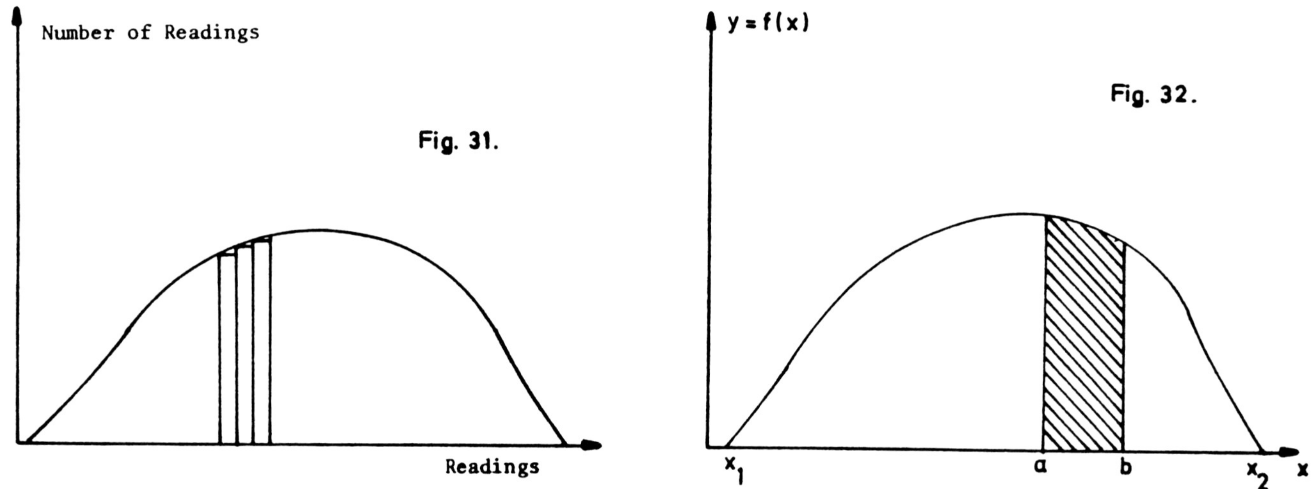

Let us look back at Fig. 30 and ask the following question. Which of the rectangles in this diagram has the largest area? The answer is the rectangles associated with the interval between 2.6 to 2.65. Since most readings are between 2.6 to 2.65, can we conclude that the *true* value must be in this range? It is really difficult to answer this question since there are no readings between 2.65 to 2.7, *per se*, and one may argue that if we keep trying more and more experiments we may get similar data indicating that the *true* value is in the range of 2.65 to 2.7, *etc.*

Since it is not feasible to make an infinite number of measurements, we introduce the following parameter which indicates how likely the *true* value of an experiment, *for the given set of data*, is to occur in a certain interval.

$$y = \frac{\text{area of particular rectangles of a given interval}}{\text{total areas of all rectangles of all intervals}} ,$$

$$= \frac{(\text{number of readings in a given interval}) \times (\text{width of an interval})}{(\text{total number of trials}) \times (\text{width of an interval})} ,$$

$$= \frac{\text{number of readings in a given interval}}{\text{total number of trials}} \, .$$

This parameter in its last form makes sense only for a finite number of trials (discrete random variables). One can easily see that $0 \leq y \leq 1$. Also adding all these measures of likelihood of all possible intervals is equal to unity.

In the limiting case that the number of trials becomes infinite and all possible readings have been developed, Fig. 31 approaches, in general, a continuous curve in which the readings or actual results of the experiment are shown by x (a continuous set of numbers) and the vertical axis represents $y=f(x)$. In Fig. 32 we see that y can be thought of as smoothed upper boundary to the rectangles in Fig. 31. *(Experts please bear with us!)*

Naturally if we are able to express mathematically the form of $f(x)$ for a given experiment without going through an exhaustive set of trials, we have done a great job. In reality it is extremely difficult to anticipate what mathematical form this graph will have for a particular experiment. The function $f(x)$ is called the *probability density function*. As we described earlier, the chances that a *true* reading lies in a given interval equals the area of rectangle under the curve of Fig. 32 and associated with that interval to the total area of all rectangles for that experiment. We call this a *probability measure* of the *true* value being in the given area and call this *true* value a random variable. By convention the total area of all rectangles equals 1.00, since we already have assumed that the *true* value is somewhere in this entire area. Thus we may say that the probability that the true reading be between two values a and b is:

$$\Pr(a \leq x \leq b) = \int_a^b f(x)\, dx \, , \quad \text{where} \int_{x_1}^{x_2} f(x)\, dx = 1 \, .$$

In certain cases we may ask: what is the probability of the *true* value being less than a chosen value - say x? This is the case in which $a = x_1$ and $b = x$, a variable. The answer is

$$F(x) = \int_{x_1}^x f(t)\, dt \, .$$

This function is known as the *cumulative distributive function.*

Thus in the limiting case and as the number of trials becomes infinitely large associated with each random variable there is a graph of a probability density function which indicates how likely (or unlikely), a given value is close to the *true* value. The real issue, however, is that we simply do not know, *a priori*, what the mathematical form of this probability density function is for most given experiments. Certainly we cannot afford to go over an infinite number of trials for each experiment to find out that curve. In statistical theory, however, an extremely well-behaved function called the *Gaussian distribution function* is known that can be considered to be the best approximation for the probability density function of *many* real and practical cases. This function is as follows:

$$f(x) = (1 / \sqrt{2\pi}\, \sigma)\, \exp[-(x-m)^2 / 2\sigma^2] \, , \quad -\infty < x < \infty.$$

Here m is called the *mean value* and σ is the *standard deviation*. A graph of this function for several different values of σ is shown in Fig. 33a.

Without getting into too much detail we observe that $f(x)$ is maximum at $x = m$. Further we can show that for any Gaussian distribution function the following is true:

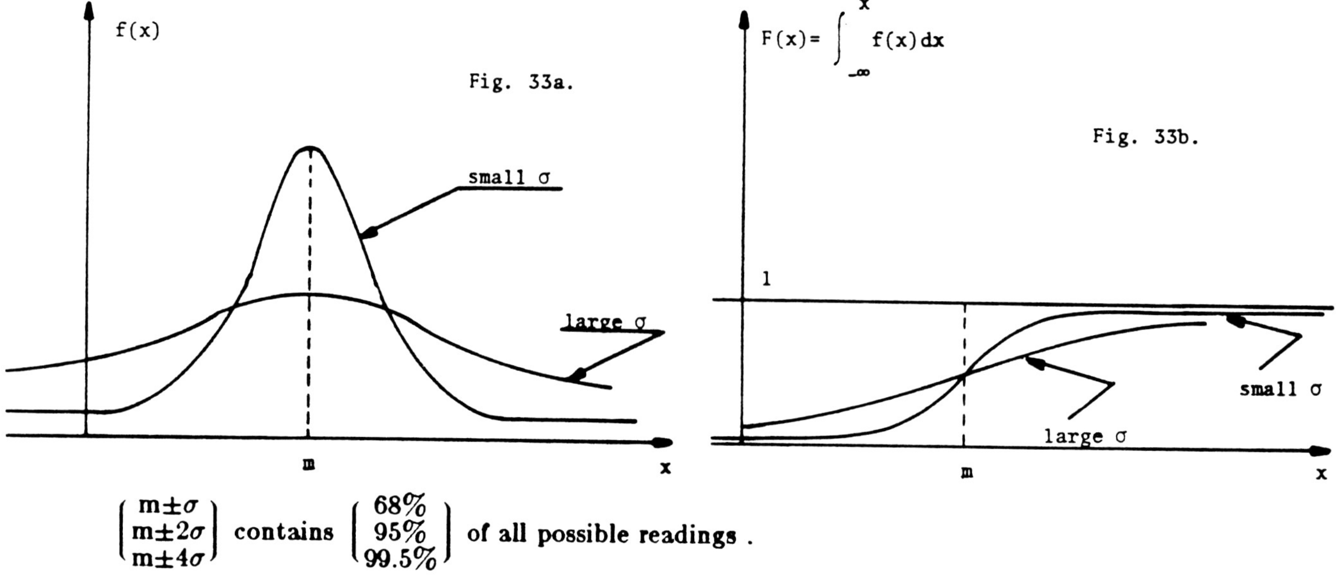

$$\begin{pmatrix} m\pm\sigma \\ m\pm2\sigma \\ m\pm4\sigma \end{pmatrix} \text{ contains } \begin{pmatrix} 68\% \\ 95\% \\ 99.5\% \end{pmatrix} \text{ of all possible readings .}$$

In fact, because of this attractive property of the Gaussian distribution function and because physical experiments do not require an infinite number of readings, we are confident that the Gaussian distribution function is a good choice to represent most random variables of interest to us.

The real issue is that we can not assume *every* random variable is Gaussian. There are tests such as the χ^2 (chi-square) statistical test that provide some indication on how good such approximation is. But we must recall that there are other equally important statistical functions that are useful and must be evaluated for possible curve fitting on a given statistical data of an experiment. To name some of these statistical functions we recall the Uniform, Discrete Uniform, Exponential, Gamma, Weibull, Log-normal, Beta, Triangular, Bernoulli, Geometric, Binomial, Negative Binomial, and Poisson. Each of these functions have been studied thoroughly by statisticians and can be applied to a given set of data to anticipate the probability density function of an experiment. We are, however, content with what we have learned so far. That is, with a reasonable assumption we may consider the *true* probability distribution of a random variable to be a Gaussian distribution with mean and standard deviation (of sample) as developed from the experiment. This choice can always be refined if additional statistical information can be obtained. At any rate, the definitions of the probability density function (pdf) and the cumulative distribution function (cdf) are independent of any particular form of the statistical function. In fact, we always have f(x) = dF(x)/dx .

The two quantities of the *mean* or *expected value* m and *standard deviation* σ (or variance $\nu=\sigma^2$) of a random variable play an extremely important role in any statistical analysis. The mean is a measure of the central tendency of a random variable and the variance is a measure of the dispersion of a random variable about its mean. We will be looking into methods of the dispersion of a random variable about its mean. We will be looking into methods of generating these numbers in our future experiments. Here we present a way of *estimating* these quantities for a given set of trials consisting of n readings $x_1, x_2, ..., x_n$. We define m as a simple arithmetic average, i.e.,

$$m = \sum_{l=1}^{n} x_l / n .$$

The standard deviation σ of the entire population is given by

$$\sigma = \sqrt{(\sum_{l=1}^{n} x_l^2 - nm^2) / n} .$$

An estimate of σ called s (standard deviation of sample) standard deviation of sample is given as

$$s = \sqrt{(\sum_{i=1}^{n} x_k^2 - nm^2) / (n-1)} .$$

For our set of data, having n=30 we obtain

$$\sum_{i=1}^{n} x_i = 77.0504, \ m = 2.5683, \ \sigma = 0.2378, \ s = 0.2419.$$

Suppose we decide that our results are sufficiently close to a Gaussian distribution function with m=2.5683 and s=0.2418 (notice we chose the standard deviation of the sample). Then the probability that the *true* result will be in the range of m$\pm$ s is calculated as follows :

$$Pr(m{-}s < x < m + s) = \int_{m-s}^{m+s} \frac{1}{\sqrt{2\pi}s} \exp[-(x{-}m)^2/2s^2] \, dx .$$

Letting u=(x–m)/s, then dx=sdu, and

$$Pr(m{-}s < x < m{+}s) = \int_{-1}^{1} \exp[-u^2/2] s du \, / \, \sqrt{2\pi}s,$$

$$= 2\int_{0}^{1} \frac{1}{\sqrt{2\pi}} \, \exp[-u^2/2] \, du = 0.69268. \quad (\text{Of course!})$$

This number, of course, is larger than our first estimate of $Pr= 2 \times \dfrac{5}{30} = \dfrac{10}{30} = \dfrac{1}{3}$, based on the total area of the rectangles (around 2.5683$\pm$0.2419, i.e., 2.5 to 2.7), because we made a further assumption that our random variable is a Gaussian random variable.

0-9.4 REMARKS

We finally close this section with some remarks on how to calculate the overall accuracy of the result of an experiment. In general, the results or response of an experiment is a multivariable function of several readings or assumed quantities.

$$R^\circ = R^\circ (c_1, c_2,...., c_n) .$$

Here R° is a nominal (or *true*) value of the result of an experiment and $c_1, c_2,...,c_n$ are the readings or values of n components producing this response R°. If each of these c_i's is off by $\pm\Delta c_i$, then the measured response becomes

$$R = R(c_1 \pm \Delta c_1 , c_2 \pm \Delta c_2 ,...., c_n \pm \Delta c_n),$$

$$= R^\circ + \Delta R^\circ.$$

We choose $\pm\Delta c_i$ for the obvious reason that we do not know in which direction these errors are. To calculate ΔR°, we may use the standard Taylor series expansion that results in

$$R^\circ + \Delta R^\circ = R^\circ + \frac{\partial R^\circ}{\partial c_1}(\pm\Delta c_1) + ... + \frac{\partial R^\circ}{\partial c_n}(\pm\Delta c_n) + \frac{1}{2}[\frac{\partial^2 R^\circ}{\partial c_1^2}(\Delta c_1)^2 + ...\,] + ... \, .$$

We may assume the Δc_i's are sufficiently small to justify dropping $(\Delta c_i)^2$ and all higher-order terms. Thus

$$\Delta R^{\circ} = \frac{\partial R^{\circ}}{\partial c_1} \, (\pm \Delta c_1) + \ldots + \left(\frac{\partial R^{\circ}}{\partial c_n}\right) (\pm \Delta c_n) \; .$$

We may choose an absolute value of errors to avoid the $(\pm)$; also we may choose the largest $\left| \dfrac{\partial R^{\circ}}{\partial c_1} \Delta c_1 \right|$ as the worst possible error. Then we get

$$|\Delta R^{\circ}| \leq \sum_{i=1}^{n} \left| \frac{\partial R^{\circ}}{\partial c_1} \Delta c_i \right| \leq n \left| \frac{\partial R^{\circ}}{\partial c_j} \Delta c_j \right| \; .$$

This quantity gives us an indication of what the results may be. Further discussion on this subject requires knowledge of the theory of sensitivity which will be briefly studied in Section III-2.1. We will be looking into some applications of error analysis in future experiments.

WORKSHEET

PART
ONE

FAMILIARIZATION WITH BASIC ELECTRONIC INSTRUMENTATION

PART ONE

FAMILIARIZATION WITH BASIC ELECTRONIC INSTRUMENTATION

I-0. INTRODUCTION

In this part we will be concentrating on the basic operations of the units that are being used in electrical measurements in everyday laboratory work. These units include power supplies, D'Arsonval meters, digital multimeters, function generators and oscilloscopes. Our main objective is to learn their basic operations and/or functions. Since we do not wish to be biased toward any particular manufacturer, we thus mainly study the common and generic properties of these units. The list of our experiments in this part is as follows.

EXPERIMENT I-1: FAMILIARIZATION WITH LABORATORY INSTRUMENTS

EXPERIMENT I-2: FAMILIARIZATION WITH ELEMENTARY OSCILLOSCOPE AND FUNCTION GENERATORS

EXPERIMENT I-3: AC MEASUREMENTS

EXPERIMENT I-4: OSCILLOSCOPE CONTINUED

EXPERIMENT I-5: POWER MEASUREMENTS

EXPERIMENT I-6: POWER MEASUREMENTS IN POLYPHASE CIRCUITS

For additional information on these experiments the reader may wish to consult the references at the end of this section.

Recalling Section 0-5, it is expected that each experimenter observes the following common procedures that are useful for each experiment.

0. *Prior to coming* to the laboratory to make sure all the pre-laboratory work such as reading the text of the experiment, performing various homework, preparation of charts and/or graph sheets, if needed, are done.

1. *Upon arriving* at the laboratory to make sure all the necessary parts and units for the day's experiment are available.

2. *While organizing* yourself to start setting up the experiment to make sure all the power switches on each device are off and that the various meters are set to their highest scale or maximum range.

3. Finally, upon *leaving* the laboratory, to make sure to reverse all the steps taken during the experiment and put the laboratory station in such shape that the next group can easily repeat the above steps.

Perhaps it is also advisable to review Part Zero before going to the laboratory. Make sure all the abbreviations are remembered.

I-1. REFERENCES

[0]   W.H. Hoyt, Jr., and J.E. Kemmerly, Engineering Circuit Analysis. New York: McGraw-Hill, Second Ed., 1971.

[1]   K.W. Sessions and W.A. Fischer, Understanding Oscilloscopes and Display Waveforms. New York: Wiley, 1978.

[2]   F.G. Stremler, ECE 300 Laboratory Notes. Madison, WI: The University of Wisconsin-Madison, 1978.

[3] A.T. Tiedemann, ECE 210/211 Experiment Notes. Madison, WI: The University of Wisconsin-Madison, 1976.

[4] B.D. Wedlock and J.K. Roberge, Electric Component and Measurements. Englewood Cliffs, NJ: Prentice-Hall, 1969.

[5] S. Wolf, Guide to Electronic Measurements and Laboratory Practice. Englewood Cliffs, NJ: Prentice-Hall, 1983.

I-2. TABLE OF PREFIX

The following table of prefixes and units as we have used them in this textbook, are generally common in electrical engineering.

Symbol	Prefix	Unit
T	tetra	10^{12}
G	giga	10^{9}
M	mega	10^{6}
K	kilo	10^{3}
h	hecto	10^{2}
dk	deka	10
d	deci	10^{-1}
c	centi	10^{-2}
m	milli	10^{-3}
μ	micro	10^{-6}
n	nano	10^{-9}
p	pico	10^{-12}

EXPERIMENT I-1

FAMILIARIZATION WITH LABORATORY INSTRUMENTS

EQUIPMENT:

ELECTRICAL EQUIPMENT PACKAGE (EEP)
VOLTMETER (*SUCH AS* VACUUM-TUBE VOLTMETER: VTVM)
VOLT-OHM-MILLIAMETER (VOM)
AUTO TRANSFORMER WITH CALIBRATED DIAL
RHEOSTAT (250^+ mA and 250^+ Ω)
RESISTORS (1 kΩ, 10 kΩ /2 (i.e., two of each), 100 kΩ: ALL 1% PRECISION)

Table of Contents   Page No.

1. INTRODUCTION

Perhaps this experiment is the hardest experiment to write about. No matter where we start to explain we face questions that have not been addressed before. In this and the next few experiments we study the basic properties of various instruments commonly used in most electrical engineering laboratories. In so doing we interchangeably use different devices to describe the functions of one or more devices. For example, to talk about a *power supply* we need a *voltmeter* and to show how a voltmeter works we need a power supply or a *function generator* . Therefore we urge our readers to be patient and try to follow the given instructions as closely as possible. We will, however, start our work from basic applications of a power supply, and make our way up to the applications and potentials of *oscilloscopes* in modern measurement techniques in the next several experiments. Even though we later introduce some experiments on how to measure various resistances, *etc.*, here we overlook this fact and we use several components to facilitate our work as though we have prior knowledge of these components.

2. PURPOSE

The purpose of this experiment is to acquaint students with the basic operations of a Power Supply (PS), a Digital Multimeter (DMM), a Volt-Ohm-Milliameter (VOM), and a voltmeter *such as* the Vacuum-Tube Voltmeter (VTVM). We will, however, interchangeably use the VTVM and/or just V as an abbreviation for a voltmeter. In general, the Electrical Equipment Package (EEP) (cf., Section 0-3.2), is capable of handling the job of a typical VOM and/or VTVM. The VOM and the VTVM are often the handiest instruments in most laboratories where for moderately accurate measurements are needed. But essential to any laboratory is the availability of an electrical source or signal. The easiest way to generate an electrical signal is to use the power supply unit of the EEP. In this experiment we briefly describe the functions and/or structures of: a typical dual power supply (DPS or PS); VOM and VTVM; and a DMM.

3. DUAL POWER SUPPLY (DPS) or (PS)

Perhaps the most familiar PS to us is a simple carbon battery that we have seen in a host of home appliances. We have also learned that these batteries, depending on the application, have a finite life time and in most cases they are relatively expensive. Another easily available power supply that we mostly take for granted is the city electricity line that provides electricity in the form of a sinusoidal *alternating current* (AC). The nice thing about this power supply is its availability. It continuously gives us a signal of relatively consistent characteristics. The difficulty of using this source is the fact that most of today's electronic devices need *direct current* (DC). Furthermore most of our laboratory testing is performed at low level voltages for reasons of safety and practicality. Since a durable, high performance, and efficient battery is not easily available, we appeal to our PS unit which operates from regular commercial power lines and produces for our usage a variable DC voltage level.

One might well ask, what are the properties of an ideal PS? We appeal to our intuition and experience with batteries to answer this question. We have learned that regular batteries die after a few hours or days of constant use. Therefore, by durable we mean the battery must have a long life time. We have also seen that a portable radio perhaps needs a stronger battery than does a desk clock. This refers to the amount of voltage or current that one requires compared with the other. Thus by high performance we mean for a given voltage the battery should supply enough current as needed for an application or in other words we want to have some control over the current drawn from a PS. Finally, we want this operation to be consistent and usable over a wide range of applications - this is called efficiency. These ideas are the same as those requirements for a durable, high performance, and an efficient PS. Most commercial PS in the market are based on these considerations.

In most electronic circuits using various integrated circuit components we need two separate levels of positive and negative voltages. The *Dual Power Supply* (DPS) provides us with such capabilities. In each DPS two power supplies are housed that each produces a maximum voltage level of either plus or minus sign. All power supplies are designed such that their output voltage does not drastically change as load changes. In other words, the power supplies compensate for current changes needed in each application. EXP. II-5 and EXP. VI-2 are concerned with the actual construction of a power supply circuit.

4. THE VOM AND VTVM

The VOM and VTVM are two multipurpose measuring devices which are used to measure DC voltage and current, AC voltage, and resistance. Both instruments use a D'Arsonval meter movement to provide an indication of the magnitude of the variable being measured. At present we will only use these two instruments to measure DC voltage and current. Their other functions will be considered later.

Before the magnitude of a quantity (that is to be measured) can be stated with any degree of certainty, the instrument being used to make the measurement must be *calibrated* by application of known inputs. We assume this has already been done for us so please do not tamper with the measuring devices. Also remember that when reading a signal always assume its magnitude is the largest that the device can measure. Therefore set the device at the highest scale. If the resulting reading are not satisfactory, then gradually lower the scale of the device. Starting with the lowest scale first often causes serious damage to the measuring device.

4.1. FAMILIARIZATION WITH THE VOM

Multimeters are made with precision resistors and a calibrated meter movement, and no external provision is made for calibration adjustment. Still it is desirable to check the accuracy of the meter from time to time. Measure several DC voltages from the PS voltage reference source (VRS) on the three lowest VOM ranges and determine the accuracy of these scales.

WARNING: When not in use, the Function Selector Switch of the VOM should be kept on a high DC-volt scale to avoid:

1. Draining the batteries when on an ohms position.
2. Burning out the rectifiers when on an AC position.
3. Catastrophic damage should a subsequent user carelessly begin to use it without first checking that the Function Selector Switch is at an appropriate setting.

4.1.1. CURRENT RANGE EXTENSION

A D'Arsonval meter movement can be modeled as a coil resistance R_m in series with an ideal (zero resistance) ammeter, Fig. 1. A typical value of R_m is listed on the face of the laboratory meter (say 50 Ω for Heath EVM-18) with the full scale current sensitivity (say 1 mA). This meter movement can then be calibrated and used directly to measure DC currents from 0 to 1 mA in magnitude. To extend the range of this basic meter movement (i.e., to measure currents of larger magnitude), a parallel shunt resistance R_s can be used to bypass some of the current around the meter movement.

The value of R_s is computed from the known value of meter resistance R_m, current for full scale deflection I_{fs}, and current scale deflection of the extended meter $\overline{I}_{fs}$.

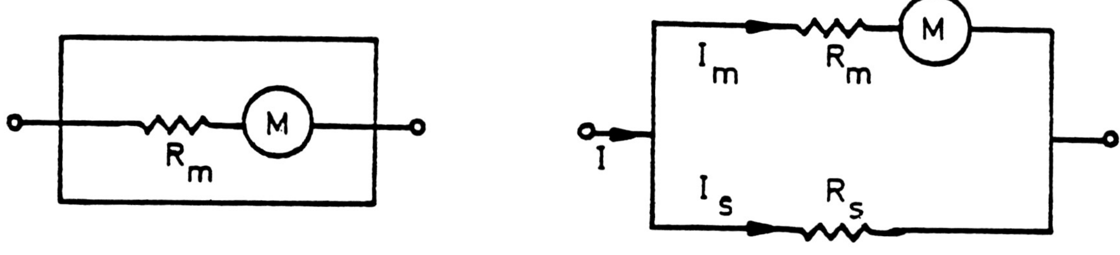

Fig. 1.   An Ideal Ammeter   Fig. 2.

From Fig. 2, the relationship between currents and resistances in the circuit is

$$I_m R_m = I_s R_s , \quad I = I_s + I_m .$$

Then we obtain:

$$R_s = R_m (I_m / I_s) = R_m / [(I / I_m) - 1].$$

To construct an extended meter with a full scale current $\overline{I}_{fs}$ from the basic meter when I_{fs} and R_m are known the required value of shunt resistance is

$$R_s = R_m / [(\overline{I}_{fs} / I_{fs}) - 1].$$

As an example, suppose we have the following VOM which has a full range of 50 microamperes (i.e., 50×10^{-6} A) and a resistance of about 1800 ohms. This is connected in series with a resistor that is adjusted by the manufacturer so that the series combination is 5000 ohms. Thus, the indicating instrument itself can be used as either a microammeter with a full-scale current of 50 microamperes or a millivoltmeter with a full-scale voltage of 250 millivolts. A special terminal is provided for these ranges on the instrument, and this terminal is illustrated in Fig. 3. We can shunt the indicating instrument with various resistors to obtain other ranges as shown in Fig. 3. The R_s is chosen so that $(I - 50\mu A) R_s$ equals 250 millivolts. (Fig. 3, is a schematic diagram of Simpson Model 260 VOM,[2]).

4.2. FAMILIARIZATION WITH THE OHMMETER [2]

A simplified ohmmeter circuit is shown in Fig. 4. This circuit is called a "series Ohmmeter" because the unknown resistance R_x and the battery (built into the instrument) are in series.

The operation of the series ohmmeter is as follows. The terminals are first shorted and R is adjusted to yield full-scale current I_o on the indicating instrument; this point is marked "0" on the ohms scale. What value of added external resistance will result in a half-scale current? Obviously, a value equal to the Thevenin (cf., EXP. II-3) resistance, R_{th}, as seen looking into the instrument from the terminals. (The *range* of an ohmmeter is specified by the *mid-scale* value.) The scale is further calibrated by noting that if $R_x = 2 R_{th}$, $I = I_o / 3$; if $R_x = R_{th} / 2$, $I = 2 I_o / 3$; *etc.*

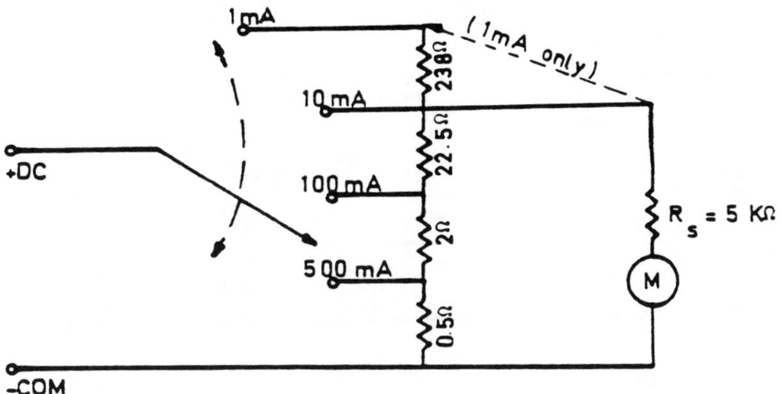

Fig. 3. Milliammeter [2].

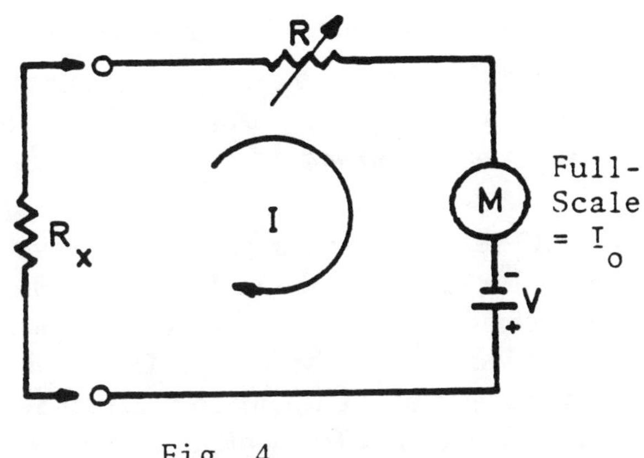

Fig. 4.

Why must we go through the process of shorting the terminals and adjusting R? If we had a perfectly stable power supply, adjustment of R would not be necessary, but batteries change in two ways: (1) their internal resistance increases; and (2) their voltage decreases. The adjustment of the resistor R in Fig. 4, can correct for battery resistance changes. However, it cannot correct for voltage changes because if the battery voltage decreases, R must be decreased in order to achieve full-scale current with the terminals shorted. But this means that we have changed the Thevenin resistance, and our calibration is wrong! Another disadvantage of the circuit of Fig. 4, is that it is necessary to change the voltage value each time we want to change the range (i.e., the mid-scale value) of the ohmmeter. The ohmmeter circuit in the following sample multimeter illustrates a compromise; only two sources are required for three ranges, and the variable resistor ("ZERO OHMS") corrects fairly well for changes in both battery resistance and voltage.

CAUTION: NEVER USE AN OHMMETER TO MEASURE THE RESISTANCE OF A "LIVE" CIRCUIT.

To avoid connecting a VOM to a live circuit accidently while it is in the "resistance" mode, *place* the VOM in the highest *voltage* setting when it is not in use.

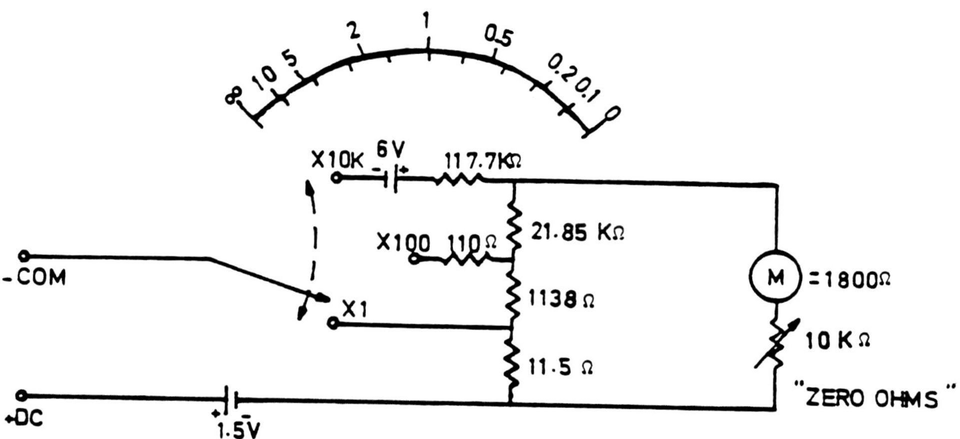

Fig. 5. A Schematic Diagram of a Multiscale Ohmmeter, (Simpson Model 260M, [2]).

4.3. FAMILIARIZATION WITH THE VOLTMETER

A current measuring device is a voltmeter when its resistance is known. *Locate* the operating controls and note the several scales on the meter face of your VTVM.

Before turning the equipment on, *check* to see that the meter reads zero. If not, carefully turn the set screw located on most of the meter faces, using a screw-driver, to obtain a zero setting. *This adjustment is only made with the meter turned off and is required very infrequently and it generally has been done for us.*

Turn the voltmeter on. If you are using a vacuum-tube voltmeter *set* the Function Selector Switch on the DC+ position. Also *set* the probe to DC, and *connect* it to the ground lead. Wait a minute until the meter needle comes to a steady position. *Adjust* the meter to zero with the Zero Adjust Control. *Set* the Function Selector switch to DC– and check the zero. When properly adjusted the meter needle remains stationary when the function switch is changed from DC– to DC+ or *vice versa*. *This adjustment is required frequently.* Leave the VTVM on. If you are using any other voltmeter you should carefully review its operations manual to learn about that device.

4.3.1. VOLTAGE RANGE EXTENSION

Though the D'Arsonval meter movement is a current-sensitive device, it can be used to measure voltage. The voltage for full scale deflection of the meter movement is

$$V_{fs} = I_{fs} R_m .$$

To extend the voltage range of the basic meter, a series dropping resistor R_d is used as shown in Fig. 6.

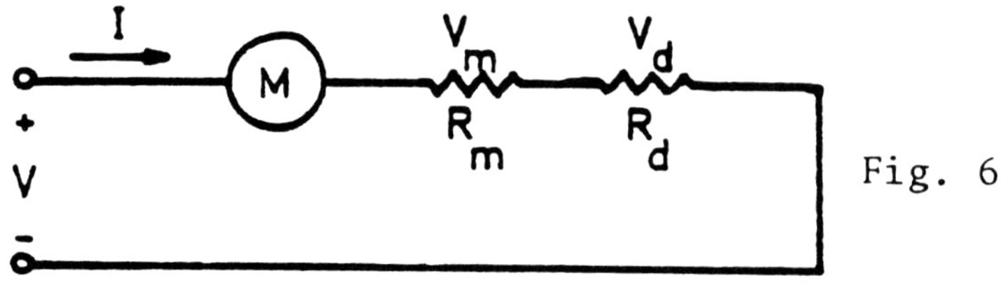

Fig. 6.

From this circuit we have $V = V_m + V_d = V_m + I R_d$. To determine the value of R_d for full scale deflection of a voltage $\bar{V}_{fs}$, given V_{fs} and R_m, we have

$$R_d = (V - V_m) / I = (\overline{V}_{fs} - V_{fs}) / I_{fs}.$$

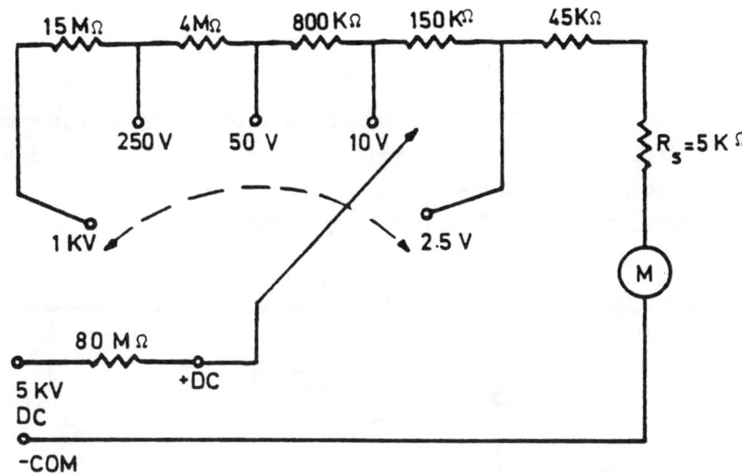

Fig. 7.   A Schematic Diagram of a Multiscale DC Voltmeter [2].

4.4. REMARKS ON R_M AND I_{FS}

One must realize that the values of R_m and I_{fs} as used above are generally speaking different from those indicated on the meter face. We can determine these values but we will postpone this sort of measurement until future experiments when more experience has been gained.

4.5. USE OF METERS IN ELECTRICAL CIRCUITS

Ammeters are used by inserting the ammeter in the branch of the circuit in which the current is to be measured. Voltmeters are used by connecting the voltmeter to the nodes between which the voltage is to be measured. These applications are shown in Fig. 8.

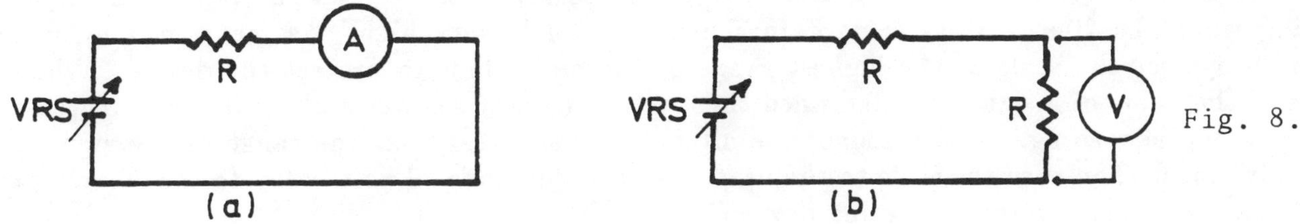

Any meter connected in a circuit for measurement purposes becomes part of that circuit. The currents and voltages present in the circuit are then not the same as those which would be present without the meter.

An ammeter is always connected in series with the circuit element through which the current is to be measured. The voltage across the element is then less than the voltage which would be present without the ammeter by an amount equal to the voltage drop across the ammeter. The voltage across the ammeter may be minimized by using an ammeter with as *small* a value of resistance as possible.

A voltmeter is always connected in parallel with a circuit element for measurement of its voltage. The resistance of that portion of the circuit is lowered by the presence of the meter; the meter is said to load the circuit. The voltage measured is then not the same voltage that exists across the element when the voltmeter is removed. The difference between the voltage across the element with and without the voltmeter in the circuit can be reduced by using a voltmeter with as *high* a resistance as possible.

4.6. LOADING

Since actual voltmeters have finite resistance, rather than the infinite resistance associated with an ideal voltmeter, the effect of this finite resistance must be considered when making a measurement. Assume that it is required to measure the voltage across resistance R_B in the circuits of Fig. 9. The resistance in these circuits are 1% precision resistors. Recall VRS means voltage reference source (PS).

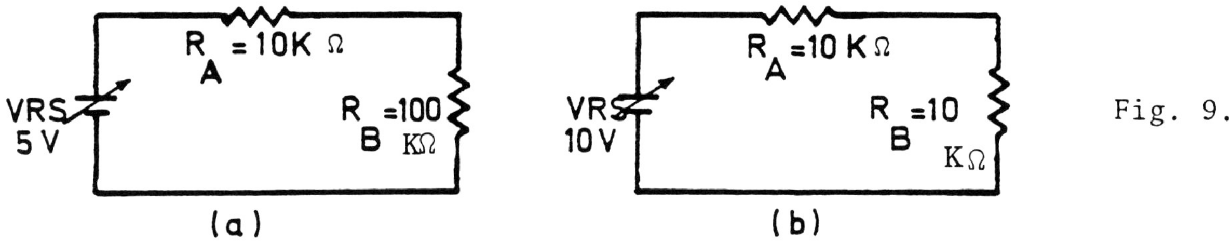

(Optional)   Using the values indicated, calculate the voltage across resistor R_B in circuits (a) and (b). Then measure V_B in circuits (a) and (b) at two separate scales of the voltmeter. Compare the calculated values of V_B that would exist if the resistance of the appropriate voltmeter is considered as part of the circuit. Compare these values with those measured earlier.

4.7. CALIBRATION AND METER ERRORS

The accuracy of a meter depends on many factors such as loading and the magnitude of the signals to be read but above all it depends on the movement of the meter itself. This measure of accuracy is denoted as a percentage of full-scale deflection; therefore at smaller deflection, the meter reading is less accurate than at the maximum deflection of the meter. For example if a meter is accurate to about 3% of full scale and is used on the 80-volt scale to read 20 volts, then the actual reading would be 20±2.4 volts or 12% inaccuracy which is very high. As mentioned before, one must always set the scale at the highest range of the meter first to protect the device. When the range of the value of a signal is determined then we must select a lower scale in order to increase the accuracy of the reading. In the above case after it became clear that the range of the signal is 20 volts, one must then turn the scale to the next closest scale above 20 volts - say to the 30 volt range. This concept is true for any measuring device.

By comparison with a highly accurate meter we can always calibrate our meters. But this is not expected from students. Just be aware that in many cases we may have to draw a calibration curve for a meter that is being used. This curve shows at each reading how much correction must be added to or subtracted from the actual readings. A calibration curve can be obtained for *any* measuring device.

5. DIGITAL MULTIMETER (DMM)

A digital Multimeter is a compact device which combines the common features of several measuring devices housed in one unit. Most commercial DMM's are capable of performing at least the following measurements:

DC voltage and current measurements   ,

AC voltage and current measurements   ,

dB measurements   ,

Resistance measurements   ,

Temperature measurements   .

The structure and use of a DMM, as far as its various ranges and placement in a circuit for testing purposes, is the same as for a typical voltmeter or current-meter. Instructions on how to use a particular DMM must be obtained from the manufacturer.

6. LABORATORY PROCEDURE

6.1. PRE-LABORATORY WORK

It is expected that you will be given sufficient instructions (perhaps directly from the manufacturer of your EEP), regarding the equipment you will be using in the laboratory. Before arriving at the laboratory please read the appropriate instructions very carefully. Also look at your instrument specifications to have some general idea about the types of measuring devices you will be using.

6.2. LABORATORY WORK

6.2.1. EEP-POWER SUPPLY FAMILIARIZATION

Here we denote the output of a power supply as VRS which stands for voltage reference source. To study the terminal characteristics of a PS we construct the circuit shown in Fig. 10, on a given breadboard.

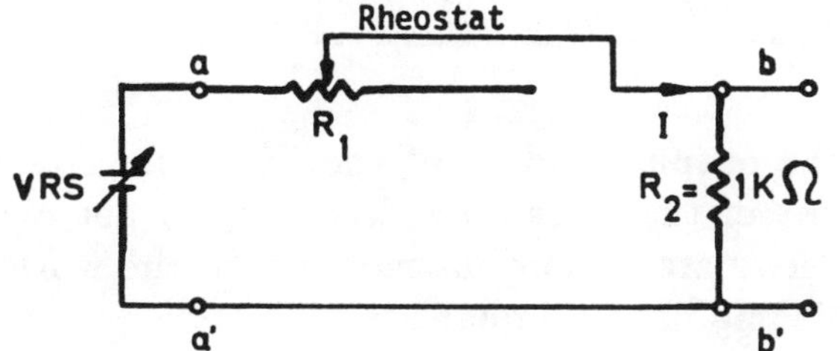

PLEASE RECORD ALL YOUR DATA

1. *Set* your VRS at +10 volts, using the PS calibrated dial indicating its output. *Read* no load output of PS by removing R_2, and *use* DMM to read $V_{aa'} = V_{bb'}$. *Adjust* the dial of PS to read exactly +10 volts on DMM, if needed. Here R_1 can be any small value.

2. *Repeat* step 1, for –10 volts VRS.

3. *Connect* R_2 to the circuit of step 1. Now , *read* $V_{aa'}$ with DMM. Do you still have +10 volts?

4. *Gradually* increase R_1 and *read* $V_{bb'}$. Since you know R_2, what is I_o for each setting of R_1. Do *not* adjust the current limiter switch of the PS. Then *measure* $V_{aa'}$.

5. *Repeat* Step 4, five times. *Plot* $V_{aa'}$ versus I.

6.2.2. A GENERAL PURPOSE AC POWER SUPPLY

Another power supply available to us is the circuit shown in Fig. 11.

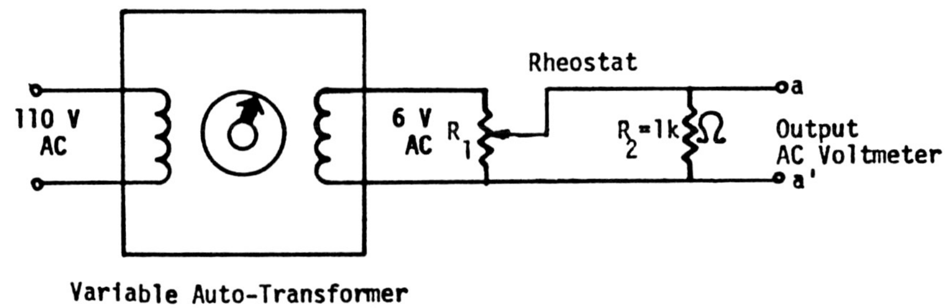

Fig. 11. A Simple AC Power Supply.

We may set the variable auto-transformer dial at a low level, first, then *VERY CAREFULLY* change the value of Rheostat R_1 to read several output levels at *port* aa$'$. Since this is a high voltage experiment and the dial may be set accidentally at a *higher* level we urge you to consult your instructor before testing this circuit.

6.2.3. VOM AND VTVM FAMILIARIZATION

Upon starting your work in the laboratory, *study* the function of various switches on VOM and VTVM and make a few measurements using PS voltage reference source (VRS). You may use the circuit of Fig. 12, setting both VOM and VTVM at DC.

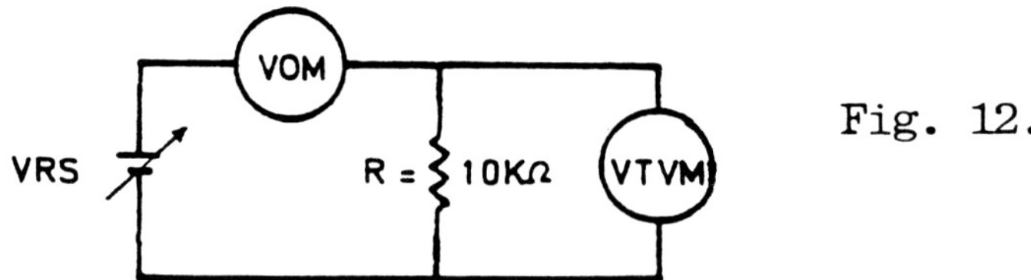

Compare your reading with the output reading of the PS dial. You will get your first lesson immediately! You will see that most of these readings, literally, do not agree with each other. Do not panic and just remember that there are no two instruments in the world which will yield *exactly* the same reading for measuring the same physical quantity.

What you should do is choose one frame of reference as your basis and make all your measurements in that basis. You will, of course, get different readings if you use DMM instead of the VTVM or VOM. Therefore, from now on remember to use only one measuring device consistently for collecting your data. You may, of course, get close results or almost identical results if you use different devices. But there is always the possibility of error if you use different units during any given experiment.

7. PROBLEM SET

You may work on these problems before the laboratory but you must submit as many as you can with your report next period.

1. Explain, by describing the effects on the meters and on the circuits being measured, why a voltmeter should never be connected in series, or an ammeter in parallel, with the circuit being measured.

2. If a 10 mA meter movement has an internal resistance of 4 Ω, calculate the shunt resistance required to increase its range to:

    a)   50 mA,

    b)   250 mA.

3. If a 1 mA, 100 Ω meter movement is inserted into a circuit branch whose resistance is 500 Ω, find the error in the indicated current values due to the insertion of the meter, assuming the voltage across the branch is not affected by insertion of the meter. If the movement is converted into an ammeter whose full scale range is 10 mA, find the resulting error.

4. Two resistors with values of 12,000 Ω and 6,000 Ω are connected in series across a 50 V voltage source. A voltmeter with a 1,000 Ω/V rating is used on its 40 V scale to measure the voltage across the 12,000 Ω resistor. Calculate the error in the measured voltage value due to the loading effect of the voltmeter on the circuit.

8. FOR YOUR REPORT

Document your observation and the solution to the above problem set in our standard report form.

EXPERIMENT I-2

**FAMILIARIZATION WITH ELEMENTARY OSCILLOSCOPE
AND FUNCTION GENERATOR**

EQUIPMENT:

    EEP

    EXTRA FUNCTION GENERATOR FOR LISSAJOUS FIGURES, OR

    RESISTOR (2 kΩ), and CAPACITOR (0.1 μF), OR

    PHASE SHIFT NETWORK (Fig. 20b.)

Table of Contents   Page No.

1. PURPOSE

In this experiment we are primarily interested in exploring the basic operations and/or functions of an oscilloscope which is the "heart" of any electrical engineering laboratory. Essential to such an exploration is some basic knowledge of the operating conditions of a function generator (FG) other than the power supply we studied before. As in the preceding experiment we again confront the same difficulty that we faced earlier. That is we can not explain how a function generator works unless we use an oscilloscope and *vice versa* . Therefore we will study these two units simultaneously. The students should realize, however, that it is extremely instructive to read the manufacturer's manual for each unit they are going to use prior to coming to the laboratory.

2. STRUCTURE OF A GENERAL PURPOSE OSCILLOSCOPE: AN OVERVIEW

In the simplest description an oscilloscope is an electronic measuring device that displays electrical voltages in graphic form.

Any general purpose oscilloscope consists of four related equally important parts as follows:

2A. *Display*

2B. *Vertical Amplifier*

2C. *Horizontal or Time-Base Amplifier*

2D. *Triggering Function*

We describe briefly each of these parts for a generic and/or a basic general purpose oscilloscope.

2A. Display

The main function of an oscilloscope is to display electrical signal voltages on a phosphorescent screen. This display results from the motion of an *electron beam* generated by an *electron gun* . This gun is located at one end of a tube which is surrounded by controlled electrostatic fields and closes with the front panel phosphorescent screen of the oscilloscope. This tube is called the *Cathode Ray Tube* (CRT). Other important elements associated with the CRT are *vertical and horizontal deflection plates.* Each of these plates controls the motion of the electron beam generated by the electron gun. The phosphorescent screen of the CRT is made of a phosphor coating that produces a light spot when an electron beam strikes it. The output of the electron gun is limited by a *control grid* that sets a limit on the flow of electrons according to the applied voltage. The *INTENSITY CONTROL SWITCH* on the CRT panel is for manually adjusting the flow of electrons striking the phosphorescent screen. This is done by letting the electron beam pass through two nodes that focus and speed up the electrons to the screen. Permanent damage may occur to the screen if the *intensity switch* reach which controls the beam brightness on the screen is set so brightly that a circle form around the spot. Each of the deflection plates, depending on its potential, controls the motion of the electron beam. This control is usually done by bending the electron beam that becomes attracted toward the positive plate of each of these deflection plates. When no voltage is applied to either set of deflection plates the electron beam passes straight to the center of the screen. Any variation from this point is the consequence of the level of applied voltages to each of the deflection plates. In most applications the voltage to be measured must be amplified before it is applied to either of these deflection plates. This amplification is generally performed internally by the vertical amplifier section of the oscilloscope. In addition to the *intensity control switch* that we describe above there is a *BEAM FINDER SWITCH* associated with each display part that brings the off-screen spots to the CRT display screen. The *power switch* that turns the oscilloscope on and off is also an element of this display part.

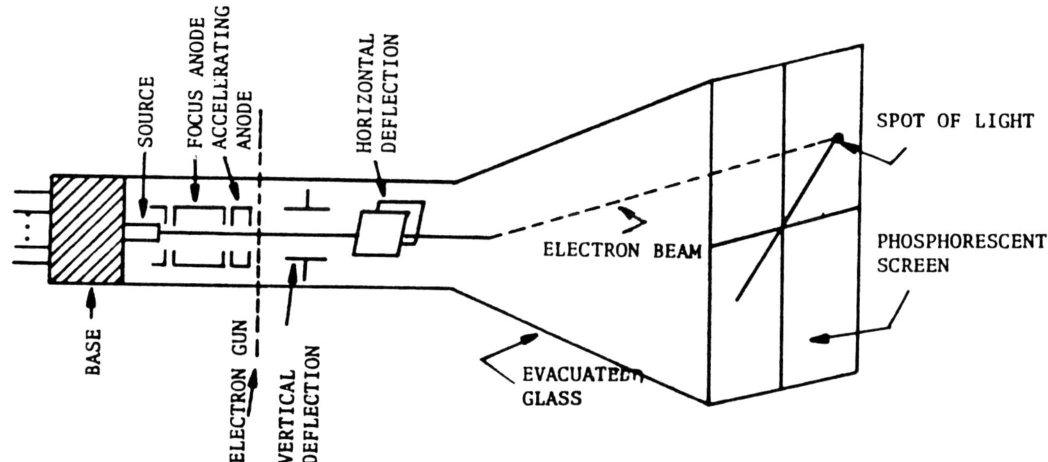

Fig. 1. Schematic Diagram of a Cathode Ray Tube (CRT).

2B. *Vertical Amplifier*

Each general purpose oscilloscope has at least one vertical amplifier. The most commonly used oscilloscope today is the dual beam (trace) oscilloscope which has two vertical amplifiers and is thus capable of displaying two signals on the screen simultaneously. Each vertical amplifier (y-axis) receives a signal and amplifies it, if needed, to display on the screen. The vertical amplifier controls consists of at least four elements. The most important one is the *SENSITIVITY CONTROL SWITCH*. This switch adjusts the necessary amplifications of the signal for displaying. Found on this control switch is the multiple of *VOLTS/DIV*ision, indicating the amplitude of the signal (which is proportional to the number of divisions on the vertical axis of the screen). Next is the *INPUT* connector (generally in the form of a BNC connector) to receive the external signal (directly from the source or through a *PROBE*) to the oscilloscope. We will study the probe later. The third important element is the *COUPLING SWITCH*. This switch indicates how the input voltage must be connected to the vertical amplifier. This switch consists of three positions marked *AC*, *GROUND*, and *DC*. If the switch is on *AC* then only the AC component of the input signal will reach the vertical amplifiers, i.e., the DC component will be blocked. If the switch is on *DC* then both AC *and* DC components of the signal will reach the vertical amplifier, i.e., no DC blocking will take place. If the switch is on *GROUND*, then only one straight line will appear on the screen showing the position of the reference line. We usually must check the position of the reference line before making any voltage measurement. Finally, the fourth important element is the *POSITION SWITCH* . This switch is a signal searcher that brings the signal to the screen and indicates if changes are needed for the sizes of *VOLTS/DIV* on the amplitude display. Figs. 2 to 4 indicate the function of *VOLTS/DIV*, *COUPLING SWITCH*, and the *POSITION SWITCH*.

Dual trace or two-channel oscilloscopes have two identical vertical amplifiers and can display two applied signals either simultaneously, or alone, or in an alternating mode. This type of oscilloscope is also capable of subtracting the signals applied on one channel from that applied on the other, provided that the *VOLTS/DIV* settings on both channels are the same. Provisions have also been made in most of these oscilloscopes to apply signals to both channels in order to display one signal versus the other. This is also called the X-Y mode.

In many applications it is much safer to connect a given signal to the oscilloscope through a device called a *PROBE* . This device generally consists of a simple RC (or RLC) network and protects the oscilloscope against possible damage in the form of high input signal amplitude and/or parasitic signals. If the probe is used then the appropriate scale factor must be considered in the measurement since, in reality, a probe is a step-down circuit. One major source of measurement

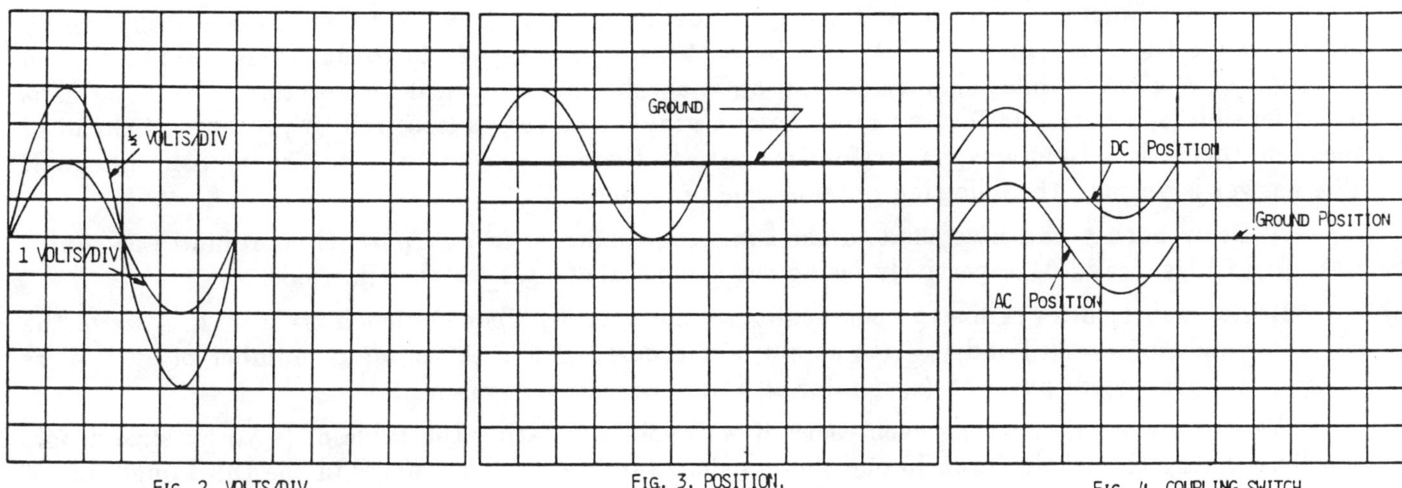

FIG. 2. VOLTS/DIV.   FIG. 3. POSITION.   FIG. 4. COUPLING SWITCH.

error is attributing a wrong value to the *gain* of a given probe. This error may also be due to incorrectly compensating the probe. Further information regarding the probe must be sought in the operating manual.

2C. Horizontal or Time-Base Amplifier

To display a signal we clearly need a two-dimensional coordinate system. The vertical amplifier is responsible for displaying the y-axis or the amplitude of the signal while the x-axis is responsible for the running variable which is time. Before we further present the function of an oscilloscope and the method of displaying a waveform we entertain our readers with the following thought.

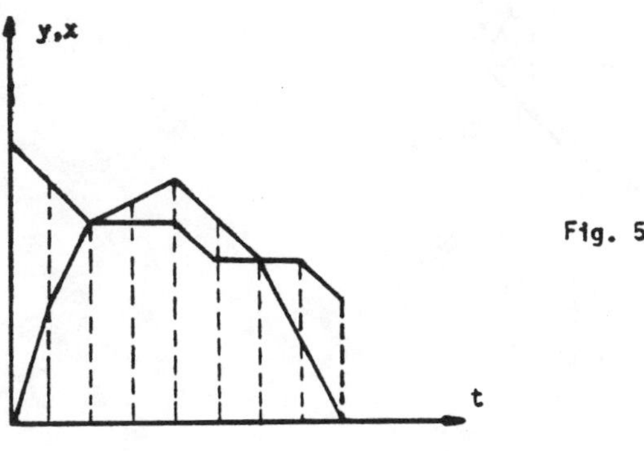

Fig. 5.

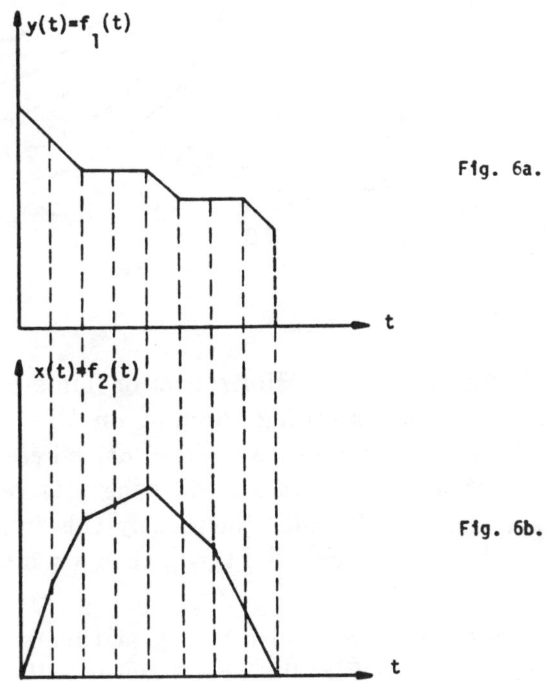

Fig. 6a.

Fig. 6b.

Given two functions $y = f_1(t)$ and $x = f_2(t)$, how do we draw $y = g(x)$? We assume both x and y are given for the *same interval* of t. Perhaps the *simplest* answer is to find t as $t = f_2^{-1}(x)$, then $y = f_1[f_2^{-1}(x)] \triangleq g(x)$. While this formal solution is a valid mathematical formulation in reality this is not a feasible approach. In fact we often have a graph of x versus t as well as y versus t instead of these mathematical functions. Therefore we ask, is there any other way to accomplish this goal? The answer is yes and the following provides the procedure.

1) Draw both y and x versus t in the same coordinate or in two parallel coordinates as shown in Figs. 5 and 6, respectively, using the same scale factor for the time axis.

2) At each t, draw a line perpendicular to the t-axes. Then for each t_i, i=1,2,..., *write* the values of $x_i = x(t_i)$ and $y_i = y(t_i)$ as a set of ordered pairs. After many number of such pairs are generated, put each (x_i, y_i), for all i, in a coordinate system.

We can make this drawing even easier if we avoid generating the table of (x_i, y_i) by straightforwardly drawing the whole graph in one shot as shown in Fig. 7. The curves in the third quadrant of Fig. 7 are directed arcs of concentric circles whose radii are the points t_1, t_2, *etc.*

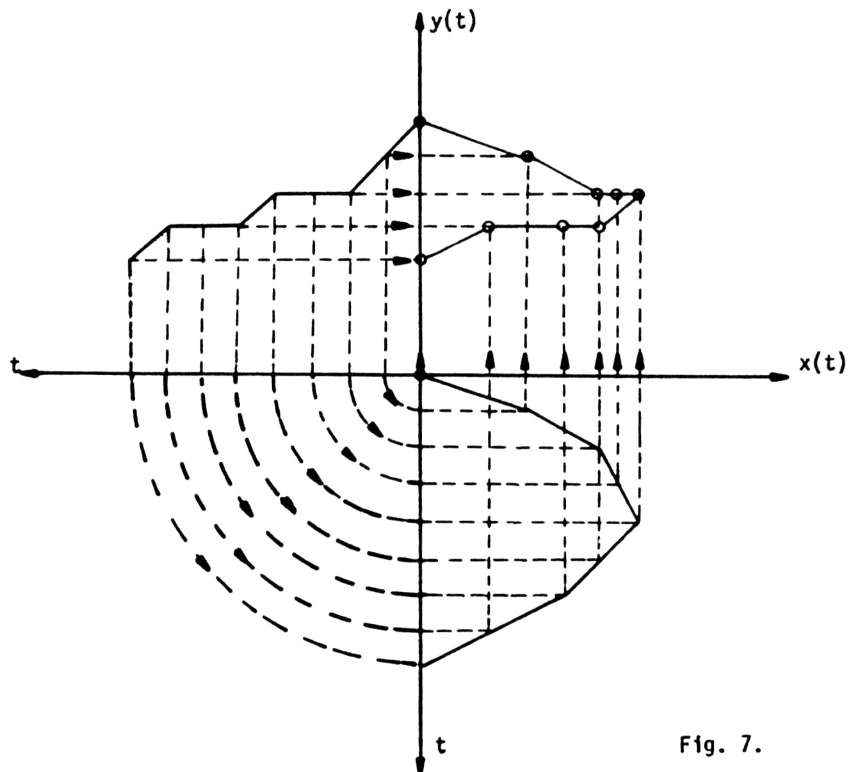

Going back to our Horizontal or Time-Base Amplifier discussions, one may wonder what is the relevance of the preceding thought on this subject. It becomes obvious if we choose x(t) to be a linear function of time, say $x(t) = \alpha t$, where α is a constant. Then the drawing y(t) versus x(t) becomes actually the same as drawing y(t) versus αt. But in the practical cases we do not have to generate $x(t) = \alpha t$ for all t (incidently this function is called a ramp function). In other words, if we want to draw y(t), for $t \in [t_0, t_f]$, then we may only use x(t) for the interval $t \in [t_0, t_f]$, as shown in Fig. 8.

In reality we may wish to repeat this drawing several times, then we need a train of these functions as shown in Fig. 9. This wave is called a *sawtooth* signal. But in practice we do not have a perfect 90 degree line at each multiple of t_f. The actual time-base signal is shown in Fig. 10. The

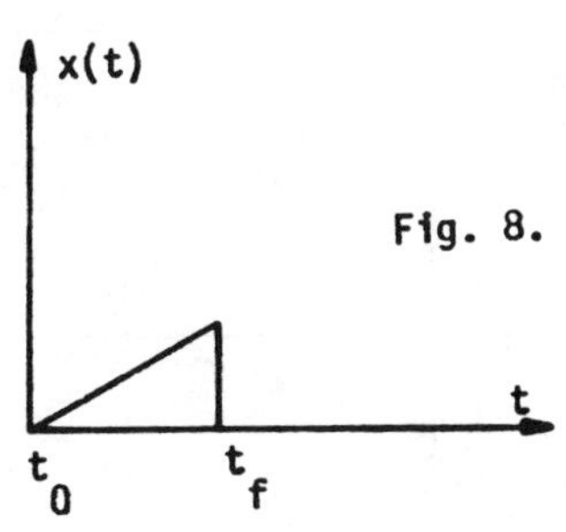

Fig. 8.

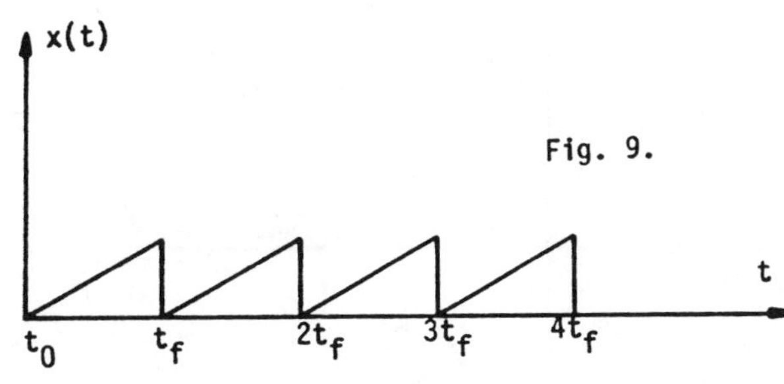

Fig. 9.

time δ is called *flyback* or retrace time [1].

During this very short time the beam is shot off and the beam is sent back to the left side of the screen. The increasing portion of this wave is a constant rate signal with the rate of multiple SEC/DIV. The rate can be adjusted by a switch that has discrete calibrated setting and also contains a concentric control for further sweep control between the settings. This SEC/DIV is actually the rate at which the beam sweeps pass the screen.

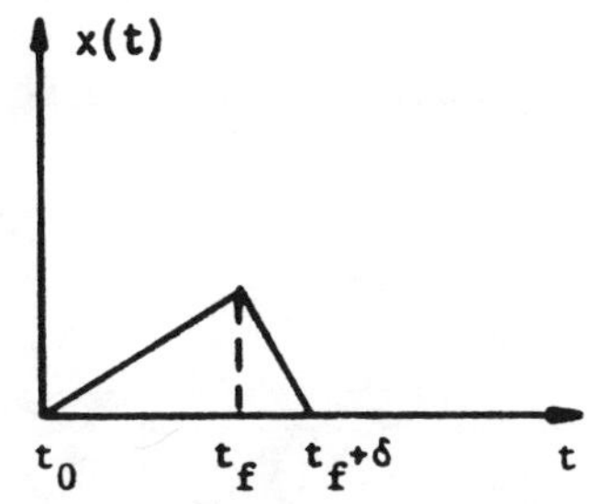

Fig. 10. Time-Base Signal
(Sweep Waveform)

Now, suppose we have the signal of Fig. 9, for the time-base and the y-axes is given by Fig. 11. What is y(t) versus x(t)? If we apply y(t) to the vertical amplifier and x(t) of Fig. 9, to the time-base, then with an *appropriate setting* we get y versus x on the screen as shown in Fig. 12.

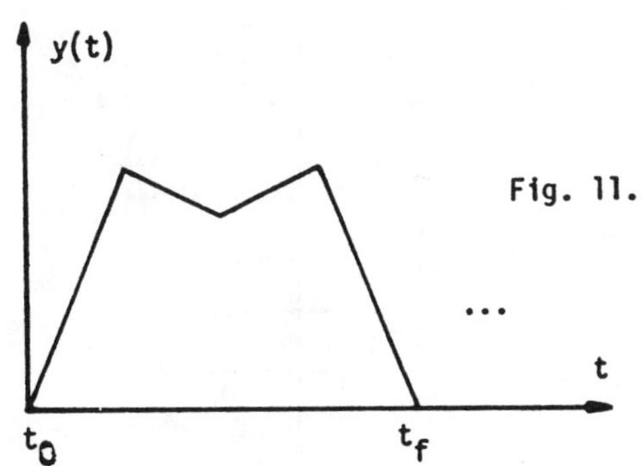

Fig. 11.

It is clear that under the above *appropriate setting* no matter how many identical y(t) signals we have we will get one perfect *stable* display on the screen. We can also display more of the input wave if we slightly change x(t) as depicted by Fig. 12b. But what about the two cases of Fig. 13, which may be considered as *inappropriate settings* ?

Here one may wonder what caused x(t) to start at a later time. Then, as a result, only portions of each y(t) are shown that are superimposed on each other and none of these waves are actually a complete replica of the applied signal on the y-axis. From the viewer's point of view this is an

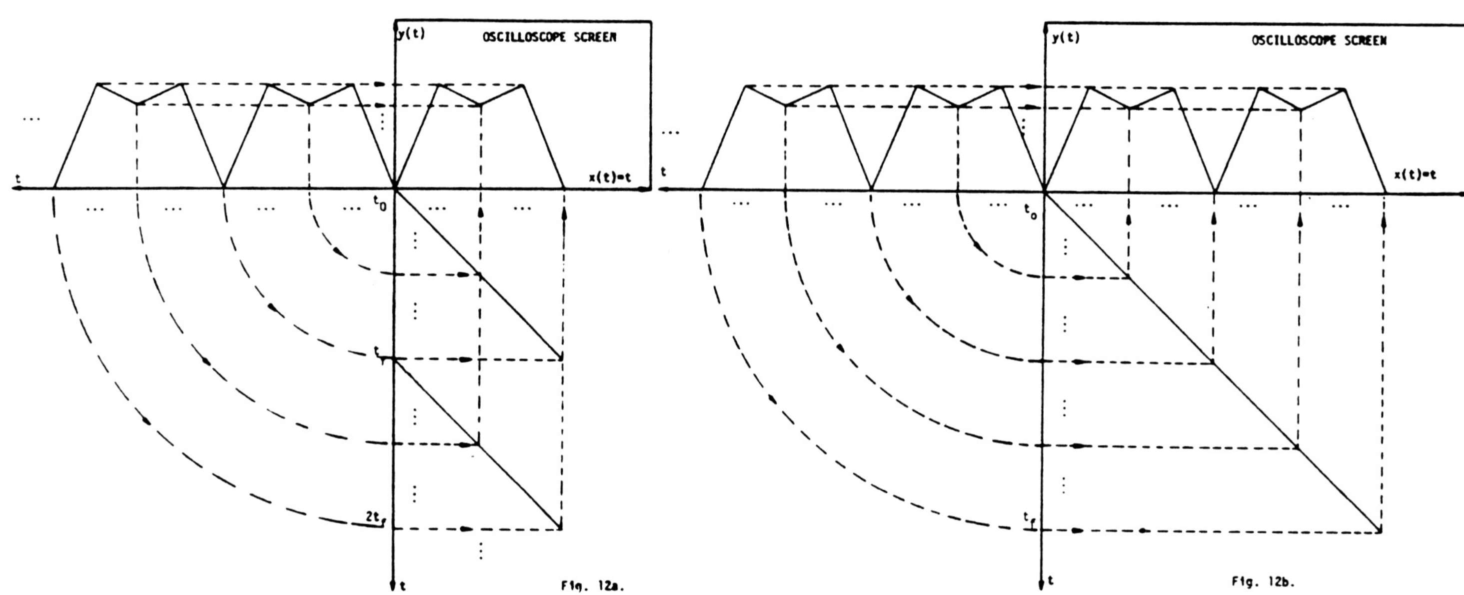

Fig. 12a.　　　　　Fig. 12b.

unstable picture. The reason for this and what can be done to avoid it is the subject of the next section.

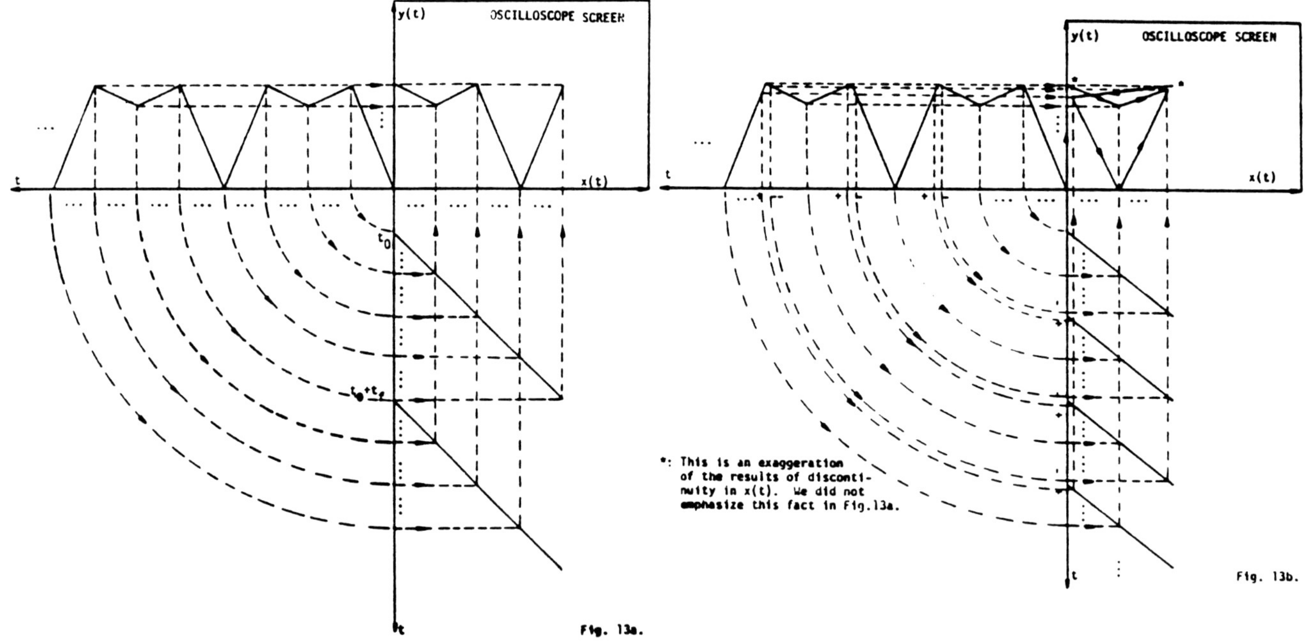

Fig. 13a.　　　　　Fig. 13b.

2D. *Triggering Function*

From Fig. 13, we intuitively realize that in order for the display to become stable $x(t)$ must be generated *in accordance with* $y(t)$ in the following two senses: (1) The indicated time interval ($t_f - t_0$) must cover that of $y(t)$; and (2) the starting point t_0 must be the same or earlier than $y(t)$. This operation is called the triggering function and it is responsible for generating a stable display on the screen. If t_0 of $x(t)$ starts after $y(t)$ starts then we will still get a picture on the screen. Triggering is said to occur on the *positive* or *negative slopes* of the actual wave depending on when t_0 begins and its relation to the actual y(t). There is a control switch on each oscilloscope to indicate this operation. Thus, in short, the triggering circuit is responsible for the *proper* functioning of the time-base signals. This is done in several different ways. The most popular modes for triggering functions are: *AUTOmatic; NORmal;* and *TV* . In *AUTO* mode the triggering signal is at such a level that deflection occurs without difficulty. This mode is used in most applications since necessary adjustments, if needed, are done automatically. *NORM* does not have this self-correcting capability, but under the right circumstances both *NORM* and *AUTO* act the same. *TV* is used for triggering on television signals. Where does the triggering circuit get its source? There are several different ways to acquire this. The main types are: *EXTernal* (that is done externally); *LINE* (that some portion of the line voltage which the oscilloscope is connected to is used); and *X-Y* (that shows a display of two signals applied to the two inputs of the oscilloscope). We can summarize the triggering function and time-base sweep generator as follows, [3]:

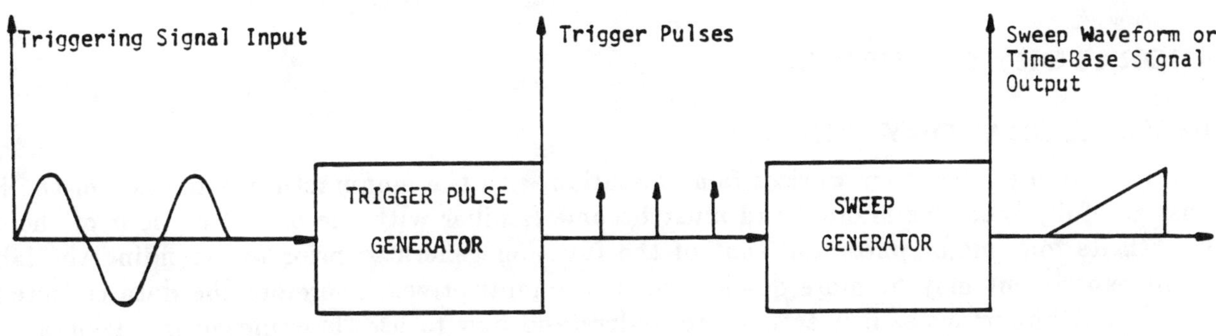

Fig. 14.

The above was a concise description of a general purpose dual oscilloscope. There are many different special purpose oscilloscopes on the market that work basically just as the above general purpose oscilloscope but have many additional features. To name a few we recall the following set:

      DIGITAL OSCILLOSCOPES   ,
      STORAGE OSCILLOSCOPES   ,
      SAMPLING OSCILLOSCOPES   ,
      LOGIC ANALYZERS   ,
      SPECTRUM ANALYZERS   ,
      CURVE TRACERS   .

The interested readers must persuade their instructors to give them lectures on these units.

Further information must be gathered from the manufacturers' manuals for the specific units that are being used at each institution.

3. FUNCTION GENERATOR

This is a very important piece of test equipment for any electrical laboratory. We have studied the power supply which was considered to be a function generator in a restricted sense. In general, a function generator produces a variety of waveforms. Some typical waveforms are sinusoidal, square, and triangular over a wide range of frequency and amplitudes. More sophisticated function generators will give, in addition to the above waveforms, pulse and ramp waveforms. To use a function generator one must be careful that all the necessary arrangements as specified in the manufacturer's manual are fulfilled. In addition, the following general remarks must be kept in mind when using any function generator. All connections from the function generator to the circuit should be as short as possible and well shielded. Coaxial cables are ideal for this application. Also one must be alert to possible reflections arising from mismatched impedances of the circuit and the output of the function generator together with its output cables. This can be understood as follows. Suppose a signal is sent from the generator through a cable or transmission line and it reaches a circuit whose impedance is different from the impedance of the transmission line. Then reflection will take place and the reflected signal will go back to the generator. Depending on the value of these two mismatched impedances either the reflected wave will be added to or subtracted from the original signal and as the result distortion will take place. Therefore, in these cases it is essential to use matching filters to avoid this situation. Many other details with regard to the operation of a given function generator must be studied that goes beyond the scope of the present discussion. We urge readers to consult the manufacturer's manual of the unit they are going to use prior to coming to the laboratory.

4. LABORATORY PROCEDURE

4.1. PRE-LABORATORY WORK

Essential to any laboratory work is familiarization with the manufacturer's manual of the EEP. One must carefully read this manual and must become familiar with the basic functions of the oscilloscope with its four major parts and that of the function generator prior to attending the laboratory. This experiment may be more qualitative than quantitative. Therefore the data collected are not so critical or important as it is actually to understand how to use these measuring devices.

4.2. LABORATORY WORK

As mentioned in Section 2, an oscilloscope is used to display electrical voltages. This powerful device shows us the actual shape of a waveform, thus enabling us to use all the powerful mathematical tools at our disposal to analyze the given signal. Here we study some basic operations of an oscilloscope.

4.2.1. SIGNAL DISPLAYING AND AMPLITUDE MEASUREMENTS

Directly *connect* the sinusoidal output of the function generator to the vertical input of the oscilloscope. *Set* the time base switch at a setting such that a stable picture is seen on the screen. In order to have a perfect display, first *set* Input Coupling (*AC*-Ground-*DC*) Switch to Ground to put the reference line in the center of screen, then switch back to *DC* position. Next, *check* the *VOLTS/DIV* of the Sensitivity Control Switch to see whether the amplitude of the waveform is within the screen, if not decrease the *VOLTS/DIV* settings. What is the peak-to-peak voltage amplitude of the waveform? This can be determined by multiplying the number of divisions (peak-to-peak) of the waveform by the *VOLTS/DIV* setting. *Read* several points of this waveform and draw the entire waveform on the chart of Fig. 15a. *Record* the frequency of this waveform from the function generator frequency selector switch. *Repeat* these measurements for two other signal frequencies of the function generator. *Use* the chart of Figs. 15b, 15c and 15d, for the drawings.

4.2.2. FREQUENCY MEASUREMENTS

In the previous case we used the output signal of the function generator as our source. Clearly from the function generator frequency selector switch we can determine the frequency of this signal. We can also find this number from the display on the oscilloscope's screen as follows. Simply *adjust* the time-base selector switch such that one complete cycle of the waveform is displayed on the screen. Then *read* the number of horizontal divisions that covers this cycle. *Multiply* this number by the *SEC/DIV* setting on the time-base selector switch to get the time interval for one cycle. The frequency of this waveform is the inverse of this number. *Measure* the frequency of the waveforms drawn in the previous case and compare your measurements with the actual number shown on each graph as was determined previously from the frequency selector switch in each case above.

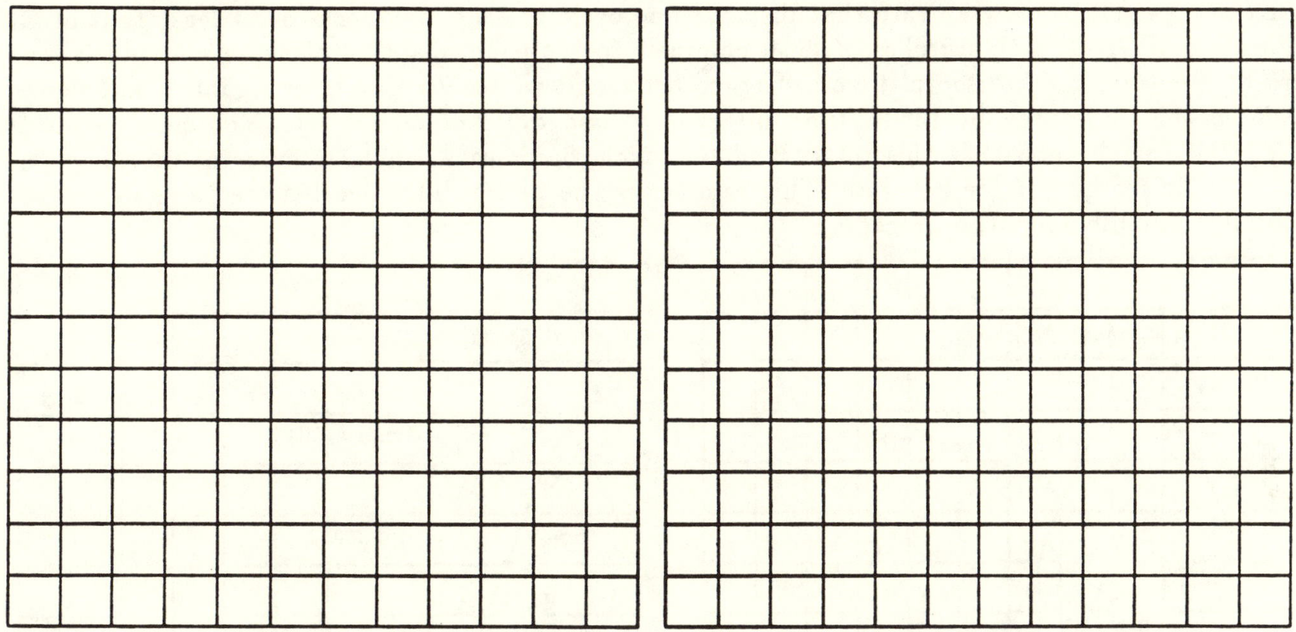

Fig. 15a.   Fig. 15b.

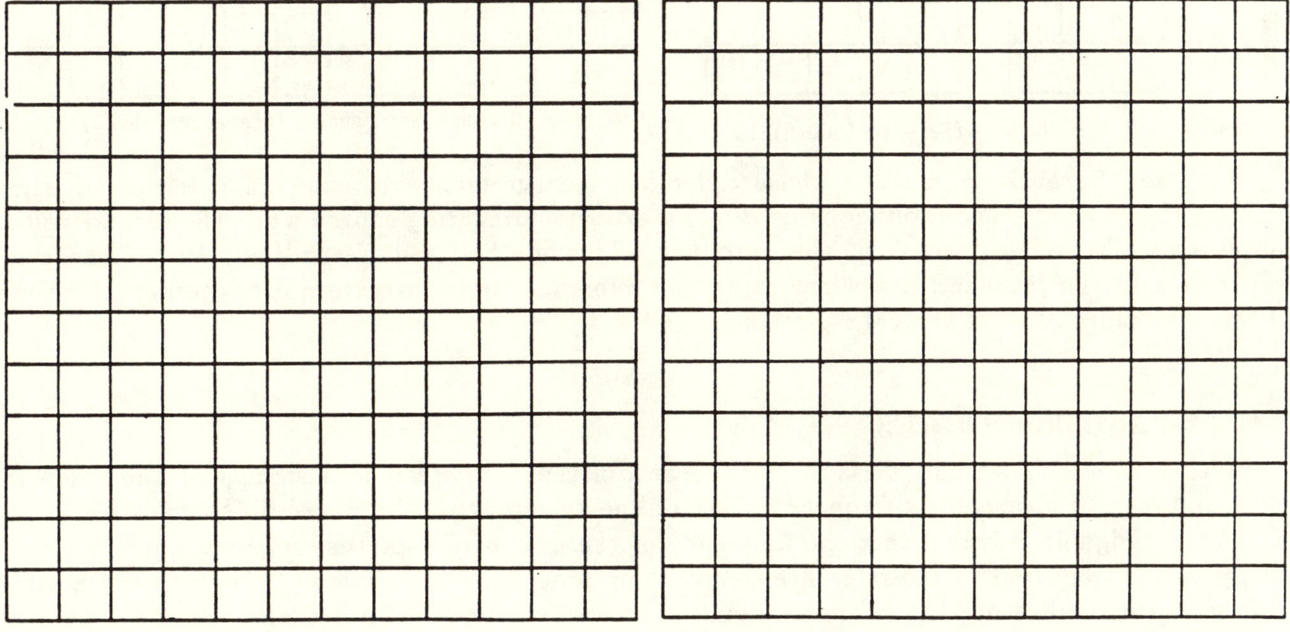

Fig. 15c.   Fig. 15d.

4.2.3. PHASE MEASUREMENTS

Another important application of an oscilloscope is to measure the phase difference between two signals A and B. Here we will present only one of several ways of doing this. Suppose the frequency of both signals A and B are the same. Then we may *apply* signal A to channel one (CH1) and signal B to channel two (CH2). We must make sure that the oscilloscope is in its *DUAL* mode and the *Ground* reference of both channels coincide. *Set INPUT COUPLING* switch to *AC* and the *SEC/DIV* switch to a sweep rate that displays one cycle of each waveform or at least that of CH1. *Adjust* the *VOLTS/DIV* switches of both channels to get equal displays of signals A and B on the screen. *Measure* the horizontal time difference between two *similar* points of signal A and signal B. *Multiply* this difference by (360°/time interval of one cycle of signal A). You may adjust the *SEC/DIV* switch such that this interval of one cycle becomes a whole number of divisions. This makes your readings more accurate. This number is the phase difference between signals A and B. Figs. 16a and 16b show this procedure.

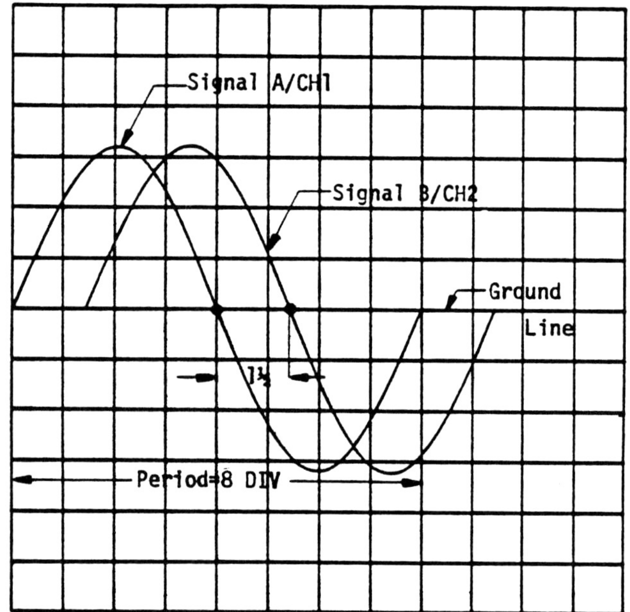

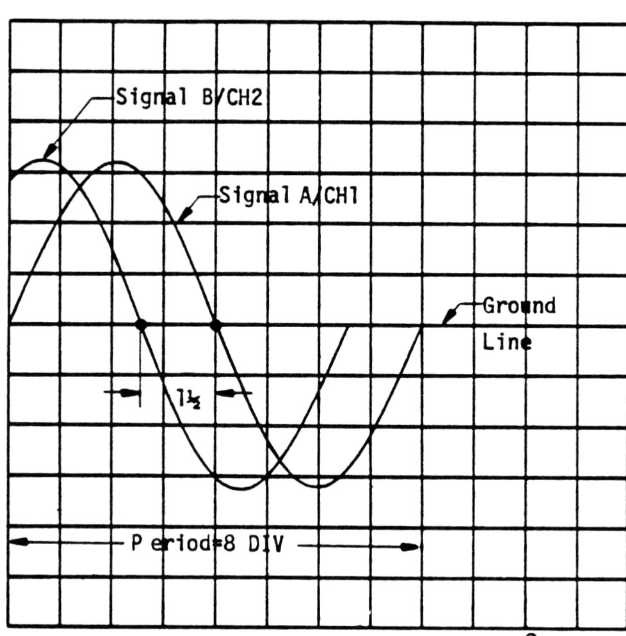

Fig. 16a. B Lags A : Phase Difference=(360°/8)*1½=67.5° Fig. 16b. B Leads A : Phase Difference=(360°/8)*1½=67.5°

One can alternatively make a phase difference measurement by using an external triggering function. Here the reference voltage is used as an external triggering source while the second voltage is applied on the vertical input of the amplifier. This method requires adjustment of the slope, sweep rate and the level settings when using the reference signal as external triggering. One may find this way more difficult than the previous case.

4.2.4. LISSAJOUS FIGURES

In this approach we can measure the phase difference between two signals of the same frequency and we can measure the frequency of an unknown signal and its phase difference with respect to a reference signal, provided that the frequency of the unknown signal is an exact multiple or submultiple of the frequency of the reference signal. The procedure is as follows. *Apply* to the horizontal input of the oscilloscope the reference signal A whose frequency is known and signal B whose frequency is an exact multiple or submultiple of the frequency of signal A to the vertical input of the oscilloscope. *Adjust the POSITION* switches to put the figures in the center of the screen. The following are a few typical cases:

1. Suppose signal A is $x(t) = A \sin \omega t$ and signal B is $y(t) = B \sin (\omega t + \theta)$. Then the Lissajous figure is shown in Fig. 17. It can be easily shown that the angle is $\theta = \sin^{-1}(\frac{\alpha}{\beta})$. As an exercise, the student should try to derive this relationship.

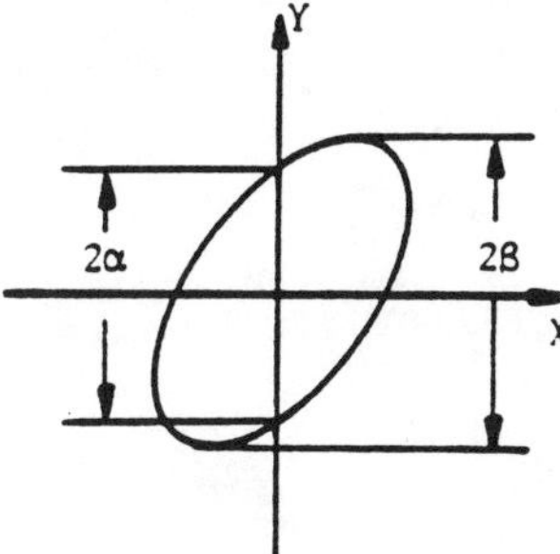

Fig. 17. Determining Phase Difference from Lissajous Figure.

2. Suppose both signals A and B have the same amplitude and are in phase. But the frequency of signal B is an exact multiple or submultiple of signal A. Then the Lissajous figures, which can be generated by the previous method, give us the following typical forms:

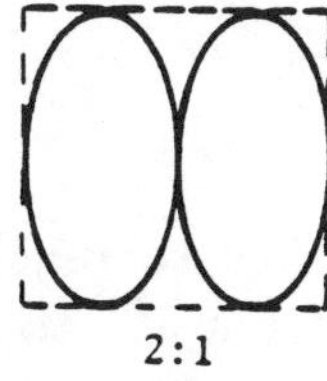
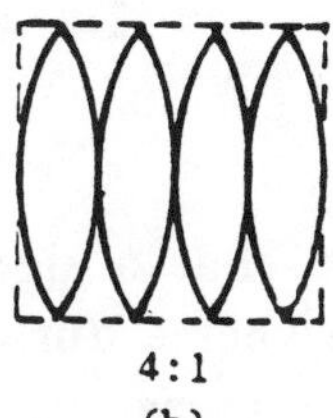
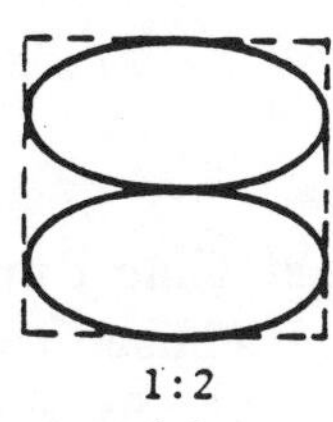

2:1
(a)

4:1
(b)

1:2
(c)

Fig. 18. Lissajous Figures

The signal frequency on the horizontal input is the reference value. If one draws a square that encloses any of these Lissajous figures, then the ratio of the frequency of the number of points of contact of this square on top to that on the side gives the ratio of the frequency of the unknown signal to that of the reference signal. For example, in case (a), the unknown frequency of signal B (vertical) is twice the frequency of known signal A.

3. Suppose both signals A and B have the same amplitude and the frequency of signal B is an exact multiple or submultiple of frequency of signal A but signals A and B are in phase. Some typical Lissajous figures for this case are illustrated below with the various phase differences indicated.

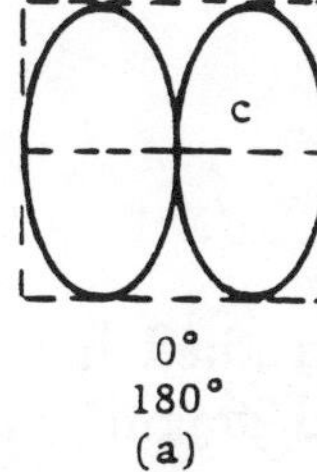
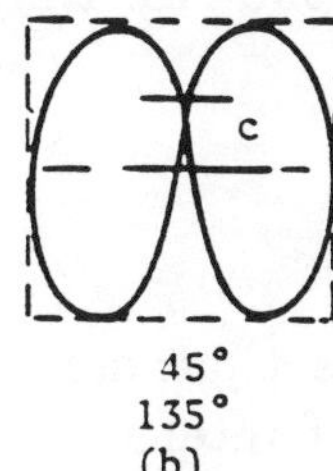
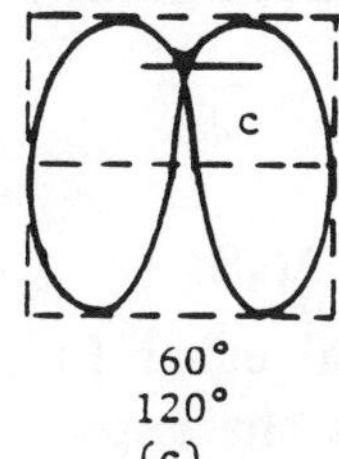

Fig. 19.

0°
180°
(a)

45°
135°
(b)

60°
120°
(c)

The phase difference is determined from the relative position of the crossing point A compared to the center point C. The total distance from the center point to the top side of the square corresponds to $+90°$ and that to the bottom side is $-90°$. The frequency ratio is the same as that in case 2, above.

Using an extra function generator *investigate* these properties for several different signals at various frequency and/or phase difference relations according to the above instructions. If an extra function generator is not available then the following circuits can be built instead to generate the signal B with a phase difference [3]. (Analyze these circuits.)

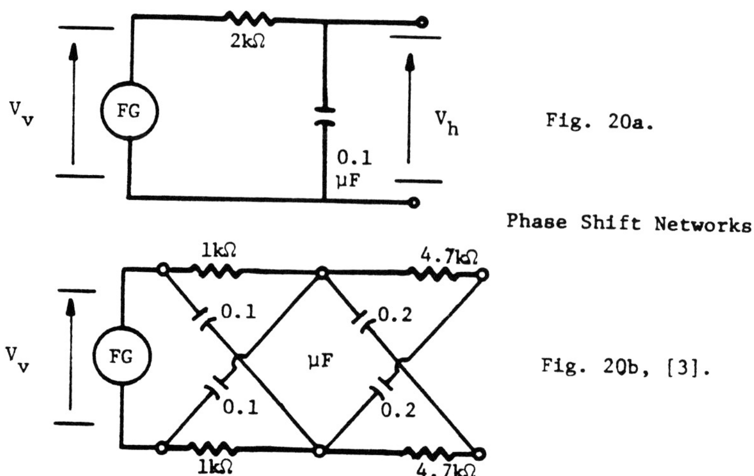

Fig. 20a.

Phase Shift Networks.

Fig. 20b, [3].

4.2.5. RISE TIME MEASUREMENTS

In general the rise time refers to the first time duration that a signal takes to reach its final value. Sometimes the definition for this time is chosen to be either a duration from 5% to 95%, or a duration from 10% to 90%, of the final value. At any rate this measurement is performed in exactly the same manner as measuring the time duration between any two points of a given display on the screen. It is, of course, better to enlarge the display around its beginnings to distinguish these two points. Most of today's oscilloscopes have this capability.

Also since the rise time of the oscilloscope is a nonzero quantity (in other words, the oscilloscope does not respond instantaneously to an input) the above measurement is only an approximation to the true rise time of the signal. The actual rise time is measured from the following equation:

$$R_{t(\text{true})}^2 = \left[R_{t(\text{measured})}^2 - R_{t(\text{oscilloscope})}^2 \right].$$

Here $R_{t(\cdot)}$ stands for the corresponding rise time as indicated by subscript in $(\cdot)$.

Using the above relation, *find* the best approximation to the rise time of your oscilloscope by measuring the known rise time of a signal.

4.2.6. MISCELLANEOUS OBSERVATIONS

Each oscilloscope has, most likely, several other functions that must be studied individually. To name just a few, certain oscilloscopes have the capability of intensifying and magnifying a portion of the sweep as we please. This enables us to study a section of the sweep more carefully for any relevant measurements. Also we must realize that the oscilloscope, like any measuring device when in use, becomes a part of the circuit. Its non-ideal characteristics must always be realized and accounted for. In the following we give some idea of how the non-infinite input impedance of an oscilloscope effects the input signals.

A typical oscilloscope input circuit is shown in Fig. 21.

Here $C_2 \ll C_1$, where C_1 is the coupling capacitor and $(C_2 \| R)$ is the input impedance of the oscilloscope. The V_s and R_s are source parameters and we may choose $R_s = 0$. An *AC* model of this circuit for $R_s = 0$ is

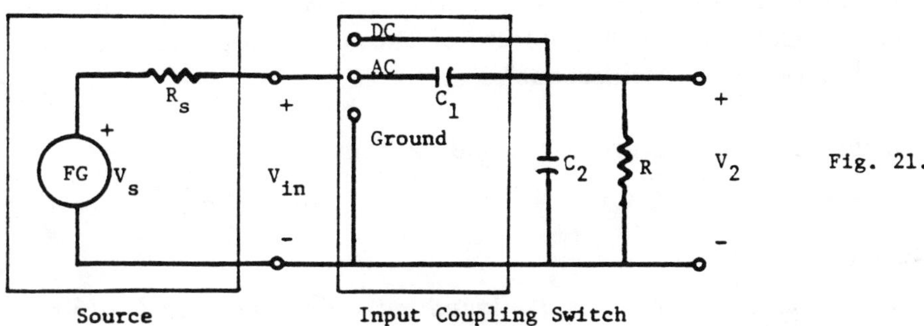

$$V_2(s) / V_s(s) = [sC_1 / (C_1 + C_2)] / [s + 1 / R(C_1 + C_2)] \cdot$$

For $V_2(s)$ a step input (i.e., constant source), we have $V_s = A / s$, when A is the size of this step input. Then

$$V_2(s) = [AC_1 / (C_1 + C_2)] / [s + 1 / R(C_1 + C_2)],$$

and

$$v_2(t) = [AC_1 / (C_1 + C_2)] \exp [-t / R(C_1 + C_2)].$$

Letting $C_2 \ll C_1$, we get

$$v_2(t) \simeq A \exp (-t / RC_1).$$

This is a decaying exponential instead of a perfectly straight line as a step input should be. The only way this function becomes a straight line is when the input resistance $R \rightarrow \infty$.

5. PROBLEM SET

1. What are some of the advantages oscilloscopes posses over other electronic measuring instruments?

2. If the vertical sensitivity switch of an oscilloscope is set to

     a) 50 mV/DIV, b) 1 V/DIV,

   what will be the peak-to-peak distance of the display shown on an oscilloscope if a sine wave is applied whose amplitude is 0.25 V?

3. Calculate the elapsed time between two points on a waveform separated by 3 cm if the sweep time of the oscilloscope is:

     a) 2 s/cm, b) 1 ms/cm, c) 50 μs/cm.

4. The waveform shown in Fig. 22 is applied to the vertical input of an oscilloscope.

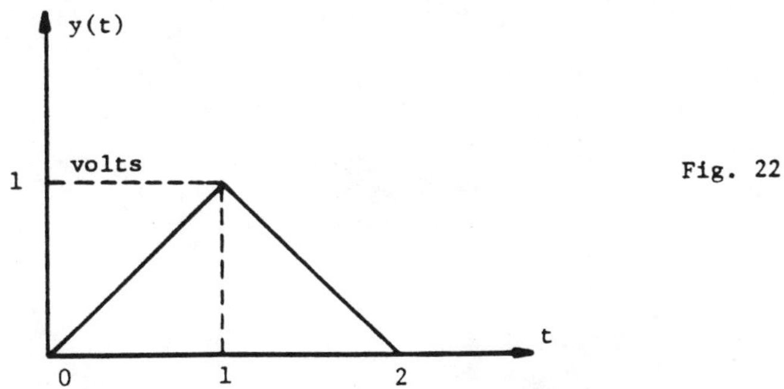

What is the waveform that must be applied to the horizontal input if $y(t)$ can be exactly shown on the oscilloscope screen?

5. How to measure the rise time of an oscilloscope? What is the frequency response of an oscilloscope? Investigate the loading effects of the oscilloscope on a given circuit. What is the input impedance of your oscilloscope?

Hint: Consult the manufacturer's manual for the EEP.

6. FOR YOUR REPORT

Following our standard report form, document your data and observations; submit with answers to the above problem set.

EXPERIMENT I-3

AC MEASUREMENTS

EQUIPMENT:

EEP
VTVM
VOM

Table of Contents   Page No.

1. PURPOSE

This experiment will complement the previous two experiments. Its main purpose is to acquaint the student with the operation and applications of the EEP with particular emphasis on function generators and the basic operation of the VOM and VTVM in measuring AC voltages. The use of a "transfer" instrument for calibration, and frequency response of the meters are also considered.

2. INTRODUCTION

Fundamental electrical standards are of the DC type. The fact that no separate standards are required for AC (alternating or in general periodic) quantities implies the existence of measuring devices that are equally effective on both DC and AC. These devices can be calibrated with the DC standards and are thereby calibrated for AC use as well. Such instruments are called "transfer standards". The electrodynamometer meter movement is a good example of a transfer instrument useful for both DC and low frequency (50-400Hz) AC measurements.

The oscilloscope is inherently a transfer instrument. Within the limitations of its frequency response (which can represent a very large bandwidth) the vertical deflection of the electron beam is proportional to the instantaneous voltage applied at the vertical input terminals. A DC calibration of this deflection system therefore constitutes the only calibration required. Application of an AC voltage to the vertical input permits immediate determination of the peak-to-peak (P-P) value of the AC signal. If the waveform is given, then the knowledge of the P-P value allows immediate calculation of any other characteristics of the AC signal, such as the average value or the root-mean-square value as described next. Our main objective is to make AC measurements using all measuring devices available. But first we must review a few definitions.

3. SOME DEFINITIONS

3.1. MEASURE OF SINE-WAVE AMPLITUDE

Sine-wave voltage and current amplitudes are commonly described in four ways: (1) the peak value, (2) the peak-to-peak value, (3) the average value, (4) the RMS (root-mean-square) value. The particular description used depends on the application. Measuring instruments such as vacuum-tube voltmeters and oscilloscopes respond to the peak-to-peak value which for a sine-wave is merely twice the peak value. Moving coil meters respond to the average value (effective rectified DC value) and the RMS value is mostly used for the measurement of power in an AC circuit. Our digital multimeter does a whole lot more than either the VOM or the VTVM.

Consider the following figure. We will summarize the definitions of these commonly used measures.

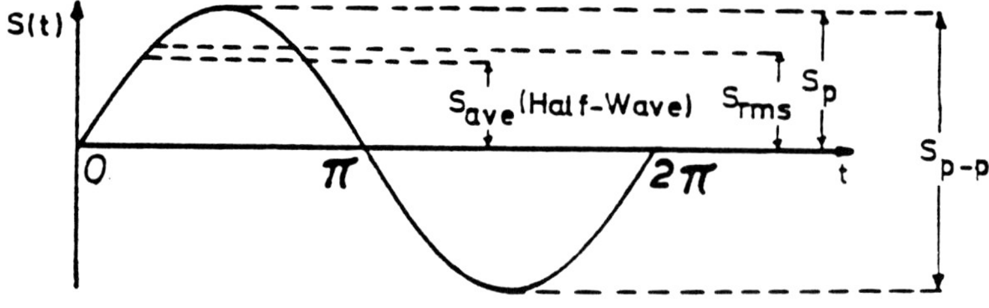

Fig. 1. An Electrical Signal S(t).

3.1.1. AVERAGE VALUE

The average value of any periodical signal is defined by

$$S_{ave} = \bar{S} = \frac{1}{T} \int_0^T S(\tau)\, d\tau,$$

where T is the period of the waveform. In general, we may define an average value for any desired interval. For a sinusoidal signal where $S(t) = S_p \sin\omega t$ we have $S_{ave} = 0$. In practice the sine-wave must be rectified (full wave) to obtain meter readings corresponding to the average value definition.

3.1.2. THE RMS (ROOT-MEAN-SQUARE) VALUE

The RMS value is the effective value needed in power calculations. The RMS value of a sine-wave current will produce the same heating (power) in a resistor as an identical DC current, i.e., 1 RMS AC ampere produces the same amount of heat in a resistor in a given time as 1 ampere of DC current. The RMS value of any periodical signal is obtained by taking the square root of the mean (average) of the squared values of that signal. This operation is illustrated in Fig. 2, for a sinusoidal wave. The square of a sine-wave ($I_p^2 \sin^2\omega t$) is plotted directly below the sine-wave. The result is another "sine-wave" which is completely above zero and twice the frequency. The mean or average value of this "sine-wave" is clearly $I_p^2 / 2$. On taking the square root

$$I_{RMS} = \sqrt{I_p^2/2} = I_p/\sqrt{2} = 0.707 I_p \cdot$$

In general, given a periodic signal S(t), the RMS becomes

$$S_{RMS} = \sqrt{\frac{1}{T} \int_0^T S^2(\tau)\, d\tau} \cdot$$

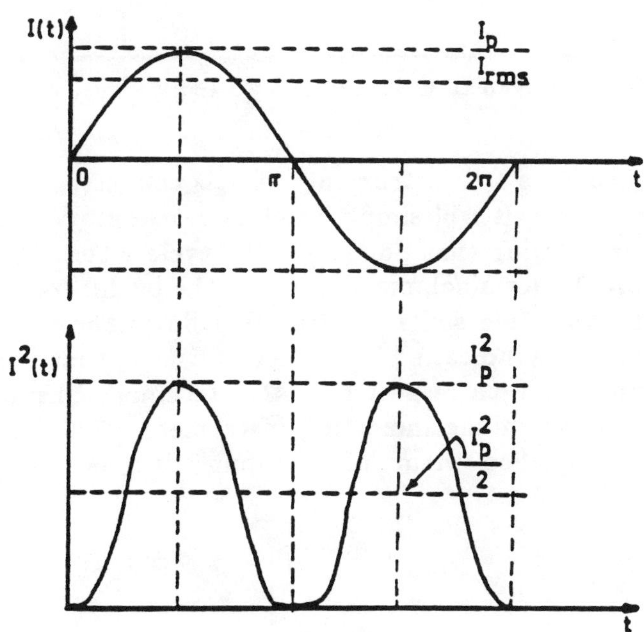

Fig. 2.   A Sinusoidal Signal.

Analysis for both current and voltages waves are exactly the same and we will not repeat the results again. Thus in general, for a periodic waveform such as $S(t) = S_P \sin(\omega t)$, we have

$$S_{ave} = \bar{S} = 0 \,,$$

$$S_{RMS} = \frac{S_p}{\sqrt{2}} = 0.707\, S_p \iff S_p = \sqrt{2}\, S_{RMS} = 1.414\, S_{RMS}\,.$$

For the full wave rectified S(t) we have $S_{ave} = \dfrac{2}{\pi}\, S_p = 0.637\, S_p$.

Note: If the wave form is not known, the use of a suitable sweep on the oscilloscope horizontal deflection plates will permit a "display" of the AC signal which can be photographed and graphically measured or integrated to get the desired information.

4. AC VOLTMETER[2]

The set of series resistors in the AC voltmeter circuit are similar to those for the DC voltmeter except that they are only one fourth as large. Thus the AC voltmeter specification is 5000 ohms per volt, Fig. 3.

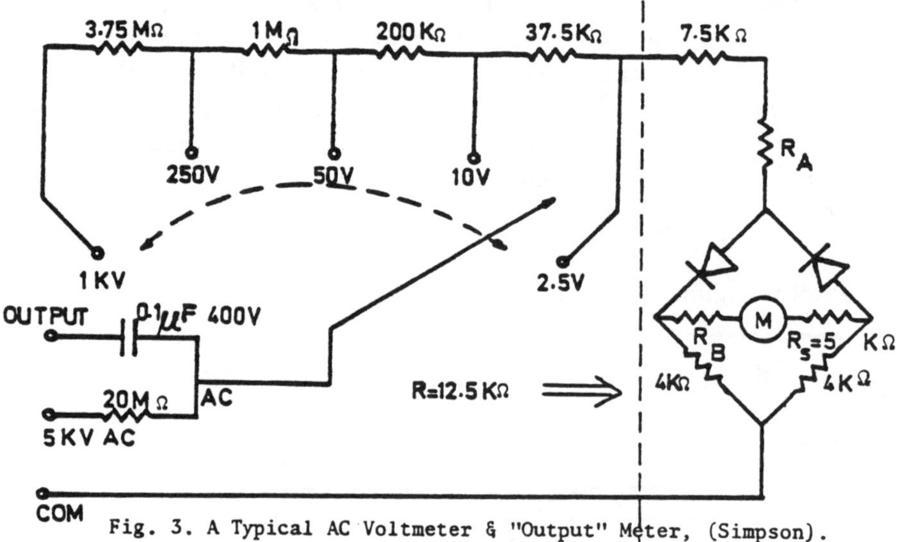

Fig. 3. A Typical AC Voltmeter & "Output" Meter, (Simpson).

The indicating instrument is a DC instrument. If it is connected across an AC signal having a frequency higher than a few hertz, it will simply tend to read an average value of zero. Now consider the effect of the diodes. During that portion of the cycle when the *"AC"* terminal is positive with respect to the *"COMMON"* terminal, current enters the bridge-like circuit from the top, passes through the left-hand diode, and then splits -- part going down through the 4K ohms resistor and part going *left to right* through the indicating instrument. During that portion of the cycle when the *"COMMON"* terminal is positive with respect to *"AC"* terminal, current enters from the bottom, but still passes *left to right* through the indicating instrument. The voltage across the indicating instrument is thus a full-wave-rectified version of the input signal as shown in Fig. 4b.

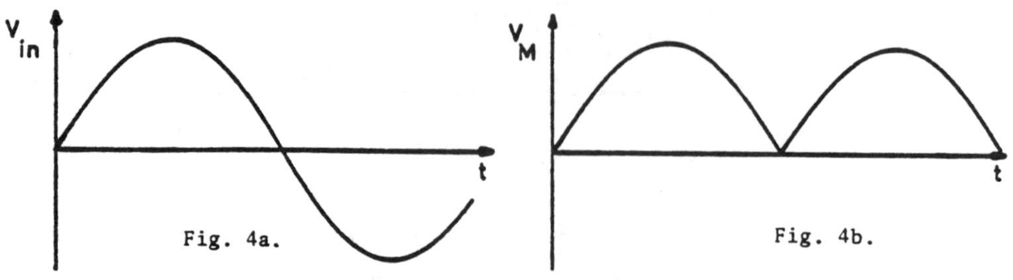

The voltage across the instrument now has a non-zero average value, and the indicating instrument responds to this average value of the voltage across (or current through) it. The average value of a full-wave-rectified sine-wave is $(2/\pi)$ (PEAK) which is approximately 0.637 (PEAK); however, we want the instrument to read the RMS value which is $(1/\sqrt{2})$(PEAK) or approximately 0.707 (PEAK). Thus, the resistor values are chosen to make the instrument *read* 1.11 times the average value of the full-wave-rectified input $(1.11 \simeq 0.707/0.637)$. Resistors R_A and R_B are adjusted by the manufacturer to achieve this calibration with the particular pair of diodes in each instrument.

The instrument always reads 1.11 times the average value of the full-wave-rectified input voltage. Suppose we connect a 1.5 volt battery to the 2.5-volt AC voltage range; the instrument will read 1.67 volts $(=1.11 \times 1.5)$. If we connect a 10 volt square wave (+10 volts for half the period and − 10 volts for the other half) to the AC voltmeter, the reading would be 11.1 volts. If the waveform is +10 volts for one fourth of the period and zero for the remaining three fourths, the reading will be 2.78 volts.

Thus, it is *imperative* that we know the waveform that is being measured with the AC voltmeter. The reading is the RMS value only if the input is a sine wave. The reading is always 1.11 times the average value of the full-wave-rectified input.

The resistance of the diodes varies, of course, with the value of their current and voltages. (The slope of the i versus v characteristic is not a straight line --- particularly, near zero.) This resistance change is negligible compared with the overall AC voltmeter resistance for higher ranges, but most multimeters include a special scale for the lowest AC range.

5. OUTPUT METER[2]

In many electronic circuits, the voltage of interest is superimposed on a DC voltage that is chosen to make the electronic devices operate in a linear region. Often, however, we would like to measure only the variational part of the voltage-- not the DC. Thus, we need some method of removing, or *"blocking"*, the DC portion of the voltage. A series capacitor is inserted when the *"OUTPUT"* terminal is used. If the signal being measured has no DC component (that is, has an average value over a period equal to zero), the meter reading is the same on both *"AC"* and *"OUT-PUT"*, although the presence of the capacitor when using the *"OUTPUT"* terminal affects the low-frequency response.

An example or two may help to clarify the effect of the series capacitor. If a 1.5 volt battery is connected, the "output meter" reads zero-- after a small, quick deflection while the capacitor is being charged. The square wave used as an example earlier would yield the same reading as when measured on *"AC"*--11.1 volts--because it has a zero average value. On the other hand, the waveform with a +10 volts for one fourth the period and zero for the remainder would read 4.17 volts because the *"OUTPUT METER"* (1) blocks the DC--which is the same as shifting the waveform to make the average value equal to zero--resulting in a waveform that is +7.5 volts for one fourth the period and − 2.5 volts for the remainder; (2) full-wave rectifies--that is, "folds up" the negative portion; (3) takes the average value--which is 3.75 volts in our example; and (4) *reads* 1.11 $\times$ 3.75 = 4.17 volts.

The measuring process of most electronic voltmeters is identical to the *"OUTPUT METER"* function of the multimeter; that is, DC blocking followed by rectification, averaging, and multiplication by 1.11 to obtain the reading. A word of caution at this point: Many electronic voltmeters have terminals labeled *"OUTPUT"*, but these are for an entirely different purpose, and external signals should *not* be connected to them.

6. LABORATORY PROCEDURE

6.1. PRE-LABORATORY WORK

Before coming to the laboratory review the manufacturer's manual and the last two experiments and make sure you are familiar with the function generator. Then go over the previous pages of this experiment and solve the following two problems and hand them in with your report of the last experiment to your instructor.

OH! CAUGHT YOU !? If you are one of those students who walk into the laboratory without reading her/his laboratory work, then perhaps you have not done these problems and you have nothing to show. Make sure - do not get caught again!

Problem 1:

From Fig. 5, determine the reading that will result using (a) the DC voltmeter, (b) the AC voltmeter, and (c) the "output meter"; also determine the RMS value. Assume that the frequency is high enough so that the capacitor can be considered a short circuit to everything except DC.

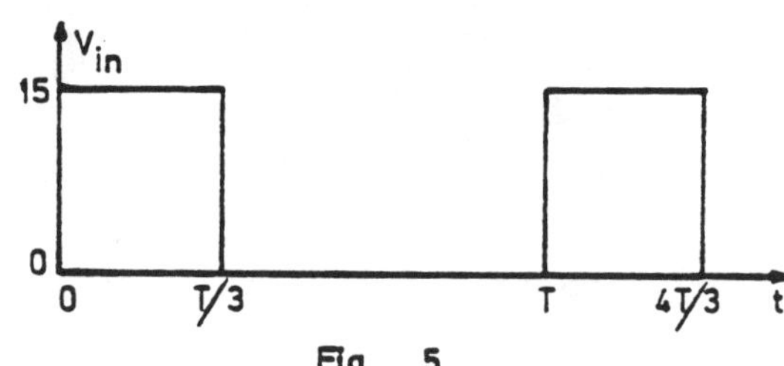

Ans.:   (a)   5 V; (b)   5.55 V;   (c)   7.4 V.

Problem 2:

If the voltage waveform in Fig. 5, is across 600 ohms, compute the AC voltmeter reading in Problem 1 in dBm, cf., Section *0-2U.*

Ans.: 17.1 dBm

6.2. LABORATORY WORK

1. *Connect* the oscilloscope (DC input) to the function generator (use a BNC male plug on both ends), and *adjust* the function generator for an 8 V, P-P sine wave at 1 kHz with an average value of zero. *Connect* the two multimeters (set for AC) in parallel with the oscilloscope input and adjust the function generator as necessary to maintain the 8 V, P-P calibration. *Measure* the function generator output voltage with the VOM and DMM on both *DC VOLTS* and *AC VOLTS.* Also *measure* on *OUTPUT* from the VOM and dB on the DMM. *Record* these data.

2. *Adjust* the offset control on the function generator to obtain an average of 4 V, as measured on the oscilloscope, for the waveform in (1) above. *Repeat* the measurements of (1) in the above.

3. *Repeat* (1) and (2) in the above for a square wave.

4. *Repeat* (1) and (2) in the above for a triangular wave.

5. *Adjust* the function generator for a sine wave output of 20 V, P-P at 1 kHz with zero average value. *Measure* the voltage on the VOM on both *AC VOLTS* and *OUTPUT.* *Use* the 10 V range. The readings should be identical---*make note* of any discrepancies. *Decrease* the frequency of the function generator until the reading on *OUTPUT* is 5% lower than that on *AC VOLTS* (taking into account the discrepancies noted above). *Record* the frequency. Then *decrease* the voltages of the function generator and *repeat* the measurement using the 2.5-volt scale.

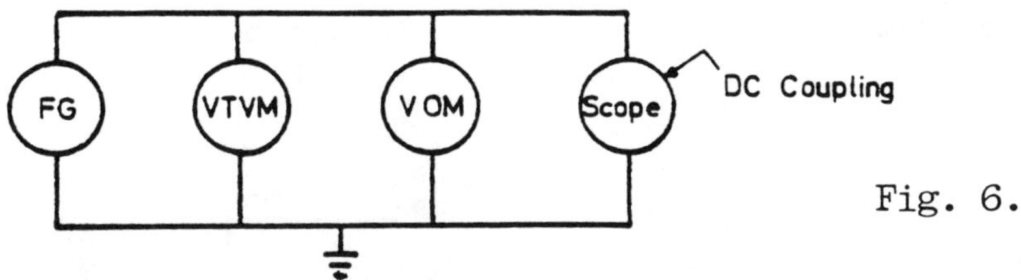

Fig. 6.

6. *Connect* the circuit shown in Fig. 6. *Measure* and plot on semilog paper the frequency response of the meters and oscilloscope. *Set* the sine-wave generator to provide a 5V RMS signal as measured by the VTVM. *Plot* the magnitude of each meter readings. RMS Volts for the VTVM and VOM, Volts peak for the oscilloscope, versus log frequency (20 Hz to 700 kHz), on the logarithmic scale.

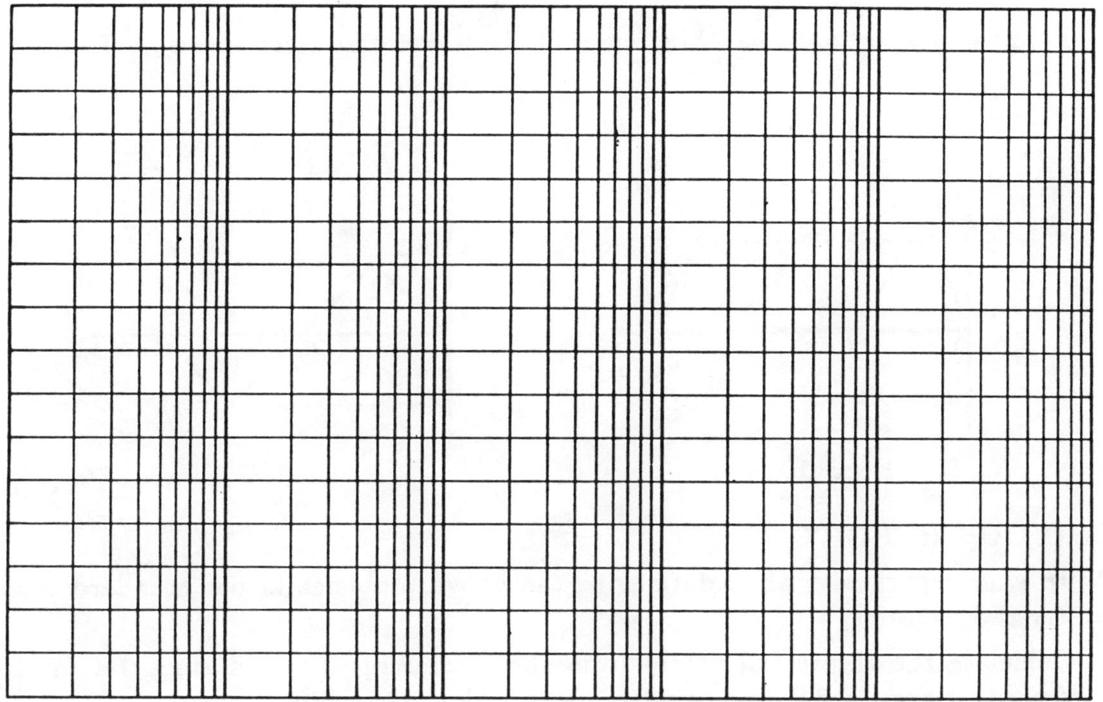

Fig. 7.

In general the sine-wave magnitude will vary as a function of frequency. The VTVM will measure the voltage with an accuracy of $\pm$ 5% full scale over a frequency range from 25 Hz to 1 MHz. *Use* the VTVM as a standard and *adjust* the sine-wave generator output voltage to keep the VTVM reading constant at 5V RMS. At what frequencies is the output voltage 0.707 of its maximum value for the VOM and the oscilloscope?

7. *Change* the sine-wave to a 400 Hz square-wave at the FG terminals. *Adjust* the P-P amplitude to 10 volts as measured by the oscilloscope. Read the P-P and RMS values of the square-wave signal with the VTVM (use appropriate scales). *Measure* the square-wave signal with the VOM (Caution: Use correct scale). *Explain* the discrepancy between the VOM and VTVM readings. *The VOM AC scales are accurate only when used to read pure AC sinusoidal signals.* This is called effect of waveform on VOM.

7. PROBLEM SET

1. Repeat the calculations of the two pre-laboratory problems for the waveform shown in Fig. 8, and Fig. 9.

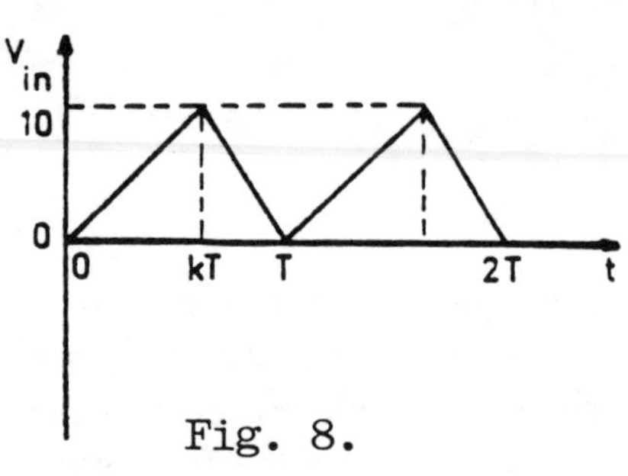

Fig. 8.

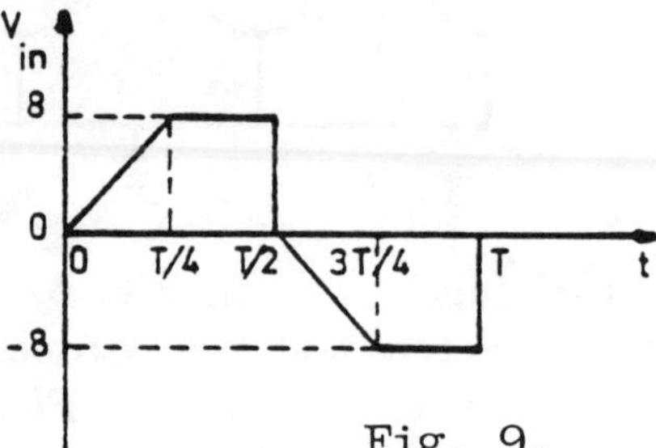

Fig. 9.

2. Find the average and the RMS values of the waveforms given below:

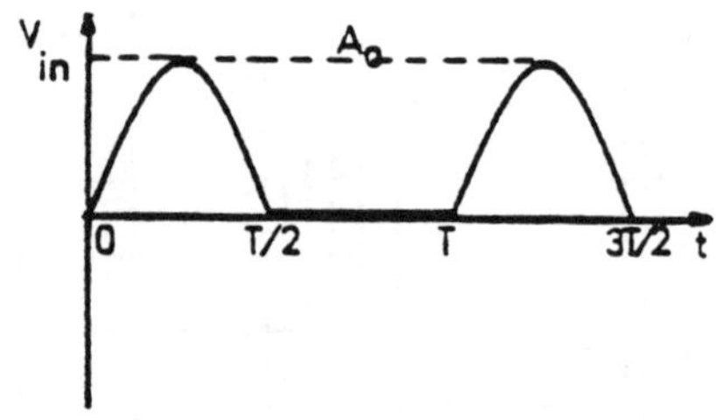

Fig. 10.

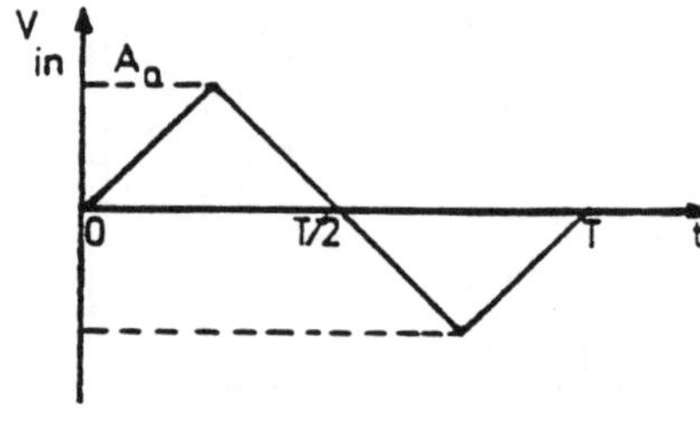

Fig. 11.

8. FOR YOUR REPORT

Write your report with the solutions to the above problems in our standard report form. In addition try the following.

1. Calculate the theoretical values from the oscilloscope data in steps 1-4 in the laboratory procedure Section 6.2, and compare these with the measured values.

2. Assume that AC voltmeter has an input resistance of 5 kΩ per volt, determine the frequency of the sine-wave for which the "output meter" reading is 5% lower than AC voltmeter reading (a) on the 2.5-volt range and (b) on the 10-volt range. Compare these quantities with the measured values.

EXPERIMENT I-4

OSCILLOSCOPE CONTINUED

EQUIPMENT:

    EEP
    CURVE TRACER
    RECORDERS
    RESISTORS (1 kΩ, 10 kΩ)
    DIODE (p-n JUNCTION)

Table of Contents   Page No.

1. PURPOSE

The purpose of this experiment is two-fold. First, to review those materials from prior experiments, in particular EXP. I-2, that have not been fully explored. Second to learn some new applications of the oscilloscope and become acquainted with *recorders* and *curve tracers*.

2. INTRODUCTION

Today you have, hopefully, received some comments from your instructor regarding your report of EXP. I-2. These comments will be helpful to show what you have actually done wrong or you could not finish due to whatever reasons. It is your right to perform those parts by yourself. Do not let your partner push you around in this course. Also do not be intimidated by, say, those partners who know everything and do all the work for you. It is your time and money that is at stake. Of course, if you are lucky enough not to have any questions or problems, then you may continue to the next part. Otherwise repeat every part you did not complete before going further.

2.1. CURVE TRACER

Perhaps it is clear by now that our main objective in becoming familiar with various electrical measuring devices is for the sake of learning how to produce some characteristics of a device or circuit from which one can analyze its electrical behavior. A common characteristic of say a two-terminal device is its V-I characteristic. Here we look into an instrument called a curve tracer, which in general can be used to find the characteristics of a three-terminal device. The main structure of this unit consists of the following parts

1. CRT (with power and position switches)   ,
2. Vertical Amplifier   ,
3. Horizontal Amplifier   ,
4. Input Circuit with Limiter Switches   ,
5. Step Generator   ,
6. Device Test Panel   .

The function of the first three parts is *similar* to that of the oscilloscope. The main function of part 4, is to control the amount of input voltages applied to the device under study. This is generally done by a control multi-scale switch and an input limiting variable resistor switch that drops a large part of the input voltage. This part has several other switches to check on polarities or other relevant properties of the device.

Part 5 is responsible for producing the multiple characteristics of the device. This is mainly useful to display the characteristics of a three-terminal device, in which we draw a typical curve of one parameter (say, the output current) versus another (say, the output voltage), while the third parameter (say, the input current) is kept constant. We may repeat this for several different values of the third parameter. This operation is performed internally by the *Step Generator* circuit. Finally, part 6 is where we interconnect the device for testing or producing its characteristics. We will fully utilize this instrument later when we study three-terminal devices such as transistors.

2.2. RECORDERS

The recorder is an instrument for duplicating certain applications of the curve tracer or for that matter the oscilloscope. Recorders are found in many electrical measurement laboratories.

These recording instruments are devices that draw one quantity versus another. While there are several different recording instruments with different applications available, in general all of these instruments contain two parts. One is a measuring part that transforms whatever it measures into electrical signals. The other part is a recording part that displays this signal in graphical form on special papers. In the operation of the *Strip-Chart Recorder* the paper rolls as a function of time while the pen moves up and down on a perpendicular axis. The motion of pen is proportional to the amplitude of the electrical signal produced by the measuring part of the recorder.

Another major type of recorder is *X-Y Plotter (Recorder)*. Here, the pen's motion is a function of the two variables x and y that are both functions of time. Just like an oscilloscope, many of these instruments have a time base section that generates a signal or sweep which can be used instead of x. Thus an electrical signal, as it is generated from the measuring part of the recorder, can be drawn or plotted versus time.

The main difference between these recorders and any oscilloscope is the speed of recording. These units are generally used for low frequency signals only. Recorders are generally useless at higher-frequency applications.

3. LABORATORY PROCEDURE

3.1. PRE-LABORATORY WORK

Review those materials discussed in the earlier experiments prior to attending the laboratory. Prepare a list of the oscilloscope functions and/or operations that you have not understood so far. Bring also your earlier reports to the laboratory.

3.2. LABORATORY WORK

If a curve tracer is available, then use this instrument to find the V-I characteristics of a resistor and a p-n junction diode. If a curve tracer is not available, then follow the procedure described below to draw this V-I characteristic. Here we use an oscilloscope to complete the experiment.

First, *construct* the circuit of Fig. 1.

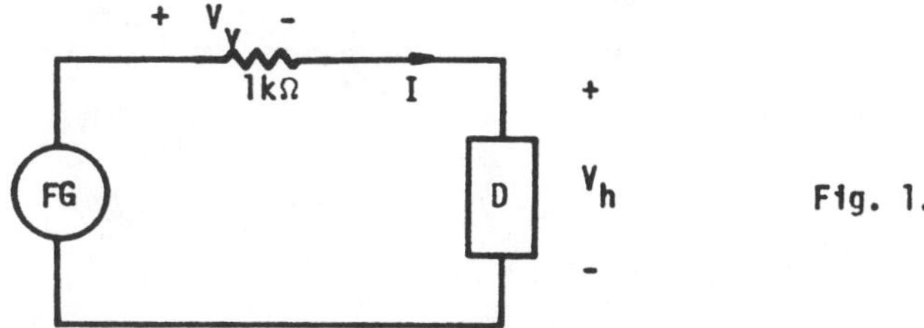

To draw the V-I characteristic of the device D, we need to apply the voltage V_h across the device to the horizontal input amplifier (CH1) and to apply the current I, through the device, to the vertical input amplifier (CH2). Since I is proportional to the voltage across a 1 kΩ resistor, we can represent this current by voltage drop V_v. Hence we apply V_v to the vertical input amplifier. *Watch for the scale on the vertical amplifier!*

Here, then is the laboratory work: *Construct* the circuit of Fig. 1, with D first as a 10 kΩ resistor and then as a p-n junction diode. In each case, *apply* V_h to the horizontal input and V_v to the vertical input of the oscilloscope. With proper oscilloscope triggering you will see the V-I characteristics of these two devices. *Draw* your results in the chart of Fig. 2, with proper scales. Please be aware that we are *not* analyzing these device characteristics at this time, since that is the subject of future experiments.

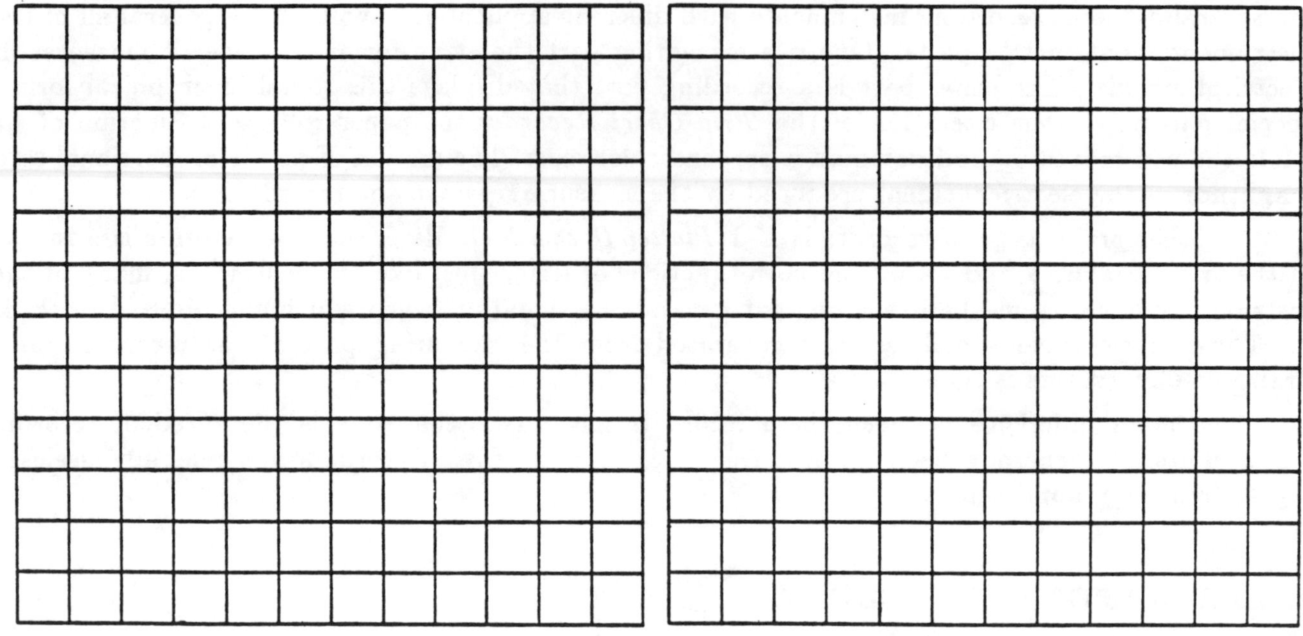

Fig. 2.

Repeat the above experiment using any one of the recorders available. Be careful about the frequency of the applied signal from the function generator since most recorders can only be used in certain frequency range.

4. FOR YOUR REPORT

Prepare your report in our standard form including the data as generated in the above.

EXPERIMENT I-5

POWER MEASUREMENTS

EQUIPMENT:

EEP (WITH PROBE)
WATTMETER (10 A/220 V)
AC-VOLTMETER (110 V)
AC-VOM (5 A)
ISOLATION TRANSFORMER (0.75 KVA, 115 V/115 V AC)
RESISTOR (36 Ω and 200 WATTS)
INDUCTOR (110 to 130 mH WITH Q $\geq$ 10 AT 60 Hz)
CAPACITOR (68 μF)
CAPACITANCE BOX (TO COVER C $= 1/L\,\omega_o^2$, $\omega_o = 377\,$rad/sec)

Table of Contents   Page No.

1. PURPOSE

The purpose of this experiment is to acquaint the student with the basic operation of a watt meter and to measure the consumed power in an electrical load. A similar power measurement using an ordinary voltmeter, a current-meter, and an oscilloscope (to measure the phase difference between voltage and current by Lissajous figures) is to be arranged and the data of these two separate methods compared.

2. RESONANT CIRCUITS

Before the concept of consumed power in a general RLC network (electrical load) is presented, we will review the *phenomenon* of *resonance* in an RLC network.

Consider the following two circuits:

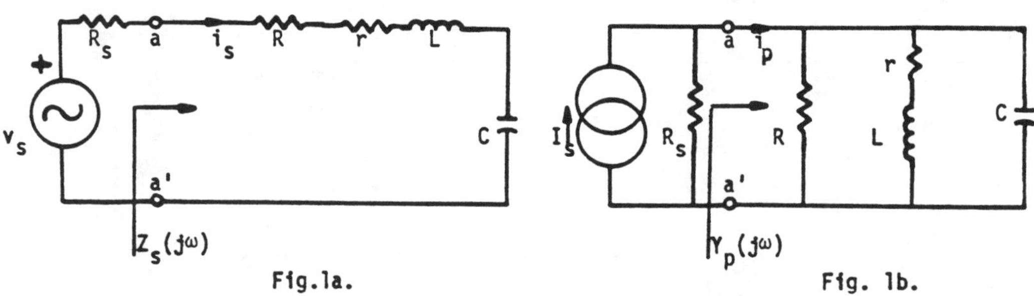

Fig.1a.   Fig. 1b.

Here r represents the resistance of wiring, coils, and capacitors. In either of these circuits we have an input *immittance* (*impedance* or ad*mittance*) that is a function of all circuit elements as well as the frequency $s = j\omega$. If such a circuit becomes a pure resistance at a given frequency then we say that a resonance phenomenon has taken place. We can easily show that either of the above circuits can exhibit this phenomenon. The immittance functions are as follows:

$$Z_s(j\omega) = R + r + j\left(\omega L - \frac{1}{\omega C}\right), \text{ and}$$

$$Y_p(j\omega) = \frac{1}{R} + \frac{1}{r + j\omega L} + j\omega C.$$

For the time being let r=0 (i.e., ideal R, L and C). Then we see that both of these two immittance functions become a pure resistance (or conductance) at an angular frequency of $\omega_o = 1/\sqrt{LC}$. Indeed the values of these immittance functions are $Z_s(j\omega_o) = R$ and $Y_p(j\omega_o) = 1/R$. Incidently these values actually correspond to the maximum values of the amplitudes of these complex functions.

If we normalized these amplitude functions by these maximum values and draw the new normalized amplitude of the immittance functions versus frequency we get a typical curve as shown in Fig. 2. From this graph we immediately see that as expected the maximum occurs at ω_o with $\omega_o = \sqrt{\omega_1 \omega_2}$, ω_1 and ω_2 are depicted in the graph of Fig. 2. The quantity

$$\Delta\omega = \omega_2 - \omega_1 ,$$

is called the *bandwidth* of the corresponding circuit.

Here we are concerned with a new concept. We may wonder how sharp this curve can be and how can sharpness be related to the other circuit parameters. The answer to this question lies in the so-called *quality factor* known as Q. It is clear that the maximum is definitely a function of the resistance elements. But what about ω_1 and ω_2? It is well known that this sharpness is related to the concept of energy in a circuit [0]. Specifically, one may define Q as follows [0]:

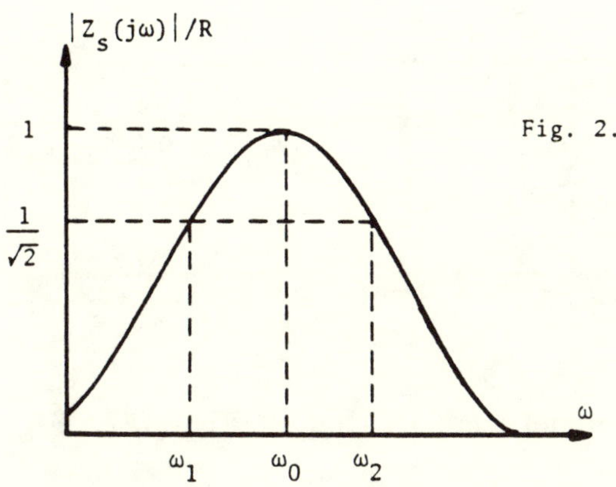

Fig. 2.

$$Q = 2\pi \; \frac{\text{maximum energy stored}}{\text{total dissipated energy per cycle}} \; .$$

To compute Q for the circuit of Fig. 1a, with r=0, and $i_s = I_s \cos\omega_o t$, we have

$$v_L = L \, di_s / dt = - I_s \, \omega_o \, L \, \sin\omega_o t \; , \text{ and}$$

$$v_C = \int_0^t i_s \, dt / C = (I_s / \omega_o C) \sin\omega_o t \; .$$

At resonance we must have $\omega_o = 1 / \sqrt{LC}$ which is the same as $v_L + v_C = 0$. The total instantaneous stored energy in this circuit is as follows:

$$W_L = \frac{1}{2}Li_s^2 = \frac{1}{2}LI_s^2 \cos^2\omega_o t \; , \text{ and}$$

$$W_C = \frac{1}{2}Cv_C^2 = \frac{1}{2} (I_s^2 / \omega_o C) \sin^2 \omega_o t \; .$$

$$W_L + W_C = \frac{1}{2}I_s^2(L \cos^2\omega_o t + \sin^2\omega_o t / \omega_o C) \; .$$

Since $\quad \omega_o LC = 1$, thus

$$W_L + W_C = \frac{1}{2}I_s^2 L \; .$$

Total energy dissipated per cycle (of period T) is

$$W_R = \frac{1}{2} I_s^2 RT \; , \text{ and therfore}$$

$$Q_s = 2\pi \, (\frac{1}{2}I_s^2 L) \, / \, (\frac{1}{2}I_s^2 RT) = \omega_o L/R = 1/\omega_o CR = \sqrt{L/C} \, R \; .$$

Similar algebra for the circuit of Fig. 1b, with r=0, results in

$$Q_p = \omega_o CR = R \, / \, \omega_o L = R\sqrt{C/L} \; .$$

In actual situations we have non-perfect elements such as inductors and capacitors. Thus we will always encounter these non-perfect elements to analyze. It is interesting to note that this concept of Q makes analysis of some of these circuits easy. To give an idea of what we mean, consider the circuits of Fig. 3, and Fig. 4. For the circuit of Fig. 3, we have $Q_s = \omega_o L_s / R_s$. We can show that if Q_s is sufficiently large around ω_o we can replace the circuit of Fig. 3, by that of Fig. 4, provided $L_p = L_s$ and $R_p = Q_s^2 \, r_s$.

The advantage of this transformation is that it makes analysis in most practical cases very simple. After all it is clear that we can easily measure the Q and ω_o for a given electrical load from

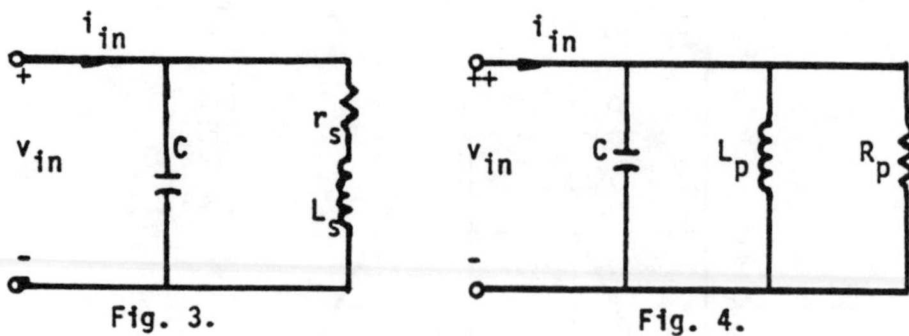

Fig. 3.   Fig. 4.

which we can approximate a model such as that of Fig. 3, or Fig. 4. Thus we can continue our analysis much faster than before.

3. BASIC NOTION OF ELECTRICAL POWER

The electrical load of every house or user can generally be modeled by a simple parallel or series RLC network. The electricity that we use is a sinusoidal waveform with the frequency of 60 Hz and an amplitude of approximately 110 volts AC. The current that we use is also a sinusoidal waveform of the same frequency. Because of the RLC load, there is a phase difference between the current and the voltage waveforms. Independent of any particular assumption on the electrical characteristics of the current and the voltage waveforms of any two-terminal network, the instantaneous   power delivered to any such network is

$$p(t) = v(t)\, i(t) .$$

This relation is developed as follows:

$$\text{power} = (\text{work}) \,/\, (\text{time}) .$$

In the electrical sense, work is equal to (charge)$\times$(voltage). Thus rearranging the above relation results in

$$\text{power} = (\text{voltage})\times(\text{charge} \,/\, \text{time}) = v(t)\, i(t) .$$

If both voltage and current have the same frequency and these are given with respect to the same common reference, then we have

$$v(t) = V_m \sin(\omega t + \theta_v) , \text{ and } i(t) = I_m \sin(\omega t + \theta_I) .$$

Thus the expression for *consumed power* becomes

$$p(t) = V_m I_m \sin(\omega t + \theta_v) \sin(\omega t + \theta_I) .$$

This of course, is a periodic function of time that may not necessarily give us the best interpretation of the consumed power. In general, we are thus interested in an average value of this quantity which is as follows. (We also recall that the average value of a well-posed periodic function can be calculated in any period of the function.)

$$P_{ave} = \frac{\omega}{\pi} \int_0^{\pi/\omega} p(t)dt = V_m I_m \cos(\theta_v - \theta_I) \,/\, 2 .$$

We recall that: $V_{RMS} = V_m \,/\, \sqrt{2}$, $I_{RMS} = I_m \,/\, \sqrt{2}$, and $\theta = \theta_v - \theta_I$. Therefore,

$$P_{ave} = V_{RMS}\, I_{RMS} \cos\theta ,$$

and $\cos\theta$ is known as the *power factor* (PF).

From the above equation we see that in order to measure the power we need to know the RMS values of both voltage and current and the phase difference between i(t) and v(t). These three measurements are done by a wattmeter.

The concept of consumed or dissipative power always refers to the actual power loss in the resistive elements. There is another term that relates the concept of power to the non-resistive elements such as perfect inductors. This is called the *reactive power* and is equal to

$$P_{var} = V_{RMS}\, I_{RMS} \sin\theta \ .$$

(Here var stands for volts-amperes-reactive.) There is one more term relating to this concept, called the *apparent power*, P_A which is

$$P_A = \sqrt{P_{ave}^2 + P_{var}^2} = V_{RMS}\, I_{RMS} \ .$$

Since our meter is only calibrated to measure P_{ave}, the power companies are very keen in bringing the PF of each user close to 1, i.e., $\theta \simeq 0°$ and $\sin\theta = 0$. This *improvement* of PF is generally done by putting in parallel a capacitor in front of each user who has a large reactive load.

4. WATTMETERS

Generally speaking, each wattmeter has two separate coils: one is a current coil which has a very low resistance; the other one is called the voltage coil which has a much higher resistance than the current coil. The torque produced on the meter pointer is proportional to the instantaneous product of the *currents* flowing in these two coils. The mechanical inertia of the meter movement causes a deflection which is proportional to the average of this torque.

Each wattmeter has four connections (or three cf:, Fig. 5). Two for the current coil and two for the voltage coil. The current coil must be connected in series and the voltage coil must be connected in parallel. The electric companies, generally speaking, will put this meter at the very beginning of the electrical lines of each user in order to measure every bit of electrical loss in the consumer's place. Also as the Fig. 5, suggests we must have a meaningful relationship between the current and the voltage coils in order to have a correct reading. If a positive current is flowing into the ($\pm$) terminal (cf., Fig. 5) of the current coil and the positive voltage is also at the ($\pm$) terminal of the voltage coil, then the power reading is correct. Otherwise our reading would be in error [2].

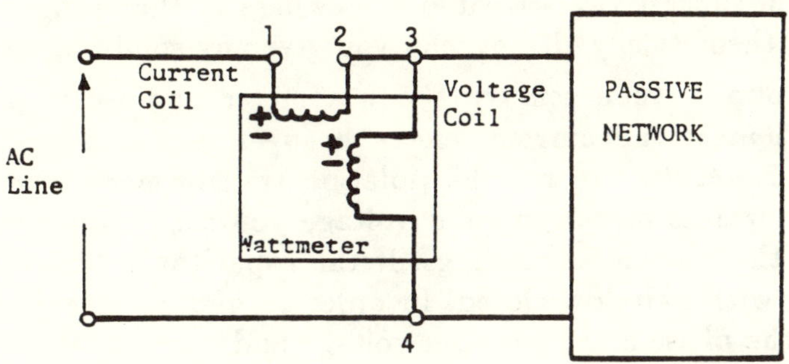

Fig. 5. A Typical Wattmeter Connection to a Passive Network.

5. LABORATORY PROCEDURE

5.1. PRE-LABORATORY WORK

Review the section on power measurement from your circuit textbook. Also review the functions of the VTVM and VOM from EXP. I-1. Next analyze the circuits given in Fig. 6. These circuits will be made available to you for your experiment.

5.2. LABORATORY WORK

The circuit of Fig. 6 is to be used in your work. Therefore, please do not change or tamper with it. Remember, this is a live, 60 Hz, 110 volts AC circuit. You must only work at the terminals (ab), (a′ b′) as well as c,c′ as indicated for you on the circuit board. Please exercise every precaution to avoid a serious shock.

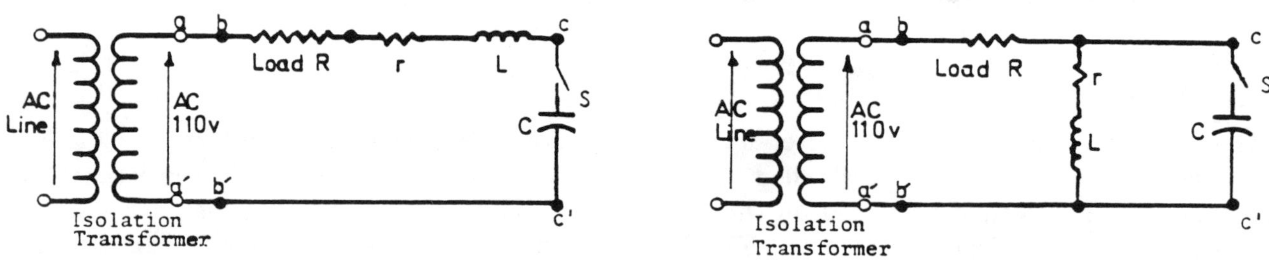

Fig. 6. Two RLC Networks.

5.2.1. POWER MEASUREMENT

Connect terminals number 1 to 4 of the wattmeter to the terminals (ab), (a′ b′) of the circuit of Fig. 6, as described next. But first, *disconnect* the isolation transformer from the load network as well as from the line. *Connect* terminal number 1 of the wattmeter to node a, terminals 2 and 3 to node b, and terminal 4 to node b′. *Connect* nodes a′ to b′. Now, *reconnect* the isolation transformer to the circuit and line. *Record* your reading. *Open* switch S and connect the capacitance box to the terminals c,c′. *Record* the wattmeter readings for several different settings of this capacitance box. When is the power reading maximum? *Record* the value of the corresponding capacitance as C_M. You may not get an exact setting on your capacitance box for this value, thus you may only have a range of values for C_M which suggest C_M is located somewhere in that range. Is this value the same as what you were expecting theoretically? If not, can you give any solid reason?

Disconnect the line voltage, then remove the wattmeter and isolation transformer. *Set* the capacitance box at C_M. *Connect* the current meter between nodes a and b and AC-voltmeter between nodes b and b′. *Connect* a′ to b′, the isolation transformer to both circuit and the line voltage. What are the RMS values of current and voltage you are reading on these meters? *Relate* your readings with that of the wattmeter readings. If the capacitance boxes are not available, then *use* the regular load circuit with switch S closed. In order to find a correct value for the dissipated power, we must determine the phase angle between voltage and current. This is done by putting voltage $V_{bb'}$ on the horizontal input amplifier of an oscilloscope (or CH1), and the voltage across the resistance R on the vertical input amplifier of the oscilloscope (or CH2). Using the probe is highly recommended. Having both RMS values of current and voltage and their phase difference, we can determine the calculated power which is dissipated in this load network. This value should then be compared with your previous readings.

If time permits repeat this experiment for the parallel circuit of Fig. 6.

6. PROBLEM SET

1. Prove the equivalence of circuits of Fig. 3 and Fig. 4, for $Q_s \gg 1$.
   (Hint: First find the exact V_{in} for each of these two circuits.)

2. Consider the following two power measurement arrangements for a given load.

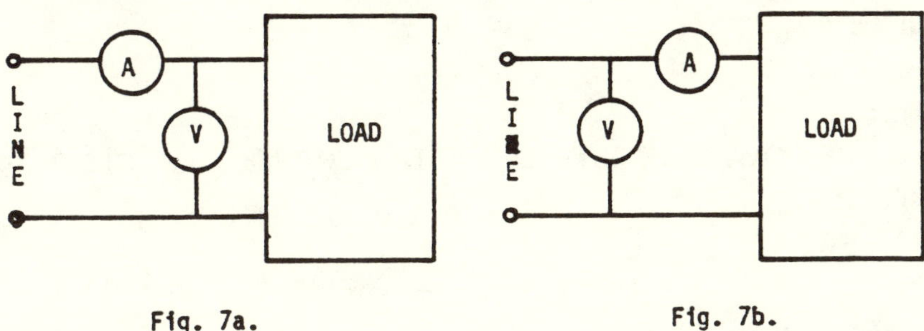

Fig. 7a.   Fig. 7b.

Either of these two circuits give a tentative arrangement for setting the wiring of a wattmeter using an ordinary ammeter and voltmeter. Discuss which of these arrangements will yield the more accurate measurements of power. Perhaps both are correct if used under specific conditions. Then, if so, what are those conditions?

7. FOR YOUR REPORT

Write your report in our standard format. Be sure to discuss your data and give reasons for any discrepancies.

EXPERIMENT I-6

POWER MEASUREMENT IN POLYPHASE CIRCUITS

EQUIPMENT:

> EEP
> AC VOLTMETER
> VOMs (2)
> WATTMETER
> LOAD BOX (HIGH POWER RESISTORS/3, INDUCTORS/3, CAPACITORS/3)
> 3-POLE SWITCH
> SINGLE-POLE SWITCH
> JACK BOXES (2)
> HIGH VOLTAGE CONNECTORS

Table of Contents Page No.

1. PURPOSE

The purpose of this experiment is to acquaint the student with the basic concepts of polyphase circuits and various methods of measuring power dissipated in these systems.

2. INTRODUCTION

The electricity produced by an electric power company is in the form of balanced sinusoidal *polyphase* sources due to the economical advantages of producing and transmitting electricity in this fashion. Also the inevitable industrial applications of electrically driven rotating machines and motors require the use of polyphase power sources to generate a constant torque which will cause the least vibration and thus yield the maximum mechanical durability for these machines. In the following experiment we will study some basic properties of several commonly used polyphase systems.

2.1. SINGLE-PHASE SOURCES

perhaps it sounds strange that we now study a single-phase source. Nevertheless we realize that most of our daily applications involves using this type of power source. These are referred to a perfect sinusoidal waveform of 60 Hz frequency having an amplitude of about 110 volts AC. While it may be produced using a regular single-phase generator, in reality it will most likely be generated as a part of a multi-phase source. At any rate, a single-phase source can be represented as a source with perfect sinusoidal waveform as shown in Fig. 1. As is evident from Fig. 1, it is convenient to use a phasor diagram to represent this source.

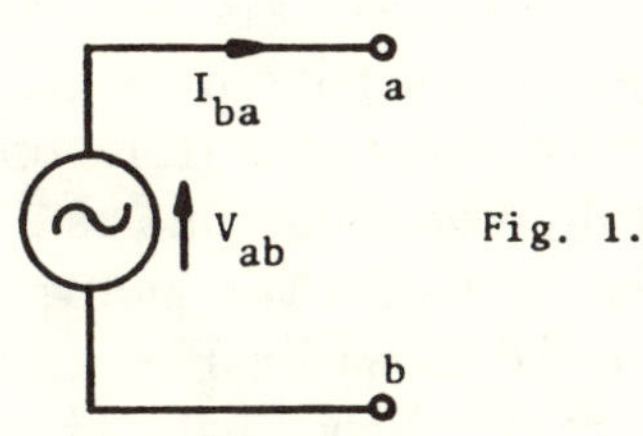

Fig. 1.

2.2. SINGLE-PHASE THREE-WIRE SOURCES (SPTWS)

On many occasions the current and/or the voltage of a regular single-phase source is not enough for a particular application. In these and the similar cases we may use a single-phase three-wire source which is the combination of two identical regular single-phase sources in series. This SPTWS has many applications in most home appliances. A typical diagram for this source is given by Fig. 2. If the source is connected to two identical loads, then the diagram becomes that of Fig. 3. Here $V_{an} = -V_{bn}$, by definition. Thus $V_{an} + V_{bn} = 0$. The current through each load is $I_{na} = \dfrac{V_{an}}{Z}$ and $-I_{bn} = \dfrac{V_{bn}}{Z}$ or $I_{bn} = -\dfrac{V_{bn}}{Z} = \dfrac{V_{an}}{Z} = I_{na}$.

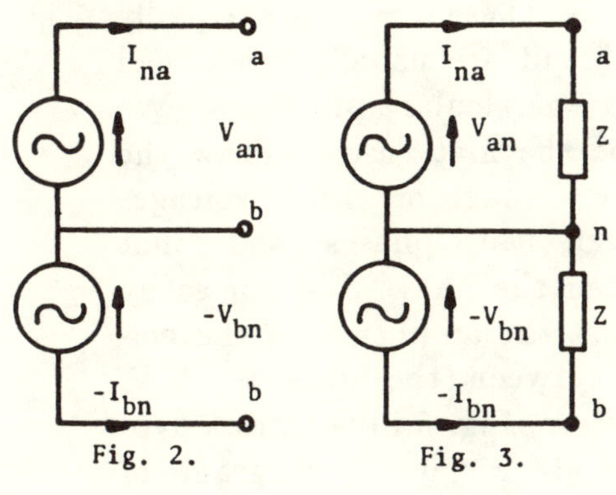

Fig. 2.   Fig. 3.

In this condition the current through the center line is zero, since $I = I_{na} + (-I_{bn}) = I_{na} - I_{na} = 0$. In a practical situation, of course, we have neither two identical sources nor two identical loads. Thus some unbalanced current always exists in the center line. One other source of such unbalance are the impedances of the top and bottom wirings which are not always equal. In general, we expect a non-zero current in the center line.

2.3. TWO-PHASE THREE-WIRE SOURCES (TPTWS)

This is the more general case of a SPTWS. Here we have a source diagram similar to that of Fig. 2, but with $V_{an} = \pm jV_{bn}$. The main application of this source is in the two-phase servomotors that are designed such that the magnetic field of one coil is perpendicular to the magnetic field of the other coil because of j in $V_{an} = \pm jV_{bn}$. Here the voltage between the top and bottom lines is as follows:

$$V_{ab} = V_{an} + V_{nb} = V_{an} - V_{bn} = V_{an}(1\pm j)$$
$$= \sqrt{2}\, V_{an} \angle (\pm 45°).$$

The current can be easily computed for any case under study.

2.4. THREE-PHASE BALANCED SOURCES (TPBS)

Three-phase balanced sources refers to a system that in general has three hot wires called *lines* and one *neutral* wire. These sources are generated such that the voltage between each two consecutive lines are the same in amplitude and with 120° phase difference from one to the next, respectively. We generally select one line as reference for phase measurements. This choice makes the analysis easier. A typical model for this system is shown in Fig. 4. This is called a Y (or star) connection. Here we arbitrarily choose $V_{an} = |V_{an}| \angle (0°)$. Then we may have either:

$$V_{bn} = V \angle (-120°), \text{ and } V_{cn} = V \angle (-240°), \text{ or}$$
$$V_{bn} = V \angle (120°), \text{ and } V_{cn} = V \angle (240°).$$

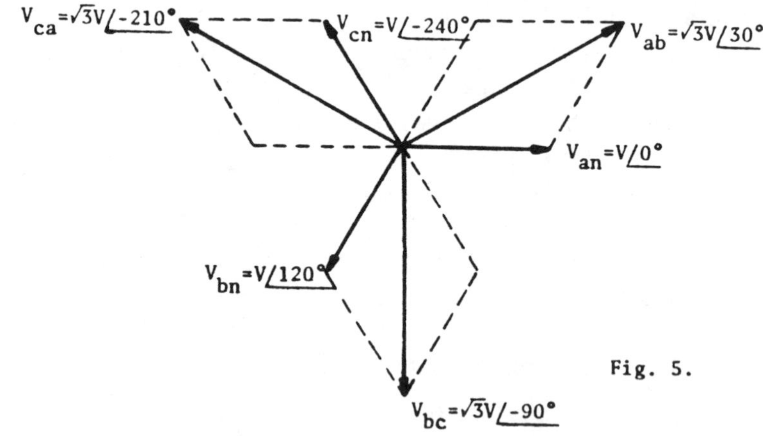

(Recall the positive direction of rotation is counterclockwise.) Each of these cases are equally useful and are usually predefined for a particular problem. We choose the first case to draw the phasor diagram of voltages between the phases and that between the lines. We can easily see that the amplitude of the voltage between the lines is $\sqrt{3}\,V$. Therefore, Fig. 5 illustrates a typical voltage phasor diagram of TPBS.

2.5. APPLICATIONS OF TPBS

There are two different ways to connect a load to a TPBS. These are known as the Y-connection and Δ-connection, respectively. A discussion of both applications follows.

These sets of loads may be balanced depending on whether the three loads in each phase are equal or different. To see some applications of the TPBS we connect these sources to any of the above sets of loads in the following manner. First we connect the source of Fig. 4, to the load of Fig. 6. This arrangement is called the Y-Y connection, Fig. 8. Here we drew the load slightly

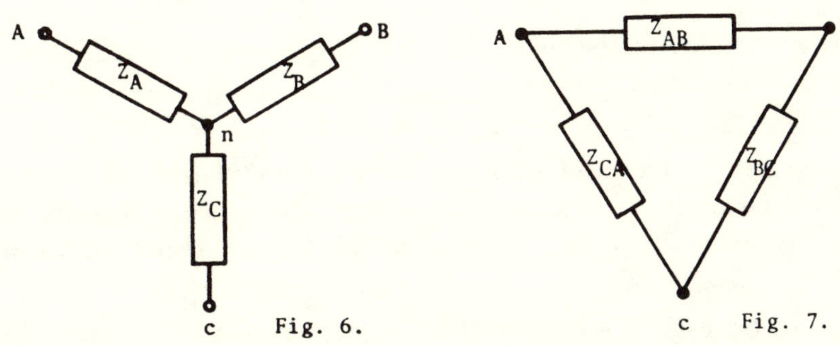

Fig. 6.   Fig. 7.

different from that of Fig. 6, to see the ease of analysis of this arrangement. Independent of what the sources and the loads may be (i.e., balanced or not), we have the following set of equations for line currents.

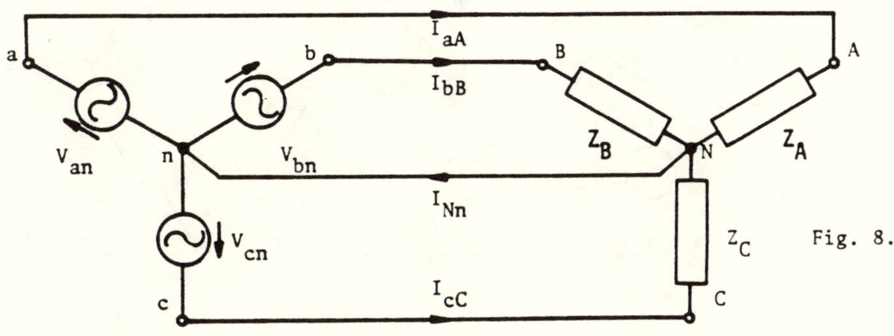

Fig. 8.

$$I_{aA} = V_{an}/Z_A \ , \quad I_{bB} = V_{bn}/Z_B \ , \quad \text{and} \quad I_{cC} = V_{cn}/Z_C \ .$$

Let $V_{an}=V \angle (0°)$, $V_{bn}=V \angle (-120°)$, and $V_{cn} = V \angle (-240°)$, then we have

$$I_{aA} = (V/ \ |Z_A|) \angle (- \phi_A) \quad ,$$

$$I_{bB} = (V/ \ |Z_B|) \angle (-120° - \phi_B) \ ,$$

$$I_{cC} = (V/ \ |Z_C|) \angle (-240° - \phi_C) \ .$$

For a set of balanced loads with $Z_A = Z_B = Z_C = |Z| \angle (\phi_Z)$, we have

$$I_{aA} = (V \ / \ |Z|) \angle (-\phi_Z) \triangleq I_l \angle (-\phi_Z) \ , \text{ similarly,}$$

$$I_{bB} = I_l \angle (-120° - \phi_Z) \quad ,$$

$$I_{cC} = I_l \angle (-240° - \phi_Z) \quad .$$

Clearly we always have

$$I_{Nn} = I_{aA} + I_{bB} + I_{cC} \ .$$

For a set of balanced loads this current becomes

$$I_{Nn} = I_l [\ \angle (-\phi_Z) + \angle (-120° - \phi_Z) + \angle (-240° - \phi_Z)] \equiv 0 \ .$$

Only for this set of balanced loads and balanced sources (and, of course, a set of balanced connecting wirings), this neutral current I_{Nn} is equal to zero. Only in this special case we may ignore the neutral wire. But in general we always have some I_{Nn} even for the seemingly balanced operation. We do not, however, need to use a very low resistance wire due to its cost, since the power lost in this wire is small enough to justify a less expensive wire.

The analysis we presented in the above is, of course, very elementary. Nevertheless it gives some idea as to what must be done in similar situations. There are a host of computerized techniques to deal with most of the realistic problems in which the balanced operation is not the case.

To analyze a load given by Fig. 7, we must study the ways that the source is connected to this load. Here the easiest way to analyze this problem is the case in which the source and the load are connected in the form of Δ-Δ. If the source given as a Y-connection TPBS, without the neutral wire, then we have the circuit of Fig. 9.

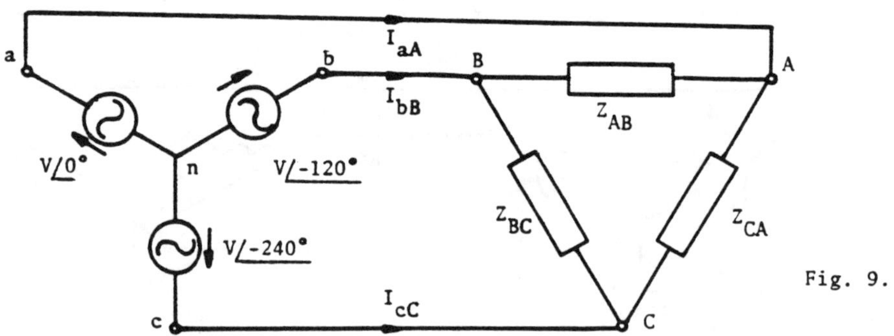

Fig. 9.

This circuit, for the purpose of analysis, can be transformed into the following circuit in which the equivalent sources in terms of line voltage between abc are substituted for the sources given in the Y-form.

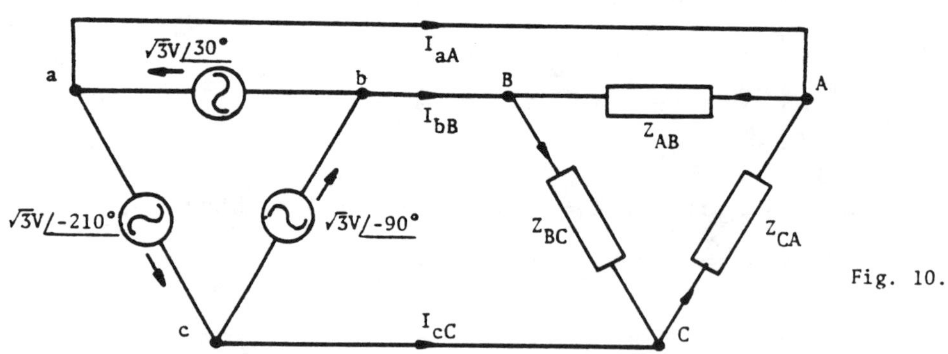

Fig. 10.

Here we can easily show that

$$I_{AB} = V_{ab}/Z_{AB} = (\sqrt{3}\, V \angle (30°))/Z_{AB} \quad ,$$

$$I_{BC} = V_{bc}/Z_{BC} = (\sqrt{3}\, V \angle (-90°))/Z_{BC} \quad ,$$

$$I_{CA} = V_{ca}/Z_{CA} = (\sqrt{3}\, V \angle (-210°))/Z_{CA} \quad .$$

if $Z_{AB} = Z_{BC} = Z_{CA} = |Z| \angle (\phi_z)$, a set of balanced loads, then each phase current becomes

$$I_{AB} = (\sqrt{3}\, V/|Z|) \angle (30° - \phi_z) \triangleq I_p \angle (30° - \phi_z) \quad ,$$

$$I_{BC} = I_p \angle (-90° - \phi_z) \quad ,\text{and} \quad I_{CA} = I_p \angle (-210° - \phi_z) \quad .$$

Clearly in this special case we have $|I_{AB}| = |I_{BC}| = |I_{CA}|$.

To compute the currents through the source lines we have, for example, $I_{aA} = I_{AB} - I_{CA}$, and similarly for I_{bB} and I_{cC} . We can easily show that

$$I_{aA} = I_p [\angle (30° - \phi_z) - \angle (-210° - \phi_z)] = \sqrt{3} \, I_p \angle (- \phi_z) \quad ,$$

$$I_{bB} = \sqrt{3} \, I_p \angle (-120° - \phi_z)] , \text{ and } I_{cC} = \sqrt{3} \, I_p \angle (-210° - \phi_z) \quad .$$

2.6. LOAD TRANSFORMATIONS

An obvious question that may be raised now is under what conditions is the load of Fig. 6 (from the terminal points of A, B, C) equivalent to that of Fig. 7. This transformation has many applications.

To answer this question we see that the equivalent impedance between any two similar points in each load must be equal. Thus we have the following set of equations:

$$Z_A + Z_B = Z_{AB} || (Z_{BC} + Z_{CA}) \quad ,$$

$$Z_B + Z_C = Z_{BC} || (Z_{CA} + Z_{AB}) \quad ,$$

$$Z_C + Z_A = Z_{CA} || (Z_{AB} + Z_{BC}) \quad .$$

To solve for Z_A , Z_B , Z_C results in

$$Z_A = Z_{AB} Z_{CA} / (Z_{AB} + Z_{BC} + Z_{CA}) \quad ,$$

$$Z_B = Z_{BC} Z_{AB} / (Z_{AB} + Z_{BC} + Z_{CA}) \quad ,$$

$$Z_C = Z_{CA} Z_{BC} / (Z_{AB} + Z_{BC} + Z_{CA}) \quad ,$$

Similarly, we can show that

$$Z_{AB} = Z_A + Z_B + Z_A Z_B / Z_C \quad ,$$

$$Z_{BC} = Z_B + Z_C + Z_B Z_C / Z_A \quad ,$$

$$Z_{CA} = Z_C + Z_A + Z_C Z_A / Z_B \quad .$$

2.7. POWER MEASUREMENTS: THEORETICAL

To measure the power dissipated in a polyphase system we must measure the power dissipated in each element of this system. This measurement requires complete analysis of the system, i.e., complete determination of the voltage across and the current through each and every element. Having found all these quantities, the power is then measured in terms of an average value of the instantaneous power of each element k, namely $p_k(t) = v_k(t) i_k(t)$. Since most of these instantaneous powers are periodic the average value can be calculated for any given cycle. Instead of making the measurements for all the elements, however, we show below that in reality we need to make one measurement less than the total number of phases. We then show how to use wattmeters to make these measurements in the next section.

Consider the circuit of Fig. 8. Suppose $Z_A = Z_B = Z_C = |Z| \angle (\phi_z)$. Let us assume

$$V_{an} = V_m \cos \omega t \quad , \quad i_{aA}(t) = I_m \cos (\omega t - \phi_z) \quad .$$

If we replace each of these V_m and I_m by their RMS values we have

$$p_{Z_A}(t) = 2 V_{RMS} I_{RMS} \cos \omega t \cos (\omega t - \phi_z) = V_{RMS} I_{RMS} \cos \phi_z + V_{RMS} I_{RMS} \cos (2\omega t - \phi_z) \quad .$$

Similarly

$$p_{Z_B}(t) = V_{RMS}\, I_{RMS}\cos\phi_z + V_{RMS}\, I_{RMS}\cos(2\omega t - 240\degree - \phi_z)\ ,\ \text{and}$$

$$p_{Z_C}(t) = V_{RMS}\, I_{RMS}\cos\phi_z + V_{RMS}\, I_{RMS}\cos(2\omega t - 120\degree - \phi_z)$$

Then the total instantaneous power becomes

$$p(t) = 3\, V_{RMS}\, I_{RMS}\cos\phi_z\ .$$

Thus

$$P_{ave} = \frac{1}{T}\int_0^T p(t)\,dt = 3\, V_{RMS}\, I_{RMS}\cos\phi_z\ .$$

This is a constant quantity which is three times the dissipated power in each phase. We may equivalently show this dissipated power in terms of the line voltage and/or current quantity if we needed to do so.

Note: The analysis presented here is very elementary. Much more detailed and careful computations are needed to measure this power in a general polyphase system.

2.8. POWER MEASUREMENTS USING WATTMETERS

One obvious way of measuring power in a three-phase system is to connect three wattmeters to each load and measure the consumed power as we did for a single-load in the previous experiment. This is theoretically correct but we can show that we actually do not need to do this for each load. But most importantly in a real or practical situation we have no direct access to the load. In other words, we know how the lines are connected to the load only externally. Therefore real measurements will not be so easy to make as the theoretical measurements discussed earlier. In-

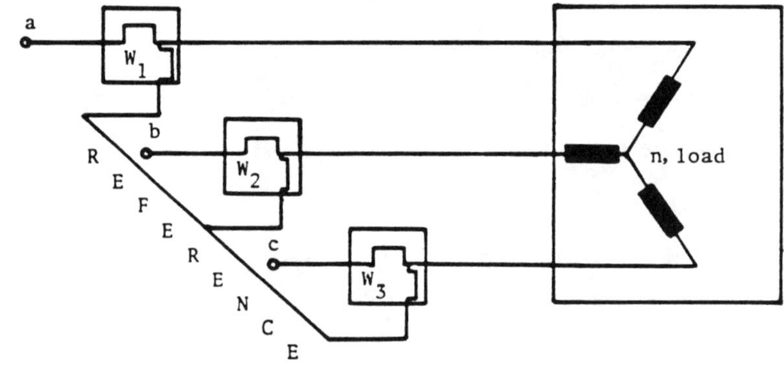

Fig. 11.

stead we put the wattmeters on the line connecting the load as shown in Fig. 11. Here we assume the load is connected in the Y-form. Then the sum of the three readings becomes the sum of the total power consumed in the load. A quick look at *this wiring* shows that we can have the reference voltage point anywhere that we please. To prove this fact we notice that if the reference voltage is v_x, then

$$p_a(t) = v_{ax}(t)\, i_a(t)\ ,\quad p_b(t) = v_{bx}(t)\, i_b(t)\ ,\quad \text{and}\quad p_c(t) = v_{cx}(t)\, i_c(t)\ .$$

Or

$$P_{a,ave} = \frac{1}{T}\int_0^T v_{ax}(t)\, i_a(t)\,dt\ ,\quad P_{b,ave} = \frac{1}{T}\int_0^T v_{bx}(t)\, i_b(t)\,dt\ ,\quad P_{c,ave} = \frac{1}{T}\int_0^T v_{cx}(t)\, i_c(t)\,dt\ .$$

We may choose

$$v_{ax} = v_{an} + v_{nx}\ ,\quad v_{bx} = v_{bn} + v_{nx}\ ,\quad \text{and}\quad v_{cx} = v_{cn} + v_{nx}\ .$$

Thus

$$P_{a,ave} + P_{b,ave} + P_{c,ave} = \frac{1}{T}\int_0^T (v_{an}\,i_a + v_{bn}\,i_b + v_{cn}\,i_c)\,dt + \frac{1}{T}\int_0^T v_{nx}\,(i_a + i_b + i_c)\,dt \ .$$

Since $i_a + i_b + i_c = 0$, here the load is a supernode. Thus the above arrangement is independent of the reference voltage value v_{nx} [0]. Therefore we can choose this reference voltage to be any value we wish. That is, we may choose as a reference, one of the line voltages. If so, such an arrangement yields the voltage reading of one of the wattmeters to be zero which means we can eliminate that wattmeter. Fig. 12 shows how two wattmeters are used to measure the consumed power of a load.

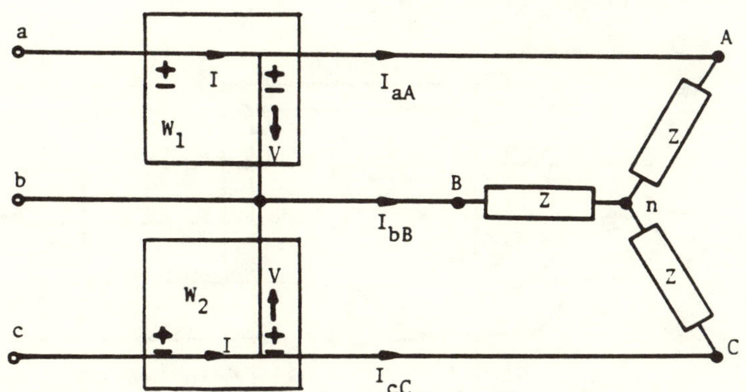

Fig. 12. Use of Two Wattmeters to Measure Total Average Dissipated Power in a Three-Phase Source.

3. LABORATORY PROCEDURE

3.1. PRE-LABORATORY WORK

Review the texts of both this experiment and that of EXP. I-5. Solve the problem of the next section and submit your complete solution with the report of EXP. I-5 which is due today.

3.2. LABORATORY WORK

To perform this experiment extra attention is required due to the high voltage which is present.

CAUTION: *BEFORE DOING ANY TESTING TURN OFF ALL POWER SWITCHES. ASK YOUR INSTRUCTOR TO CHECK YOUR CIRCUIT BEFORE TESTING IT.*

The measurement concept in this experiment is straightforward. Here several elements are mounted on a board(s), with proper binding posts, as found in Fig. 13. As will be seen we can make several different arrangements with these loads for our various experiments. Also it is helpful to have at least two boards with two separate values for RLC elements if enough boards are available.

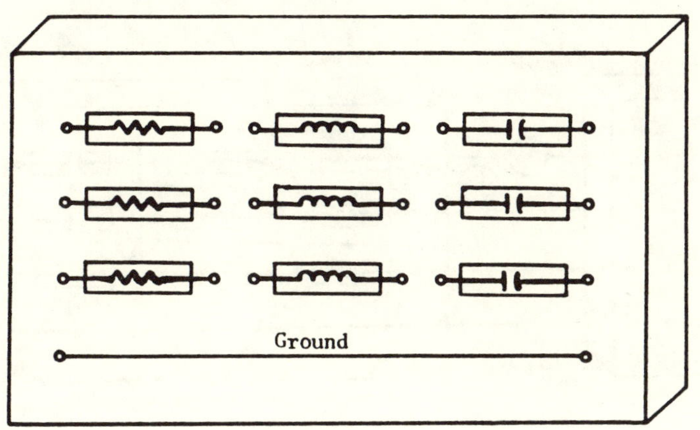

Fig. 13. Load Box

3.2.1. BALANCED THREE-PHASE Y-CONNECTED LOAD

Consider the three-phase, four-wire system of Fig. 14. This system is assumed to be balanced regardless of the type of load. Suppose we also choose, in each phase, a balanced load ($R + 1/j\omega C$) that is Y-connected. Here, with an ordinary wattmeter and voltage and current-meters, we are able to analyze the system as follows:

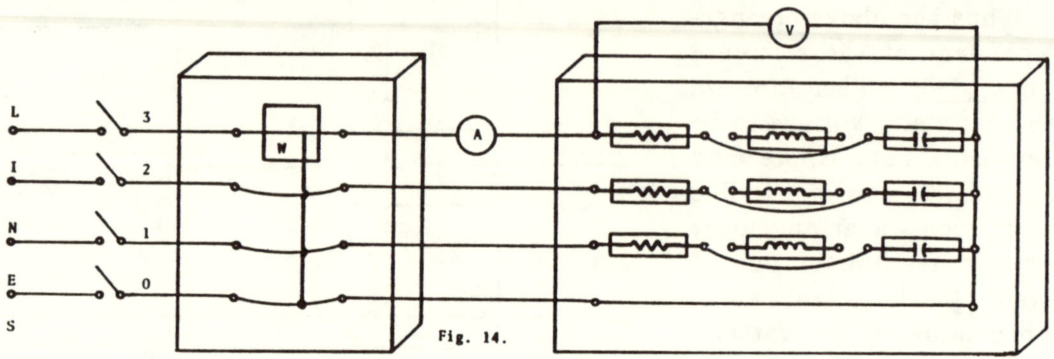

Fig. 14.

CAUTION: BEFORE REMOVING THE MEASURING DEVICES OPEN ALL SWITCHES.

Connect the wattmeter, current-meter, and voltmeter to line number 3 as shown in Fig. 14. *Close* the three-pole switch and the neutral switch. *Record* the above three readings. *Remove* very carefully the voltmeter and read the voltage across the resistive element of the load. *Open* the switches and *set* the above meters on line number 2 and subsequently on line number 1. *Close* the switches and *repeat*, recording the readings for line number 2 and subsequently for line number 1 as you did for line number 3. *Record* all these data in a table. *Draw* a complete phasor diagram for this system. What is the phase difference between voltage and current of each phase? What are the line currents and voltages?   *Compute* the impedance of each phase load from your measurements and *compare* your results with the actual values of these elements as given to you on the load box. *Use* the configuration of Fig. 12 for power measuring with two wattmeters (open the neutral switch) and *measure* the total dissipated power in this system. Is this value the same as that calculated using three wattmeters?   *Discuss* all the discrepancies you have observed in this experiment.

Repeat the above measurements while the neutral switch remains open.

3.2.2. BALANCED THREE-PHASE Δ-CONNECTED LOAD

Consider the load given by Fig. 15. Here we choose ($R + j\omega L$) as the load in each phase. The connection is obviously in the Δ-form. Here we only use one current meter at the time, but that, of course, slows the measurement procedure.

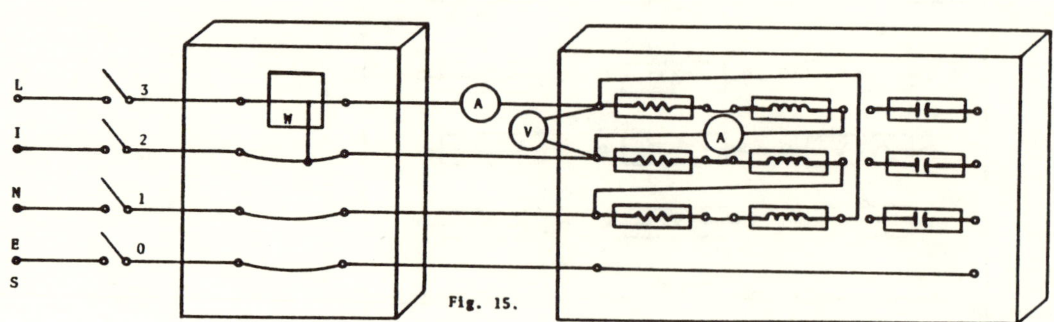

Fig. 15.

CAUTION: BEFORE REMOVING THE MEASURING DEVICES OPEN ALL SWITCHES.

Connect the wattmeter, current-meter and voltmeter to line number 3 as shown in Fig. 15. *Close* the three-pole switch. *Record* these readings. Using the current of line 1 and voltages of lines 1 and 2, *repeat* the above step by inserting the wattmeter between lines 1 and 2. *Record* the total dissipated power by this method. *Insert* the wattmeter in each phase load. *Add* the three readings of the wattmeter and *compare* it with the readings of two-wattmeter method. *Measure* the line currents and voltages. *Measure* each phase load impedance. *Compare* your results with what you expect theoretically. *Discuss* the effects of the actual element values of the load box in your results.

3.2.3. UNBALANCED LOAD

Make up an unbalanced load by, say, choosing R to be in phase one, L to be in phase two, and C to be in phase three (or pick any other feasible arrangement). *Connect* in either Y- or Δ-configuration. Then *repeat* the steps of either Section 3.2.1, or Section 3.2.2, which ever is appropriate.

4. PROBLEM SET

For the circuit in Fig. 16, if $V_{ab} = 100\sqrt{3}\,\angle\,(30°)$, $V_{bc} = 100\sqrt{3}\,\angle\,(-90°)$, and $V_{ca} = 100\sqrt{3}\,\angle\,(-210°)$, then find the currents through each load and the total dissipated power in this system. What is the total reactive power in this system?

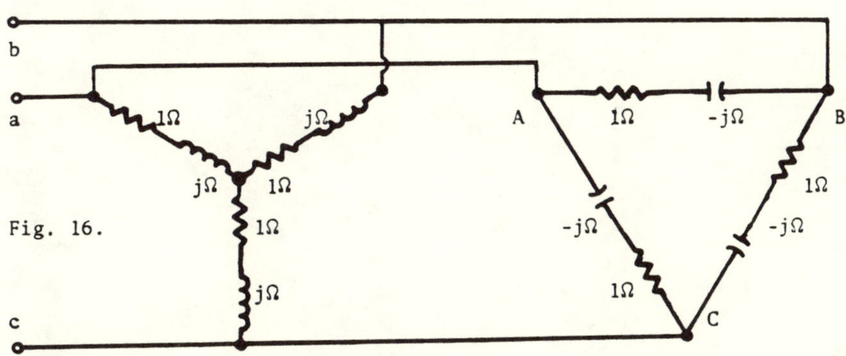

5. FOR YOUR REPORT

Submit your data with whatever conclusions and/or discrepancies you have found in your experiment in the standard form.

WORKSHEET

WORKSHEET

WORKSHEET

PART TWO

FAMILIARIZATION WITH
SOME BASIC CIRCUIT
COMPONENTS AND THEOREMS--
MAGNETIC MATERIAL AND HYSTERESIS

PART TWO

FAMILIARIZATION WITH SOME BASIC CIRCUIT COMPONENTS AND

THEOREMS--MAGNETIC MATERIAL AND HYSTERESIS

II-0. INTRODUCTION

In this part we will study some basic measurement techniques in order to find the element values of some linear circuit components. These techniques are based on the fundamental theorems of circuit theory. These theorems, such as the Thevenin, Norton and Tellegen theorems, are also reviewed. We then study the basic wave-shaping properties and/or applications of diodes especially for AC signal rectification, and some simple RC filter design. The fundamentals of linear passive network synthesis are reviewed next. Finally, we study some basic properties of magnetic materials and hysteresis. The list of our experiments in this part is as follows.

EXPERIMENT II-1: RESISTANCE MEASUREMENTS

EXPERIMENT II-2: DIODES: CHARACTERISTICS AND APPLICATIONS

EXPERIMENT II-3: THEVENIN'S THEOREM AND IMPEDANCE MATCHING

EXPERIMENT II-4: WAVE-SHAPING CIRCUITS: SIMPLE RC FILTERS

EXPERIMENT II-5: AC TO DC POWER SUPPLIES [3]

EXPERIMENT II-6: PASSIVE NETWORK SYNTHESIS

EXPERIMENT II-7: MAGNETIC MATERIAL AND HYSTERESIS [3]

For additional information on these experiments the reader may consult the references of this part.

Please review page 41 of this textbook for common laboratory procedures.

II-1. REFERENCES

[A] Textbooks:

[0] D.J. Comer, Modern Electronic Circuit Design. Reading, MA: Addison-Wesley Pub. Co., 1976.

[1] C.E. Lowman, Magnetic Recording. New York: McGraw-Hill, 1972.

[2] J. Millman, Microelectronics. New York: McGraw-Hill, 1979.

[3] F.G. Stremler, ECE 300 Laboratory Notes. Madison, WI: The University of Wisconsin-Madison, 1978.

[4] M.E. Van Valkenburg, Introduction to Modern Network Synthesis. New York: Wiley, 1967.

[5] M.E. Van Valkenburg, Network Analysis. Englewood Cliffs, NJ: Prentice-Hall, 3rd Edition, 1974.

[B] Manufacturer' Manuals:

[6] General Electric Company, Semiconductor Data Handbook. Syracuse, NY: GE, 3rd Edition, 1977.

[7] Motorola, Semiconductor Data Library, vol. 1. Phoenix, AZ: Motorola Company, 1974.

EXPERIMENT II-1

RESISTANCE MEASUREMENTS

EQUIPMENT:

EEP	GR DECADE RESISTANCE BOX *
VTVM **	DECADE RESISTANCE BOX
VOM	PRECISION RESISTOR (WIREWOUND RESISTOR)
OHMMETER	UNKNOWN CARBON COMPOSITION RESISTORS/4
CAPACITORS (1 μF, 20 μF)	RESISTORS (1 kΩ/2, 33 kΩ)
BRIDGE (UNIVERSAL, AC, DC)	

Table of Contents   Page No.

*   GR stands for General Radio (U.S.A.). In this and the following experiments this unit can be replaced by any other brand of variable resistance box.

**   VTVM stands for Vacuum Tube Voltmeter. Any comparable voltmeter can be used as well.

1. PURPOSE

The purpose of this experiment is to obtain the DC volt-ampere (V-I) characteristic *curves* of several resistive elements. This measurement is done by using a DMM or VTVM/VOM and DC bridges. Later we expand this DC bridge concept into an AC bridge measurement to find the capacitance of an unknown capacitive element.

2. INTRODUCTION

The first step toward useful application of any two-terminal device as a circuit component is the determination of the volt-ampere (V-I) characteristics of that device. This characteristic is then used to obtain the value of the parameters for small signal models of the device, or is used directly in graphical design techniques to analyze the circuit in conjunction with other elements. The DC characteristics of a device are determined from point-by-point DC measurements of voltage

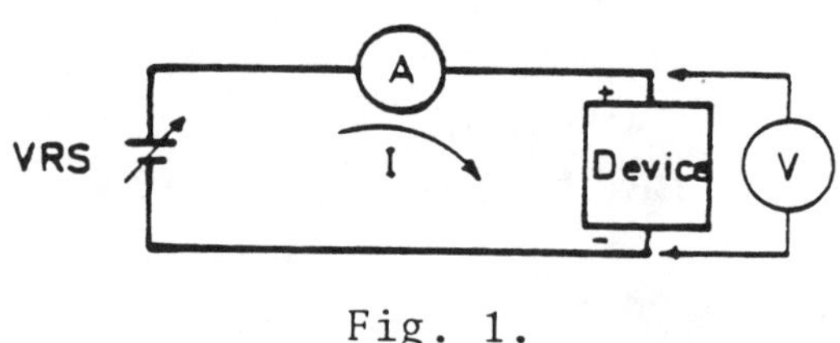

Fig. 1.

and current. A two-terminal device can be completely characterized by a series of measurements of the voltage across and the current through the device. A simple circuit to make this measurement is given by Fig. 1. One can also use a curve tracer to find this characteristic as has already been done in EXP. I-4.

In general, for an unknown device, we should expect a nonlinear relation between V and I. For example, if we look into the (V-I) characteristic of a diode, we get a nonlinear curve. From experience, however, one knows that the (V-I) characteristic of a resistor is a straight line, i.e., the relation between V and I is linear. This, of course, was expected from Ohm's law. When it is known that a component can be considered as a fixed resistance, easier methods exist to determine the value of that resistance. With the component in an active circuit, a single measurement of the voltage across and current through the component allows its resistance to be calculated by Ohm's law. When the component is not in a circuit, instruments such as ohmmeters, a DMM or DC bridges can be used to determine the resistance.

2.1. SOME REMARKS ON RESISTANCE MEASUREMENTS

The ideal resistor is a two-terminal circuit element in which Ohm's law ($v = Ri$) holds instantaneously. R is called the resistance of the element and its dimensional unit is the Ohm (Ω). The inverse of a resistive element, $G = 1/R$, yields the conductance of that circuit element. The dimensional unit of conductance is the Siemens (S) or Mho (Ω)$^{-1}$.

To measure the resistance value of both low and high resistances we must be very careful about the various kinds of errors that may occur. For low resistance we have to worry about the leads from the resistive element to the remainder of the circuit (cf., Fig. 2).

Reading = R+2r.

Fig. 2.

For high resistance we have to look into the leakage paths between the resistance terminals (cf., Fig. 3).

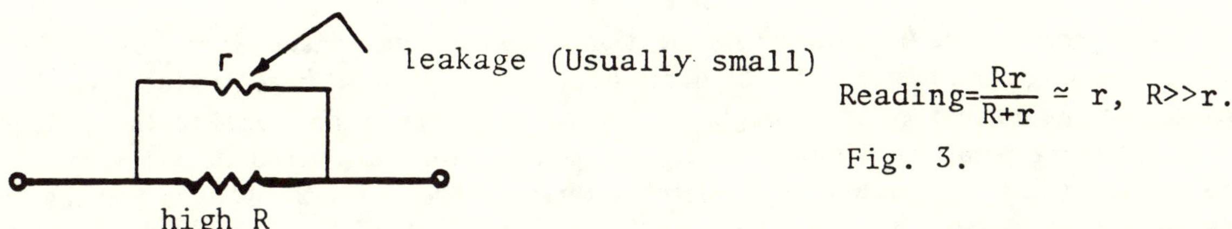

$$\text{Reading} = \frac{Rr}{R+r} \simeq r, \; R \gg r.$$

Fig. 3.

These leakage paths are always unwanted and their causes, such as moisture or humidity, must be eliminated. *LEAKAGE PATHS CAN ALSO RESULT FROM THE RESISTOR BODY BY HANDLING. THUS WHILE READING DO NOT TOUCH YOUR CIRCUIT.*

In industrial manufacturing applications we have no time to measure each resistor as we want to use them. Therefore, a universal color code has been developed which gives an approximate value for the resistors. This coding is as follows:

Here $d_2 > d_1$ so that there will be no ambiguities on which side to start reading.

Fig. 4.

The resistance value is $R = \overline{AB} \times 10^C \pm D\%$ where :

Black $= 0$	Green $= 5$	Gold (C–only)	$= -1$
Brown $= 1$	Blue $= 6$	Silver (C–only)	$= -2$
Red $= 2$	Violet $= 7$		
Orange$= 3$	Gray $= 8$	Tolerance (D–only) $=$	$\begin{cases} 20\% & \text{No band} \\ 10\% & \text{Silver} \\ 5\% & \text{Gold.} \end{cases}$
Yellow $= 4$	White $= 9$		

Suppose for a particular case we have the following strips: A=red, B=violet, C=orange, and D=gold. Then the value of the resistance is

$$R = 27 \times 10^3 \, \Omega \pm 5\% , \text{ i.e., } 25650\Omega < R < 28350\Omega .$$

The main reason behind using this color code is that we can always read the value of resistors as they are mounted on the circuit in any given direction and without taking them out. There is no other way of writing these values such that we can read them from all sides of the resistor.

There are, of course, some limitations, on the resistor values. Some of these limitations are as follows:

(a) *Temperature variation:* All resistor values change with temperature. In most large scale and complex applications a cooling system is built into the circuit to rectify this situation. The variation of resistance with temperature is specified by a temperature coefficient (TC) and is given by the manufacturer. Resistors can be made to have both positive and negative temperature coefficients. There are, of course, cases in which we would like to have a temperature dependent device such as a thermistor to be used for temperature measurements (cf., EXP. VI-12).

(b) *Current and voltage limitations:* As mentioned above, temperature is one of the fundamental limitations in using the resistor. On the other hand, the additional heat produced in the resistor is proportional to the current going through the resistor which is proportional to the voltage across the resistor. Thus we associate a power rating for each resistor, expressed in watts, which simply says if the power rating of a resistor at a reference temperature -- say 25 °C -- is P, then the maximum voltage and/or current that should be applied to this resistor is $V_{Max} = \sqrt{PR}$ and $I_{Max} = \sqrt{P/R}$. Exceeding these values will generally derate the resistance.

(c) *Frequency limitations:* At high frequency the performance of a resistor will change because of many new factors such as stray capacitances and lead inductances. Therefore, bear in mind that the calculated or measured R may change as frequency changes.

There are, in general, four types of resistors; namely, (1) Carbon composition available in the range of 1 Ω to 22 MΩ and a maximum power of 2 watts; (2) wirewound, 1 Ω to 100 kΩ, and 200 Watts; (3) metal film, 0.1 Ω to 150 MΩ and 1 watt; and (4) carbon film, 10 Ω to 100 MΩ and 2 watts maximum. Resistors are also made of semiconductors.

3. LABORATORY PROCEDURE

Note: *DO NOT EXCEED VOLTAGES AND CURRENTS SPECIFIED IN THE PROCEDURES OR THE COMPONENT UNDER TEST MAY BE DESTROYED DUE TO EXCESSIVE POWER DISSIPATION.*

3.1. CONSTRUCTION OF V-I CHARACTERISTIC

Precision Resistor: Using the circuit of Fig. 5, *measure* the volt-ampere (V-I) characteristics of the wire wound resistor supplied. *Measure* values of voltage and current over a voltage range from 0 to + 5 volts in 0.5 volt increments. *Plot* the values on graph of Fig. 6, in the first quadrant of a rectangular coordinate system. *Reverse* the leads of the voltage reference source (VRS)

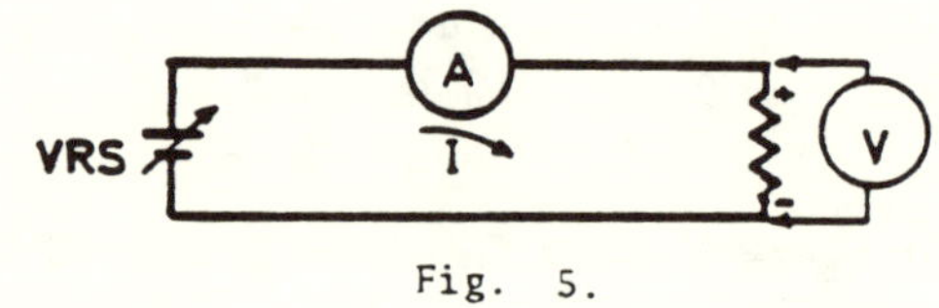

Fig. 5.

and also of the current meter in the circuit, and again *measure* the voltage and current over a range of voltage from 0 to + 5 volts in 0.5 volt increments. In what quadrant should these values be plotted? From this graph *determine* the value of the resistance. How does this value compare with that indicated on the resistor case?

3.2. SIMPLE SERIES OHMMETER

The circuit for a simple series ohmmeter is given by Fig. 7. When the test leads are "shorted" ($R_X = 0$) the 1 kΩ potentiometer can be adjusted so that the current through the circuit deflects the meter full-scale. Let this value of current be I_1 ; then,

$$I_1 (R_m + R_f + R_a) = 1.5 \text{ volts.}$$

If an unknown resistance R_X is inserted in the circuit between the test leads a new value of current I_2 exists, where

$$I_2 (R_m + R_f + R_a) + I_2 R_X = 1.5 \text{ volts.}$$

Solving these two equations yields

$$R_X = (R_m + R_f + R_a) (I_1 - I_2) / I_2 .$$

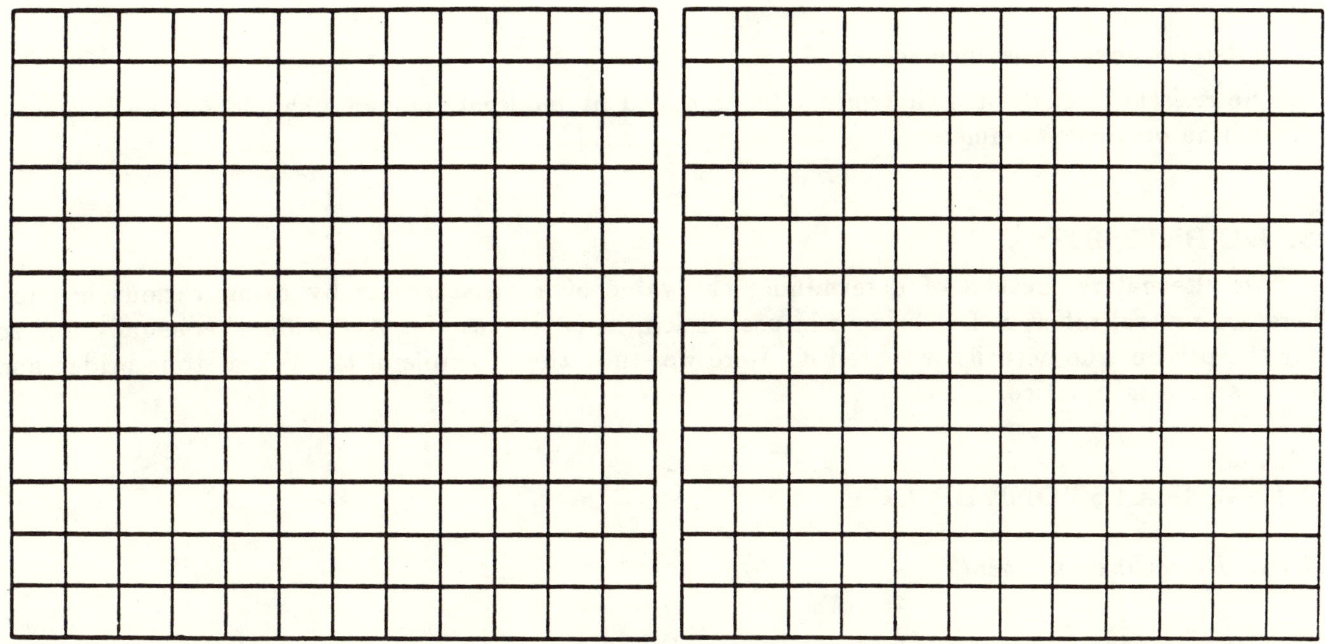

Fig. 6.

With a VRS output of 1.5 volts and a 1 mA full-scale meter movement the sum of the resistances $R_m + R_f + R_a$ must be set to 1500 Ω for full scale deflection. Then it follows that $R_X = 1500\ \Omega\ ((I_1 / I_2) - 1)$. When $I_2 = I_1/2$ (mid-scale deflection), then $R_X = 1500\ \Omega$. When $I_2 = I_1/3$, then $R_X = 3000\ \Omega$. The scale is nonlinear with higher resistance readings bunched at the lower end.

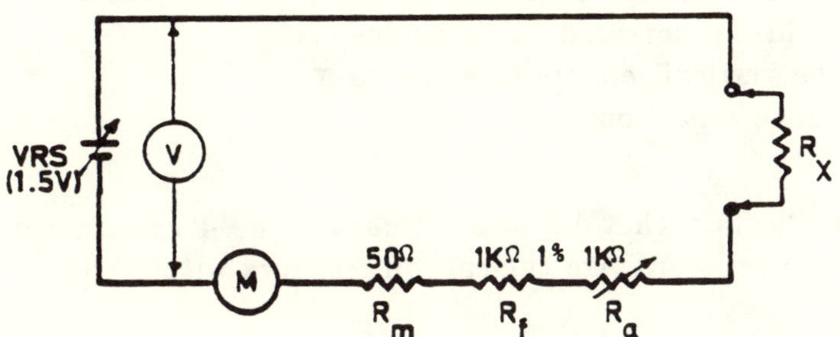

Fig. 7. A Simple Ohmmeter.

Construct the circuit of Fig. 7, and *measure* resistance values of 100 Ω, 1 kΩ and 10 kΩ from the GR decade resistance box.

Note: At each step in the procedure, the VRS terminal voltage must be checked (and readjusted if necessary) to maintain 1.5 V reading on the voltage meter.

Caution: If using a VTVM as a voltmeter insure that the switch on the VTVM probe is in the DC position.

What changes would have to be made to this circuit to provide ranges of 150 Ω and 15 kΩ mid-scale? The difficulty of providing multiple resistance ranges limits this circuit's use in practical ohmmeters.

REMARKS:

1. An ohmmeter is *never* used to determine a resistance while the circuit containing the resistance is in operation. The value of a resistance measured in a circuit by an ohmmeter may be affected by other circuit elements in the circuit. In general, a resistor should be removed from a circuit

to determine its resistance accurately.

2. The resistance of devices that might be damaged by moderate currents should not be measured with an ordinary ohmmeter.

3.3. DC BRIDGES

An alternative method of determining the value of a resistance is by using various bridges. There are several different DC bridges in use. Each, as is the case for AC bridges, is usually named after the person who actually invented it. Here we study the principle of the Wheatstone bridge and how it is used in practice.

3.3.1. WHEATSTONE BRIDGE

3.3.1A. Direct measurement

Construct the Wheatstone bridge shown in Fig. 8. The oscilloscope with coupling input set at DC will be used as a null detector. R_X is the resistance to be determined. When the voltage indicated by the oscilloscope equals 0 volts the bridge is said to be balanced. *Start* with the oscilloscope's vertical sensitivity set to 10 V/cm. The R_3 is used to balance the bridge. As the bridge approaches balance the vertical sensitivity setting of the oscilloscope should be increased. The bridge should be balanced with the vertical sensitivity equal to 10 mV/cm. Under the balance conditions

$$R_X = R_2\,R_3\,/\,R_1 .$$

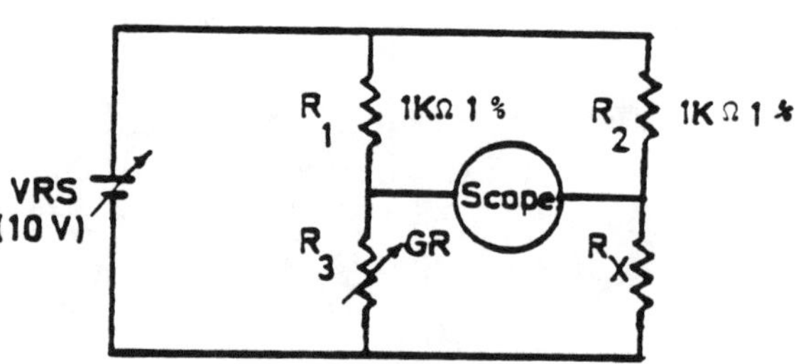

Fig. 8.   Wheatstone Bridge.

This relation results from the fact that $V_{ab} = 0$. *Measure* the 4 carbon composition resistors supplied and the 10 kΩ precision resistor. Are the carbon resistors within their tolerance?

3.3.1B. Substitution method

The substitution method involves a comparison between the unknown resistance and an accurately calibrated resistor. The bridge is first balanced with the unknown resistance as R_X. If the accurately calibrated resistance is adjusted to restore a null reading, then R_X equals the value indicated on the accurately calibrated resistance. *Use* one *decade resistance* for R_3 and GR decade resistance as the accurately calibrated resistor. *Measure* the 4 carbon composition resistors and the 10 kΩ precision resistor. *Compare* the measurements with those in part *3.3.1A*.

3.4. AC BRIDGES

The concept of AC bridges is the same as that of the DC bridges. Here, instead of resistive elements, four impedances are located in the four branches of the bridge. The general configurations of this and one special case, called the Wien bridge, are presented in Fig. 9.

Here we may use the oscilloscope for the null detector. The balance condition is again the same as that of the DC bridge which is $V_{ab} = 0$. If this condition is met, then we have $Z_X = Z_2\,Z_3/Z_1$.

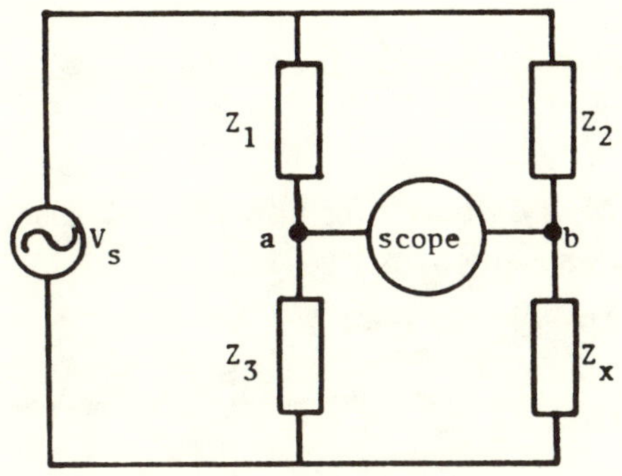

Fig. 9. A General AC Bridge.

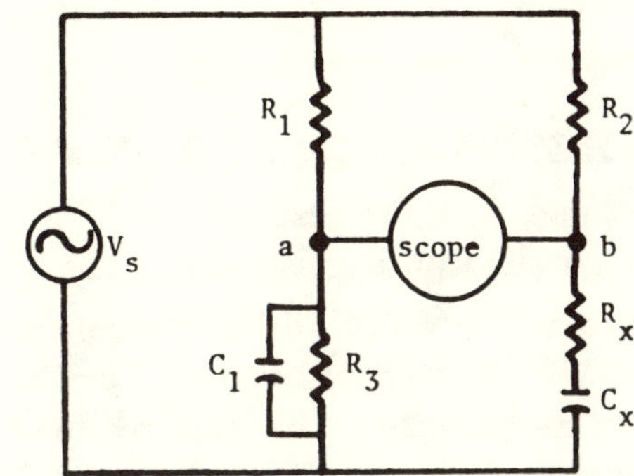

Fig. 10. A Wien Bridge.

Hypothetically let $Z_1 = 1/sC_1$, $Z_2 = R_2$, $Z_3 = R_3$, and $Z_x = sL_x$. Then in the *steady state* and for a sinusoidal input we get

$$sL_x = R_2 R_3 sC_1 \Rightarrow L_x = R_2 R_3 C_1 .$$

Similarly, $\qquad C_x = L_1 / R_2 R_3 .$

These conditions are seldom met in practical problems since one cannot obtain a pure inductor and/or capacitor to use in these bridges. In reality we must use other bridges, such as that of Fig. 10, to find capacitances and/or inductances.

For Fig. 10, we have $Z_1 = R_1$, $Z_2 = R_2$, $Z_3 = R_3 / (1+j\omega R_3 C_3)$, and $Z_x = R_x + 1/j\omega C_x$. Using the balance condition results in

$$C_x = (C_3 + 1 / \omega^2 R_3^2 C_3) R_1 / R_2 , \text{ and}$$

$$R_x = (1 + \omega^2 R_3^2 C_3^2) R_2 R_3 / R_1 .$$

Of course, these are functions of ω, the balanced frequency, which may be difficult to obtain. These equations can also be reversed to use this bridge as a frequency meter.

An alternative method to measure the value of an unknown capacitor involves using the circuit of Fig. 11, which simply is a voltage divider. It is easy to show that

$$V_c / V_s = C_x / (C + C_x) .$$

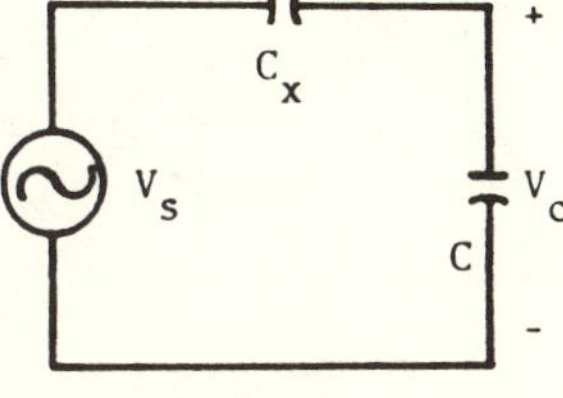

Fig. 11.

A careful monitoring of this ratio with a known C, or possibly a variable C, will result in the value of C_x. This is a reasonably accurate method to perform this experiment.

3.5. COMMENTS ON BRIDGE SENSITIVITY

The concept of bridge sensitivity goes back to the years of 1833-1845, as described in the classical textbook of B. Hague (Alternative Current Bridge Methods--revised by T.R. Foord, Pitman Press, 1971). This is perhaps the most interesting result in the early history of electrical instrumentation. This historical reference was brought to the attention of this writer by Emeritus Professor T.J. Higgins of the University of Wisconsin-Madison. In fact, the original work of Wheatstone had considered this problem. We will discuss sensitivity problems of the networks in Section III-2.1. Here, for the sake of record, we point out that to measure various resistances, etc., accurately, via these bridges, we must be very concerned with the accuracy of our standard or known values. Inaccuracies in these standard values would be amplified by further multiplications, and would result in less accurate readings for the unknown element values. These problems are referred to as bridge sensitivity problems.

3.6. INPUT RESISTANCE OF THE VTVM

Measure the DC input resistance of the VTVM by observing the rate of discharge of a 1 μF capacitor through the input resistance of the VTVM as shown in Fig. 12.

Place contact S *ON* and adjust the reference supply so that the VTVM reads 10 volts. *Turn* contact S to *OFF* position and *measure* the time required for the capacitor voltage to decay to 3.68 volts or 36.8 percent of the initial value. *Repeat* the measurement ten times. Calculate the mean and use this as the most probable value of the period. This period is equal to the time constant τ of the circuit where $\tau = RC$. Hence the input resistance, R, of the VTVM is τ/C.

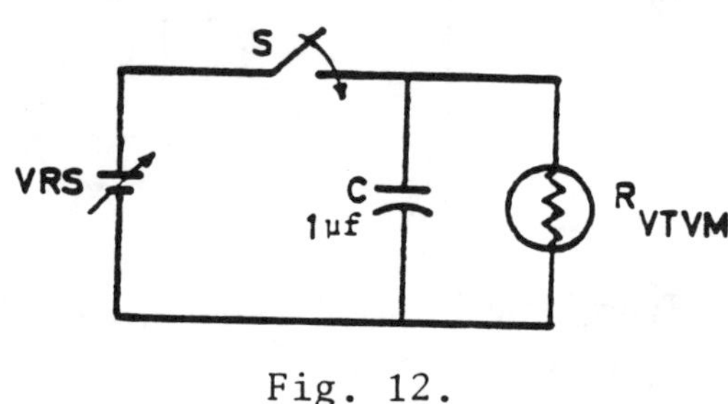

Fig. 12.

Repeat the above procedure to measure the DC input resistance of the oscilloscope and compare with the value given by the manufacturer. *Use* a 20 μF capacitor instead of the 1 μF capacitor. *Make* 10 measurements. *Calculate* the mean of these measurements. These results illustrate loading effects of the measuring devices on the circuits.

4. PROBLEM SET

1. Consider the circuit of Fig. 13. Let the VRS be set for 1 volt (assume the voltmeter has a 2% error in this range). Suppose that the milliammeter A is used in the range of 5 mA with 2% error in this range. How accurately can a 100 Ω resistor be measured in this setting?

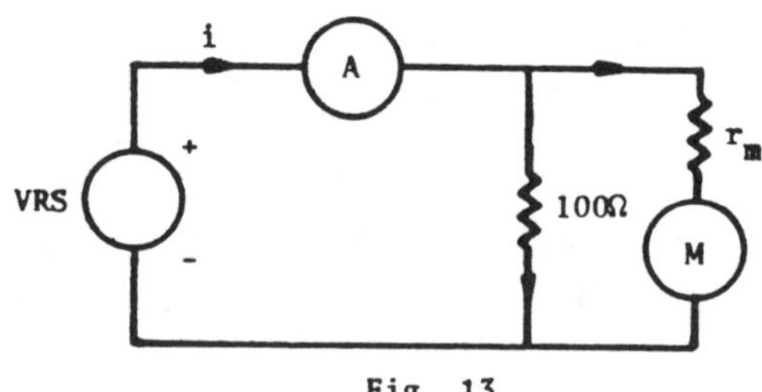

Fig. 13.

2. Give the color code used to identify the following composition resistors:
   (a) 6800 Ω $\pm$ 20%   (c) 3.9 Ω $\pm$ 10%

   (b) 510 Ω $\pm$ 5%   (d) 100 Ω $\pm$ 20% .

3. Two resistors with unmarked values are received. However, it is known that one is a 250 Ω, 5 W resistor and the other is a 50 Ω, 25 W resistor. How can the 50 Ω, 25 W resistor be identified by inspection?

4. Set up a circuit to measure the input resistance and input capacitance of an oscilloscope.

5. Set up a circuit to measure output impedance of a function generator.

5. FOR YOUR REPORT

Submit the solutions to the above problem set and your data in our standard report form.

6. APPENDIX

6.1. STANDARD RESISTANCE VALUES

The nominal values in Ohms of carbon resistors that are available in power ratings of 1/4, 1/2, 1 and 2 W are as follows. These values may be multiplied by 10^n, $0 \le n \le 6$.

$\pm$ 5%	2.7	3.0	3.3	3.6	3.9	4.3	4.7	5.1	5.6	6.2	6.8	7.5
$\pm$ 10%	2.7		3.3		3.9		4.7		5.6		6.8	
$\pm$ 5%	8.2	9.1	10	11	12	13	15	16	18	20	22	24
$\pm$ 10%	8.2		10		12		15		18		22	

6.2. STANDARD CAPACITANCE VALUES

The nominal values of several different capacitor types are given in the following tables.

Ceramic-disk capacitors, 10% tolerance, pF				
3.3	24	91	330	800
5	25	100	350	820
6	27	120	360	1000
6.8	30	130	390	2200
7.5	39	150	400	2500
8	47	180	470	2700
10	50	200	500	3000
12	51	220	510	3300
15	56	240	560	4300
18	68	250	600	4700
20	75	270	680	6800
22	82	300	750	8200

Electrolytic capacitors for bypass application, μF	
250	2000
500	3000
1000	4000
1500	5000

Tantalum capacitors, 10% tolerance, μF		
0.0047	0.010	0.022
0.0056	0.012	0.027
0.0068	0.015	0.033
0.0082	0.018	0.039

All the above values may be multiplied by 10^n, where n = 0, 1, 2, 3, 4, 5 (up 330 μF) .

EXPERIMENT II-2

DIODES: CHARACTERISTICS AND APPLICATIONS

EQUIPMENT:

EEP	p-n DIODES (2)
VOM	ZENER DIODE
VTVM	SILICON DIODE
OHMMETER	RESISTORS (1 kΩ, 10 kΩ, 33 kΩ)
CURVE TRACER	PHOTOCONDUCTOR
CAPACITOR (0.1 μF)	BLACK PAPER

Table of Contents   Page No.

1. PURPOSE

The purpose of this experiment is to acquaint the student with the (V-I) characteristic of a diode from which one concludes a diode is a nonlinear element. The method of generating a local small-signal model to represent a given diode from this characteristic is also presented. Finally, some basic wave shaping properties of circuits involving diodes are included. Additional applications of circuits with diodes are covered in EXP. II-5.

2. INTRODUCTION

An ideal diode is an element which acts exactly like an ideal switch, i.e., when it closes at a certain voltage it allows infinite current to flow. This is not, of course, the case for a real diode, however. In the following we construct the (V-I) characteristic of several diodes, then continue to study additional properties of diodes. We do not, however, plan to study the structure of diodes or how they are built. This information can be found in many standard electronics textbooks such as [2].

3. LABORATORY PROCEDURE

3.1. CONSTRUCTION OF (V-I) CHARACTERISTIC

In the following, several diodes are provided that are to be used to determine their (V-I) characteristics. One can always use a curve tracer to get these results, however, we will do this the hard way!

3.1A. *p-n Diode*

Using the circuit of Fig. 1, and with the cathode connected to the minus (–) terminal of the voltage reference source (VRS), *measure* values of voltage and current over a *current* range from 0 to 1 mA in 0.1 mA increments. *Plot* these data in rectangular coordinates (use the first quadrant of Fig. 3a). *Reverse* the leads of the VRS and of the current meter and *measure* values of voltage and current over a voltage range of 0 to 10 V in 1 volt increments. *Plot* these data on the same graph and in the appropriate quadrant. Can a single value of resistance be determined for this device? If not, could the diode be considered as having two different resistances, one for each direction of voltage polarity across the diode? If this is possible, what values of resistance would you select?

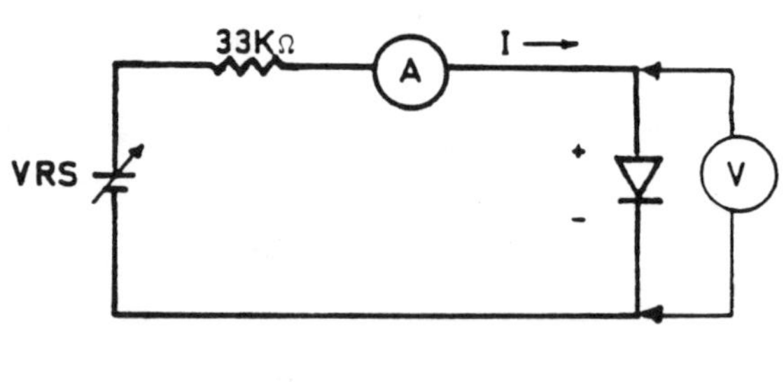

Fig. 1.

tances, one for each direction of voltage polarity across the diode? If this is possible, what values of resistance would you select?

3.1B. *Zener Diode*

Using the circuit of Fig. 2, and with the cathode connected to the minus (–) terminal of the VRS, *measure* values of voltage and current over a *current* range from 0 to 1 mA in 0.1 mA increments. *Plot* these data on the rectangular coordinate system of Fig. 3b, and in

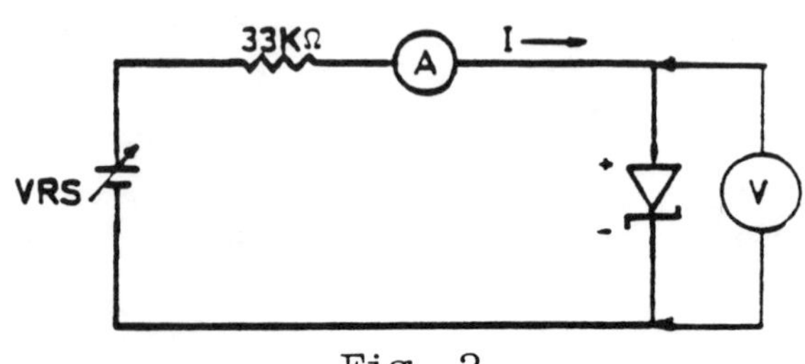

Fig. 2.

the first quadrant. *Switch* the leads of VRS and of the current meter and *measure* the values of voltage and current over a current range from 0 to 1 mA in 0.1 mA increments. Note that a voltage of approximately 6 volts magnitude must be applied before a measurable current for this connection is obtained. Beyond 6 volts slight increases in the VRS voltage will cause large current increases. How does this diode differ from the previous diode considered? This diode usually is used for voltage regulation. (Why?)

There are many other kinds of diodes for various other applications, but we can not go through them all. One of these is a photodiode.

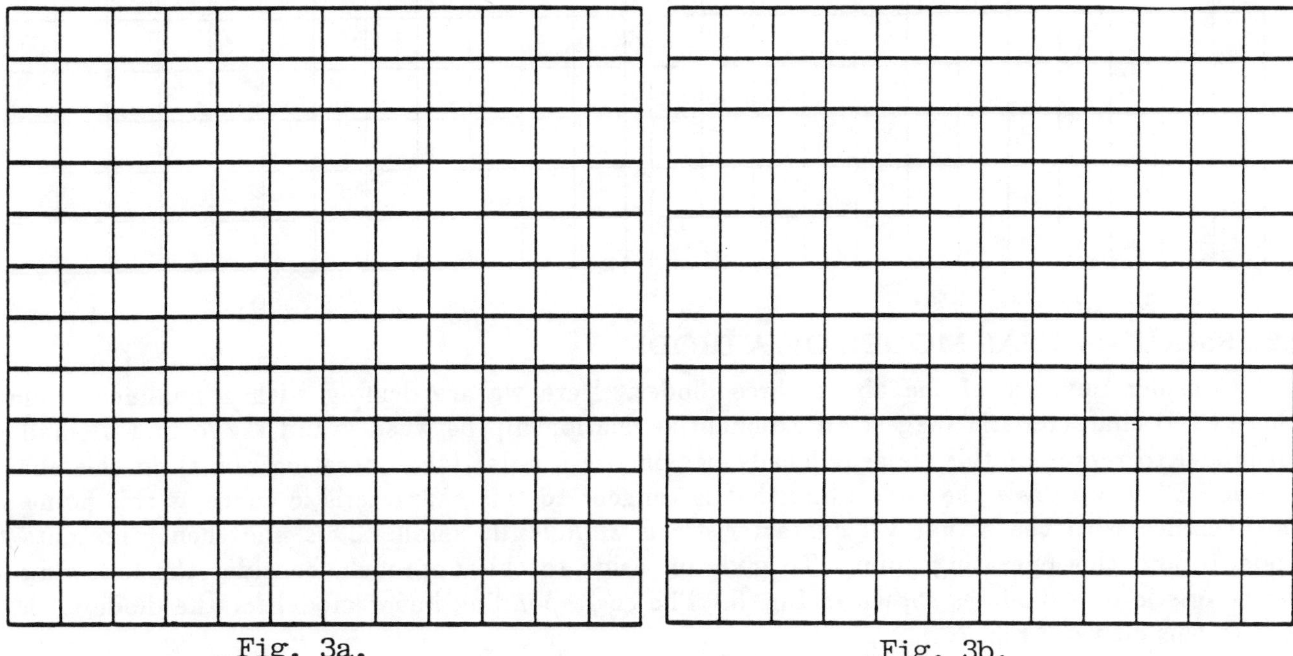

Fig. 3a.   Fig. 3b.

3.1C. Photoconductor

Using the circuit shown in Fig. 4, *measure* the volt-ampere characteristic of the supplied photoconductor. *Place* the photoconductor in a position on the board such that its top surface faces the overhead lights. *Measure* and *plot* the voltage and current values for voltages from 0 to 5 volt increments on Fig. 5a. *Reverse* the leads of the VRS and of the current meter and *repeat* the process. Next *cover* the top surface of the photoconductor with the photographic film provided, or black paper, and *determine* the volt-ampere characteristics over the same range of voltages as before. *Plot* these values on the same graph as the previous values. Next *cover* the top surface of the photoconductor with masking tape. *Repeat* the voltage and current measurements as before. *Plot* this curve on graph of Fig. 5b. What values of resistance characterize the photoconductor under each of the three conditions of light intensity. *Give* a qualitative description of the resistance property of the photoconductor.

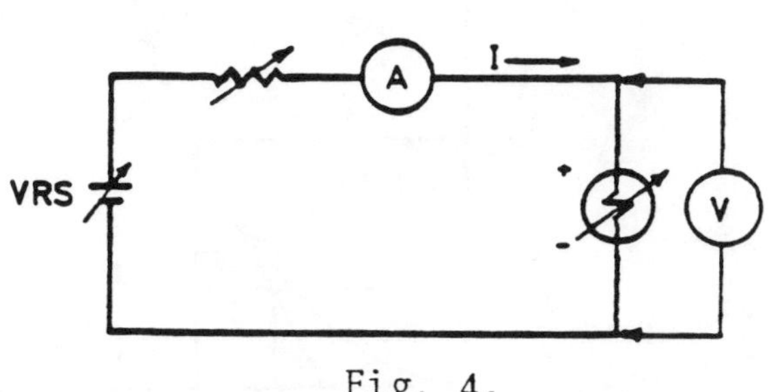

Fig. 4.

Note: As you may have observed the circuits for the last three experiments are the same and the purpose of the 33 kΩ series resistor is for the protection of the device. Thus to go from one experiment to the next just change the various diodes and continue the measurement. Of course, each time while changing the diode *you must turn off the power supply.*

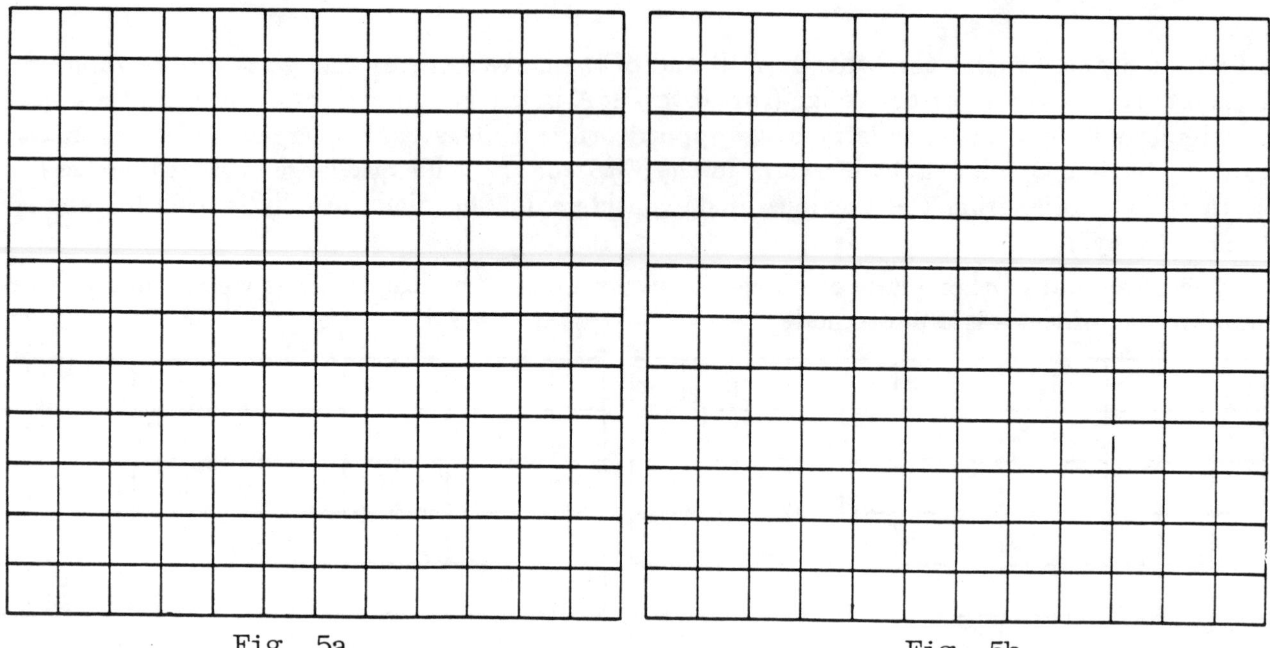

Fig. 5a.   Fig. 5b.

3.2. SMALL-SIGNAL MODEL OF A DIODE

Consider any one of the above three diodes. Here we are dealing with a nonlinear element whose (V-I) characteristic suggest an exponential relationship between v and i. To find a small-signal model to represent this element locally, we choose a point (the operating point) on this characteristic. Then we draw the *widest* line that is tangent to this characteristic curve at this point and its difference with the actual V-I characteristic is *sufficiently* small. This line then represents the diode around this operating point. To give an inside to this approach, consider the following V-I characteristic of a diode as shown in Fig. 6. The corresponding linear model for the diode is shown next to this curve in Fig. 7.

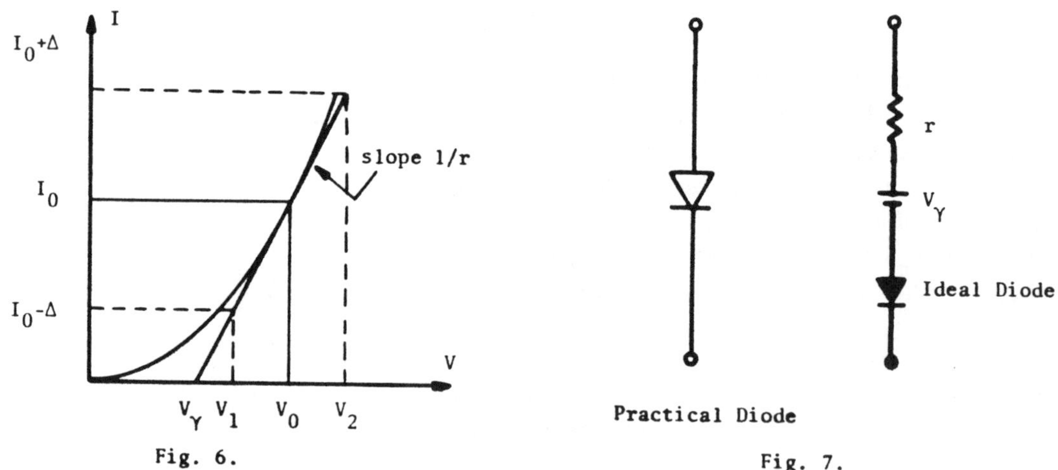

Fig. 6.   Fig. 7.

Find the best linear model of the three diodes you have studied earlier.

3.3. DIODE WAVE-SHAPING PROPERTY

There are several interesting wave-shaping properties of diodes that one can study. Some of these properties such as rectification are reported in EXP. II-5. Here we look into two other wave-shaping properties of diodes; namely, *clipping* and *clamping*. *Use* the silicon diode for the following testing.

    Construct the *Clipping* circuit of Fig. 8. *Apply* a sinusoidal waveform input whose maximum amplitude is larger than E with f=100 Hz. *Choose* E=0, at first. *Apply* the input to CH1 of the oscilloscope and use CH2 to display the output. *Draw* both input and output on Fig. 9a. *Draw* the transfer function curve (Input-Output relation) on Fig. 9b. *Increase* E to a larger level and *repeat* the above test. *Increase* the frequency to a much higher value -- say 1 MHz and *repeat* the above experiment.

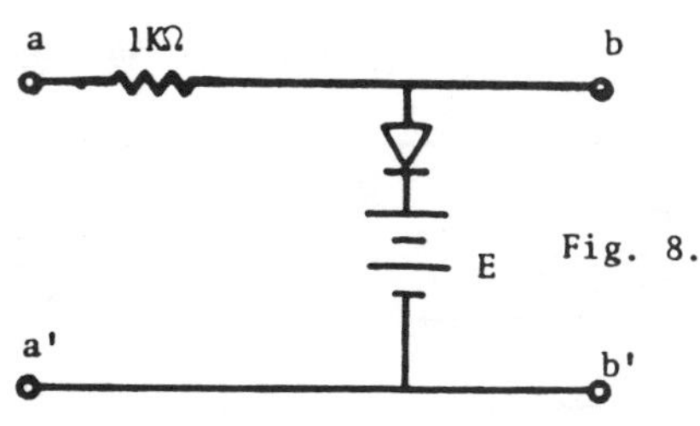
Fig. 8.

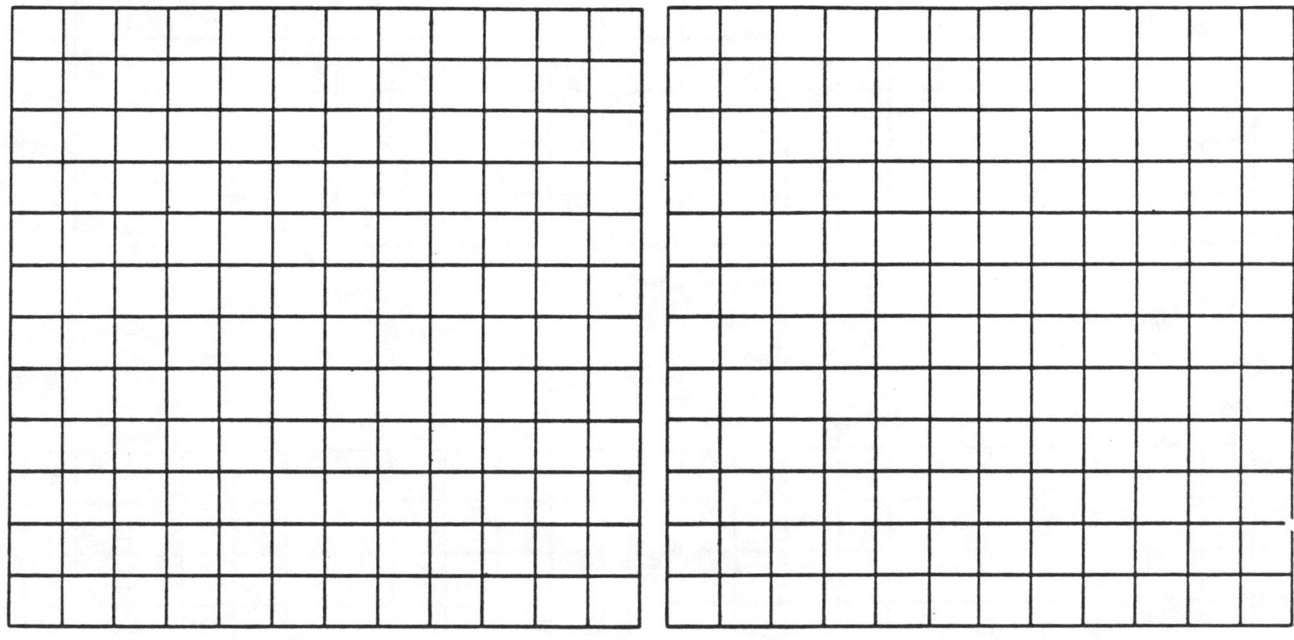

Fig. 9a. Fig. 9b.

    Next *construct* the circuit of Fig. 10, and as before, *draw* the input and output waveforms for this circuit on Fig. 12a. *Draw* the circuit's transfer function curve on Fig. 12b.

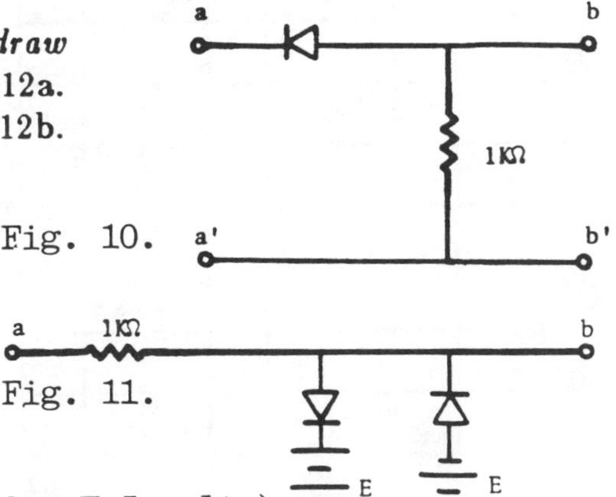
Fig. 10.

Fig. 11.

    Find the transfer function for the *limiter* circuit of Fig. 11, using the graph of Fig. 13b. But, first, *draw* its output on the chart of Fig. 13a for an input $v_{aa'} = 10 \sin \omega t$.

(Set E=5 volts)

    Construct the *Clamping* circuit of Fig. 14. *Draw* the output of this circuit on the graph of Fig. 15, for a sinusoidal input of $v_{aa'} = 10 \sin 1000t$.

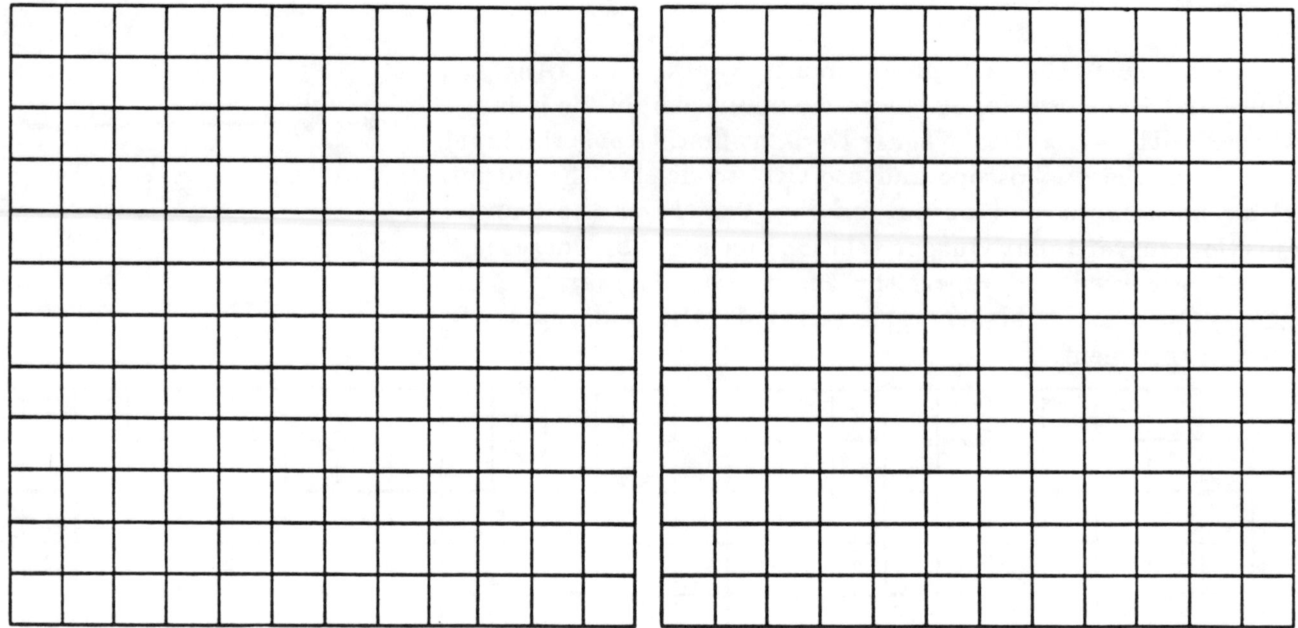

Fig. 12a.   Fig. 12b.

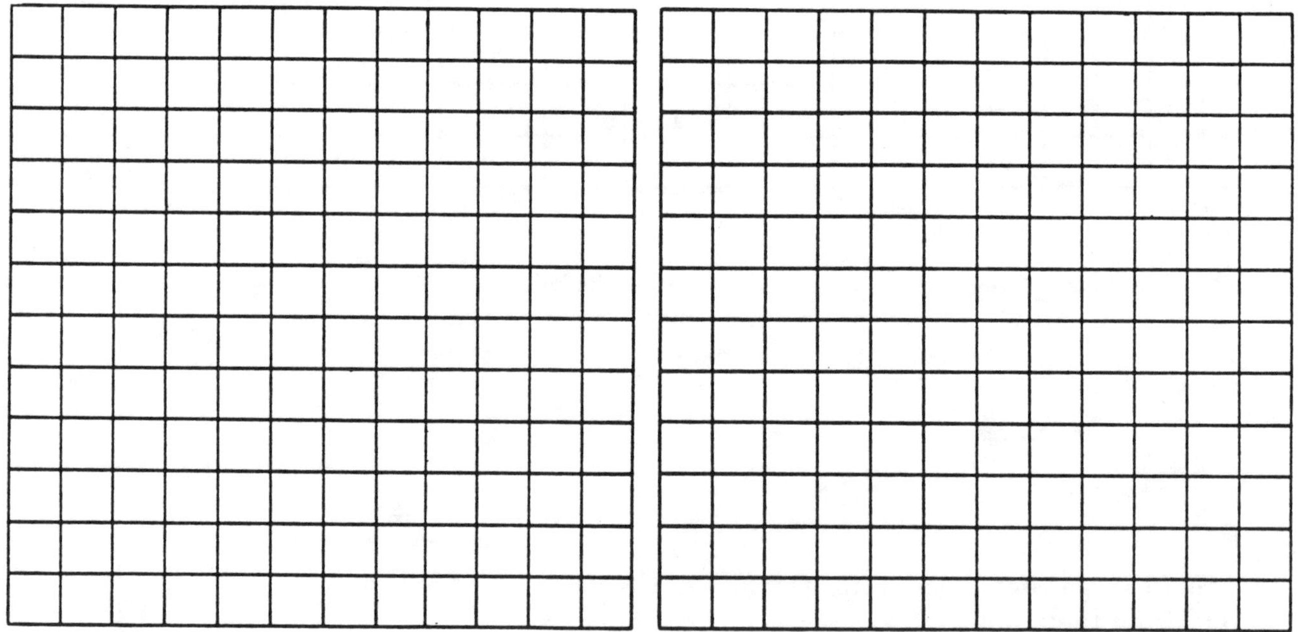

Fig. 13a.   Fig. 13b.

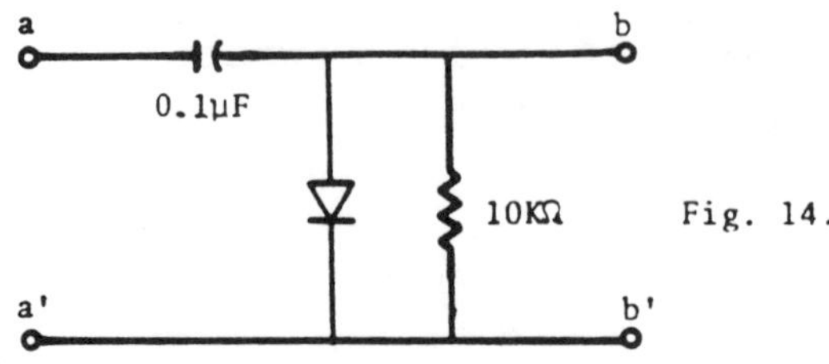

Fig. 14.

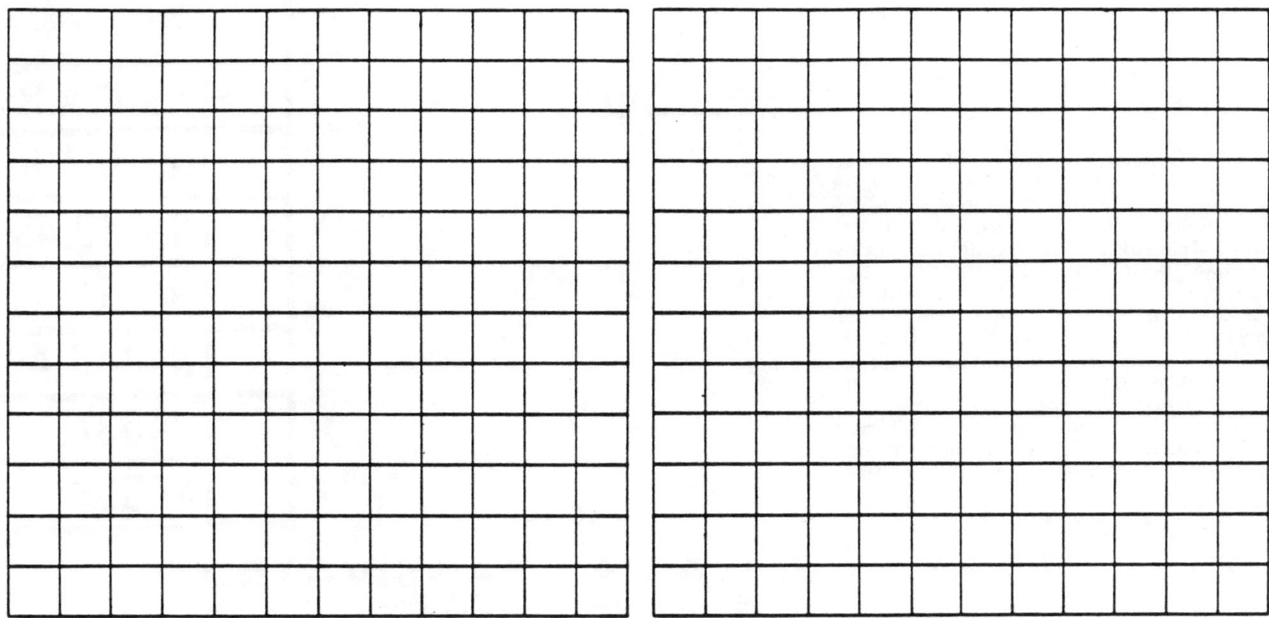

Fig. 15a.   Fig. 15b.

4. PROBLEM SET

1. What information should appear on the graphs for the volt-ampere characteristics of the two-terminal devices under study?

2. For what reasons should these data be plotted as they are taken in the laboratory?

3. In making the voltage and current measurements required in this experiment, what criterion should be used in selecting the voltage and current scales on the meters to be used, and why?

4. What is the purpose of the 33 kΩ resistor in the circuits for determining the volt-ampere characteristics of the p-n diode and Zener Diode?

5. If the two voltage sources shown by E in Fig. 11 were different or zero then what would have been the transfer function of this circuit?

5. FOR YOUR REPORT

Please submit your data with all the necessary discussions in our standard report form.

6. MANUFACTURERS' SPECIFICATION DATA SHEETS

In the following pages (cf., Figs. 16, 17 and 18), we present the specification data sheets of three widely used diodes for the sake of illustrations and review of their basic properties [6] and [7]. Please read these pages very carefully.

Silicon
Diodes

This family of General Electric silicon signal diodes are very high speed switching diodes for computer circuits and general purpose applications. These diodes incorporate an oxide passivated planar structure. This structure makes possible a diode having high conductance, fast recovery time, low leakage, and low capacitance combined with improved uniformity and reliability. These diodes are contained in two different packages; double heat sink miniature package, and milli-heat sink package.

They are electrically the same as their equivalent types in each of the two different packages (see page two for groupings of electrically equivalent types in each of the two packages).

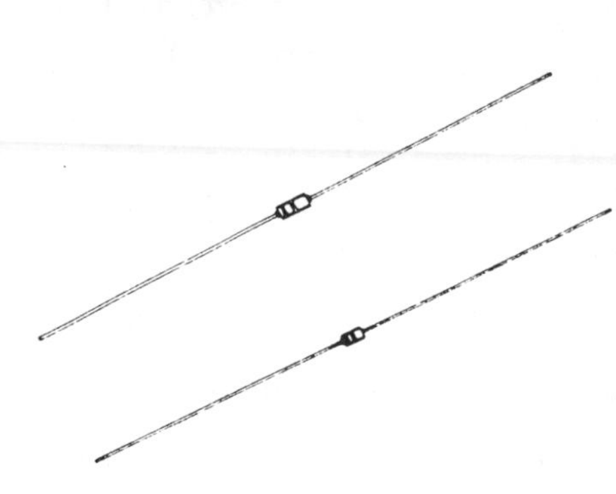

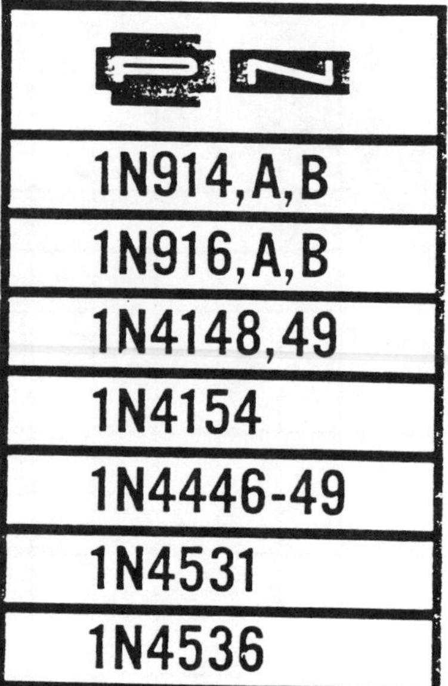

PLANAR EPITAXIAL PASSIVATED
with Controlled Conductance

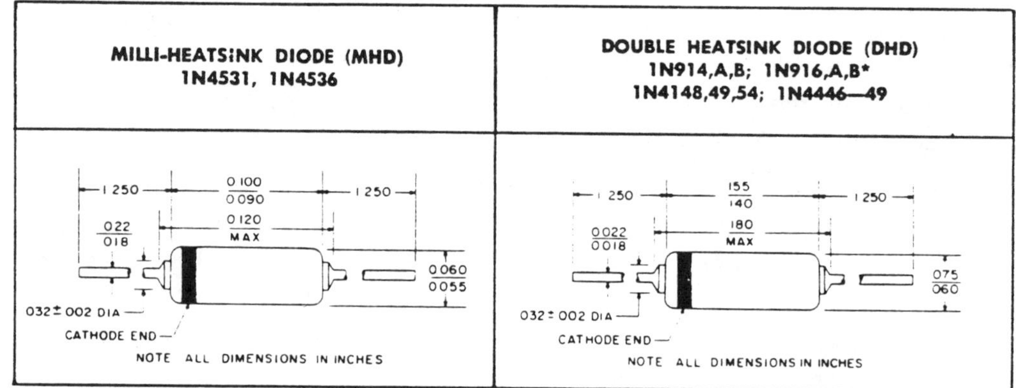

Dissipation: 500mW @ 25°C free air
Derate: 2.85mW/°C for temp. above 25°C amb. based on max. $T_J = 200°C$

Dissipation: 500mW @ 25°C free air
Derate: 2.85mW/°C for temp. above 25°C amb. based on max. $T_J = 200°C$

FEATURES	1N914 1N914A 1N914B	1N4148 1N4446 1N4448 1N4531	1N916 1N916A 1N916B	1N4149 1N4447 1N4449	1N4536 1N4154
Reverse Recovery Time of 2 nanoseconds maximum					●
Reverse Recovery Time of 4 nanoseconds maximum	●	●	●	●	
Capacitance of 2 pF maximum			●	●	
Capacitance of 4 pF maximum	●	●			●
Power Dissipation to 500 mW		●		●	●
Power Dissipation to 250 mW					
Meets all MIL-S-19500C requirements	●	●	●	●	●

HEATSINK SPACING FROM END OF DIODE BODY	STEADY STATE THERMAL RESISTANCE °C/mW (NOTE 1)		POWER DISSIPATION AT 25°C mW (NOTE 2)	
	MHD	DHD	MHD	DHD
.062"	.230	.250	760	700
.250"	.319	.319	550	550
.500"	.438	.438	400	400

NOTE 1 See Figure 7 for thermal resistance for short pulses.
NOTE 2 This power rating is based on a maximum junction temperature of 200°C.

Figure 1

Fig. 16. Silicon Diodes (Courtesy of General Electric [6]).

absolute maximum ratings: (25°C) (unless otherwise specified)

		MHD & DHD	MHD & DHD	
Voltage				
Reverse		75	25	Volts
Current				
Average Rectified		150	150	mA
Recurrent Peak Forward		450	450	mA
Forward Steady-State DC		200	200	mA
Peak Forward Surge (1 μsec. pulse)		2000	2000	mA
Power				
Dissipation		500	500	mW
Temperature				
Operating		←—65 to +200—→		°C
Storage		←—65 to +200—→		°C

electrical characteristics: (25°C) (unless otherwise specified)

Type	Minimum Breakdown Voltage @ 100μA	Forward Voltage		Maximum Reverse Current, I_R			C_o [1]	t_{rr} [2]	V_f [3]
		I_F	V_F	20V		75V			
				25°C	150°C	25°C			
	Volts	mA	V	nA	μA	μA	pF	ns	V
1N914 1N4148 1N4531	100	10	1.0	25	50	5	4	4	
1N914A 1N4446	100	20	1.0	25	50	5	4	4	
1N914B 1N4448	100	5 100	0.62–0.72 1.0	25 [4]	50	5	4	4	2.5
1N916 1N4149	100	10	1.0	25	50	5	2	4	
1N916A 1N4447	100	20	1.0	25	50	5	2	4	
1N916B 1N4449	100	5 30	0.63–0.73 1.0	25	50	5	2	4	2.5
1N4154 1N4536	35 @ 5μA	30	1.0	100 @ 25V	100 @ 25V		4	2	

*Except as noted.

NOTES   (1) Maximum Capacitance is measured on Boonton model 75A capacitance bridge at a signal level of 50 mV at $V_R = 0$

(2) Maximum Reverse Recovery Time, $I_f = 10mA$, $V_R = -6V$, $R_L = 100\Omega$, Recovery to 1.0mA (Figure 6)

(3) Maximum Forward Recovery Voltage, $-50mA$ peak square wave, 0.1 μsec. pulse width, 5 to 100 kHz repetition rate, generator rise time $(t_r) \leqq 30nsec.$

(4) Also 3μA at 20 V at 100°C

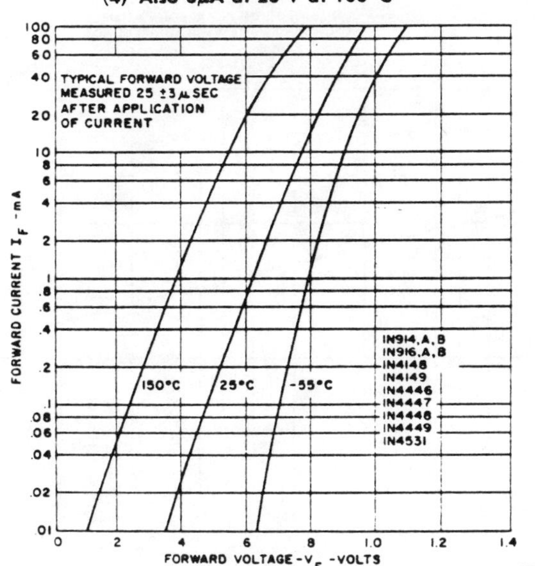

Figure 2

Fig. 16. Continued.

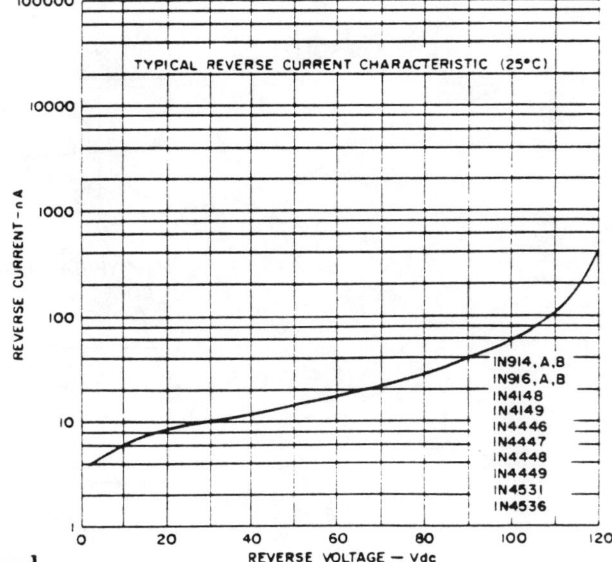

Figure 3

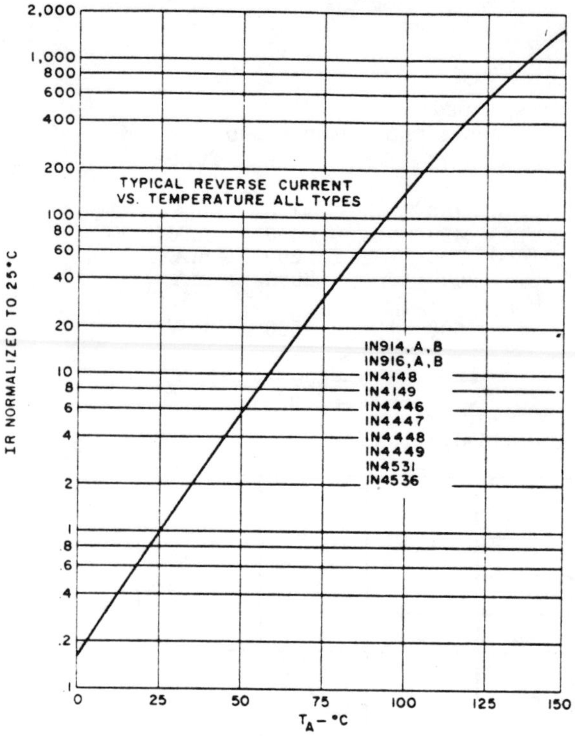

Figure 4

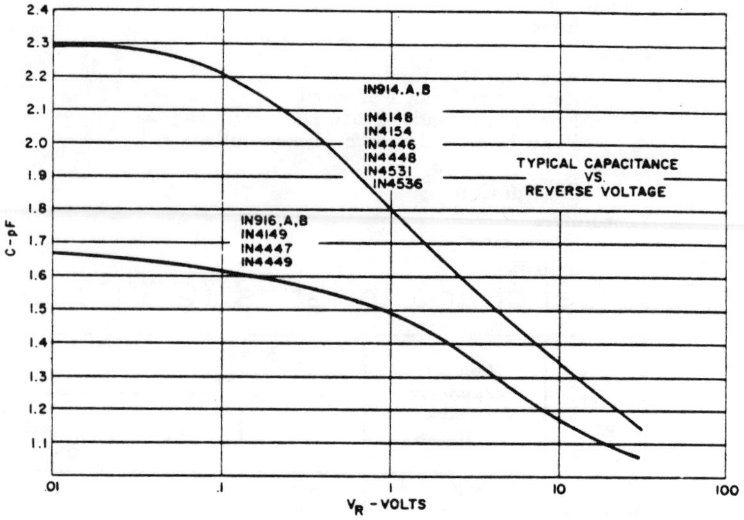

Figure 5

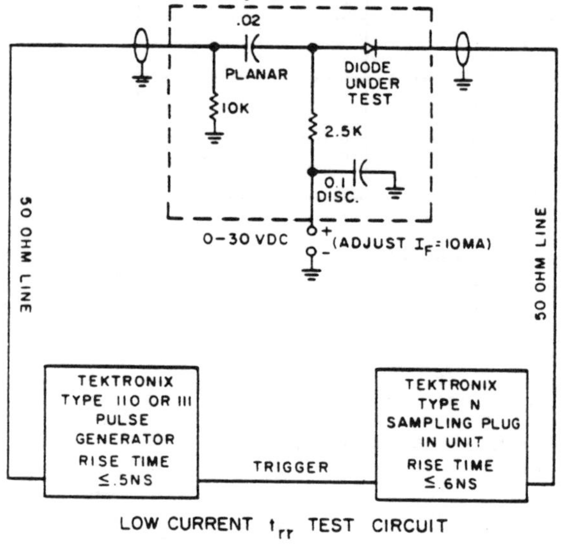

Figure 6

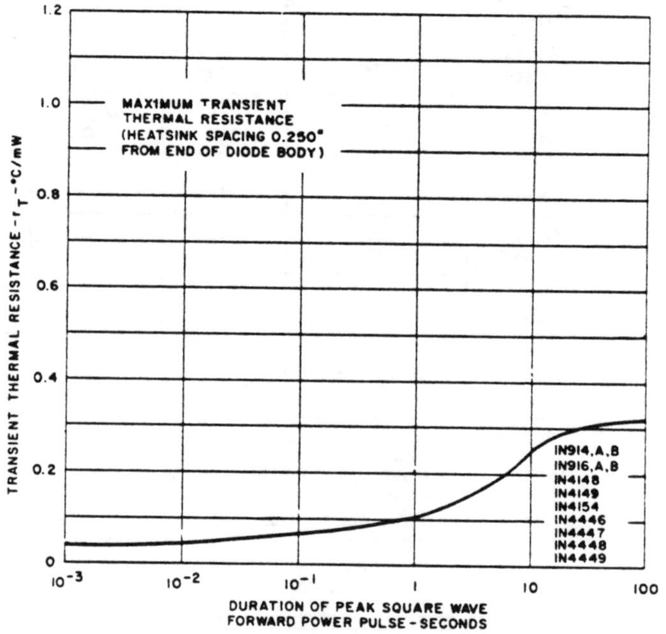

Figure 7

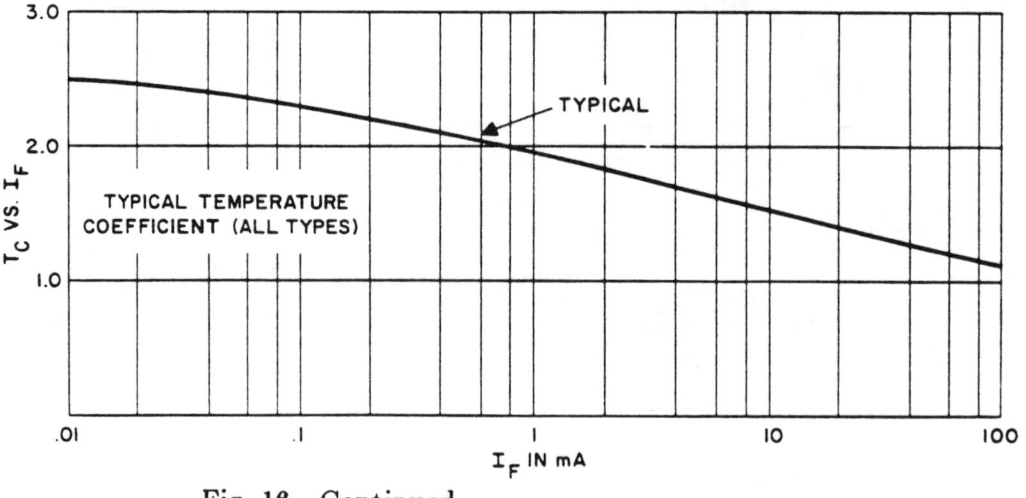

Figure 8

Fig. 16. Continued.

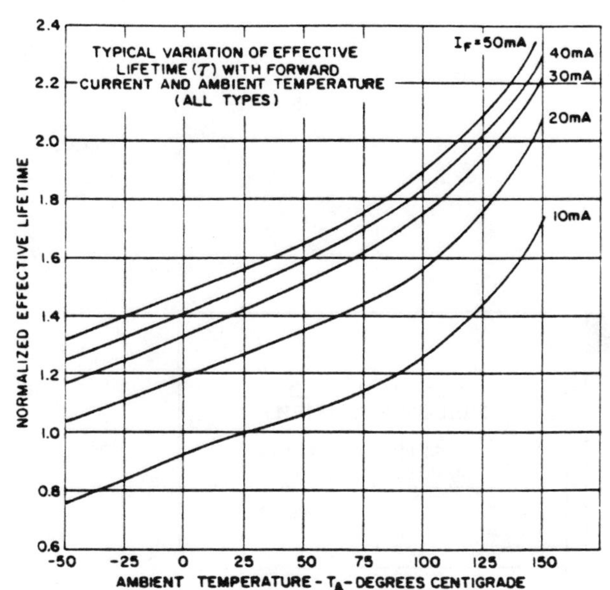

Figure 9

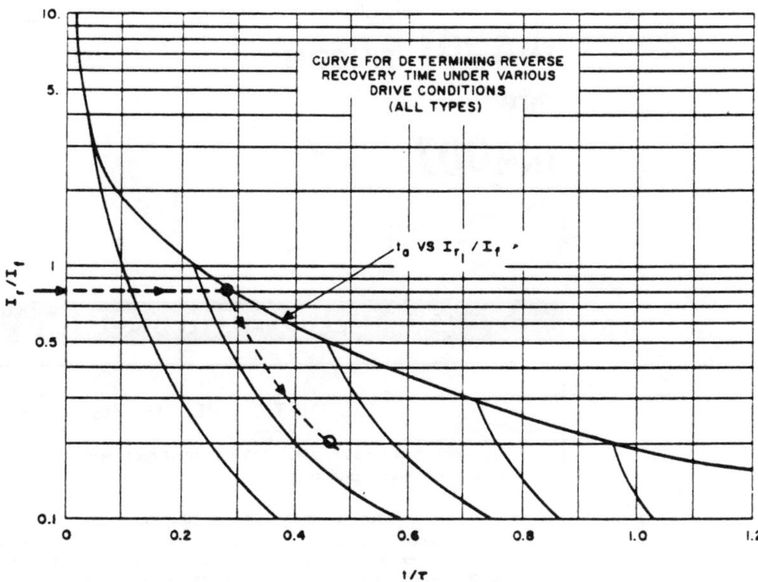

Figure 10

ESTIMATION OF REVERSE RECOVERY TIME UNDER VARIOUS DRIVE CONDITIONS

The reverse recovery time of a silicon signal diode has been shown* to be determined by a quantity called the effective lifetime, τ, and the ratio of forward and reverse current. The exact equations expressing times t_a and t_b (as defined in the sketch at right) are somewhat inconvenient for numerical evaluation, but in many cases an estimation of response time is sufficient. Figure 10 is a graphical solution to the response time equations and its use can best be illustrated by the following example:

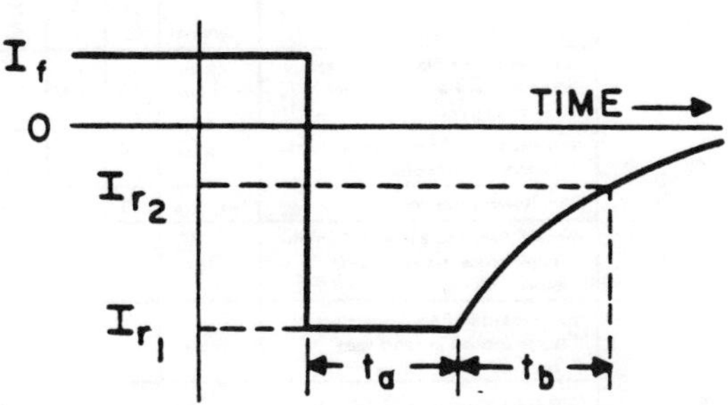

FIND: Recovery time to 5 mA reverse current when the forward current is 25 mA and the maximum reverse current is 20 mA.

SOLUTION: Enter the left side of Figure 10 at $I_{r1}/I_f = 20/25 = 0.8$ and follow horizontally until the t_a vs. I_{r1}/I_f line is reached (see dotted line). From the t/τ scale of the horizontal axis, it is seen that t_a is 0.28τ. The t_b portion of the recovery curve is estimated by moving downward parallel to the general contour lines until the $I_{r2}/I_f = 5/25 = 0.2$ line is reached. The total switching time is thus 0.46τ. The delay time, t_b, is $0.46\tau - 0.28\tau$ or 0.18τ.

The value of τ on the spec sheet should be corrected for current level. Figure 9 shows the typical variation of effective lifetime with forward current. Since the current level of the example is 25 mA, the maximum effective lifetime is approximately (6.8) (1.35) or 9.3 nsec., therefore:

$$t_a \approx (9.3)\ (.28) \approx 2.6 \text{ nsec. maximum}$$
$$t_b \approx (9.3)\ (.18) \approx 1.7 \text{ nsec. maximum}$$

Total reverse recovery time $\approx$ 4.3 nsec. maximum

Additional information on this method of diode recovery time calculation is contained in a paper entitled "Predicting Reverse Recovery Time of High Speed Semiconductor Junction Diodes" by C. H. Chen, (Publication #90.36) available on request.

---

*Ko, W. H., "The Reverse Transient Behavior of Semiconductor Junction Diodes," IRE Trans. ED-8, March 1961, pp. 123-131.

Fig. 16. Continued.

1N4001 (SILICON)
thru
1N4007

"SURMETIC" RECTIFIERS

. . . subminiature size, axial lead mounted rectifiers for general-purpose low-power applications.

Designers Data for "Worst Case" Conditions

The Designers Data Sheets permit the design of most circuits entirely from the information presented. Limit curves — representing boundaries on device characteristics — are given to facilitate "worst case" design.

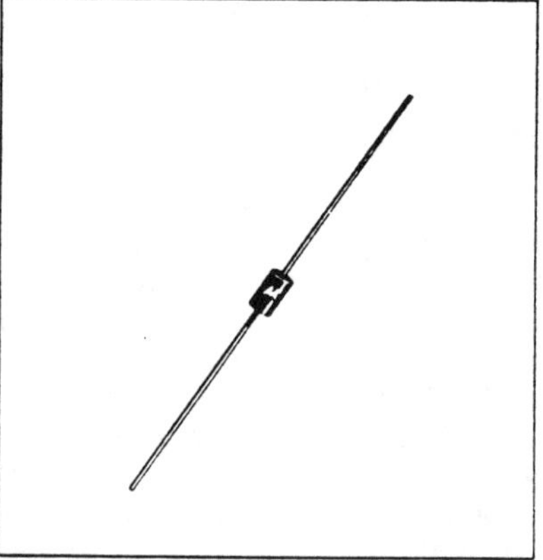

*MAXIMUM RATINGS

Rating	Symbol	1N4001	1N4002	1N4003	1N4004	1N4005	1N4006	1N4007	Unit
Peak Repetitive Reverse Voltage Working Peak Reverse Voltage DC Blocking Voltage	V_{RRM} V_{RWM} V_R	50	100	200	400	600	800	1000	Volts
Non-Repetitive Peak Reverse Voltage (halfwave, single phase, 60 Hz)	V_{RSM}	60	120	240	480	720	1000	1200	Volts
RMS Reverse Voltage	$V_{R(RMS)}$	35	70	140	280	420	560	700	Volts
Average Rectified Forward Current (single phase, resistive load, 60 Hz, see Figure 8, $T_A = 75°C$)	I_O	1.0							Amp
Non-Repetitive Peak Surge Current (surge applied at rated load conditions, see Figure 2)	I_{FSM}	30 (for 1 cycle)							Amp
Operating and Storage Junction Temperature Range	T_J, T_{stg}	-65 to +175							°C

*ELECTRICAL CHARACTERISTICS

Characteristic and Conditions	Symbol	Typ	Max	Unit
Maximum Instantaneous Forward Voltage Drop ($i_F = 1.0$ Amp, $T_J = 25°C$) Figure 1	v_F	0.93	1 1	Volts
Maximum Full-Cycle Average Forward Voltage Drop ($I_O = 1 0$ Amp, $T_L = 75°C$, 1 inch leads)	$V_{F(AV)}$	–	0.8	Volts
Maximum Reverse Current (rated dc voltage) $T_J = 25°C$ $T_J = 100°C$	I_R	 0.05 1.0	 10 50	µA
Maximum Full-Cycle Average Reverse Current ($I_O = 1.0$ Amp, $T_L = 75°C$, 1 inch leads	$I_{R(AV)}$	–	30	µA

*Indicates JEDEC Registered Data.

MECHANICAL CHARACTERISTICS

CASE: Void free, Transfer Molded
MAXIMUM LEAD TEMPERATURE FOR SOLDERING PURPOSES: 350°C, 3/8'' from case for 10 seconds at 5 lbs. tension
FINISH: All external surfaces are corrosion-resistant, leads are readily solderable
POLARITY: Cathode indicated by color band
WEIGHT: 0.40 Grams (approximately)

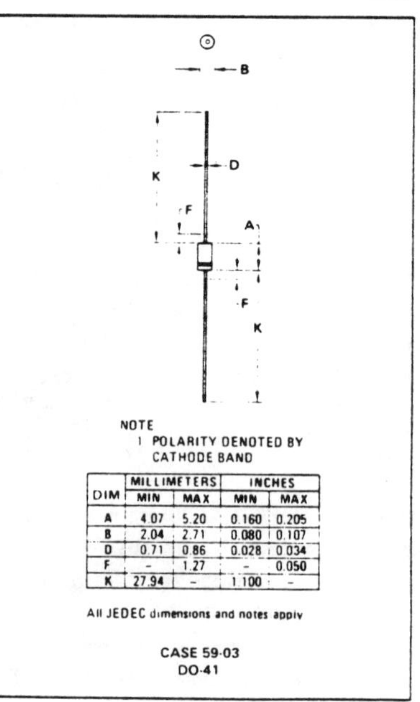

DIM	MILLIMETERS		INCHES	
	MIN	MAX	MIN	MAX
A	4.07	5.20	0.160	0.205
B	2.04	2.71	0.080	0.107
D	0.71	0.86	0.028	0 034
F	–	1.27	–	0.050
K	27.94	–	1 100	–

Fig. 17. Silicon Diodes (Courtesy of Motorola [7]).

1N4001 thru 1N4007 (continued)

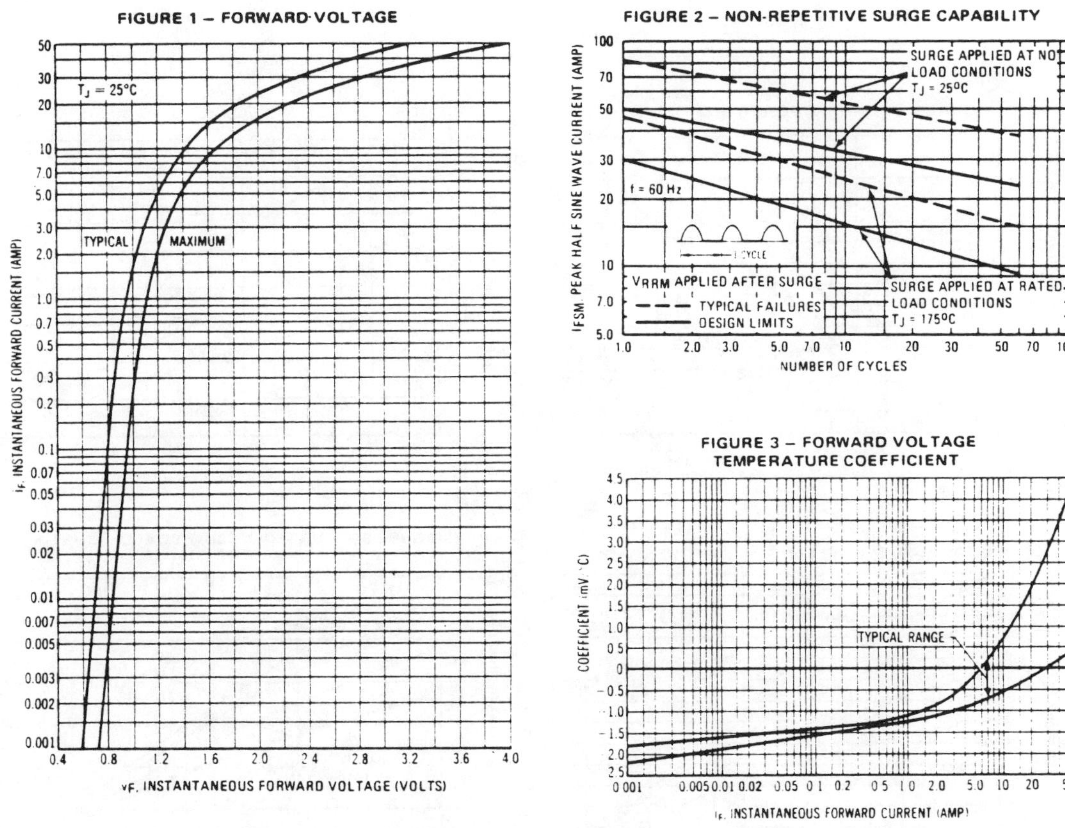

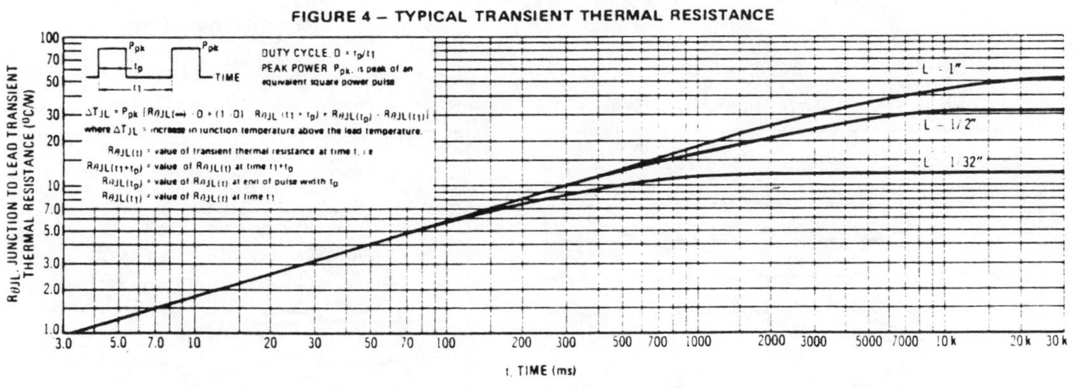

The temperature of the lead should be measured using a thermocouple placed on the lead as close as possible to the tie point. The thermal mass connected to the tie point is normally large enough so that it will not significantly respond to heat surges generated in the diode as a result of pulsed operation once steady-state conditions are achieved. Using the measured value of T_L, the junction temperature may be determined by:

$$T_J = T_L + \Delta T_{JL}$$

Fig. 17. Continued.

1N4001 thru 1N4007 (continued)

CURRENT DERATING DATA

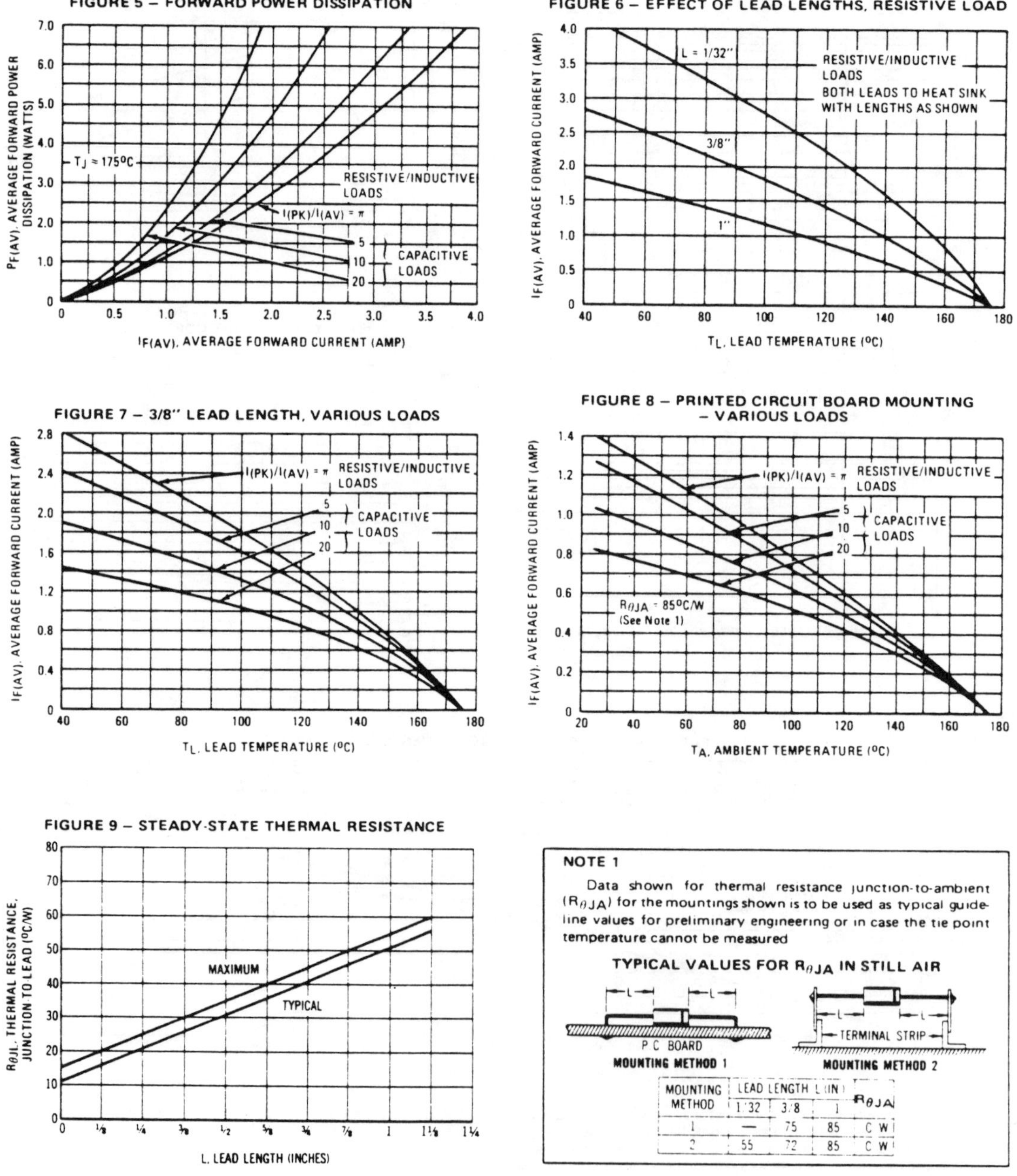

Fig. 17. Continued.

1N4001 thru 1N4007 (continued)

TYPICAL DYNAMIC CHARACTERISTICS

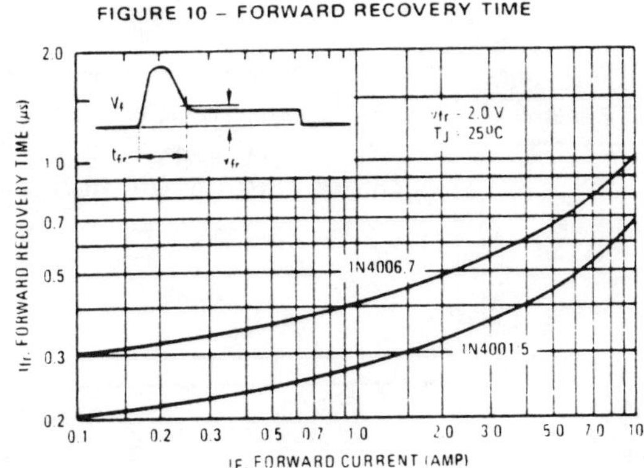

FIGURE 10 — FORWARD RECOVERY TIME

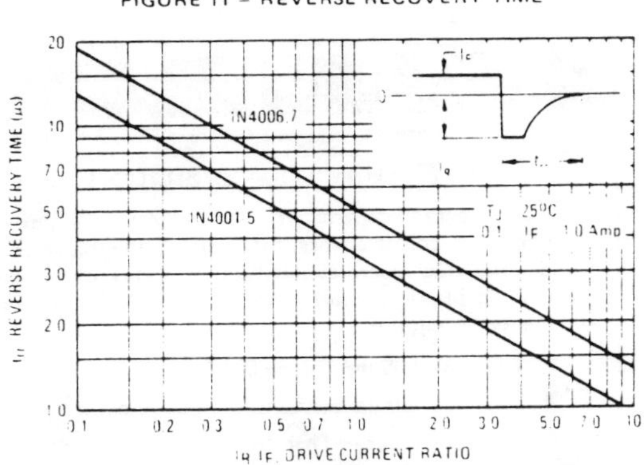

FIGURE 11 — REVERSE RECOVERY TIME

FIGURE 12 — JUNCTION CAPACITANCE

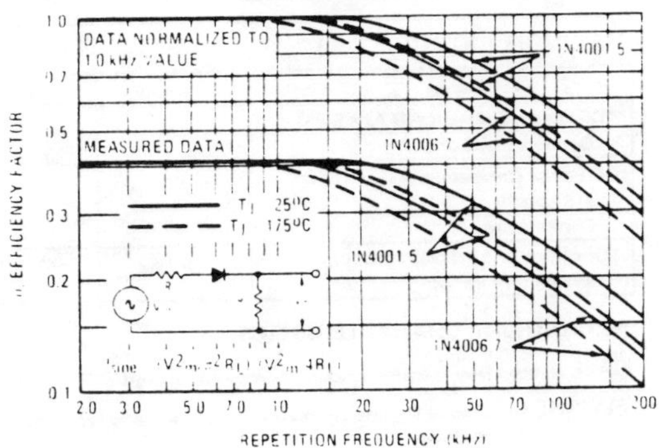

FIGURE 13 — RECTIFICATION WAVEFORM EFFICIENCY FOR SINE WAVE

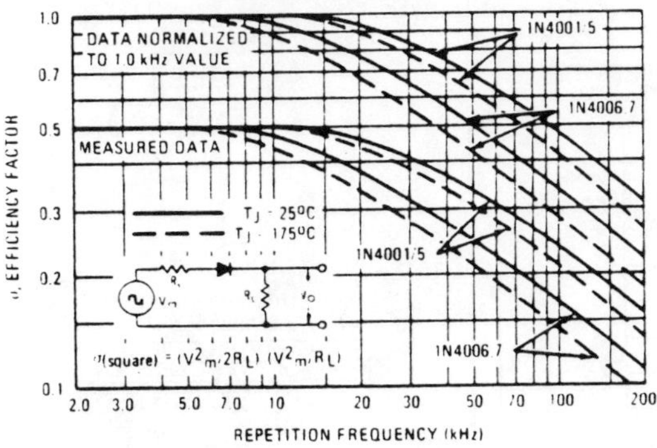

FIGURE 14 — RECTIFICATION WAVEFORM EFFICIENCY FOR SQUARE WAVE

RECTIFIER EFFICIENCY NOTE

The rectification efficiency factor η shown in Figures 13 and 14 was calculated using the formula.

$$\eta = \frac{P_{dc}}{P_{rms}} = \frac{\dfrac{V^2_{O(dc)}}{R_L}}{\dfrac{V^2_{O(rms)}}{R_L}} \cdot 100\% = \frac{V^2_{O(ac)}}{V^2_{O(ac)} + V^2_{O(dc)}} \cdot 100\% \quad (1)$$

For a sine wave input $V_m \sin(\omega t)$ to the diode, assumed lossless, the maximum theoretical efficiency factor becomes 40%, for a square wave input of amplitude V_m, the efficiency factor becomes 50%. (A full wave circuit has twice these efficiencies).

As the frequency of the input signal is increased, the reverse recovery time of the diode (Figure 11) becomes significant, resulting in an increasing ac voltage component across R_L which is opposite in polarity to the forward current thereby reducing the value of the efficiency factor η, as shown in Figures 13 and 14.

It should be emphasized that Figures 13 and 14 show waveform efficiency only, they do not account for diode losses. Data was obtained by measuring the ac component of V_O with a true rms voltmeter and the dc component with a dc voltmeter. The data was used in Equation 1 to obtain points for the Figures.

Fig. 17. Continued.

1N4728 thru 1N4764 (SILICON)
1M3.3ZS10 thru 1M200ZS10

Designers Data Sheet

1.0 WATT SURMETIC 30 SILICON ZENER DIODES

. . . a complete series of 1.0 Watt Zener Diodes with limits and operating characteristics that reflect the superior capabilities of silicon-oxide-passivated junctions. All this in an axial-lead, transfer-molded plastic package offering protection in all common environmental conditions.

- To 80 Watts Surge Rating @ 1.0 ms
- Maximum Limits Guaranteed on Six Electrical Parameters
- Package No Larger Than the Conventional 400 mW Package

Designer's Data for "Worst Case" Conditions

The Designers Data sheets permit the design of most circuits entirely from the information presented. Limit curves — representing boundaries on device characteristics — are given to facilitate "worst case" design.

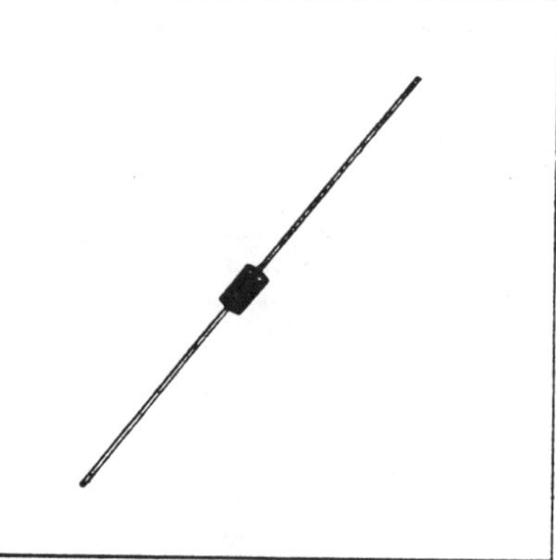

MAXIMUM RATINGS

Rating	Symbol	Value	Unit
*DC Power Dissipation @ T_A = 50°C	P_D	1.0	Watt
Derate above 50°C		6.67	mW/°C
DC Power Dissipation @ T_L = 75°C	P_D	3.0	Watts
Lead Length = 3/8"			
Derate above 75°C		24	mW/°C
*Operating and Storage Junction Temperature Range	T_J, T_{stg}	–65 to +200	°C

MECHANICAL CHARACTERISTICS

CASE: Void-free, transfer-molded, thermosetting plastic

FINISH: All external surfaces are corrosion resistant and leads are readily solderable and weldable

POLARITY: Cathode indicated by polarity band. When operated in zener mode, cathode will be positive with respect to anode

MOUNTING POSITION: Any

WEIGHT: 0.4 gram (approx).

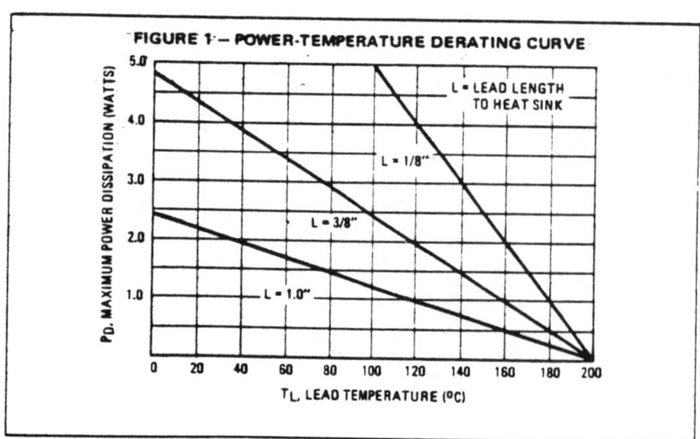

*Indicates JEDEC Registered Data

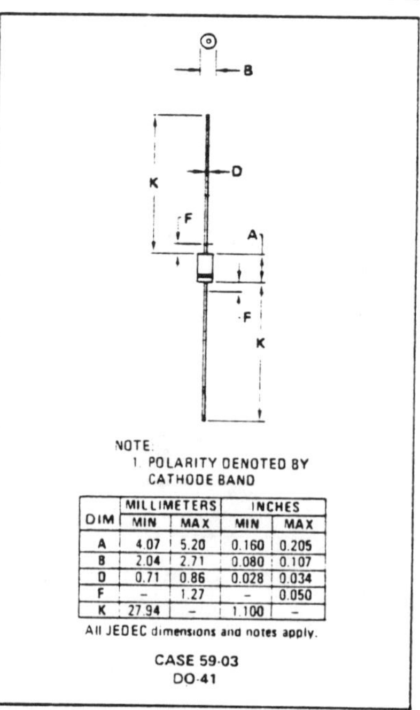

DIM	MILLIMETERS		INCHES	
	MIN	MAX	MIN	MAX
A	4.07	5.20	0.160	0.205
B	2.04	2.71	0.080	0.107
D	0.71	0.86	0.028	0.034
F	—	1.27	—	0.050
K	27.94	—	1.100	—

All JEDEC dimensions and notes apply.

CASE 59-03
DO-41

Fig. 18. Zener Regulator Diodes (Courtesy of Motorola [7]).

1N4728 thru 1N4764 (continued)
1M3.3ZS10 thru 1M200ZS10

ELECTRICAL CHARACTERISTICS (T_A = 25°C unless otherwise noted) *V_F = 1.5 V max, I_F = 200 mA for all types

JEDEC Type No. (Note 1)	Motorola Type No. (Note 2)	*Nominal Zener Voltage V_Z @ I_{ZT} Volts (Note 2 & 3)	*Test Current I_{ZT} mA	*Max Zener Impedance (Note 4)			*Leakage Current		*Surge Current @ T_A = 25°C i_r — mA (Note 5)
				Z_{ZT} @ I_{ZT} Ohms	Z_{ZK} @ I_{ZK} Ohms	I_{ZK} mA	I_R µA Max	V_R @ Volts	
1N4728	1M3.3ZS10	3.3	76	10	400	1.0	100	1.0	1380
1N4729	1M3.6ZS10	3.6	69	10	400	1.0	100	1.0	1260
1N4730	1M3.9ZS10	3.9	64	9.0	400	1.0	50	1.0	1190
1N4731	1M4.3ZS10	4.3	58·	9.0	400	1.0	10	1.0	1070
1N4732	1M4.7ZS10	4.7	53	8.0	500	1.0	10	1.0	970
1N4733	1M5.1ZS10	5.1	49	7.0	550	1.0	10	1.0	890
1N4734	1M5.6ZS10	5.6	45	5.0	600	1.0	10	2.0	810
1N4735	1M6.2ZS10	6.2	41	2.0	700	1.0	10	3.0	730
1N4736	1M6.8ZS10	6.8	37	3.5	700	1.0	10	4.0	660
1N4737	1M7.5ZS10	7.5	34	4.0	700	0.5	10	5.0	605
1N4738	1M8.2ZS10	8.2	31	4.5	700	0.5	10	6.0	550
1N4739	1M9.1ZS10	9.1	28	5.0	700	0.5	10	7.0	500
1N4740	1M10ZS10	10	25	7.0	700	0.25	10	7.6	454
1N4741	1M11ZS10	11	23	8.0	700	0.25	5.0	8.4	414
1N4742	1M12ZS10	12	21	9.0	700	0.25	5.0	9.1	380
1N4743	1M13ZS10	13	19	10	700	0.25	5.0	9.9	344
1N4744	1M15ZS10	15	17	14	700	0.25	5.0	11.4	304
1N4745	1M16ZS10	16	15.5	16	700	0.25	5.0	12.2	285
1N4746	1M18ZS10	18	14	20	750	0.25	5.0	13.7	250
1N4747	1M20ZS10	20	12.5	22	750	0.25	5.0	15.2	225
1N4748	1M22ZS10	22	11.5	23	750	0.25	5.0	16.7	205
1N4749	1M24ZS10	24	10.5	25	750	0.25	5.0	18.2	190
1N4750	1M27ZS10	27	9.5	35	750	0.25	5.0	20.6	170
1N4751	1M30ZS10	30	8.5	40	1000	0.25	5.0	22.8	150
1N4752	1M33ZS10	33	7.5	45	1000	0.25	5.0	25.1	135
1N4753	1M36ZS10	36	7.0	50	1000	0.25	5.0	27.4	125
1N4754	1M39ZS10	39	6.5	60	1000	0.25	5.0	29.7	115
1N4755	1M43ZS10	43	6.0	70	1500	0.25	5.0	32.7	110
1N4756	1M47ZS10	47	5.5	80	1500	0.25	5.0	35.8	95
1N4757	1M51ZS10	51	5.0	95	1500	0.25	5.0	38.8	90
1N4758	1M56ZS10	56	4.5	110	2000	0.25	5.0	42.6	80
1N4759	1M62ZS10	62	4.0	125	2000	0.25	5.0	47.1	70
1N4760	1M68ZS10	68	3.7	150	2000	0.25	5.0	51.7	65
1N4761	1M75ZS10	75	3.3	175	2000	0.25	5.0	56.0	60
1N4762	1M82ZS10	82	3.0	200	3000	0.25	5.0	62.2	55
1N4763	1M91ZS10	91	2.8	250	3000	0.25	5.0	69.2	50
1N4764	1M100ZS10	100	2.5	350	3000	0.25	5.0	76.0	45
—	1M110ZS10	110	2.3	450	4000	0.25	5.0	83.6	—
—	1M120ZS10	120	2.0	550	4500	0.25	5.0	91.2	—
—	1M130ZS10	130	1.9	700	5000	0.25	5.0	98.8	—
—	1M150ZS10	150	1.7	1000	6000	0.25	5.0	114.0	—
—	1M160ZS10	160	1.6	1100	6500	0.25	5.0	121.6	—
—	1M180ZS10	180	1.4	1200	7000	0.25	5.0	136.8	—
—	1M200ZS10	200	1.2	1500	8000	0.25	5.0	152.0	—

*Indicates JEDEC Registered Data

NOTE 1 – TOLERANCE AND TYPE NUMBER DESIGNATION

The JEDEC type numbers listed have a standard tolerance on the nominal zener voltage of ±10%. A standard tolerance of ±5% on individual units is also available and is indicated by suffixing "A" to the standard type number.

NOTE 2 – SPECIALS AVAILABLE INCLUDE:

(A) NOMINAL ZENER VOLTAGES BETWEEN THE VOLTAGES SHOWN AND TIGHTER VOLTAGE TOLERANCES: To designate units with zener voltages other than those assigned JEDEC numbers and/or tight voltage tolerances (±5%, ±3%, ±2%, ±1%), the Motorola type number should be used.

(B) MATCHED SETS: (Standard Tolerances are ±5.0%, ±3.0%, ±2.0%, ±1.0%).

Zener diodes can be obtained in sets consisting of two or more matched devices. The method for specifying such matched sets is similar to the one described in (A), except that two extra suffixes are added to the code number described.

These units are marked with code letters to identify the matched sets and, in addition, each unit in a set is marked with the same serial number, which is different for each set being ordered.

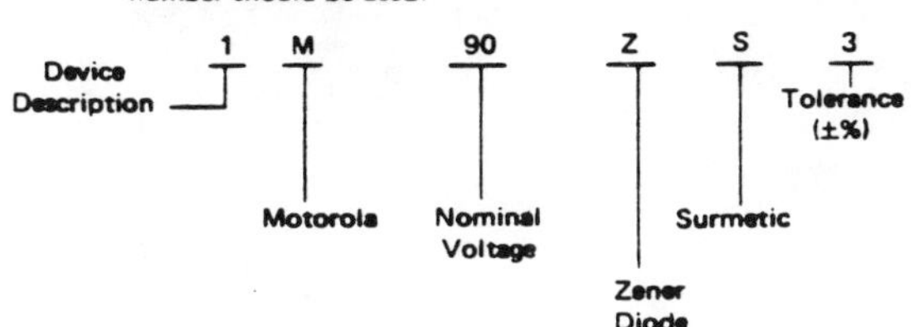

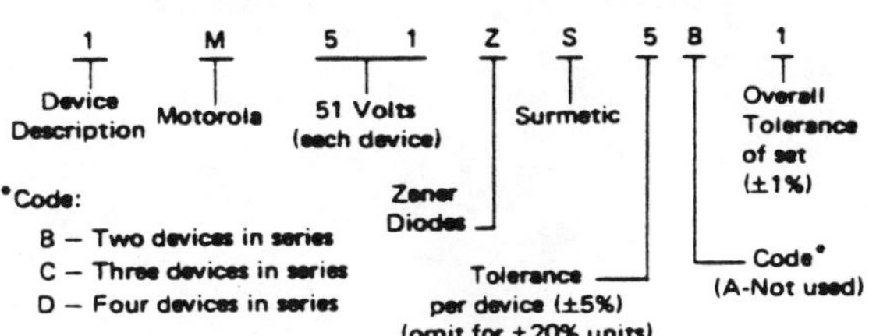

Fig. 18. Continued.

1N4728 thru 1N4764 (continued)
1M3.3ZS10 thru 1M200ZS10

(C) ZENER CLIPPERS: (Standard Tolerance ±10% and ±5%).

Special clipper diodes with opposing Zener junctions built into the device are available by using the following nomenclature:

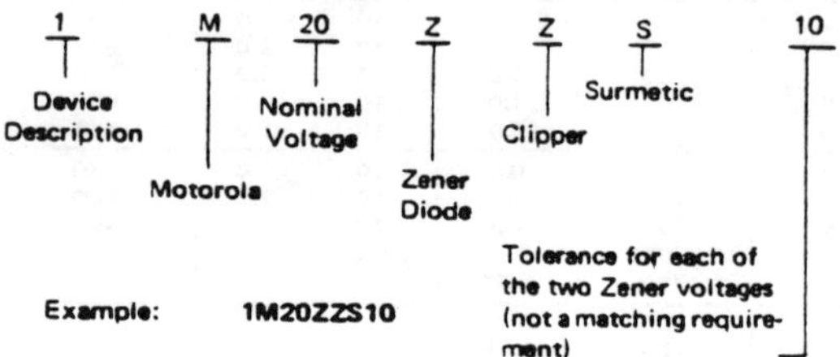

NOTE 3 — ZENER VOLTAGE (V_Z) MEASUREMENT

Motorola guarantees the zener voltage when measured at 90 seconds while maintaining the lead temperature (T_L) at 30°C ± 1°C, 3/8″ from the diode body.

NOTE 4 — ZENER IMPEDANCE (Z_Z) DERIVATION

The zener impedance is derived from the 60 cycle ac voltage, which results when an ac current having an rms value equal to 10% of the dc zener current (I_{ZT} or I_{ZK}) is superimposed on I_{ZT} or I_{ZK}.

NOTE 5 — SURGE CURRENT (i_r) NON-REPETITIVE

The rating listed in the electrical characteristics table is maximum peak, non-repetitive, reverse surge current of 1/2 square wave or equivalent sine wave pulse of 1/120 second duration superimposed on the test current, I_{ZT}, per JEDEC registration, however, actual device capability is as described in Figures 4 and 5.

APPLICATION NOTE

Since the actual voltage available from a given zener diode is temperature dependent, it is necessary to determine junction temperature under any set of operating conditions in order to calculate its value. The following procedure is recommended:

Lead Temperature, T_L, should be determined from:

$$T_L = \theta_{LA} P_D + T_A$$

θ_{LA} is the lead-to-ambient thermal resistance (°C/W) and P_D is the power dissipation. The value for θ_{LA} will vary and depends on the device mounting method. θ_{LA} is generally 30-40°C/W for the various clips and tie points in common use and for printed circuit board wiring.

The temperature of the lead can also be measured using a thermocouple placed on the lead as close as possible to the tie point. The thermal mass connected to the tie point is normally large enough so that it will not significantly respond to heat surges generated in the diode as a result of pulsed operation once steady-state conditions are achieved. Using the measured value of T_L, the junction temperature may be determined by:

$$T_J = T_L + \Delta T_{JL}$$

ΔT_{JL} is the increase in junction temperature above the lead temperature and may be found from Figure 2 for a train of power pulses (L = 3/8 inch) or from Figure 3 for dc power.

$$\Delta T_{JL} = \theta_{JL} P_D$$

For worst-case design, using expected limits of I_Z, limits of P_D and the extremes of $T_J(\Delta T_J)$ may be estimated. Changes in voltage, V_Z, can then be found from:

$$\Delta V = \theta_{VZ} \Delta T_J$$

θ_{VZ}, the zener voltage temperature coefficient, is found from Figures 6 and 7.

Under high power-pulse operation, the zener voltage will vary with time and may also be affected significantly by the zener resistance. For best regulation, keep current excursions as low as possible.

Data of Figure 2 should not be used to compute surge capability. Surge limitations are given in Figure 4. They are lower than would be expected by considering only junction temperature, as current crowding effects cause temperatures to be extremely high in small spots resulting in device degradation should the limits of Figure 4 be exceeded.

Fig. 18. Continued.

1N4728 thru 1N4764 (continued)
1M3.3ZS10 thru 1M200ZS10

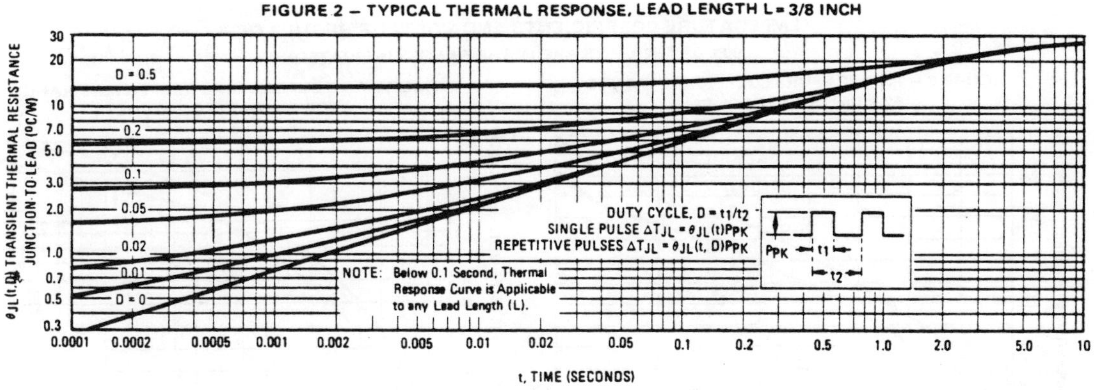

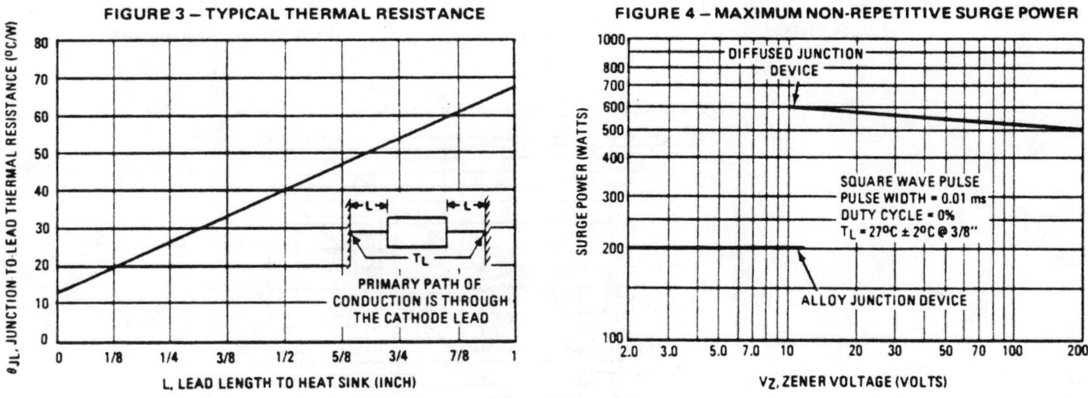

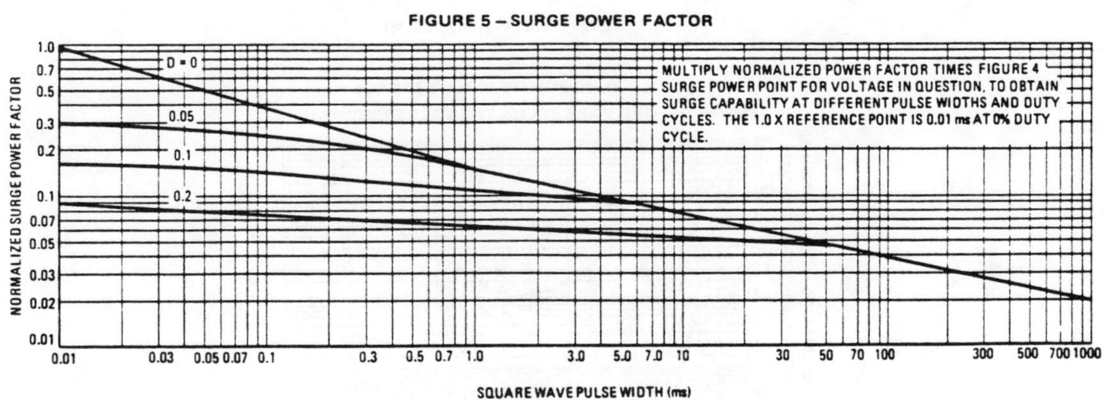

Fig. 18. Continued.

1N4728 thru 1N4764 (continued)
1M3.3ZS10 thru 1M200ZS10

TEMPERATURE COEFFICIENTS AND VOLTAGE REGULATION
(90% OF THE UNITS ARE IN THE RANGES INDICATED)

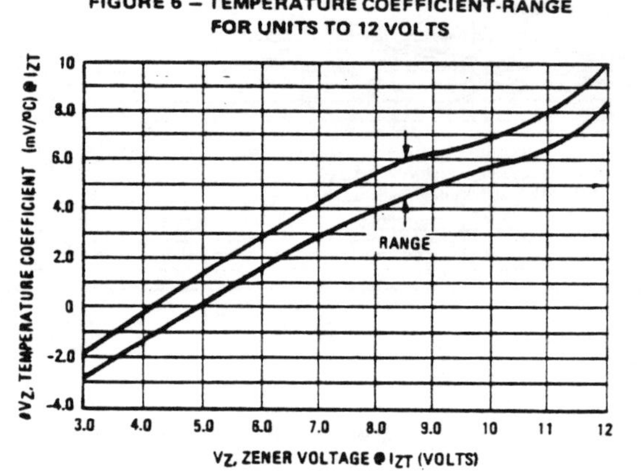

FIGURE 6 — TEMPERATURE COEFFICIENT-RANGE FOR UNITS TO 12 VOLTS

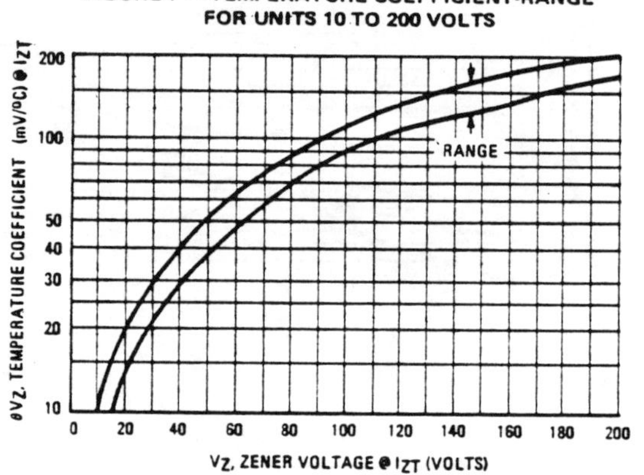

FIGURE 7 — TEMPERATURE COEFFICIENT-RANGE FOR UNITS 10 TO 200 VOLTS

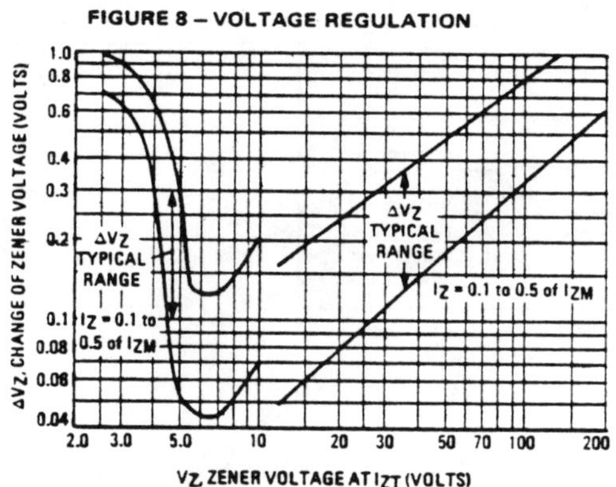

FIGURE 8 — VOLTAGE REGULATION

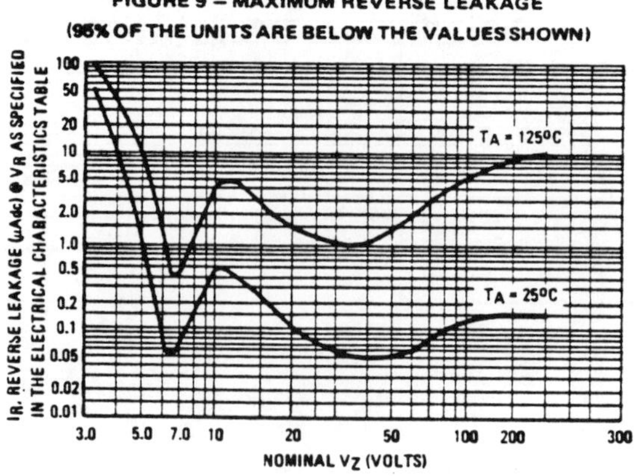

FIGURE 9 — MAXIMUM REVERSE LEAKAGE
(95% OF THE UNITS ARE BELOW THE VALUES SHOWN)

Fig. 18. Continued.

EXPERIMENT II-3

THEVENIN'S THEOREM AND IMPEDANCE MATCHING

EQUIPMENT:

> EEP
> BLACK BOX (UNKNOWN CIRCUIT!)
> GR DECADE RESISTANCE BOX
> DECADE RESISTANCE BOX

Table of Contents   Page No.

1. PURPOSE

The purpose of this experiment is to demonstrate Thevenin's (or Norton's) theorem and its application in circuit theory and to give the conditions for the maximum power transfer in resistive circuits. The maximum power transfer as applied to circuits with inductance and capacitance is also mentioned. Finally, Tellegen's theorem is included in this experiment.

2. INTRODUCTION

Thevenin's or Norton's theorem is one of the most important and powerful theorems in circuit theory. These two seemingly different theorems are indeed complementary to each other. In the following we state these two theorems.

2A. Thevenin's Theorem

Given any two terminals of a network composed of linear passive and active circuit elements and independent voltage and current sources, we can find an equivalent voltage source and an equivalent series impedance which behave exactly the same as the original network.

We emphasize that depending on where we pull these two terminals of a given network we will have different equivalent circuits. Thus Thevenin's equivalent network is a function of the selected port of the circuit and it only represents the port property.

How to measure this voltage source and series impedance?

The voltage source is equal to the potential difference between the two terminals caused by the network with no external elements connected to these terminals. The series impedance is equal to the impedance looking into the two terminal points with all independent power sources within the terminal pair set equal to zero (i.e., the voltage sources are short circuited and the current sources are open circuited).

2B. Norton's Theorem

Given any two terminals of a network composed of linear passive and active circuit elements and independent voltage and current sources, we can find an equivalent current source with an equivalent parallel impedance which behave exactly the same as the original network.

To emphasize the port dependency of these theorems consider Fig. 1. In this figure, for some-one who is looking into networks (a), (b) and (c) at port aa′, she/he sees exactly the same input voltage and input current for all the three networks. In other words, the equivalence between these networks holds only at one specific port.

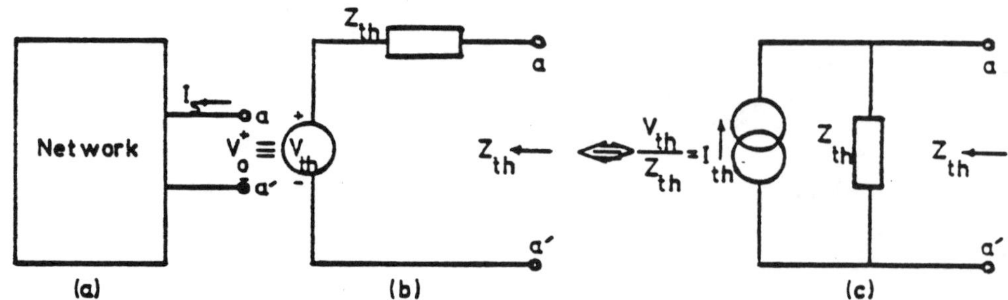

Fig. 1. Thevenin's and/or Norton's Theorem.

Another important network theorem is the maximum Power Transfer Theorem which is discussed next.

2C. The Maximum Power Transfer Theorem

This theorem states that given a voltage source V_{th} and a series resistance R_{th} , the power dissipated in R_l , which is in series with the voltage source (V_{th}, R_{th}) will be maximum when $R_l = R_{th}$.

Referring to Fig. 2, it is easy to show that if R_l is set equal to R_{th}, then the power dissipated in R_l is maximum. To prove this claim, simply start with $P_l = R_l I^2 = V_{th}^2 R_l / (R_{th} + R_l)^2$. Then find dP_l / dR_l and set that quantity equal to zero and solve for R_l. If instead of R_{th} we had $Z_{th} = X + jY$ then we can show that the optimal load impedance for the maximum power dissipated is $Z_l = X - jY$. Prove this result.

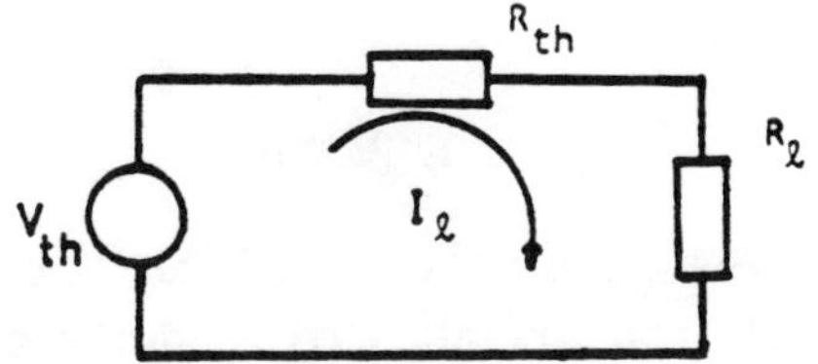

Fig. 2. Maximum Power Transfer.

Another important theorem with many applications in circuit theory is Tellegen's theorem. This theorem is as follows [5].

2D. Tellegen's Theorem

Consider a network with k branches. Choose arbitrary directions and values for currents and voltages of those elements provided that both Kirchhoff's current law and Kirchhoff's voltage law satisfy these choices. Then

$$\sum_{j=1}^{k} v_j(t)\, i_j(t) = 0 .$$

This theorem has been generalized into several directions and has been the subject of many research projects. One interesting extension is as follows. The above result also holds for voltages measured at time t_1 and currents measured at t_2 such that the sum of their products also vanish. That is,

$$\sum_{j=1}^{k} v_j(t_1)\, i_j(t_2) = 0 .$$

Another extension of this results says that if we have two networks N_1 and N_2 with the same topology and we choose a set of currents and voltages for their branches with the main conditions described earlier in the theorem, then

$$\sum_{j=1}^{k} v_j^{N_1}(t_1)\, i_j^{N_2}(t_2) = \sum_{j=1}^{k} v_j^{N_2}(t_1)\, i_j^{N_1}(t_2) = 0 .$$

These are powerful theorems from circuit theory that we must know well.

3. LABORATORY PROCEDURE

3.1. EXPERIMENTAL DETERMINATION OF THE THEVENIN EQUIVALENT

A "black box" consisting of an unknown interconnection of voltage sources and resistances has been supplied to determine its Thevenin equivalent circuit. *Because of practical limitations this "black box" will actually consist of the voltage reference source (VRS) connected to a second "aluminum box" that contains an unknown resistive network.* The "black box" and its subsystems are illustrated in Fig. 3.

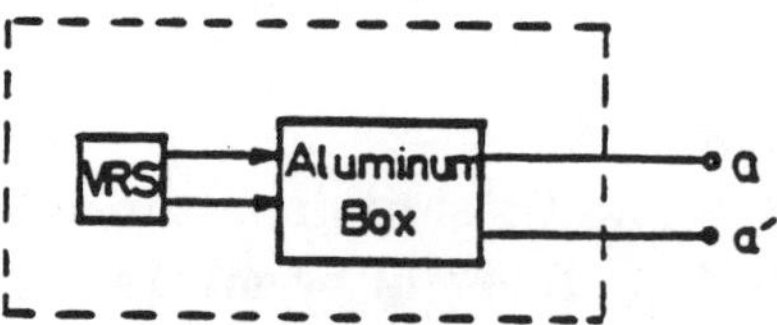

Fig. 3. "Black Box".

Considering the "black box" as an entity with only its two output terminals available for measurement, you are allowed to make any voltage and/or current measurements you wish at these terminals to determine the Thevenin equivalent V_{th} and R_{th}. You are asked *not* to open the aluminum box or measure the output of the VRS. *Use caution when applying any meters to make measurements at the output terminals. Always start with the meter on its highest voltage or current scale when making measurements.* After your measurements are completed, *determine* the Thevenin equivalent voltage, V_{th} and series resistance, R_{th}, and *record* these values. *Have your instructor initial your values of* V_{th} *and* R_{th} *and* draw *a square around them.* These values are not to be changed once initialed. It is up to you to determine what measurements and how many should be taken to determine V_{th} and R_{th}. If you desire to make measurements with a load applied at the output terminals you may use the GR Decade Resistance Box as a load. Usually three to four times is enough to give a reasonable value. By now you are aware that your first reading may not always be your most accurate one. In practice usually many readings for each experiment is collected, from which and by statistical analysis a value is chosen that best represents the *true* value of the desired quantity.

We now turn to another measurement.

3.2. THE THEVENIN EQUIVALENT VOLTAGE AND RESISTANCE OF VRS

Disconnect the VRS from the aluminum box. *Do not change* the voltage setting. *Measure* the output voltage of VRS with no load. Next *connect* a resistance of at least 1 kΩ to the output of the VRS, using the GR Decade Resistance Box. *Measure* the DC current supplied by the VRS to the external load. *Determine* the Thevenin equivalent voltage V_0 and resistance R_0 of the VRS.

3.3. THE THEVENIN EQUIVALENT OF A "BLACK BOX"

An experimental procedure analogous to the method used when calculating the Thevenin equivalent from a given circuit diagram will now be used. This procedure require that the voltage source in the "black box" be available for measurement. *Measure* the output voltage of the "black box" with no load connected using a DMM. Because of the high input impedance of the DMM this corresponds to the open circuit output voltage. Next *disconnect* the leads from the VRS to the aluminum box, at the VRS. The VRS now has no connections made at its output terminals. Now, *connect* a load of R_0 across the input terminals to the aluminum box. At the output terminals of the aluminum box measure the resistance between the output terminals. Now *remove* R_0 and use this information to determine V_{th} and R_{th}.

3.4. THE THEVENIN EQUIVALENT BY CALCULATION

If the circuit in the aluminum box is not given to you, then *disconnect* the VRS from the aluminum box. *Open* the aluminum box and examine its contents. *Draw* the circuit diagram of the "black box" including the VRS. From the circuit diagram calculate the Thevenin equivalent V_{th} and R_{th}.

3.5. COMPARISON OF THEVENIN EQUIVALENT CIRCUIT AND "BLACK BOX"

Reconstruct the "black box" as in procedure 3.1. Using the "black box" *connect* a Decade Resistance Box as the load, R_L, shown in Fig. 4. Using a DMM *measure* the voltage across R_L and current through R_L for values of R_L from $R_L = 0$ to $R_L = 10\,R_{th}$, including one measurement of $R_L = R_{th}$.

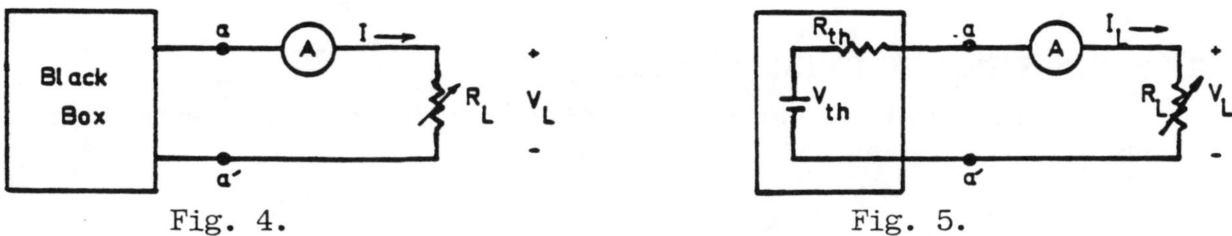

Fig. 4.   Fig. 5.

Construct the Thevenin equivalent circuit of the "black box" using the VRS in series with a GR Decade Resistance Box, Fig. 5. *Determine* the voltage setting of the VRS and the resistance of the GR Decade Resistance Box so that it simulates V_{th} and R_{th}. Repeat the above measurements by changing R_L from $R_L = 0$ to $R_L = 10\,R_{th}$, including one measurement of $R_L = R_{th}$ in the circuit of Fig. 5.

Compare the values of I_L and V_L for various R_L in the above two steps. Can the Thevenin equivalent and "black box" be distinguished by measurements made at their output terminals? If they can be distinguished, should this be the case, and what practical consideration might justify differences in output values under the same load for two circuits which are supposedly identical?

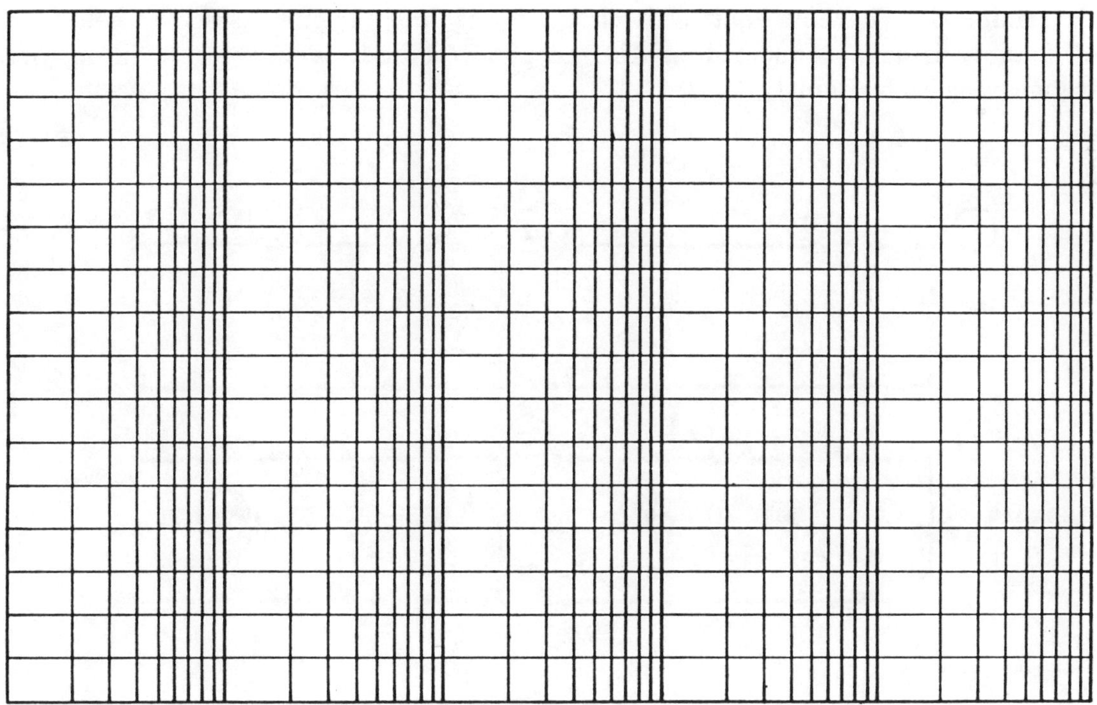

Fig. 6.

3.6. MAXIMUM POWER TRANSFER

Plot the power $P_L = V_L \times I_L$ as a function of R_L on the semilog (R_L on log scale) chart of Fig. 6. *Determine* the value of R_L for maximum power transfer. Using the Thevenin equivalent circuit calculate R_L for maximum power transfer. Is it possible to use information about optimal R_L to find V_{th} and R_{th}? If yes, how? If no, why?

3.7. TELLEGEN THEOREM VERIFICATION

Construct the following two circuits using the PS and the two resistance decade boxes. *Verify* the three forms of Tellegen's theorem.

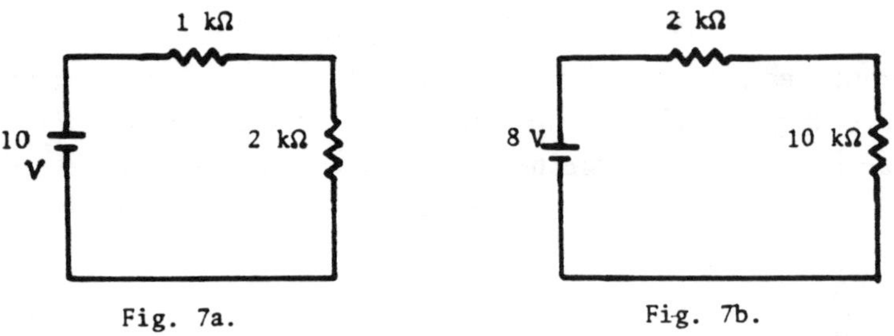

Fig. 7a.   Fig. 7b.

4. PROBLEM SET

1. Determine the Thevenin equivalent circuit for each of the circuits shown below. What value of load resistance connected at the output terminals would result in maximum power transfer for each circuit?

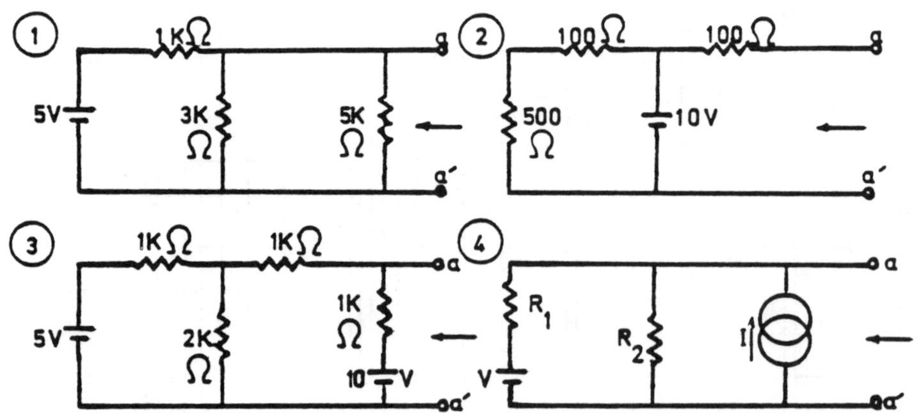

Fig. 8. Problems.

5. FOR YOUR REPORT

Submit your data with the solution to the above problems in our standard report form.

EXPERIMENT II-4

WAVE SHAPING CIRCUITS: SIMPLE RC FILTERS

EQUIPMENT:

EEP
GR DECADE RESISTOR (OR 100 kΩ POT)
CAPACITOR (0.1 μF) (OR VARIABLE CAPACITANCE BOX)
RESISTORS (6.8 kΩ, 10 kΩ/2)

Table of Contents   Page No.

1. PURPOSE

The main purpose of this experiment is to acquaint the student with the basic notion of wave shaping circuits and filter theory. As an example we choose a simple RC circuit and we will study the response of this circuit to some basic electrical waveforms.

2. INTRODUCTION

Consider a network which is represented by its Thevenin equivalent circuit. Suppose the output of this circuit is fed into another network called a filter. Depending on how this filter has been designed we may have many different kinds of output at port bb′.

Naturally, the question will always be: How do we choose our filter if we only want the output to be so and so? As a simple example, if the shape of our signal at port aa′ in Fig. 1, is a full-wave sinusoidal and if we only want to have a half-wave rectified sinusoidal, then our filter will be a simple diode. But filter designing is not always so simple as this example. As you may have already guessed, in the design of any

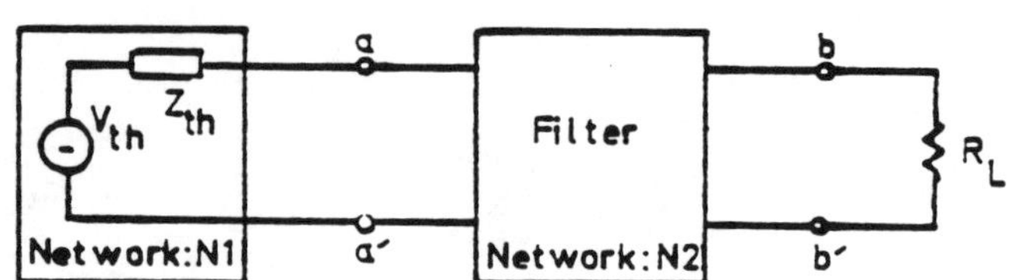

Fig. 1. Simple Representation of a Filter.

filter we have only information about the input and the output signals. From this input-output (I-O) relation we must build our filter. With the help of active network theory, general filter design has gained a tremendous lift and there are charts which can be used as design tools to build filters for many ordinary applications. More complicated and special purpose filters, of course, need additional efforts which will not concern us at this moment. An extensive review of filter design is given in EXP. II-6, and EXPs. VI-5 to 7 of this textbook.

From the above remarks we know that in order to design a filter we must have the corresponding I-O relation. In general, this I-O relation is expressed by the ratio of two rational polynomials in s, the Laplace transform variable. Depending on the numerator and denominator polynomials of this ratio, say, $T(s) = N(s)/D(s)$, we may have one of the following situations:

1. $T_l(s) = K/D(s)$, K is a constant: *LOW-PASS FILTER*

This is called a low-pass filter because if we draw $|T_l(j\omega)|$ versus ω, for frequencies below ω_l its frequency response or output is nonzero and for frequencies above ω_l the response becomes identically zero.

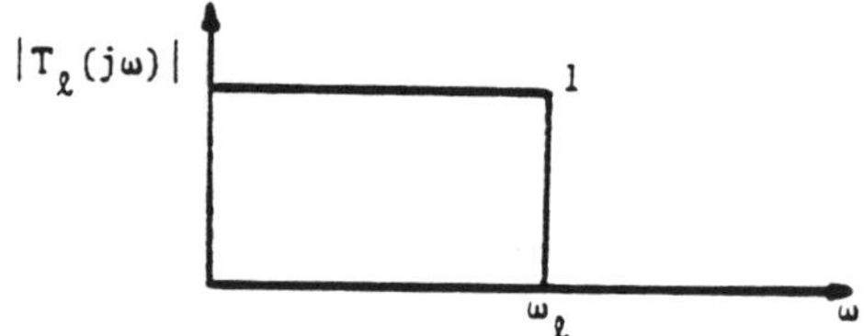

Fig. 2. An Ideal Low-Pass Filter.

2. $T_h(s) = K\, s^n / D(s)$, $n = \deg D(s)$:   *HIGH-PASS FILTER*

This is called a high-pass filter because for any frequency above ω_h its frequency response or output is nonzero and for frequencies below ω_h this response is identically zero.

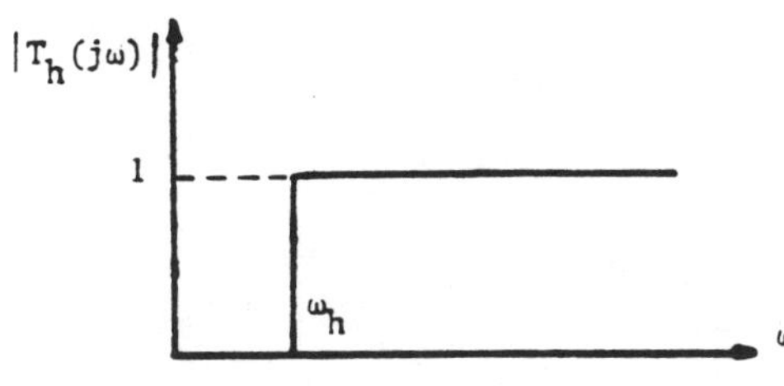

Fig. 3.   An Ideal High-Pass Filter.

3. $T_{bp}(s) = K\, s^{n/2} / D(s)$:   *BAND-PASS FILTER*

This filter is called a band-pass filter because of its frequency response which only permits signal with frequencies between $(\omega_l - \omega_h)$ go through without attenuation.

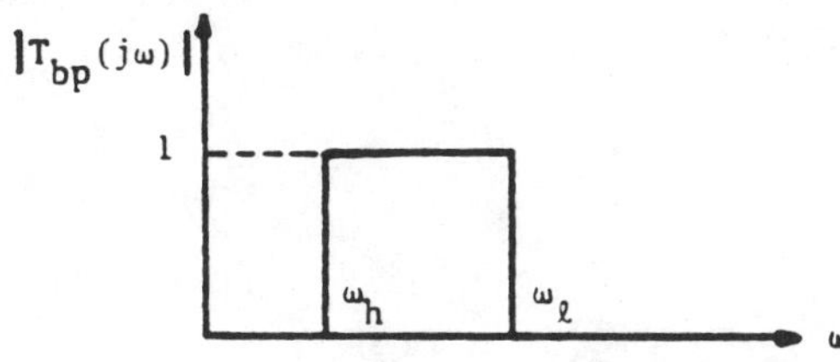

Fig. 4.   An Ideal Band-Pass Filter.

4.   *BAND-STOP FILTER*

It should now be evident that a band-pass filter can be built by cascading a low-pass filter and a high-pass filter such that $\omega_h > \omega_l$. This filter has, of course, application when we like the output to be zero for $\omega_l < \omega < \omega_h$.

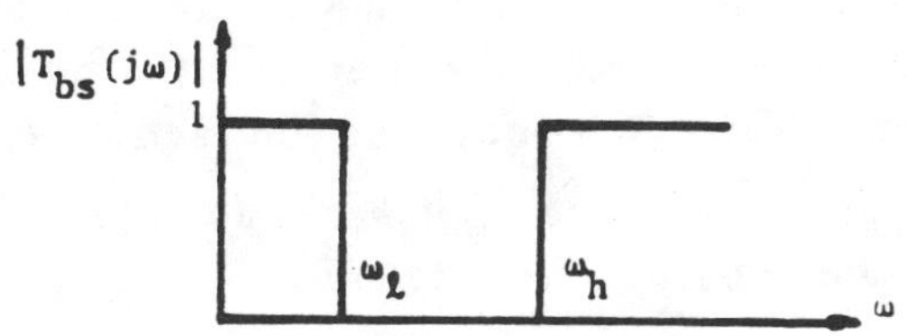

Fig. 5.   An Ideal-Band Stop Filter.

Notice that any T(s) on substituting $s = j\omega$ can be written as $T(j\omega) = |T(j\omega)| \angle T(j\omega)$. We have only drawn $|T(j\omega)|$ versus ω in the above. In many applications if $|T(j\omega)|$ becomes large we may have to use a Bode or logarithmic scale which is 20 log $|T(j\omega)|$ (dB) vs. log ω. Also we must realize that the response curves of practical filters that we design or buy if you please - are not so sharp as those shown above. In fact, we have the response of Fig. 6 instead of Fig. 2. For a practical low-pass filter, instead of maximum output 1.00 at ω_l, we have 0.707 which is 3 dB below the maximum output.

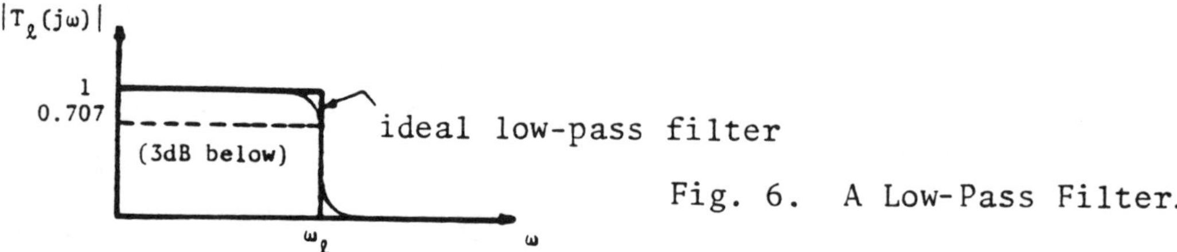

Fig. 6.   A Low-Pass Filter.

From the above remarks we see that by the appropriate choice or choices of T(s), we can modify or shape the frequency response of a network. This opportunity enables us to form any output that we like from any signal by passing that signal through an "optimal" filter. We do not want to go any further in filter theory and in the following we will only choose a few simple RC circuits and will study the wave shaping properties of these RC circuits. In particular we will see that a simple RC circuit can be adjusted such that it will be used as a signal integrator or a differentiator. For additional information please read EXP. II-6, and EXPs. VI-5 to 7.

3. LABORATORY PROCEDURE

3.1. RC CIRCUITS AS DIFFERENTIATORS AND INTEGRATORS

Build the simple RC circuit of Fig. 7. It is easy to show that $V_{out}(s) = V_{in}(s) [RCs / (1 + RCs)]$. With $T(s) = RCs / (1 + RCs)$ this circuit is a high-pass filter.

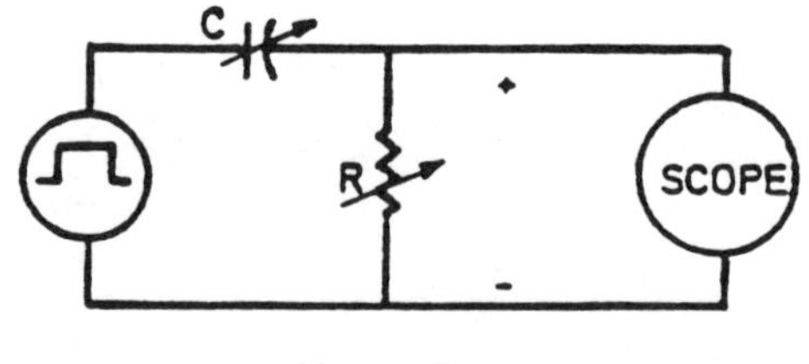

Fig. 7.

Set the square-wave generator for an output of about 1 kHz and 1 V amplitude. *Start* with a value of $C = 0.1 \ \mu F$ and vary R in steps from about 100 Ω to 100 kΩ. *Observe* the waveforms on the oscilloscope and *record* (sketch) the cases where the RC time constant relative to the period of the input signal is such that the output across the resistor is a series of (1) spikes; (2) undistorted square wave; (3) square waves with a slight tilt on the top. *Note* the conditions for each case.

For the case where the output is a series of spikes, measure the time it takes for the voltage to decay to 1/e times the initial value as compared to the time constant RC. (e=2.718)

You may have realized that for $|RCs| << 1$, this circuit acts as a differentiator. In practice we avoid differentiation due to the noise amplification. (Why?)

Draw $|T(j\omega)|$ vs. ω in your report for at least one input as ω changes.

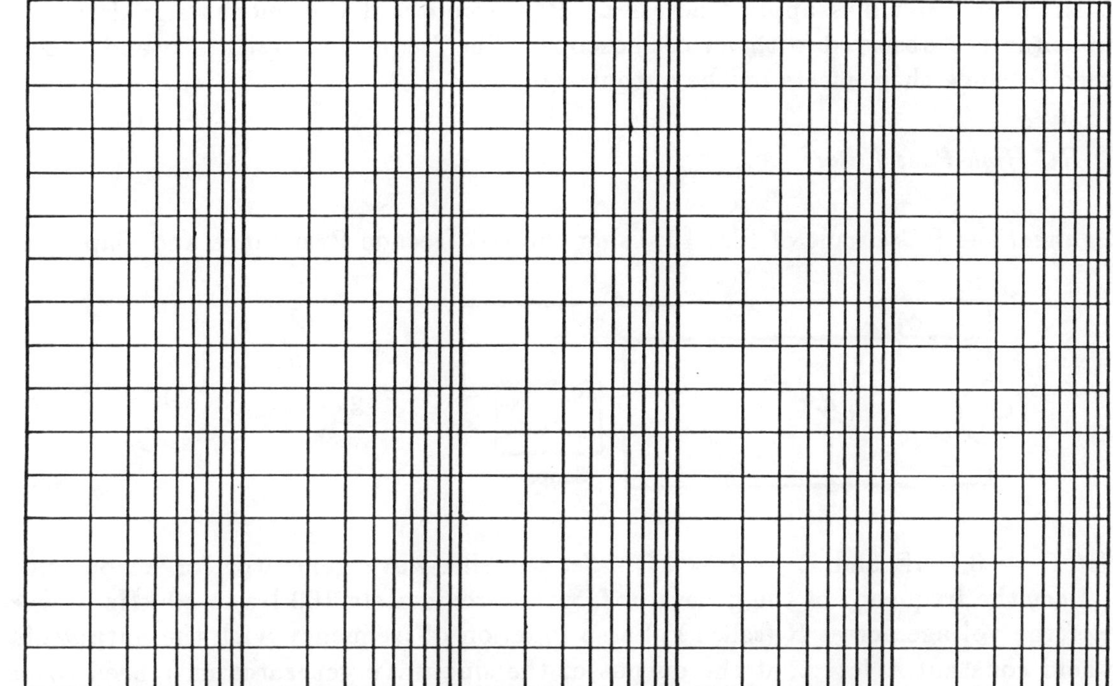

Fig. 8.

Interchange R and C and connect the oscilloscope leads across the capacitor. *Record* (sketch) the curves for the same values of R and C and signal period for which the curves were sketched previousl. *Compare* the curves.

It is easy to show that now $T(s) = 1/(1+RCs)$, which is a low-pass filter. For $|RCs| \gg 1$, this circuit acts as an integrator. Contrary to the differentiator, an integrator is a stable circuit and is used often in practice.

Draw the $|T(j\omega)|$ vs. ω in your report for at least one input as ω changes.

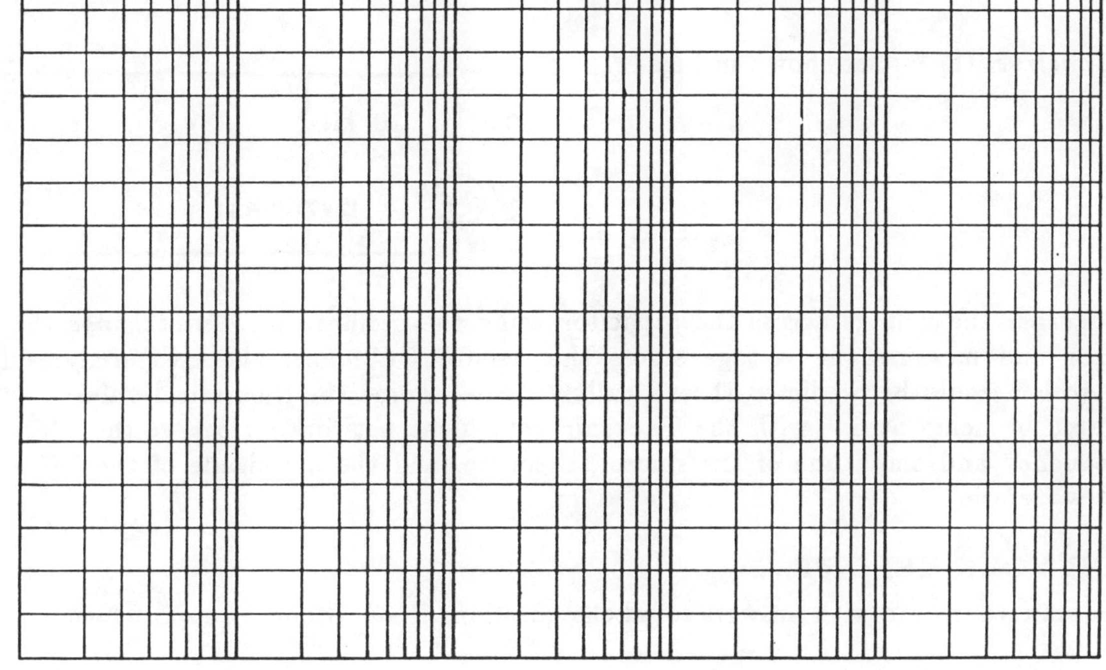

Fig. 9.

3.2. RC CIRCUITS AS FILTERS

A sinusoidal signal is applied across an RC circuit and the output signals across the resistor and capacitor are observed with an oscilloscope. The frequency, resistance and capacitance values are varied to show their effects on the output signals.

3.2A.   RC High-Pass Filter

Connect the RC circuit of Fig. 10, using the GR Decade Resistance and Capacitance Boxes for R and C.

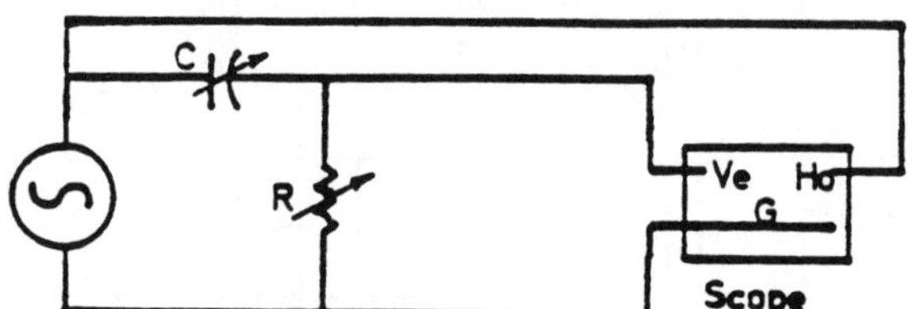

Fig. 10.   RC High-Pass Filter.

Set $C = 0.1$ μF and R $= 1.59$ kΩ. *Set* the sine-wave generator for an output V_I of 3 volts RMS. *Vary* the frequency of the generator from approximately 100 Hz to 10 kHz. *Observe* the magnitude of the voltage across R (called V_R) as a function of frequency with the output of the sine generator held constant. Note that the output of the sine-wave generator may need to be adjusted to keep the value constant at 3 volts RMS. *Sketch* V_R as a function of frequency (you may use Fig. 8). Also *measure* the phase angle between the sine-generator, V_I, and V_R, from the oscilloscope. *Determine* the frequency at which $V_R = 0.707$ V_I. What is the phase angle at this frequency?

3.2B.   RC-Low Pass Filter

Repeat 3.2A. for a low-pass RC filter; *use* Fig. 9.

3.2C.   RC Circuit with Time Varying and constant Sources

Construct the circuit shown in Fig. 11.

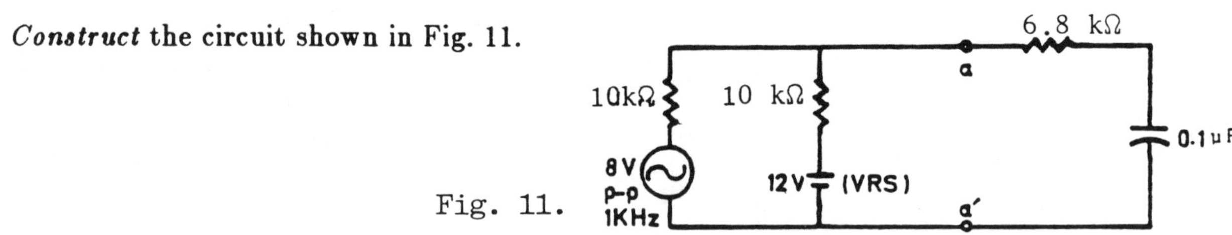

Fig. 11.

Measure the voltage across the capacitor using the oscilloscope. *Interchange* the resistor and capacitor and measure the voltage across the resistor. *Compare* the measured results with the values which would be predicted theoretically. To determine the theoretical values *replace* the signal source branches at aa′ with the Thevenin equivalent network. *Observe* the effect of changing the frequency and amplitude of the sinusoidal source, and the amplitude of the DC source on the output waveform.

4. FOR YOUR REPORT

Submit your data with answers to all the questions in our standard report form.

EXPERIMENT II-5

AC TO DC POWER SUPPLIES

EQUIPMENT:

EEP
ZENER DIODE
SILICON RECTIFIER DIODES (4)
RESISTORS (10 Ω, 33 Ω, 1 kΩ, 33 kΩ)
POT (150 Ω)
CAPACITORS (500 μF/2)
STEP-DOWN TRANSFORMER (115: 12.6V)

Table of Contents   Page No.

1. PURPOSE

The purpose of this experiment is to familiarize the student with the commonly used rectifier power supply circuits which convert AC voltage to DC voltage. Some RC filters are considered for decreasing the AC ripples.

2. INTRODUCTION

The following discussions are due to Professor F.G. Stremler of the University of Wisconsin-Madison [3] with the reference [0].

The EXP. VI-2 will complement this experiment.

2.1. SOME THEORETICAL BACKGROUND ON RECTIFICATION

There are several ways for rectification of an AC signal which, generally speaking, boil down to using circuits with several diodes in various kinds of arrangements. In the following we discuss these circuits with their theoretical justifications.

2.1.1. THE HALF-WAVE RECTIFIER

The half-wave rectifier circuit is shown in Fig. 1a. The diode conducts during the positive half-cycle of the transformer output voltage and the voltage across the resistance R follows the transformer voltage during the positive half-cycle. As a result of the internal resistance of the forward-biased diode, the voltage across R is slightly less than that supplied by the transformer. This drop of voltage for a Germanium (or Silicon) diode is approximately 0.2 V (or 0.6 V).

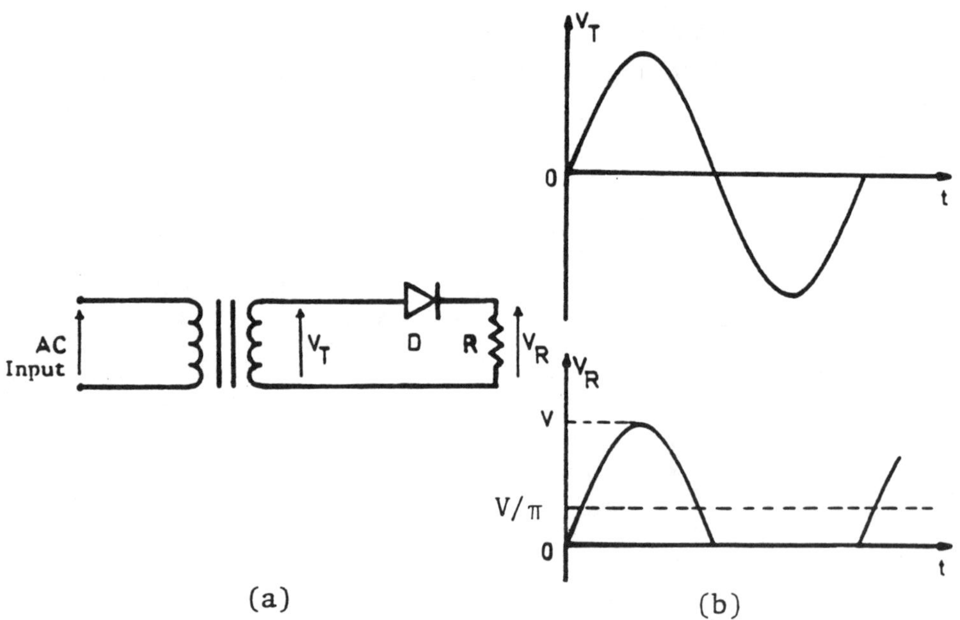

Fig. 1. Half-Wave Rectifier.

During the negative half-cycle the diode is off and no voltage is present at the output. The resulting voltage waveform is shown in Fig. 1b. The average value of this rectified waveform is

$$V_{avg} = \frac{1}{T} \int_{0}^{T/2} V \sin\omega t \, dt = V/\pi.$$

There are still some large fluctuations present, however, and usually a filter is used to smooth out these fluctuations.

2.1.2. THE FULL-WAVE RECTIFIER

A more efficient use of the alternating voltage is accomplished by rectifying both half-cycles of the transformer voltage. This can be done using two diodes as shown in Fig. 2a. Here the transformer output is center tapped and the center tap is referenced to ground. Therefore, one diode conducts on the positive half-cycle and the second diode conducts on the negative half-cycle.

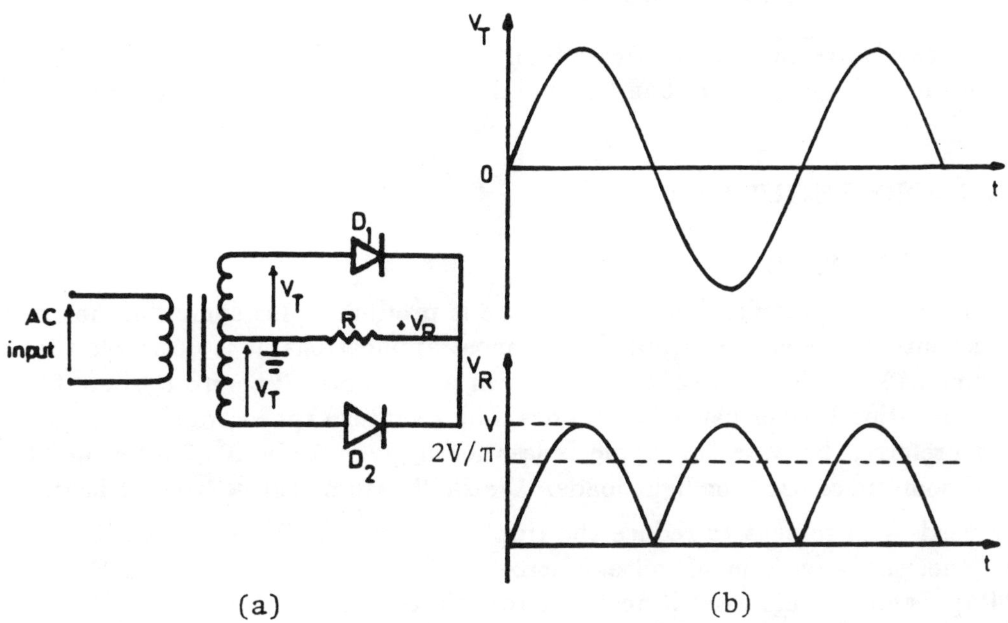

Fig. 2.   The Full-Wave Rectifier.

The resulting voltage waveform for the full-wave rectifier is shown in Fig. 2b. The average value of this rectified waveform is

$$V_{avg} = \frac{2}{T} \int_0^{T/2} V \sin \omega t \, dt = 2V/\pi.$$

The fluctuations about this average are less than for the half-wave rectifier.

2.1.3. THE BRIDGE RECTIFIER

It is possible to obtain full-wave rectification without a center tapped transformer by using the bridge rectifier shown in Fig. 3. During the positive half-cycle of the transformer output, both diodes D_1 and D_3 are forward-biased and diodes D_2 and D_4 are reverse-biased. Therefore diodes D_1 and D_3 are in series electrically with the load R and the transformer. The resulting voltage waveform across R is equal to that of the transformer output, less the voltage drops across the two diodes. In a similar way, diodes D_2 and D_4 are forward-biased on the negative half-cycle and diodes D_1 and D_3 are reverse-biased. Thus the bridge rectifier performs a full-wave rectification.

The advantage of the bridge rectifier is that it produces approximately twice the output DC voltage, compared to the two-diode full-wave rectifier, for the same transformer voltage. Also, a center-tapped transformer is not required and the peak inverse voltage ratings of the diodes can be less

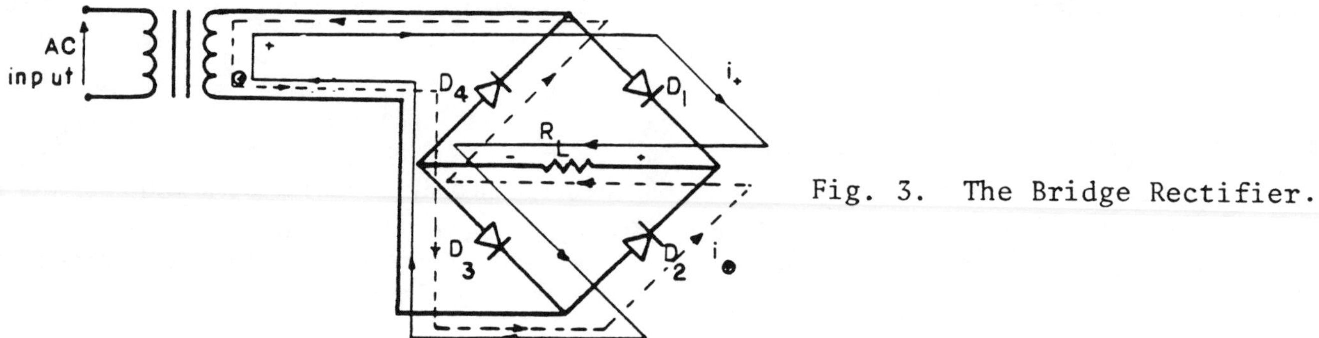

Fig. 3.   The Bridge Rectifier.

because two diodes are always in series. A disadvantage is that the forward-biased DC voltage drop across the two diodes in series is greater than the one diode in the two-diode full-wave rectifier.

2.2. FILTERING AND REGULATION

2.2.1. CAPACITOR FILTERS

In the above methods of rectification a DC voltage is produced plus some fluctuating voltage components. These fluctuations are called "ripple". For most applications, it is desirable that the ripple be minimized to produce as smooth a DC voltage waveform as possible. Therefore, some filtering is necessary. Full-wave rectification (either using two diodes and a center tapped transformer or four diodes in a bridge circuit) is preferred because the ripple is less for a given amount of filtering. However, half-wave rectification is sometimes used for light loads. We shall assume full-wave rectification here.

A commonly used arrangement to reduce the ripple is to shunt the resistance R with a large capacitor. With this done, the waveform of voltage across R becomes that shown in Fig. 4. The dashed lines show the voltage which would exist if no capacitor is used.

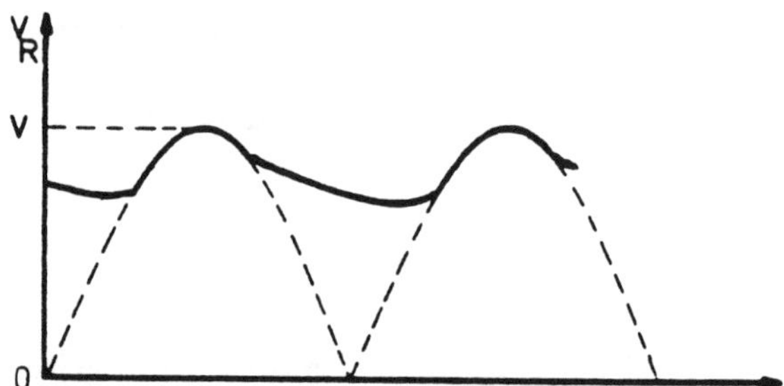

Fig. 4.   A Capacitor-Filtered Rectified Waveform.

With a capacitor, there are two components to the periodic waveform of $v_R(t)$. The positive slope region follows the transformer voltage as the source recharges the capacitor. The negative slope region is exponential as the capacitor discharges into R. Note that the average voltage is greater than it would be with no capacitors, and that with $C \to \infty$, there will be no ripple voltages. Rectifier currents exist only during the positive slope regions. This means that for a given DC bias current, the rectifier currents must have rather high peak values because the average rectifier current must equal the average load current.

A general solution to the capacitor filter problem involves solving a trancendental equation. To simplify the analysis, we shall assume that the output-voltage waveform is made up of portions of straight lines, as shown in Fig. 5. The peak value of this wave is V, the transformer voltage less the forward-biased diode voltage drop. If the capacitor discharge voltage is V_r, then, from Fig. 5, the average value of the voltage, V_{DC} is approximately

$$V_{DC} = V - (V_r / 2) \, .$$

The instantaneous ripple voltage is obtained by subtracting V_{DC} from the instantaneous load voltage. The result is shown in Fig. 6. The RMS value of this "triangular wave" is independent of the slopes or lengths of the straight lines and depends only upon the peak value.

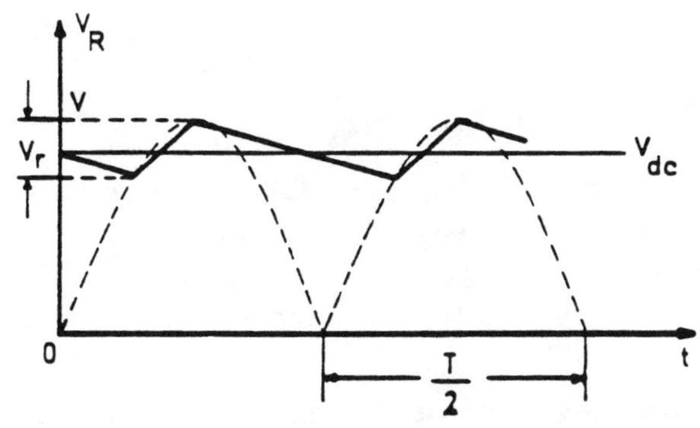

Fig. 5. The Approximate Load-Voltage Waveform in a Full-Wave Capacitor-Filtered Rectifier.

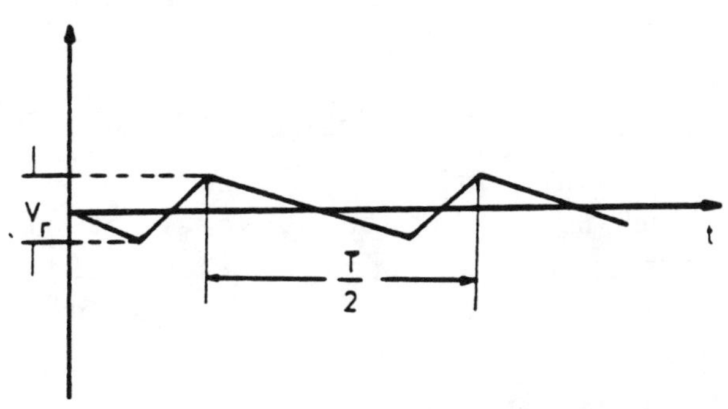

Fig. 6. The Approximate Waveform of the Ripple Voltage in a Full-Wave Capacitor Filtered Rectifier.

Calculation for the RMS ripple voltage yields

$$V'_{RMS} = \frac{V_r}{2\sqrt{3}} \, .$$

The ripple factor is given by

$$r = \frac{V'_{RMS}}{V_{DC}} = \frac{V_r}{2\sqrt{3}\, V_{DC}} \, .$$

To decrease the ripple for a given input capacitor, a resistor and second capacitor are sometimes used, as shown in Fig. 7. This configuration is called a "π-type" filter.

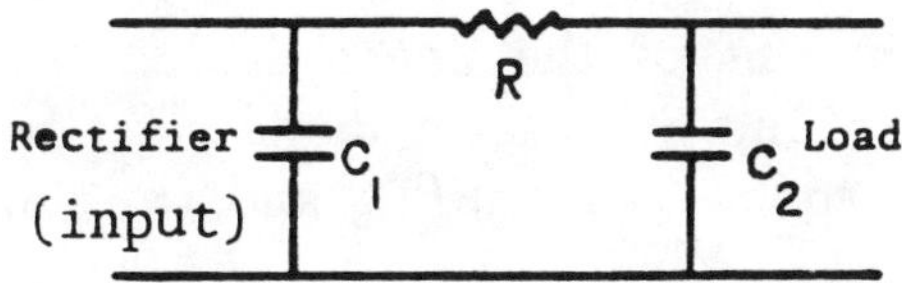

Fig. 7. The Capacitor-Input RC π-Type Filter.

The regulation of a power supply is an index which shows how the DC output voltage varies as it is loaded. Designating the no-load voltage by V_m and V_l as the rated full-load voltage, the percent regulation is defined as

$$\text{Percent Regulation} = \frac{V_m - V_l}{V_l} \times 100.$$

The regulation of a capacitor filter is relatively poor. The larger the capacitance the better will be the regulation. The ripple of a capacitor filter varies directly with the load current and inversely with the capacitance.

In order to keep the ripple low and to ensure good regulation, very large capacitors (of the order of tens or hundreds of microfarads) must be used. The most common type of capacitor for this rectifier application is the electrolytic capacitor. These capacitors are polarized and care must be taken to insert them into the circuit with the terminal marked + to the positive side of the output.

The desirable features of rectifiers employing capacitor input filters are the small ripple and the high voltage at light load. The no-load voltage is equal, theoretically, to the maximum transformer voltage. The disadvantages of this system are the relatively poor regulation, the high ripple at large loads, and the peaked currents that the diodes must pass.

2.2.2. POWER SUPPLY REGULATORS

The power supply previously discussed has several undesirable features. One of these is its poor regulation which means that the DC voltage at the load terminals is a function of the load current. This effect in turn may be thought of in terms of the output or Thevenin resistance of the power supply which does not change in a linear fashion with current. A quantity of interest is the incremental output resistance, which may be defined as $\Delta V_L / \Delta I_L$.

Another poor feature of the power supply is the ripple voltage appearing with the DC voltage. This will produce the so-called "hum" in audio amplifiers, blurring of the display on an oscilloscope, and extraneous signals in instrumentation systems.

A third undesired property is not really that of power supply itself but rather that due to poor regulation of the AC voltage source. This causes the DC voltage to vary in proportion to the AC source voltage. Some method of reducing these effects is needed in order that, within the voltage-ampere range of the system, the system may be made to appear as an ideal battery (i.e., a pure DC voltage independent of current and having no dependence upon AC line regulation).

A simple method of voltage regulation uses an avalanche (Zener) diode connected as shown in Fig. 8a. Fig. 8b is a graph of a typical current-voltage characteristic of a Zener diode. When the diode is forward-biased, the forward current increases with applied voltage and is then limited by the parameters of the circuit. When the diode is reverse-biased, a small reverse current I_s, called the saturation current, flows. The I_s remains relatively constant despite an increase in reverse bias until the Zener breakdown region is reached. Here the reverse current starts rising rapidly as a result of avalanche effects. Finally, Zener breakdown (a sharp increase in current) occurs. As long as the current through the diode is limited by the external circuit to a level within its power-handling capabilities, the diode continues to function normally. The breakdown voltage in a Zener diode is dependent on diode construction and material. Zener diodes are used as voltage regulators and as voltage reference standards.

As long as I_Z exceeds the reverse breakdown current of the Zener diode, V_L is constant and independent of R_L, V, and R_S. For a wide range of R_L values, the Zener diode must be capable of conducting a maximum current whose DC current value equals $(V - V_L) / R_S$ and thus dissipate a power equal to

$$P = V_L (V - V_L) / R_S .$$

For example, if $V = 10$ volts, $V_L = 5$ volts, $I_{L(Max)} = 100$ mA, and $R_S = 50\ \Omega$, then

$$P = V_L (V - V_L) / R_S = \frac{5(5)}{50} = 0.5 \text{ watt.}$$

The resistance, seen as the output resistance by the load, is the dynamic resistance of the avalanche diode in parallel with R_S. Usually the dynamic resistance of the diode is very small, perhaps on the order of 1 ohm or less, so the Zener regulator circuit is fairly good except for its lack of adjustability and limited range of load current.

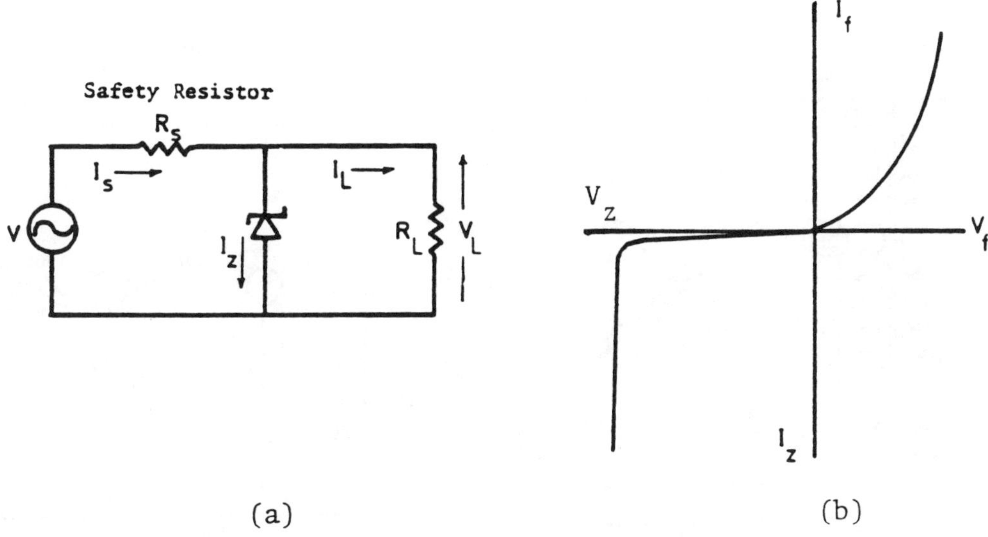

Fig. 8. (a) Zener Shunt Regulation, (b) A Zenere Diode Characteristic.

To provide for adjustability and improved regulation over a wide range of conditions, the difference between a Zener-regulated voltage and the load voltage can be sensed, amplified, and used to control the voltage drop from the supply (usually with a transistor). For many voltage ranges, this type of regulator is available as an integrated circuit requiring only a few (if any) external components. For further information on this matter please refer to EXP. VI-2.

3. LABORATORY PROCEDURE

It is expected that the circuit of Fig. 9 will be made available for your testing.

The circuit shown in Fig. 9 can be connected in various ways as designated in the following chart. (The 10 Ω resistor at the output of the rectifier circuit is for diode protection here and would not be used in actual equipment.)

Switch	Rectifier			Filter	
	half-wave rectifier	full-wave rectifier	bridge rectifier	capacitor filter	π filter
S_2			close		
S_3			close		
S_4			close		
S_5		close	close		
S_6				close	close
S_7					close

In steps 1-3 below, *sketch* the oscilloscope display of the observed waveform in chart of Fig. 10. *Measure* the peak voltage with respect to ground as accurately as possible and *indicate* it on your sketch. (*Trigger* the oscilloscope on LINE.) Also *measure* the DC voltage referenced to ground with a digital voltmeter (DVM). *Note: The steps in the laboratory procedure are cumulative; do not change switch settings from a previous step unless you are instructed to do so.*

1. *Connect* the circuit shown in Fig. 9, for a half-wave rectifier. *Observe* the waveform (with no filtering) at point A.

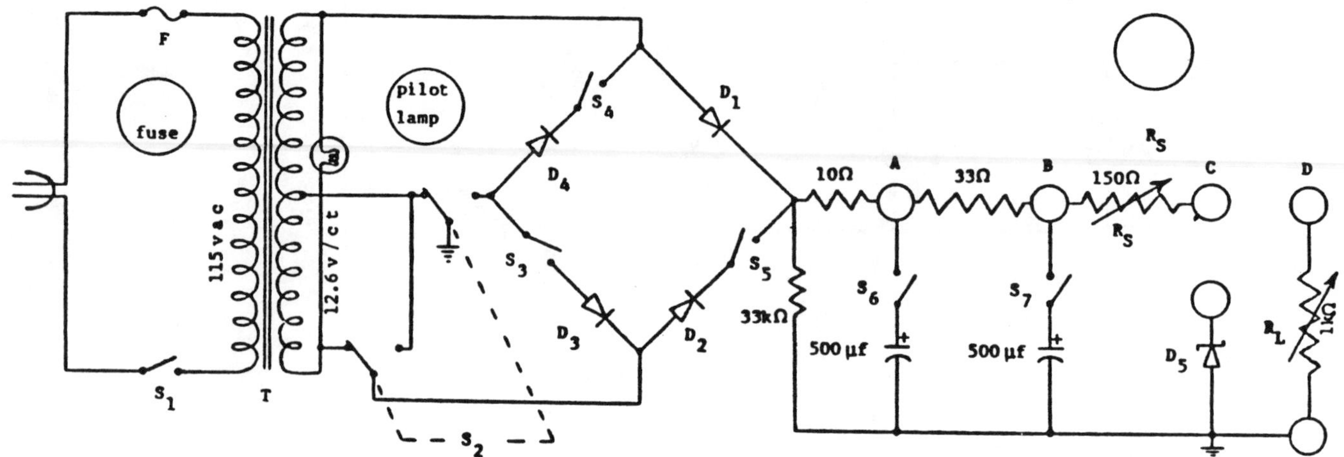

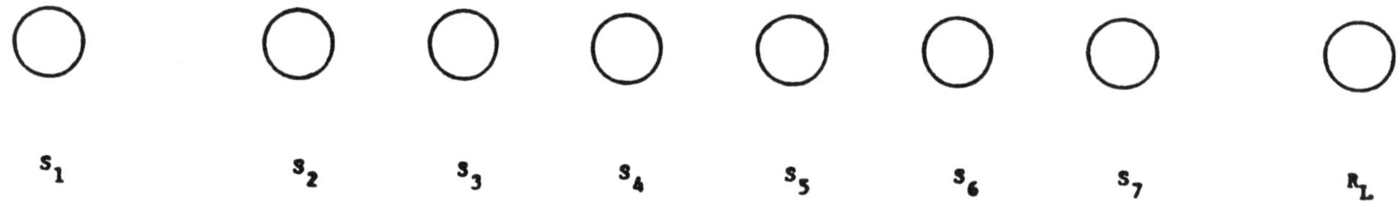

All switches are shown in position away from circuit diagram.

Fig. 9, [3].

2. *Close* switch S_5 to connect as a full-wave rectifier; *repeat* above measurements.

3. In addition, *close* switches S_2, S_3, and S_4 to produce a bridge rectifier; *repeat* the measurements (Note that now only one-half of the transformer secondary is in the circuit).

In steps 4-6 below, *close* switch S_6 to connect a capacitor filter at the output of the rectifier. *Connect* point B to load resistor R_L through a 0-100 mA DC milliammeter. *Record* the DC voltage at point B versus load current at 10 mA increments over 0-100 mA. Also *measure* P-P ripple voltage at point B with an oscilloscope.

4. *Connect* for the half-wave rectifier and *make* measurements.

5. *Connect* for the full-wave rectifier and *make* measurements.

6. *Connect* for the bridge rectifier and *make* measurements.

7. *Sketch* the ripple voltage waveform obtained in steps 4-6 above in chart of Fig. 10.

8. Now *close* switch S_7 to place a π-filter at the output of the rectifier. *Repeat* the measurements of steps 4-7 above.

9. *Connect* the 0-100 mA DC milliammeter from point C in series with the Zener diode and *adjust* R_S to give 50 mA through the Zener diode (or as close to 50 mA as possible with your circuit; make note of actual current). *Leave* R_S set at this value, disconnect milliammeter, and jumper the Zener diode to point C for the next step.

10. *Connect* R_L to point C through the 0-100 mA DC milliammeter. *Record* the DC voltage and P-P ripple voltage at point C versus load current at 10 mA increments over 0-100 mA for the half-wave rectifier.

11. *Repeat* steps 9 and 10 for the full-wave rectifier.

12. *Repeat* steps 9 and 10 for the bridge rectifier.

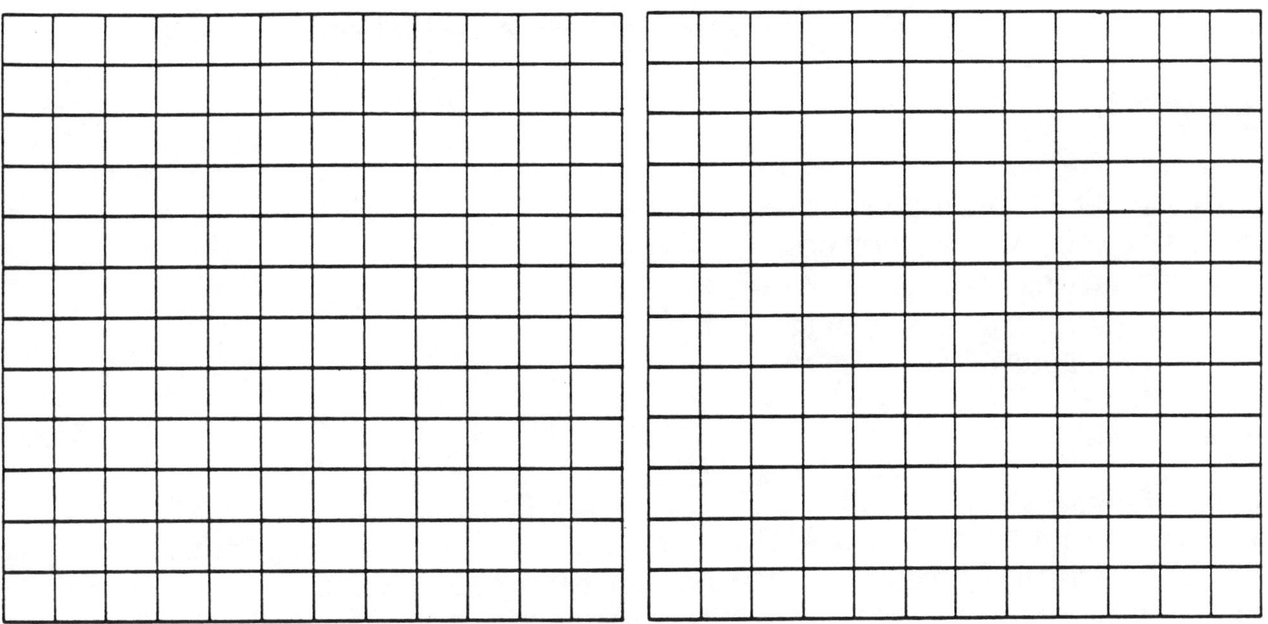

Fig. 10.

4. PROBLEM SET

1. Compute the average value of each of the waveforms in steps 1-3 in the laboratory procedure using the oscilloscope data and compare with the measured DC values.

2. Is the triangular approximation for the capacitor-input filter waveform also valid for the π-filter? Explain.

3. Is the primary function of the second capacitor in the π-filter for improved regulation or lower ripple?

4. The ratings of the 1N4731 Zener diode used in this experiment are 4.3 V at 50 mA. On the basis of your observations in this experiment, what would you suggest as a good "rule-of-thumb" specification for the maximum allowable load current in terms of rated Zener current for 5% regulation?

5. FOR YOUR REPORT

Answer as many problems as you possibly can. The report should be organized in our standard report form and it must include the following items:

1. Plot the V-I load characteristic for each type of rectifier circuit. Each plot should have three curves: (a) capacitor filter only; (b) π-filter; (c) π-filter plus Zener regulator.

2. If the rated load for this supply is 30 mA, what is the percent regulation for each condition in (1) above?

3. Calculate the P-P ripple voltage divided by the DC load voltage at a load current of 30 mA and express in percent for each of the conditions in (1) above.

4. Compare the half-wave, full-wave, and bridge rectifiers in terms of DC voltage output, percent ripple, and regulation.

5. Compute the incremental output resistance for each rectifier circuit at a load current of 30 mA both with and without the Zener regulator; compare.

EXPERIMENT II-6

PASSIVE NETWORK SYNTHESIS

EQUIPMENT:

EEP
GR DECADE RESISTANCE BOX
DECADE CAPACITANCE BOX
RESISTORS (VALUES OPEN)
CAPACITORS (VALUES OPEN)
INDUCTORS (VALUES OPEN)

Table of Contents

1. PURPOSE

The purpose of this experiment is to acquaint the student with basic concept of passive network synthesis.

2. INTRODUCTION

Up to now we have been only concerned with the analysis of circuits that are driven by input signals. To build a circuit (or filter) that produces a specific response to a given input (which is the inverse of the above concept), on the other hand, has many applications in circuit design. In this experiment we will study the basics of this concept using only passive elements. Experiments VI-5 to 7 complement the present experiment.

In general, the design requirements or constraints are formulated in terms of an input-output relationship. This relationship is generally expressed in frequency domain by various transfer functions between port variables of a given one-port network (or several transfer functions between port variables of many ports of a multi-port network). These transfer functions are to be carefully studied before the final design is chosen. Indeed, given one set of constraints (transfer function), we may have a number of different circuits that yield the same output signal when a given input signal is applied to them. In EXP. VI-5, we have described several other issues regarding the overall design of a filter that must be considered before the final choice is made. The main reasons for these concerns are, of course, the expenses of building as well as maintaining the operation of the ultimate circuit. Based on these simple considerations a large body of research concerning various techniques and algorithms to design a circuit from a set of given transfer functions has appeared in the literature since the early 1930's. These results primarily classify the types of transfer functions from which the corresponding synthesis procedures have been developed.

In general, there are two major approaches to these synthesis problems; namely, passive network synthesis and active network synthesis. As these names imply, each category uses the corresponding network elements in the final design. But because of the bulkiness of the inductors and/or transformers used in passive network synthesis, this approach has lost its place to active network synthesis, in particular to RC-active networks at low frequencies. To synthesize with only passive network elements, however, certain properties, as will be described, must be met. These limitations can largely be removed using active network components. We will postpone the discussions on active network synthesis to EXPs. VI-5 to 7. Nevertheless, we point out that the vast applications of active network synthesis have practically changed the entire picture of the conventional network synthesis techniques. This change has been strengthened by the accessibility of these active network components. We must, however, emphasize that the passive network synthesis is more systematic and /or theoretical than the active network synthesis. We strongly believe that students must be at least acquainted with some basic and general ideas behind passive network synthesis. Here we study various requirements that are to be satisfied if a transfer function is going to be synthesized with only passive network elements. *Notice that we concentrate only on the one-port passive network.*

3. SYNTHESIS OF THE ONE-PORT PASSIVE NETWORK

Before we present the basic ideas behind passive network synthesis, recall the immittance functions of some basic network elements. The resistance of a resistor is R, the impedance of an inductor is sL, and the admittance of a capacitor is sC. Thus, if we are asked to design a transfer function $T(s) = V(s)/I(s) = 2$ ohms, then we choose a simple resistor of 2 ohms to represent this transfer function. Similarly, if $T(s) = V(s)/I(s) = 2s$, then an inductor of 2 H is sufficient to represent this transfer function. Finally, if $T(s) = I(s)/V(s) = 2s$, then a capacitor of 2 F represents this transfer function. *In reality, however, we have much more complicated transfer functions to work with than these simple elementary cases.* In fact, in most problems the given transfer function is not easily decomposed into such simple and nice terms. Incidently, we call these *noticeable terms. To make a given transfer function representable by several noticeable terms we must execute procedures that are called synthesis algorithms. Actually, that is what network synthesis is all about.* These algorithms are the subject of separate courses. Nevertheless, we present several major results that at least enable us to know where to look for answers to these very important design questions.

Considering the above three elementary examples, can we state some common features of these transfer functions?

Recall that these transfer functions were R, sL, and sC where s $= \sigma + j\omega$ is the Laplace variable. We immediately see that these are rational functions of s with constant coefficients. In general, of course, these transfer functions are rational quotient polynomials of s. Furthermore, substituting s $= \sigma + j\omega$ in these transfer function results in complex numbers that have positive real parts for all $\sigma \geq 0$. This is a remarkable property of these transfer functions. Thus the question that will be immediately raised is the following. Does every transfer function have this positive realness property? (Notice that we are dealing with one-port networks, therefore there is only one transfer function to work with.) The answer is definitely "yes" if the transfer function has come from a passive circuit or equivalently if the transfer function is supposed to be synthesized with only passive elements. This celebrated result was first proved by Otto Brune in his doctoral thesis at M.I.T. in early 1931 [4], and it became the cornerstone of the systematic study of network synthesis. You may verify this result by choosing any RLC network and checking the real part of its driving-point immittance function.

To test the positivity of any given transfer function is not, however, a trivial task. Indeed, a large effort must be devoted to this testing, and this has been the subject of many studies over the years.

We now summarize the basic properties of positive real functions as appear in our synthesis questions. We must check any given transfer function T(s) $=$ p(s)/q(s) against these conditions before any attempt to synthesize T(s) with passive network elements is made. These conditions are well known and can be found for example in the classical textbook of M.E. Van Valkenburg [4].

Given T(s) $=$ p(s)/q(s), where p and q are both rational functions of s, the Laplace variable, then T(s) is positive real function if the following are true:

1. The coefficients of p and q are all positive and thus as a result

    a. T(s) is real for all real values of s, including s $= 0$.

    b. The complex poles and zeros of T(s) occur in complex conjugate pairs.

2. The poles and zeros of T(s) have non-positive real parts.

3. The imaginary-axis poles (and zeros) of T(s) are simple with positive residues. (Note residues of zeros are found from 1/T(s).)

4. The degrees of the highest and the lowest terms of p(s) and q(s) differ by only one.

If the transfer function satisfies all the above requirements then it is *fully* within the realm of positive real functions and can be synthesized with passive elements. The above checks are possible by inspection expect those regarding poles and zeros. For higher-order polynomials (which is often the case in practice) we must turn to special techniques to check these requirements without actually finding the roots of the corresponding polynomials. This special study is well beyond the scope of this textbook. Nevertheless, due to many advances including the availability of computers, we can test to see (for a moderate sized transfer function) whether these requirements on poles and zeros are met or not by directly computing them. We do not elaborate on this testing further.

Suppose a given transfer function passes the above tests and becomes a good candidate to be synthesized by the passive network elements. Then what steps should we take to carry out this process? In the following we very briefly go over some of the major procedures in passive network synthesis.

3.1. SYNTHESIS OF IMMITTANCE FUNCTIONS WITH LC ELEMENTS

Suppose a positive real impedance function $Z(s)$ is given to be synthesized. If this transfer function can be decomposed into a set of αs (an inductor with $L = \alpha$ H) or $1/\beta s$ (a capacitor with $C = \beta$ F), or $(\alpha s + 1/\beta s)$ (an inductor $L = \alpha$ H in series with a capacitor $C = \beta$ F - or if instead of $Z(s)$ we had $Y(s)$, then an inductor $L = \beta$ H in parallel with a capacitor $C = \alpha$ F). Then we gradually build our circuit by consecutively removing each of these elements from the original impedance function. This means we start with $Z(s)$, then we write $Z(s) = Z_1(s) + Z_2(s)$, where $Z_1(s)$ is one of the above three cases. We then continue the same procedure with $Z_2(s)$ until we exhaust our options. This is a perfectly valid idea but there is much more into it than it sounds. Many exercises are needed to be able to complete this process without errors. The same technique can be used to synthesize $Y(s) = 1/Z(s)$. The circuit corresponding to $Z(s)$ is called Foster's first form and that corresponding to $Y(s)$ is called Foster's second form [4].

Example 1: Consider an impedance function $Z(s) = (2s^2 + 1)/s(s^2 + 1)$. Find Foster's first and second forms.

The first step is to check and see the positive-realness property holds.

1. The coefficients of p and q are all positive - check. Also, 1a and 1b are satisfied - check.

2. The poles and zeros of $Z(s)$ have non-positive real parts - check. (Because the poles are s=0, s=$\pm$j, and the zeros are s=$\pm$j $\sqrt{2}$.)

3. The imaginary-axis poles of $Z(s)$ are simple with positive residues - check. (Because the imaginary-axis poles are s=$\pm j\omega$ this is simple, i.e., we do not have more than one pole, say, at s=$+j\omega$ or s=$-j\omega$. Also the residues are positive since

$$\frac{2s^2 + 1}{s(s^2 + 1)} = \frac{1}{s} + \frac{\frac{1}{2}}{s + j} + \frac{\frac{1}{2}}{s - j} \; ; \text{ therefore residues}(1, \frac{1}{2}, \frac{1}{2}) \text{ are positive.})$$

4. The degree of the highest and the lowest terms of p(s) and q(s) differ by only one - check.

Thus $Z(s)$ is a good candidate to be synthesized by passive network elements. Now we must extract as many *noticeable terms* as we possibly can. Since in step 3 of our testing we found that

$$Z(s) = \frac{1}{s} + \frac{\frac{1}{2}}{s + j} + \frac{\frac{1}{2}}{s - j} = \frac{1}{s} + \frac{s}{s^2 + 1},$$

we immediately see that we have one *noticeable term* of $\frac{1}{s}$ free to grab. This becomes our first element. Thus we have so far the circuit of Fig. 1. Let $Z_2(s) = \frac{s}{s^2 + 1}$. This impedance function can be synthesized as follows. $Z_2(s) = 1/(s+1/s) = (s)(1/s)/(s+ 1/s)$. This is a parallel inductor of $L = 1$ H and a capacitor of $C = 1$ F. Therefore we have the circuit of Fig. 2 as our Foster's first design.

As an exercise try to find Foster's second design, i.e., start from $Y(s) = 1/Z(s)$.

Comments: Generally speaking we must look for poles at $s=0$ and $s=\infty$ as our first candidates for *noticeable terms* to be extracted from the given transfer function. Then we must go after complex conjugate poles such as $s^2 + \alpha$, where α is a constant. Suppose in this process we come across with terms that do not correspond to any of the above three cases - then what? The answer is simple. We just cannot synthesize this transfer function with only L's and C's.

There are many other ways of developing a circuit with only LC elements such as Cauer forms. This method depends on the *continued fraction expansion* of the transfer function producing a ladder LC network to represent a given transfer function. This development will result in the following general patterns depending on the properties of the transfer function at $s=0$ and $s=\infty$. Suppose we have an impedance function $Z(s)$ of an LC network. Then we may have one of the following two forms:

$$Z(s) = L_1 s + \cfrac{1}{C_2 s + \cfrac{1}{L_3 s + \cfrac{1}{C_4 s + \cfrac{1}{\ddots}}}} \text{ , or}$$

$$Z(s) = \cfrac{1}{C_1 s} + \cfrac{1}{\cfrac{1}{L_2 s} + \cfrac{1}{\cfrac{1}{C_3 s} + \cfrac{1}{\cfrac{1}{L_4 s} + \cfrac{1}{\ddots}}}} .$$

The final forms of these ladder networks is obvious. Try to build a circuit with four elements.

3.2. SYNTHESIS OF IMMITTANCE FUNCTIONS WITH RC AND RL ELEMENTS

Here again the fundamental requirement is that these functions must be positive real. The role that R plays is simply generating a constant that is added to the immittance function consisting of more RC or RL elements. But first, recall that the immittance function of an RC network is as follows:

$$Y_a(s) = \cfrac{1}{R_a + \cfrac{1}{sC_a}} = \frac{1}{R_a} \cdot \frac{s}{s + 1/R_a C_a} ,$$

$$Z_b(s) = \frac{R_b (1/sC_b)}{R_b + 1/sC_b} = \frac{1}{C_b} \cdot \frac{1}{s + 1/R_b C_b} .$$

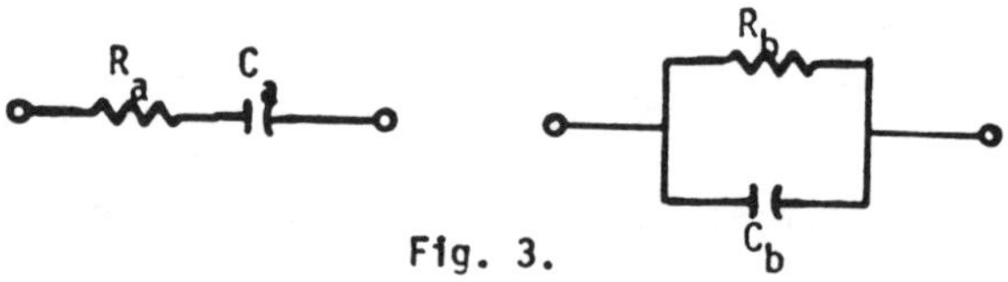

Fig. 3.

Both of these immittance functions have poles on the negative axis. Therefore the *noticeable terms* for these functions are: $(s+\alpha)$, $\alpha > 0$; constant; and $1/sC$ or sL. We may easily show the properties of RL networks as well. Thus we conclude that given, say, an impedance function $Z(s)$, we must be able to decompose this function in the following manner in order to synthesize with only RC elements:

$$Z_{RC}(s) = R_0 + \frac{1}{C_0 s} + \sum_{k=1}^{n} \frac{1}{sC_k + 1/R_k} .$$

This is called Foster's first form. Similarly, we may have

$$Y_{RC}(s) = G_0 + C_0 s + \sum_{k=1}^{n} \frac{1}{R_k + 1/C_k s} ,$$

which is called Foster's second form.

Cauer's first and second forms are also generated with the continued fraction method as we studied for the LC networks. Synthesis with RL elements is a dual method to the above scheme.

3.3. SYNTHESIS OF IMMITTANCE FUNCTIONS WITH RLC ELEMENTS

The main purpose of this experiment has been to familiarize the student with the basic concept of passive network synthesis. We very briefly, and for the sake of academic interest, state a few words about the RLC case. Passive RLC filter are useful as prototypes for active RC-filters.

If, given a positive real immittance function, we start extracting constant, poles at the origin ($\pm j\omega$; and infinity) from this transfer function, then we may exhaust the entire immittance function thus constructing the RLC network. Sometimes we may come across with a particular immittance function consisting of positive real functions but whose real values go to zero at one or several frequencies. This class of functions is called *minimum functions*, which, in addition to the above property, has the following properties: They have no poles and zeros on the imaginary axis and their values at $s=0$ and $s=\infty$ are finite and positive. Of course this requires that the degree of numerator and denominator of these functions be equal. To synthesize this particular immittance function one may use the well-known methods of Brune, Bott and Duffin as described in the classical textbook of Van Valkenburg [4]. In order to raise student's interest in this matter we will study one such method. Also to illustrate the method we will present an easy example. Consider the following transfer function:

$$Z(s) = \frac{s^2 + s + 1}{4s^2 + s + 1} .$$

Clearly this transfer function satisfies the basic properties of a positive real function. Also to check that all the conditions of these constraints are met we notice that

$$Z(j\omega) = \frac{(1 - 2\omega^2)^2 - j\, 3\omega^3}{(1 - 4\omega^2)^2 + \omega^2} .$$

Certainly Re $Z(j\omega) = (1 - 2\omega^2)^2 / [(1 - 4\omega^2)^2 + \omega^2] \geq 0$ for all ω. But interestingly at $\omega = \pm\, 1/\sqrt{2}$, Re $Z(\pm j / \sqrt{2}) = 0$. We notice that indeed $Z(s)$ is a minimum impedance function. What do we do next? Obviously the techniques we have studied before are not applicable. In other words, we can not extract poles or constants easily. Here we use Brune's method to complete the synthesis process. His approach centered around the frequency ω_1 which makes Re $Z(j\omega_1) = 0$. Clearly $Z(j\omega_1) \neq 0$. Let $Z(j\omega_1) = jX_1$ (if X_1 is positive then this circuit at ω_1 corresponds to a positive inductor and if $X_1 < 0$ at ω_1 this circuit corresponds to a negative inductor). Brune actually suggested that by removing this particular inductor, whether positive or negative, will make the remaining part of the $Z(s)$ manageable. In this example, we have $Z(j/\sqrt{2}) = -j/\sqrt{2} = (j/\sqrt{2})\,(-1) = (j\omega_1)\,(-1)$. Thus our inductor is $L=-1$. Removing this element results in

$$Z(s) = -s + Z_1(s), \text{ or}$$

$$Z_1(s) = Z(s) + s = \frac{(2s^2 + 1)\,(2s + 1)}{4s^2 + s + 1}.$$

This is what has happened, so far. We started with Fig. 4 and then we obtained the circuit of Fig. 5. Here we choose to study $1 / Z_1(s)$ since the numerator of $Z_1(s)$ is in factored form. Thus we get

$$Y_1(s) = \frac{4s^2 + s + 1}{(2s^2 + 1)\,(2s + 1)} = \frac{s}{2s^2 + 1} + \frac{1}{2s + 1}$$

$$= \frac{1}{2s + \dfrac{1}{s}} + \frac{1}{2s + 1} .$$

From our previous experience we see that the first term is the admittance of an inductor $L=2$ H in series with a capacitor of $C=1$ F. The second term is an inductor of $L=2$ H and a resistor of $R=1\ \Omega$.

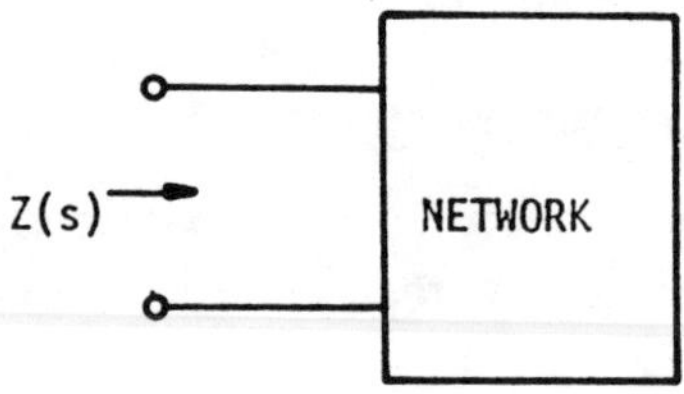

Fig. 4.

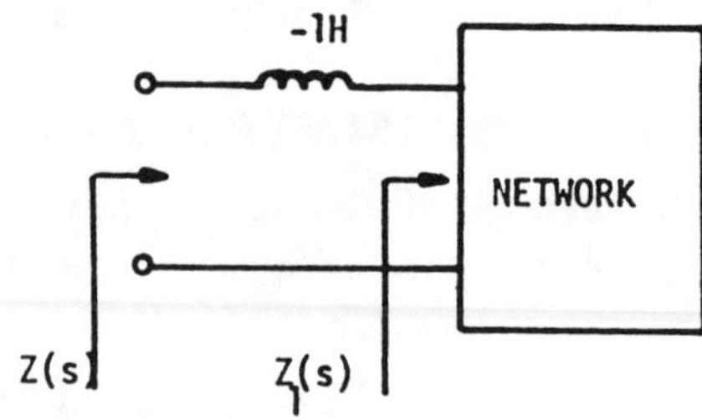

Fig. 5.

Therefore we finally obtain the circuit of Fig. 6 for our design.

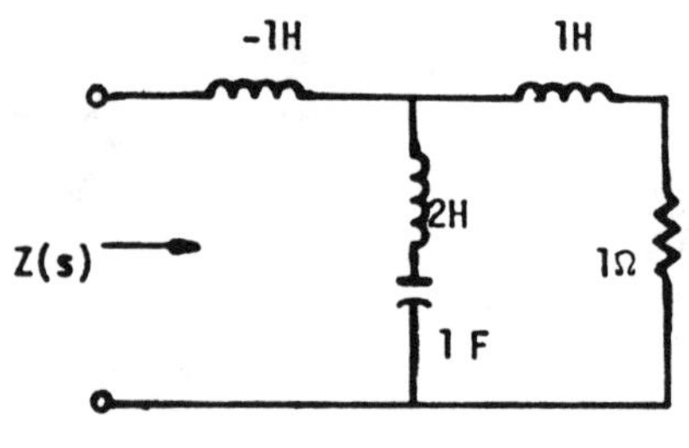

Fig. 6.

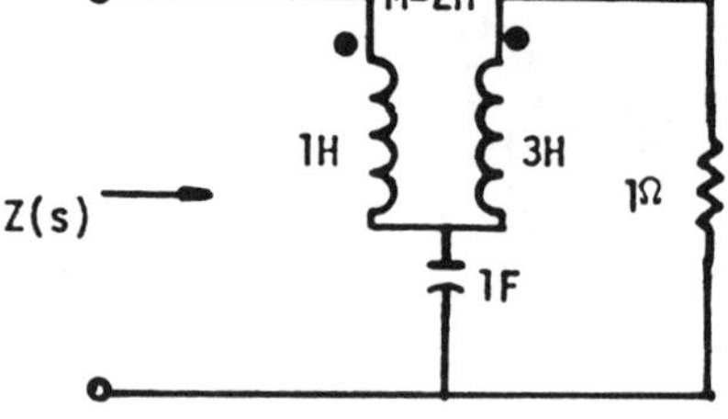

Fig. 7.

From our review in Section 0-2, we have the equivalent transformer circuit of Fig. 7 for the network of Fig. 6. Neat!

3.4. CONCLUSIONS

We have very briefly reviewed some very major results from passive network synthesis. It has not been our intention to cover every little detail entailing these problems. Nevertheless we have pointed out enough material that can be used as the starting point for further studies of this important subject in circuit design. Interested readers must consult network synthesis textbooks for additional details.

4. LABORATORY PROCEDURE

4.1. PRE-LABORATORY WORK

Carefully read the text of this experiment. Then synthesize the following transfer functions and submit them with your last report which is due today.

1. $Z(s) = \dfrac{3s + 4}{2s^2 + 5s + 3}$

2. $Z(s) = \dfrac{2s^2 + 4s + 2}{3(s + 1)}$

3. $Z(s) = \dfrac{6s^4 + 90s^3 + 7s^2 + 40s + 1}{s(9s^2 + 4)}$

4. $Z(s) = \dfrac{3s^3 + 4s^2 + 4s + 4}{6s^2 + 2s + 2}$.

4.2. LABORATORY WORK

Comments: *We believe in order to improve the quality of this textbook we must leave it up to each instructor to lead the students in selecting a numerically specific design problem that is appropriate to their laboratory facilities. This is to be our policy in most design oriented experiments in this textbook.*

Synthesize the transfer function given to you by your instructor in the laboratory. *Apply* a sinusoidal waveform to your design and *compare* its theoretical response with the practical result stemming from your designed circuit.

5. FOR YOUR REPORT

Submit your data in our standard report form.

EXPERIMENT II-7

MAGNETIC MATERIAL AND HYSTERESIS

EQUIPMENT:

EEP

ISOLATION TRANSFORMER (115 volts AC to 115 volts AC)

VARIABLE AUTO TRANSFORMER

RESISTORS (10 Ω-5 W, 100 kΩ)

CAPACITOR (10 μF)

ONE TO ONE (*) CORE TRANSFORMER WITH 1000 TURN

(*) Means different materials such as Orthonol, etc., with the corresponding
manufacturer's specification data sheets for their hysteresis loops.

Table of Contents

Page No.

1. PURPOSE

In this experiment the characteristics for various ferromagnetic core materials will be determined from their flux-current loops. These characteristics will be compared to manufacturer's catalog data.

2. INTRODUCTION

The following discussion is due to Professor F.G. Stremler of the University of Wisconsin-Madison [3] with the reference [1].

In free space, the relationship of the magnetic flux density B and the magnetizing force H can be expressed in terms of the permeability μ_0 as

$$B = \mu_0 H \,,$$

where $\mu_0 = 4\pi \times 10^{-7}$ henrys/meter $= 4\pi \times 10^{-7}$ volt-sec/ampere. In cases where the medium is other than free space, the above equation is written as

$$B = \mu H = \mu_r \mu_0 H \,,$$

here μ_r is the relative permeability. Most ferromagnetic materials have values of μ_r on the order of hundreds or thousands. Thus such materials have a high flux density for a given applied ampere-turns and thus these materials are often used for transformers and motors.

In many applications (e.g., transformer cores), ferromagnetic materials are subjected to cyclically varying values of magnetizing force H with symmetric positive and negative limits. As H varies through repeated cycles, B lags behind H, and the graph of B versus H approaches a fixed closed curve, as shown in Fig. 1. The lagging of B behind H is called *hysteresis* and the closed curve, as shown in Fig. 1, is called a *hysteresis loop*.

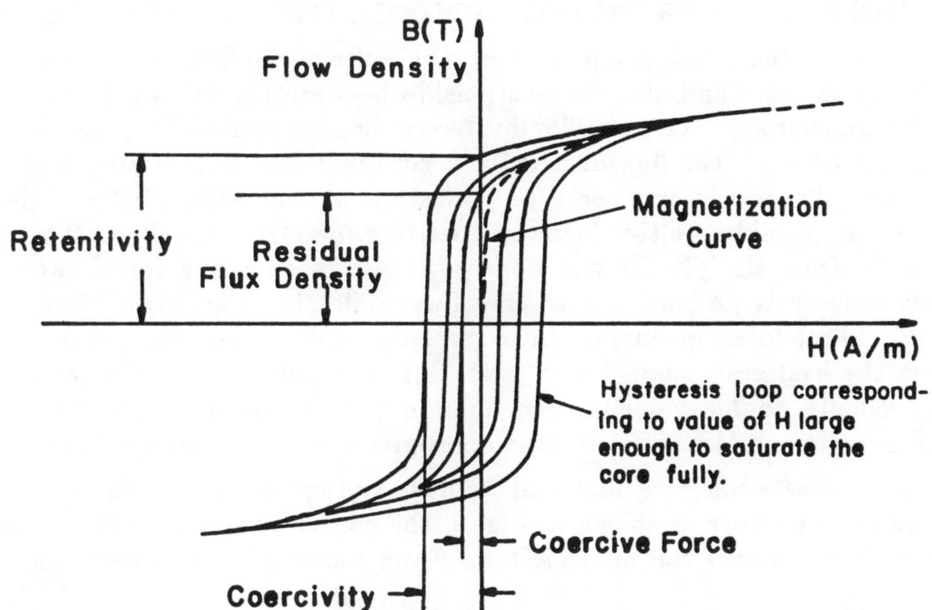

Fig. 1. Typical Hysteresis Loops and a Magnetization Curve.

As shown in Fig. 1, the flux density corresponding to a particular field intensity is not single-valued. Its value lies between certain limits depending upon the previous history of the ferromagnetic material. However, because in many situations this previous history is not known, a compromise procedure is used in making calculations by working with a single-valued curve called the *magnetization curve*. This is found by drawing a curve through the tips of a group of hysteresis loops generated by different strengths of H. Such a magnetization curve is shown in Fig. 1.

The process of magnetization and demagnetization of a ferromagnetic material in a symmetrically cyclic condition involves a storage and release of energy which is not completely reversible. As the material is magnetized during each half-cycle in the hysteresis loop, the amount of energy stored in the magnetic field exceeds that which is released upon demagnetization. The net energy density in joules per cubic meter is given by

$$W_h = K \oint H \, dB \, ,$$

where $K = 6333$ in the (rationalized) MKS system of units. Therefore the energy dissipated per cubic meter of material during one cycle is proportional to the area within the hysteresis loop.

In AC motors and transformers, we are interested in the power dissipated by the hysteresis losses in the iron. This power, P_h, can be obtained by multiplying the above W_h by the frequency, f, and the volume; that is,

$$P_h = (\text{Vol}) \, f \, W_h \, .$$

The basic hysteresis loop is obtained by slowly varying the excitation of a ferromagnetic core sample and plotting the flux density of the core versus the applied magnetizing force. However, because the ferromagnetic materials are used at power-line frequency, loops are often obtained with 60 Hz AC excitation. Strictly speaking the loops obtained are not true hysteresis loops, but are called flux-current loops or dynamic hysteresis loops. These differ from true hysteresis loops because the dynamic loop also includes the AC losses in the core, such as *Eddy-current losses*. The AC magnetization curve can be obtained from these dynamic hysteresis loops.

2.1. CORE EXCITATION FROM LOW SOURCE IMPEDANCE VOLTAGE SOURCE

It is essential that the relationship between the voltage applied to the winding on a magnetic core and the flux in the core and also the relationship between the flux in the core and the coil excitation current be understood. The familiar hysteresis or flux-current loop as shown in Fig. 1, provides a relationship between the flux density in a core and the magnetizing force. For other than very high frequency phenomena or pulse type excitation, we can say that for a given flux density B in the core, a unique, possibly multi-valued, magnetizing force H is required. This magnetizing force is obtained directly from the plot of flux density versus magnetizing force. For a given core and winding the flux density is proportional to core flux while the magnetizing force is proportional to the coil current. Thus, for a given flux in a core, the coil current must have a specific value as determined from the hysteresis loop. For a given flux, the coil current as determined from the flux-current loop is usually double valued. The ambiguity is resolved by noting whether the flux is increasing or decreasing and then selecting the appropriate current from the curve.

The relationship between core flux and applied voltage is not as direct. If a voltage e(t) is applied to a winding on a core as shown in Fig. 2, the coil voltage $e_c(t)$ will be equal to the applied voltage if the source impedance can be neglected. From Faraday's Law we obtain

$$e_c(t) = N \frac{\partial \phi}{\partial t} \, , \quad \text{or}$$

$$\Delta \phi = \frac{1}{N} \int_0^\tau e_c(t) \, dt \, ,$$

where $\Delta \phi$ is the change in flux from t=0 to $t = \tau$. The above equation can be interpreted as indicating that the flux change in the core is related to the time integral of the voltage on the core. The actual variation of the flux is obtained from the equation

$$\phi = \frac{1}{N} \int_0^t e_c(t)\, dt + K .$$

Here the integration constant K is evaluated to provide the initial value of ϕ at t=0. This is necessary when determining the steady state flux variation when the applied voltage is periodic.

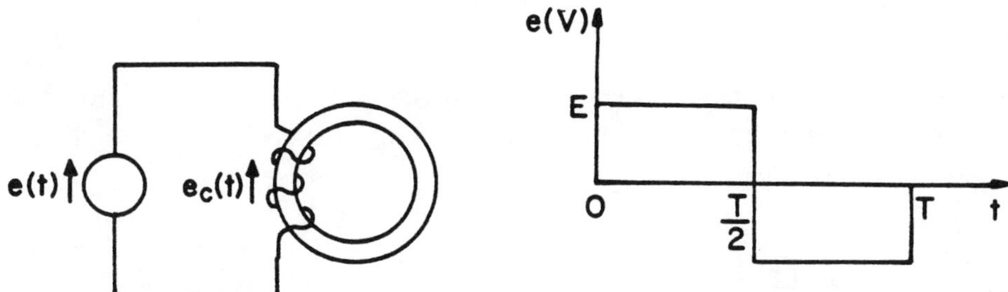

Fig. 2. Core Circuit.   Fig. 3. Square Wave Voltage.

Example 1: If the excitation e(t) in Fig. 2 is a square wave of frequency f (hertz) and amplitude E (volts), as shown in Fig. 3, then find the maximum value of flux density in the core.

The flux change during the positive half-cycle of voltage is given by

$$\Delta\phi = \frac{1}{N} \int_0^{\tau=1/2f} E\, dt = \frac{E}{2fN} .$$

During the next half-cycle the flux change is of equal magnitude but of opposite sign. If the steady state flux is assumed to be zero at t=0, then the flux variation will be as shown in Fig. 4. Such a flux variation with time would indicate that there is an average value component of flux. This is impossible in the steady state as there is no average value driving force to provide this component. Using the equation of ϕ and the integration constant K to eliminate the DC component, the steady state flux is as shown in Fig. 5. With such a flux variation, equations for $\Delta\phi$ and ϕ are satisfied with $\Delta\phi = 2\,\phi_m$ and the flux moving between ϕ_m and $-\phi_m$.

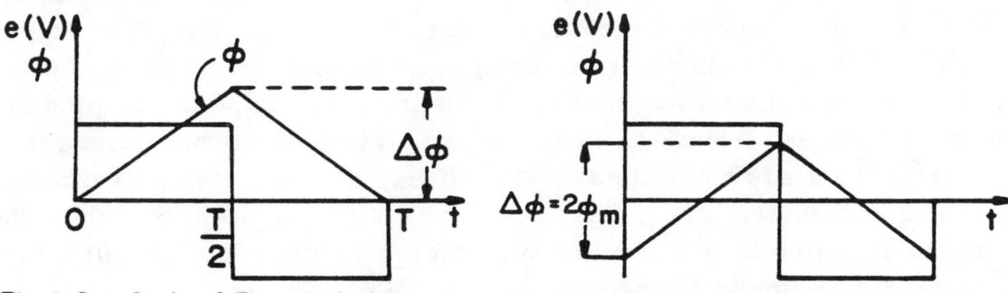

Fig. 4. One Cycle of Flux Variation.   Fig. 5. One Cycle of Flux Variation Steady State Conditions.

Substituting $\Delta\phi = 2\,\phi_m$ in $\Delta\phi = E\,/\,2fN$, yields

$$E = 4fN\,\phi_m .$$

This equation now provides the relationship between the core flux peak value and the applied voltage for a square wave excitation. Note that the voltage and flux are related by an integral, as given above.

Example 2: If the excitation e(t) in Fig. 2 is a sine wave of frequency f (hertz) and peak value E_p (volts), then find the maximum value of flux density in the core.

The relationship between a sinusoidal coil voltage and core flux is obtained in a manner similar to *Example 1*. Assume that the coil voltage is given by

$$e\,(t) = E_p \sin\omega t \; .$$

Then

$$\Delta\phi = \frac{1}{N} \int\limits_{0}^{\tau = \pi/\omega} E_p \sin\omega t \; dt = 2\,E_p \, / \, \omega N \; \cdot$$

Using a similar argument as for the square wave voltage source, the flux variation is a sinusoidal variation with a maximum variation between $+\,\phi_m$ and $-\,\phi_m$. Therefore, $\Delta\phi = 2\,\phi_m$; this yields

$$2\,\phi_m = \frac{2E_p}{\omega N} \Rightarrow \phi_m = \frac{E_p}{\omega N} \; .$$

Since $E_{RMS} = E_p \, / \, \sqrt{2}$, we obtain

$$E_{RMS} = 4.44 \; fN\phi_m \; .$$

It is important to summarize the results for the flux in a core and the applied voltage and for the flux in a core and the coil current. The instantaneous flux in a core is related to the applied voltage by an integral expression, while the instantaneous flux is related to the instantaneous current by the hysteresis or flux-current loop.

2.2. COIL CURRENT CHARACTERISTICS

Fig. 6 represents four possible idealized flux-current loops for four cores with different core materials but identical in physical dimension and coil winding. Thus, the flux density magnetizing force loops can be shown as flux-current loops and compared for the four cores since the core area and mean path length is the same for each of the cores. With a sinusoidal voltage source applied, the flux will be sinusoidal regardless of the core material. This is required to satisfy Faraday's Law. The coil current necessary to produce this sinusoidal flux is then obtained from the flux current loops as shown in Fig. 6.

Note that with a sinusoidal voltage applied to the coil, the flux will be sinusoidal no matter what the core material as long as the source impedance can be neglected. The coil current will then be the value determined from the flux-current loops. A core made of wood or the best core material would have exactly the same flux variation with the same sinusoidal voltage applied. The maximum flux value would be determined from E_{RMS} in the above. The coil current, however, would be vastly different as much larger currents would be required to establish this flux in a wood core with a relative permeability of approximately one. The wood core would probably invalidate the assumption of low source impedance in a practical case since such high currents could produce appreciable voltage drop in even small source or winding impedances.

Example 3: Consider a magnetic core connected in a circuit as shown in Fig. 7. The idealized core characteristics for this core are also shown in Fig. 7. The core material flux-current loops approximate that of 50% Iron-50% Nickel alloys such as Orthonol or Deltamax. The core is toroidal with the important dimensions given in Fig. 7. If the AC voltage applied is a sinusoidal voltage of 50 volts RMS at 60 hertz, the minimum number of turns required on the core to keep the flux below saturation can be computed using E_{RMS} and neglecting the source resistance R_S.

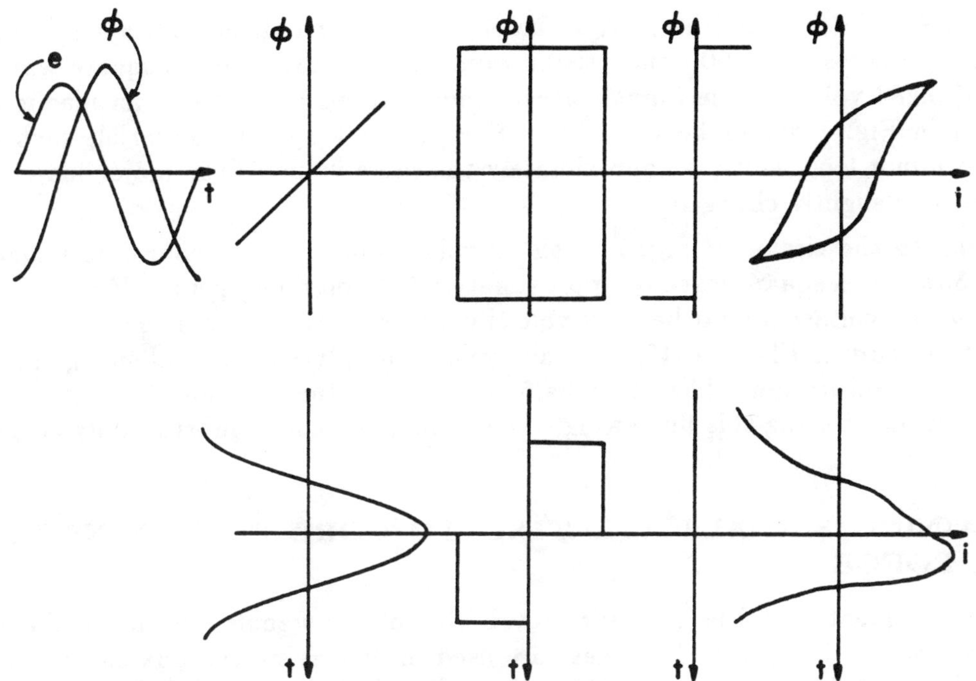

Fig.6. Exciting Current As A Function of Core Material, Four Different Cases.

$$N = \frac{E_{RMS}}{4.44 \, f \, \phi_m} = \frac{50}{(4.44)\,(60)\,(1.4)\,(0.69)\,10^{-4}} \simeq 1945 \text{ turns}.$$

With 1945 turns on the core, the exciting current is a square wave with a peak value of

$$I_p = \frac{H \, l_m}{N} = \frac{(25)\,(0.22)}{1945} = 2.83 \text{ mA} .$$

With a 2.83 mA peak exciting current, a source resistor $R_s = 100 \; \Omega$ will reduce the coil voltage by only 0.3 volts so its effects can probably be neglected. With $R_s = 1000 \; \Omega$, the voltage drop across the resistor will be on the order of 3 volts so it might possibly still be neglected. Thus, for a source resistance of approximately 100 ohms or less, the core can be considered as excited from a low impedance source. Note that this analysis is applicable only for this particular core, with this applied voltage, and with the indicated winding.

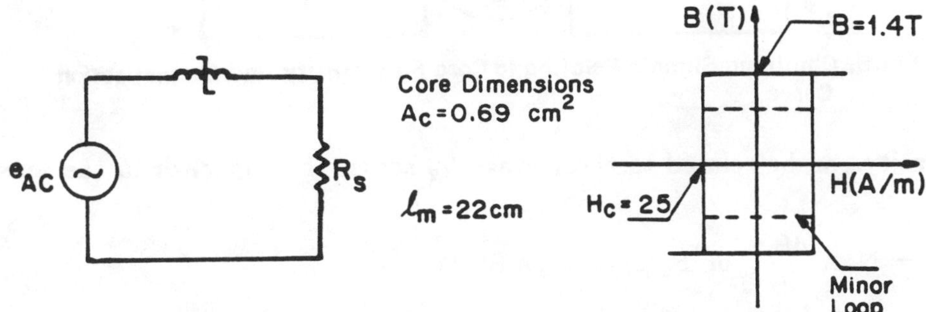

Fig. 7. Core Circuit and Idealized Core Characteristics.

With the 50 volts applied and the source resistance less than 100 ohms, substantially all the sinusoidal voltage is across the core and the core is said to absorb the voltage. This causes a sinusoidal flux variation with a consequent square wave of exciting current with a peak value of 2.83 mA. The core circuit acts as a constant current source since variation of source resistance from zero to 100 ohms has little effect on the peak value of exciting current and it stays at approximately 2.83 mA. The voltage across the 100 ohm resistor would be approximately a square wave voltage with a peak value of 0.283 volts. If the input voltage were reduced to 25 volts, an idealized minor loop as shown dotted in Fig. 7 would be traversed. For square loop core material, such as Deltamax or Orthonol, the minor loop width is approximately the same value as the major loop so that the exciting current is only slightly changed.

Returning to the circuit of Fig. 7, with 50 volts applied and a source resistance of 100 ohms, the output voltage is a square wave of approximately 0.3 volts amplitude. If the core were short circuited, the output voltage would be a 50 volt sinusoidal voltage across the source resistance. With the core in the circuit and N = 1945, one half-cycle of supply voltage will swing the flux from minus saturation to plus saturation. The other half-cycle swings the core flux from positive saturation to negative saturation. During this flux swing, the exciting current is substantially constant.

2.3. OBTAINING SIGNALS RELATING TO CORE FLUX DENSITY AND MAGNETIZING FORCE

In this experiment it will be necessary to obtain voltage signals relating to the core flux density B and the core magnetizing force H. These are used in displaying the flux-current loop for the individual core with an AC excitation. Fig. 8 shows such a circuit with coil being excited.

The magnetizing force H is obtained from the voltage drop across resistor R_1. If R_1 is made small so that $i_1 R_1 \ll e_1$, then the voltage e_1 will be sinusoidal and substantially equal to v_1. Then:

$$H = \frac{N_1 i_1}{l_m} \, , \quad V_H = i_1 R_1 \, , \quad \text{and} \quad H = \frac{N_1}{l_m R_1} V_H \, ,$$

where H = magnetizing force (A/m); l_m = core mean path length (m); N_1 = number of turns on the exciting coil; R_1 = sense resistor in exciting coil circuit (Ω); and V_H = voltage across sense resistor proportional to the magnetizing force (V).

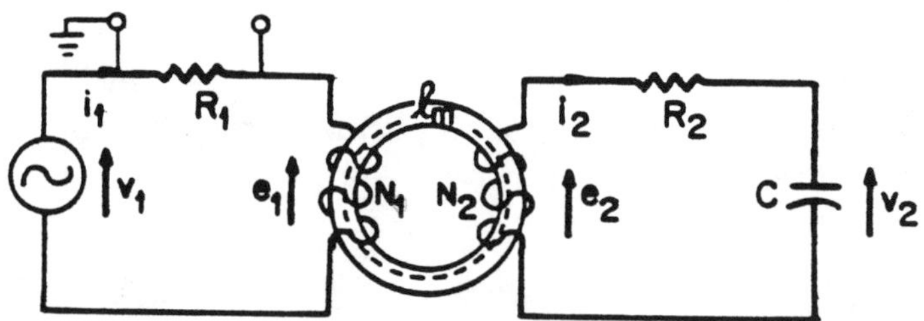

Fig. 8. Obtaining Signals Relating to Core Flux Density and Magnetization Curve.

The flux density can be related to the voltage V_2 across the capacitor in the sense circuit starting with

$$e_2 = \frac{d\lambda}{dt} = N_2 A \frac{dB}{dt} \, , \quad \text{or} \quad E_2(s) = [N_2 A B(s)] s \, .$$

Also $I_2(s) = E_2(s) / (R_2 + \frac{1}{sC})$, and $V_2(s) = I_2(s) / sC = \dfrac{[N_2 A B(s)] s}{R_2 C s + 1}$.

For s=jω, we have $V_2(j\omega) = [N_2 A B(j\omega)] j\omega / (jR_2 C\omega + 1)$. If $R_2 C\omega \gg 1$, then

$$V_2(j\omega) \simeq \frac{N_2 A}{R_2 C} B(j\omega) \, , \quad \text{or}$$

$$v_2(t) \simeq \frac{N_2 A}{R_2 C} B(t) \cdot$$

Here B = flux density in webers / m^2 (T); A = cross-sectional area of the core (m^2); C = capacitance (F); N_2 = sense coil turns; and R_2 = sense coil resistor (Ω).

The preceding two sets of equations provide us with the necessary calibration equations for the parameters of the circuit shown in Fig. 8.

2.4. APPLICATIONS TO MAGNETIC TAPE RECORDING

Although the emphasis in this experiment will be on the use of ferromagnetic materials for transformers and motors, the theory applies as well to the use of such materials as ferric oxides for magnetic recording. (Gamma ferric oxide ($Fe_2 O_3 \gamma$) is the oxide used in the manufacture of a large percentage of present tapes. The gamma distinguishes the ferromagnetic form from the nonferromagnetic alpha ferric oxide). We briefly examine this application in this section with an emphasis on audio recording.

Fundamentally, the process of magnetic recording is the magnetization of the magnetic particles in the tape. In the recording process, a current that is proportional to the incoming signal is supplied to the record head. This current passes through the head windings and produces a magnetizing force H. The magnetic tape is moved across the record head and the magnetizing force produces a magnetic flux in the tape. As the tape moves away from the record head, some of this flux remains and is called the remanent flux, ϕ_r. It is the remanent flux that stores the information in the magnetic tape.

As the recorded tape is moved across the reproduce head, the head senses the magnetic flux pattern in the tape and the remanent flux ϕ_r causes a voltage following Faraday's Law; namely,

$$e(t) = N\frac{d\phi_r}{dt} \cdot$$

In terms of frequency, this equation is

$$E(j\omega) = j\omega N \phi_r (j\omega) \cdot$$

If the frequency response of the record head were flat with frequency, then the voltage output from the reproducing head would not be constant but would increase at a 6 dB/octave (i.e., as ω) rate. In addition, there is a $90°$ phase shift between the recorded and the reproduced signal.

One function of the reproducing amplifiers is to weigh the voltage from the reproducing head with a frequency characteristic that decreases at a rate of 6 dB/octave (i.e., as $1/\omega$) with increasing frequency. However, in practice there are some deviations from the 6 dB/octave rate caused by such things as head construction, magnetic tape characteristics and the speed of tape movement across the heads. Corrections needed to straighten the reproduction frequency response to one comparable with the recording curve (plus the 6 dB/octave) for given tape characteristics and tape speed are known as *equalization*.

Various equalization standards can be used. The NAB (National Association of Broadcasters) standards used in the U.S. employ some low-frequency boost (i.e., increase in gain) and some high-frequency boost in recording. Some high-frequency boost is also used in the reproducing amplifiers. For best reproduction, these high-frequency equalization choices should match in the recording and reproducing amplifiers. Effects of mismatches, however, can be partially masked by appropriate settings of the tone controls in the reproduction process. For cassette-type audio tapes (which move at a speed of 1 7/8 ips -- inches per second), high-frequency boost is applied at frequencies either above a 120 μsec time constant or above a 70 μsec time constant. Tapes for which the recommended equalization is not given generally assume the 120 μsec choice. Note that if a tape were recorded with 70 μsec equalization and reproduced with 120 μsec equalization, then there will be too much high-frequency gain relative to the rest of the frequency components in the recorded signal. Tapes for reel-type audio recorders (which use tape speeds of 3 3/4, 7 1/2, 15 ips) generally use a 150 μsec time constant.

Next we give some attention to the details of the actual recording process. If the magnetic tape is assumed to be in an unmagnetized state and a gradually increasing magnetizing force H is applied, the value of the magnetic flux density B will follow a curve such as OP shown in Fig. 9, until the saturation flux density B_s is reached. If the magnetizing force is then reduced to zero, the flux density follows the path PQ and the tape will retain a residual flux density B_r if the magnetizing force goes to zero. In this manner, the magnetic tape is capable of storing information in the magnetic flux of the tape. The remanent flux ϕ_r is proportional to the residual flux density B_r and we have seen above that the reproducing signal is proportional to the rate of change of ϕ_r. The higher the value of B_r for a given B_s, the more efficient the tape is in storing the information.

As the magnetizing force H is cycled through symmetric positive and negative values, hysteresis loops are generated, such as those shown in dashed lines in Fig. 9. The ratio of the residual flux density to the saturation flux density, B_r / B_s, is called the "squareness factor". If $B_r / B_s = 1$, a square hysteresis loop will be generated, indicating that the tape was 100 percent efficient. Tapes commonly used in audio recording have squareness factors on the order of 0.70 to 0.75.

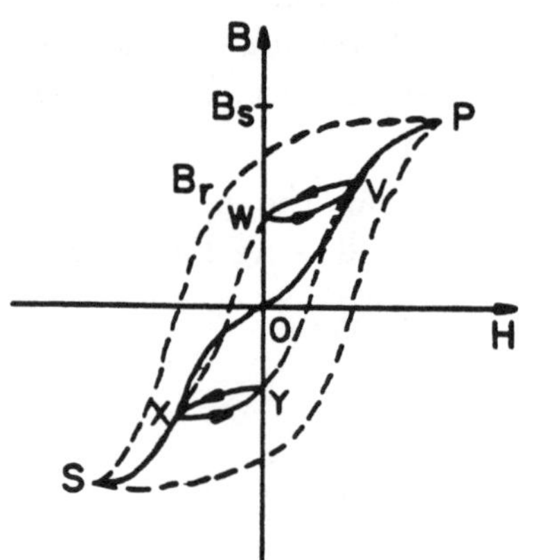

Fig.9. Hysteresis Loops for Magnetic Tape Recording.

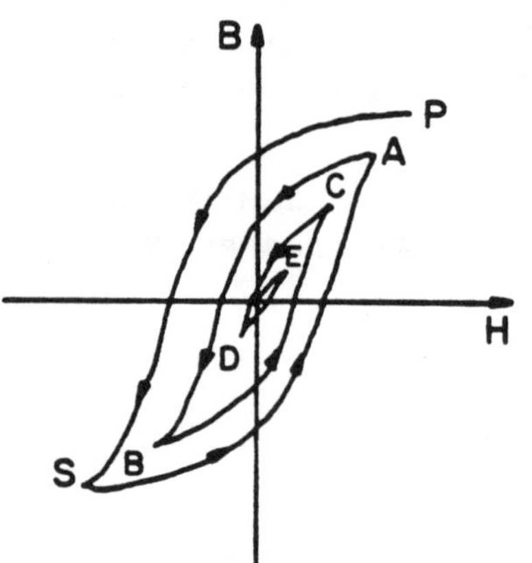

Fig.10. Hysteresis Loops for the Erase Process.

If the magnetizing force H is increased from zero to a value H_1 less than that required to saturate the magnetic tape then the magnetic flux density B will follow the curve OV in Fig. 9 for positive values of H and the curve OX for negative values. If the magnetizing force cycles between $-H_1$ and $+H_1$, the hysteresis loop XYVWX is formed. The reduction of the magnetizing force from $+H_1$ to zero and back to $+H_1$ will result in the minor hysteresis loop VWWV. Similarly, the minor hysteresis loop XYYX would be formed by cycling the magnetizing force between zero and $-H_1$.

To erase a recorded signal from magnetic tape, it is necessary to subject the tape to a slowly diminishing magnetizing force that is cycled symmetrically. The resulting hysteresis loops are shown in Fig. 10. The magnetizing force H is first cycled between the positive and negative saturation levels. Then the magnitude of the magnetizing force is slowly decreased so that the tape undergoes a series of successively smaller hysteresis loops, following the path PSABCDE, *etc*. Audio recorders generally use a special erase head supplied with a high-frequency current to provide a cyclic magnetizing force. The motion of the tape past the head provides the decreasing magnitude of the magnetizing force. Large 60 Hz electromagnets, called tape "degaussers," are used to provide a stronger magnetizing force to saturate the tape completely. These degaussers are moved over the tape and then withdrawn manually (with the electromagnet turned on) from the tape to accomplish the erasing process described above.

Because it is the residual flux density B_r that gives rise to the storage of information on the tape, it is instructive to plot B_r versus H. This is shown in Fig. 11 for a typical magnetic tape. Note that the curve is not linear. Analog recording techniques attempt to make use of the relatively linear segment between the instep and the knee of the B_r versus H curve. In contrast, digital recording techniques use only the saturated levels on the tape.

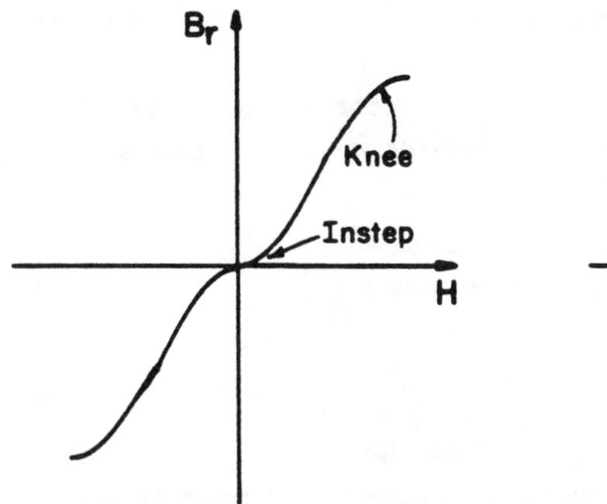

Fig.11. Plot of Residual Flux Density
Vs. Magnetizing Force for
Typical Magnetic Tape.

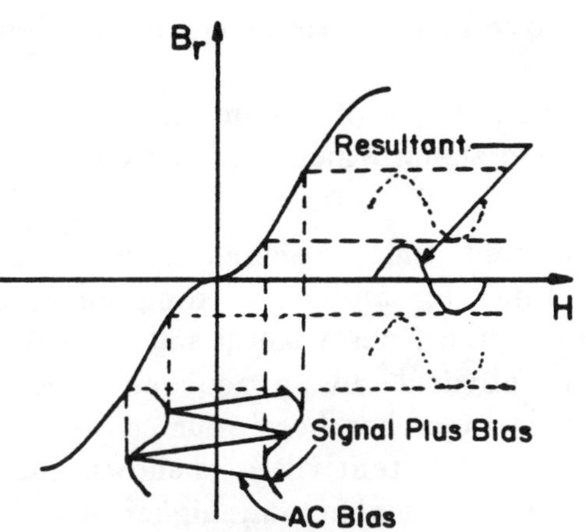

Fig.12. Magnetic Recording Using
An AC Bias.

For analog recording methods, it is necessary to add a bias to the record signal to place the signal excursions within the linear range of the B_r versus H curve. One method is to use a DC bias to accomplish this task. However, the ferric oxide on the tape is not continuous but is composed of many discrete particles, and these particles tend to polarize under the influence of the steady magnetic field resulting from the bias. The net result is that the noise level rises and so this method is not very satisfactory. In fact, the heads on good-quality tape machines should be "degaussed" occasionally to prevent any permanent magnetism from arising in the heads.

A better method of biasing is to use an AC bias, as illustrated in Fig. 12. In this method, a constant-amplitude high frequency sinusoidal waveform is added to the signal before recording on the tape. The amplitude of this waveform at the bias frequency must be sufficient to position the signal excursions within the linear range of the B_r versus H curve. This high-frequency sinusoidal waveform is removed again in the reproducing amplifiers by means of low-pass filtering. For best results, the bias frequency should be chosen to be three to five times the highest frequency to be recorded.

An optimum bias current must be selected to give good results in the recording process. Choice of this bias current should be the best balance between low distortion (too little or too much bias causes distortion), high output (the amount of magnetic induction is proportional, to a point, to the bias current), and extended high-frequency response (too large a bias current substantially reduces the high-frequency response of the tape). Different tape compositions require different amounts of bias current for optimum results. Choice of bias current is a concern only of the recording process, whereas equalization affects both recording and reproduction processes.

The majority of audio tapes that are produced currently for the consumer market are packaged in the popular audio cassette, based on the Phillips format. This cassette uses a tape width of 0.15 inches and a tape speed of 1 7/8 ips. Some typical properties of good-quality audio tape for consumer use are listed below.

coercivity: 250-500 oersteds

retentivity: 800-1300 gauss

erasure with 1,000-oersted field: $\geq$ 60 dB

distortion: $\leq$ 1% at standard record levels

print-through (transfer of a magnetic field from one layer to another within a roll of tape): $\leq$ -40 dB

signal-to-noise ratio: (ratio of maximum signal power to noise power within the passband) $\geq$ 55 dB

dynamic range (ratio of maximum signal which can be recorded at a given level of distortion to the minimum signal which can be recorded over a narrow frequency range): $\geq$ 56 dB

At the present time, research is being pushed on new metal-particle tapes for audio recording purposes. The metal particles are extracted from a metal salt by a reduction process with sodium borohydride in an aqueous solution. Interest in the metal-particle tapes has been spurred by their higher coercivity and remanence (retentivity), as shown in Fig. 13, and the increasing demand for higher and higher signal-to-noise ratios and dynamic range at slow tape speeds. From Fig. 13, it can be seen that the metal-particle tapes have about four times more energy than that of the best current ferric-oxide tapes. Drawbacks of the new metal-particle tapes are potential rust problems and the need for better tape-head designs. Bias and equalization standards are still in the development stage.

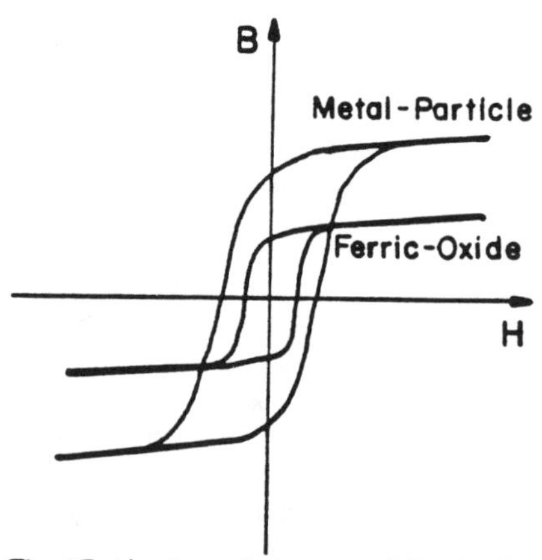

Fig. 13. Hysteresis Loops of Typical Metal-Particle and Good Ferric-Oxide Magnetic Tapes.

3. LABORATORY PROCEDURE

3.1. PRE-LABORATORY WORK

Carefully review the text of this experiment prior to coming to the laboratory.

3.2. LABORATORY WORK

3.2A. Flux-Current Loop

Set up the circuit shown in Fig. 14. *Always start from low voltage on the auto transformer;* your circuit should not require more than 30 to 40 volts. *Observe* indicated oscilloscope ground connections; *reverse* input leads on CH1 (or X channel) if necessary to obtain proper polarity in the display. For two different core materials (if provided for you), *display* the flux-current loop and *determine* the values for:

(a)   Coercivity

(b)   Retentivity

(c)   Saturation flux density

(d)   Voltage on the core when the core begins to saturate

(e)   AC magnetization curve.

Remove the 10 μF capacitor and *observe* the secondary voltage versus magnetizing force loops for your two cores. *Record* your observations. *Replace* the 10 μF capacitor. *Remove* the isolation transformer and the variable autotransformer from the circuit, and instead *substitute* a function generator. Be sure to *connect* the function generator (FG) ground to the CH1 (X-channel) ground; *set* output for a 30 Hz sinusoidal waveform (with zero DC offset). *Increase* the amplitude of the FG output until the flux-current loop just begins to show some saturation. Carefully *note* the display pattern so that it can be repeated. (If it is impossible to reach the beginning of saturation with full amplitude, use the pattern at full FG output amplitude at 30 Hz; repeatability is important). *Measure* the FG output voltage with an AC voltmeter. Next, *decrease* the FG output frequency in 2 Hz steps down to 2 Hz. At each frequency, *readjust* the FG output amplitude control as needed to return to the same point of onset of core saturation as was observed at 30 Hz. *Record* both the generator frequency and the AC voltmeter reading at each frequency. *Repeat* these measurements for your second core.

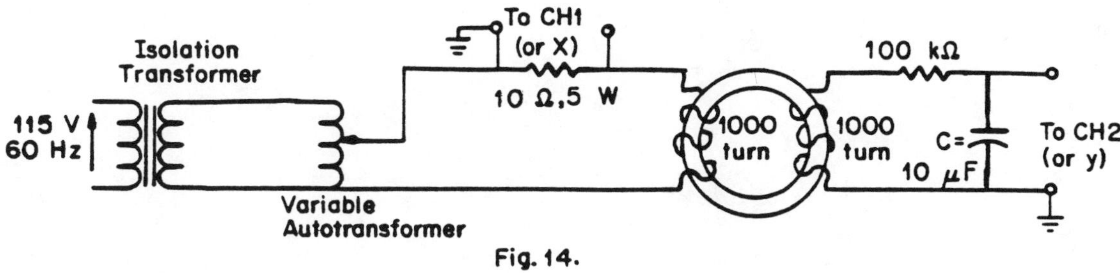

Fig. 14.

3.2B. Exciting Current

Set up the circuit shown in Fig. 15. *Observe* the relationship between the core voltage and the exciting current as the input voltage is varied. *Measure* several values below core saturation. *Record* the peak voltage across the exciting current sense resistor and the peak value of the core excitation voltage. *Repeat* for the second core.

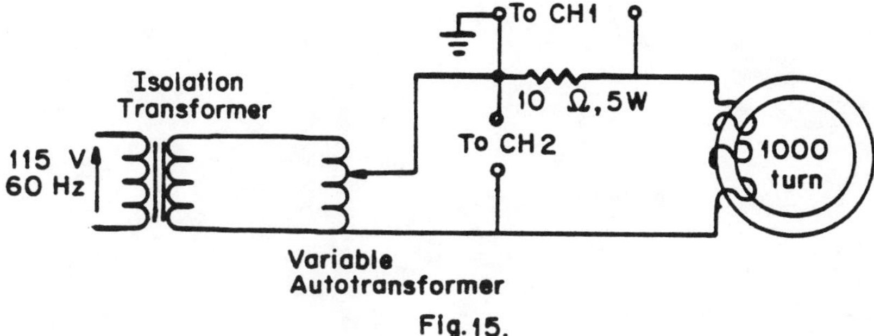

Fig. 15.

4. PROBLEM SET

1.   Explain the shape of the secondary voltage-magnetizing force loops.

2.   Why are the flux-current loops which you obtained at 60 Hz wider than those which would be obtained as DC hysteresis loops? Explain.

5. FOR YOUR REPORT

Document your data and observation in our standard report form. In this report also include the following.

1.   Compare the measured values obtained with those from the manufacturer's specification data sheets for the DC hysteresis loops of your core materials as far as:

     (a)   Coercivity

     (b)   Retentivity

     (c)   Saturation flux density

     (d)   Voltage on core when the core saturates.

2.   Plot an AC magnetization curve for each of your core materials.

3.   Plot the voltage required versus frequency to produce a given flux density (e.g., when the core just begins to saturate) for each of your two cores. Compare the shape of your plots with what you would expect from theory.

4.   Compare the peak measured exciting current with the calculated value for your two core materials at several points below saturation.

WORKSHEET

WORKSHEET

WORKSHEET

WORKSHEET

PART THREE

ACTIVE TWO-PORT NETWORKS WITH TRANSISTORS

PART THREE

ACTIVE TWO-PORT NETWORKS WITH TRANSISTORS

III-0.   INTRODUCTION

In part three, the fundamentals of two-port networks (in particular, active two-port networks with emphasis on transistors) are reviewed. The theory of two-port networks is equally applicable to circuits having *operational amplifiers* as will be discussed in the next part. The common *algebra* of three-terminal active devices is studied next, followed by a brief study of the physical structure of active three-terminal devices. It is not, however, our intention to get involved with the physical aspects of the devices in this textbook.

The table of contents of the remaining part of this review is as follows.

III-1.   REFERENCES
III-2.   FUNDAMENTALS OF TWO-PORT NETWORKS
III-2.0. DESCRIPTION OF TWO-PORT NETWORKS
III-2.1. BASIC ASPECTS OF FEEDBACK THEORY
III-2.2. GENERAL DESCRIPTION OF FEEDBACK AMPLIFIERS
III-3.   COMMON ALGEBRA OF THREE-TERMINAL DEVICES (TTD)
III-3.0. INTRODUCTION
III-3.1. ANALYTICAL APPROACH TO STUDY NONLINEAR ACTIVE TTD
III-3.2. GRAPHICAL APPROACH TO STUDY NONLINEAR TTD
III-4.   PHYSICAL STRUCTURE OF ACTIVE THREE-TERMINAL DEVICES: AN OVERVIEW
III-4.1. p-n JUNCTIONS
III-4.2. METAL-OXIDE-SEMICONDUCTOR (MOS) TRANSISTORS
III-4.3. BIPOLAR JUNCTION TRANSISTORS (BJT)
III-5.   MANUFACTURER'S SPECIFICATION DATA SHEETS

Since our main goal is to teach the relevant basic techniques as used in the electronic circuit design as well as to provide familiarization with some commonly and widely used three-terminal devices, we chose an npn transistor for the first four experiments of this part. Then we study circuits involving MOSFET and SCR in the last two experiments as though these are typical devices of their kinds. We hope this procedure avoids the confusion that some students may have at this early stage of their education as they confront several different devices. One source of confusion is how to remember the polarity of npn and/or pnp transistors -*per se*. This is why we are dealing only with npn transistors in order that students master the applications of this device in a wide range of design problems. Having learned this device thoroughly we then see no difficulty designing the counterpart of circuits used in this part with pnp transistors provided that proper polarity is observed. We must bear in mind that with the advent of the integrated circuit, transistors alone may soon become obsolete. Nevertheless we strongly believe that some elementary knowledge of this device, perhaps for the next several years, is absolutely essential.

The list of the experiments of this part is as follows.

EXPERIMENT III-1:   **BIPOLAR JUNCTION TRANSISTORS CHARACTERIS-TICS AND SMALL-SIGNAL MODEL**

EXPERIMENT III-2:   **BIPOLAR JUNCTION TRANSISTOR BIASING AND SMALL-SIGNAL MODEL PARAMETERS MEASURE-MENT**

EXPERIMENT III-3:   **BIPOLAR JUNCTION TRANSISTOR CAPACITIVE COUPLED LOAD AND MULTI-STAGE FEEDBACK AMPLIFIERS**

EXPERIMENT III-4:   **BIPOLAR JUNCTION TRANSISTORS LARGE-SIGNAL BEHAVIOR**

EXPERIMENT III-5:   **FIELD-EFFECT TRANSISTORS CHARACTERISTICS AND APPLICATIONS**

EXPERIMENT III-6:   **SILICON CONTROLLED RECTIFIERS [6]**

In future experiments we will be concentrating on advanced circuit design, in which several of the circuits that are studied in the above experiments are to be used simultaneously.

For further information on the contents of these experiments, please refer to the following references.

Please review page 41 of this textbook for common laboratory procedures.

III-1.   REFERENCES

[A] Textbooks and Articles:

[0]   S.E. Dewan and A. Straughan, Power Semiconductor Circuits. New York: John Wiley, 1975.

[1]   J.J. Ebers and J.L. Moll, "Large-signal behavior of junction transistors," Proc. IRE, vol. 42, pp. 1761-1772, Dec. 1954.

[2]   F.E. Gentry, F.W. Gutzwiller, N. Holonyak, and E.E. Van Zastrow, Principles and Applications of PNPN Devices. Englewood Cliffs, NJ : Prentice-Hall, 1964.

[3]   J. Millman, Microelectronics, digital and analog circuits and systems. New York : McGraw-Hill, 1979.

[4]   W. Shepherd, Thyristor Control of AC Circuits. London: Bradford University Press, 1975.

[5]   B.G. Streetman, Solid State Electronic Devices. Englewood Cliffs, NJ : Prentice-Hall, 1980.

[6]   F.G. Stremler, ECE 300 Laboratory Notes. Madison WI: The University of Wisconsin-Madison, 1978.

[7]   J.D. Wiley and J.E. Nordman, ECE 240-Electronic Devices: Lecture Notes. *ibid,* Dec. 1979.

[B] Manufacturers' Manuals:

[8]   General Electric Company, SCR Manual. Syracuse, NY : GE Comp., 5th Edition, 1972.

[9]   General Electric Company, Semiconductor Data Handbook. *ibid,* 1977.

[10]   J.W. Motto, Jr., Ed., Introduction to Solid-State Power Electronics. Youngwood, PA: Westinghouse Electric Corporation, 1977.

[11]   Motorola, Semiconductor Data Library (Discrete Products), vols. 1 and 2. Phoenix, AZ : Motorola Comp., 1974.

[12] Siliconix, Small Signal FET Design Catalog. Santa Clara, CA: Siliconix, 1983.

[13] Texas Instruments, Power Products Data Book. Dallas : Texas Instruments, 1985.

[14] Westinghouse Electric Corporation, SCR Designers' Handbook. Youngwood, PA: WEC, 1977.

III-2. FUNDAMENTALS OF TWO-PORT NETWORKS

III-2.0. DESCRIPTION OF TWO-PORT NETWORKS

A general description of a two-port network is that of Fig. 1. In this case four parameters are to be specified as compared to two in one-port networks. We call port aa' the input port and bb' the output port. In general, we expect the input and output variables to be related via some functional relationships. We may symbolically show this as follows.

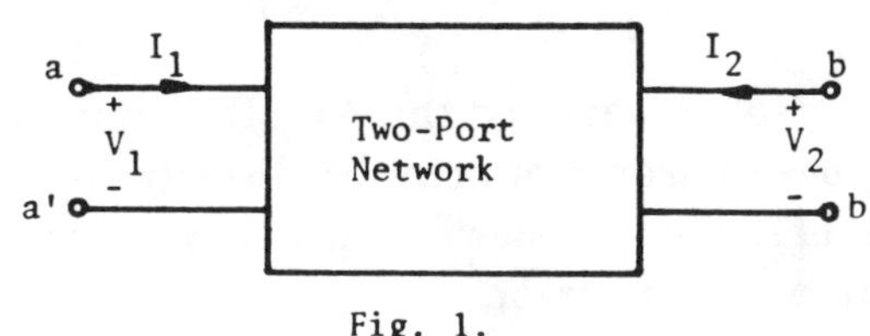

Fig. 1.

$$V_1 = f_1 (I_1 , I_2 , V_2) \ ,$$

$$I_1 = g_1 (V_1 , I_2 , V_2) \ ,$$

$$V_2 = f_2 (I_2 , I_1 , V_1) \ , \ \text{and}$$

$$I_2 = g_2 (V_2 , I_1 , V_1) \ .$$

Clearly there is a redundancy in this representation. For example, from $V_2 = f_2 (I_2 , I_1 , V_1)$ (or $I_2 = g_2 (V_2 , I_1 , V_1)$) and $V_1 = f_1 (I_1 , I_2 , V_2)$ we have $V_1 = f_1 [I_1 , I_2 , f_2 (I_2 , I_1 , V_1)]$ (or $V_1 = f_1 [I_1 , g_2 (V_2 , I_1 , V_1), V_2]$). Solving for V_1 results in

$$V_1 = V_1 (I_1 , I_2) \ \text{or} \ V_1 = V_1' (I_1 , V_2) \ , \ \text{and similarly}$$

$$I_1 = I_1 (V_1 , I_2) \ \text{or} \ I_1 = I_1' (V_1 , V_2) \ ;$$

$$V_2 = V_2 (I_2 , I_1) \ \text{or} \ V_2 = V_2' (I_2 , V_1) \ ;$$

$$I_2 = I_2 (V_2 , I_1) \ \text{or} \ I_2 = I_2' (V_2 , V_1) \ .$$

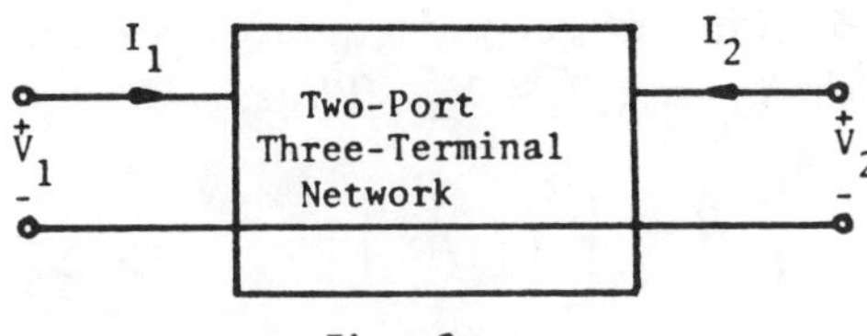

Fig. 2.

The above functional relationships are the most general representations of any two-port network. These are also valid for the two-port three-terminal networks of Fig. 2, as well.

In the case of linear two-port networks we have the linearized versions of these functionals. Depending on the choice of input and output representations of any given two-port network, we will have one of the following parameter sets.

Starting with V_1 and V_2 we get

$$V_1 = V_1 (I_1 , I_2) \triangleq (V_1 \text{ is a linear function of } I_1 , I_2) = \alpha I_1 + \beta I_2, \text{ and}$$

$$V_2 = V_2 (I_1 , I_2) \triangleq (V_2 \text{ is a linear function of } I_1 , I_2) = \alpha' I_1 + \beta' I_2.$$

Obviously the physical dimensions of these parameters $(\alpha , \alpha' , \beta , \beta')$ can be associated with impedances. Thus the natural name for these quantities is Z-parameters. Similarly, we may choose other combinations of these input and output quantities to get other sets of parameters (assuming linearity, of course). These parameter sets can be written as follows:

$$\begin{bmatrix} V_1 \\ V_2 \end{bmatrix} = \begin{pmatrix} Z_{11} & Z_{12} \\ Z_{21} & Z_{22} \end{pmatrix} \begin{bmatrix} I_1 \\ I_2 \end{bmatrix} , \ Z\text{-parameters};$$

$$\begin{bmatrix} V_1 \\ I_2 \end{bmatrix} = \begin{pmatrix} H_{11} & H_{12} \\ H_{21} & H_{22} \end{pmatrix} \begin{bmatrix} I_1 \\ V_2 \end{bmatrix} , \; H\text{--parameters};$$

$$\begin{bmatrix} I_1 \\ V_2 \end{bmatrix} = \begin{pmatrix} G_{11} & G_{12} \\ G_{21} & G_{22} \end{pmatrix} \begin{bmatrix} V_1 \\ I_2 \end{bmatrix} , \; G\text{--parameters};$$

$$\begin{bmatrix} I_1 \\ I_2 \end{bmatrix} = \begin{pmatrix} Y_{11} & Y_{12} \\ Y_{21} & Y_{22} \end{pmatrix} \begin{bmatrix} V_1 \\ V_2 \end{bmatrix} , \; Y\text{--parameters};$$

$$\begin{bmatrix} V_1 \\ I_1 \end{bmatrix} = \begin{pmatrix} A & B \\ C & D \end{pmatrix} \begin{bmatrix} V_2 \\ -I_2 \end{bmatrix} , \; ABCD\text{--parameters}.$$

Notice that for the ABCD-parameters the sign of I_2 has been changed. The physical dimensions of these parameters can be directly found from any given equation. These linearized models are also the first step or the first-order approximation for analyzing any general nonlinear two-port network such as a transistor.

Example 1: To show how these parameter sets are derived consider the circuit of Fig. 3. Derive both Z-parameters and H-parameters for this circuit.

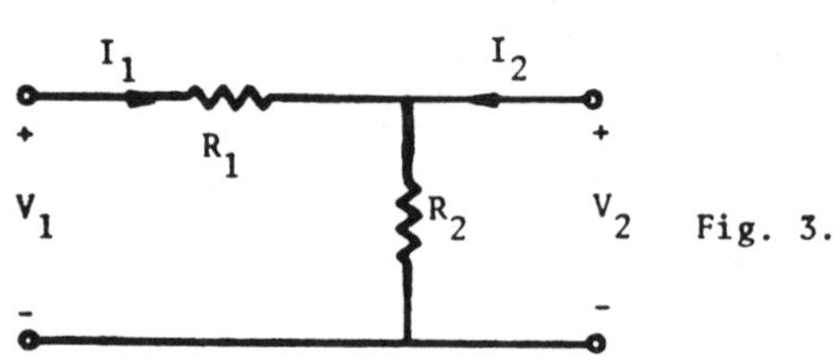

Fig. 3.

Solution: From KVL we have

$$V_1 = (R_1 + R_2)I_1 + R_2 I_2 \; ,$$

$$V_2 = R_2 I_1 + R_2 I_2 \quad .$$

Thus:

$$Z = \begin{pmatrix} R_1 + R_2 & R_2 \\ R_2 & R_2 \end{pmatrix}.$$

Also from the second equation we have $I_2 = -I_1 + V_2/R_2$. Substituting this I_2 into the first equation yields $V_1 = R_1 I_1 + V_2$. Thus:

$$H = \begin{pmatrix} R_1 & 1 \\ -1 & 1/R_2 \end{pmatrix}.$$

Problem 1:

Derive the G-parameters, the Y-parameters and the ABCD-parameters for the circuit of Fig. 3.

Problem 2:

Develop a table to relate the above five sets of parameters for a given general two-port network to each other; i.e., if the Z-parameters of a two-port network are given, then from this set of parameters find the H-parameters, the Y- parameters, *etc.*, of the same network and *vice-versa*.

An interesting aspect of this analysis is the fact that we may break any two-port network into several pieces of two-port networks that are each interconnected with themselves in some known ways. By knowing the appropriate parameters of these sub-networks and the ways these are interconnected we can express the overall network parameters almost by inspection. As we will see later, this *algebra* will make the feedback amplifier circuits much simpler to analyze than before. In the following we consider the interconnection of several two two-port networks and the corresponding overall parameters.

Suppose a two-port network N consists of two two-port networks N_1 and N_2 that are connected series to series as in Fig. 4. The implication of the ideal transformers is that the two two-port networks must be such that the current that gets into N_1 be exactly the same as that which gets into N_2. Similarly I_1 must equal I_2. If these conditions are met, then for N_1 we have:

$$\begin{bmatrix} V_1 \\ V_2 \end{bmatrix} = \begin{pmatrix} Z_{11} & Z_{12} \\ Z_{21} & Z_{22} \end{pmatrix} \begin{bmatrix} I_1 \\ I_2 \end{bmatrix} .$$

For N_2 we have:

$$\begin{bmatrix} V'_1 \\ V'_2 \end{bmatrix} = \begin{pmatrix} Z'_{11} & Z'_{12} \\ Z'_{21} & Z'_{22} \end{pmatrix} \begin{bmatrix} I'_1 \\ I'_2 \end{bmatrix} .$$

Since

$$\begin{bmatrix} V_{in} \\ V_{out} \end{bmatrix} = \begin{bmatrix} V_1 + V'_1 \\ V_2 + V'_2 \end{bmatrix} \ \& \ \begin{bmatrix} I_{in} \\ I_{out} \end{bmatrix} = \begin{bmatrix} I_1 \\ I_2 \end{bmatrix} = \begin{bmatrix} I'_1 \\ I'_2 \end{bmatrix} .$$

Thus the overall network parameters become:

$$\begin{bmatrix} V_{in} \\ V_{out} \end{bmatrix} = \begin{pmatrix} Z_{11} + Z'_{11} & Z_{12} + Z'_{12} \\ Z_{21} + Z'_{21} & Z_{22} + Z'_{22} \end{pmatrix} = \begin{bmatrix} I_{in} \\ I_{out} \end{bmatrix} .$$

We bear in mind that the above algebra is only true when we have exactly the indicated topology.

If the two networks N_1 and N_2 were interconnected parallel to parallel with valid port currents isolation (with similar reasoning as given above), we have the following overall Y-parameters for this two-port network N.

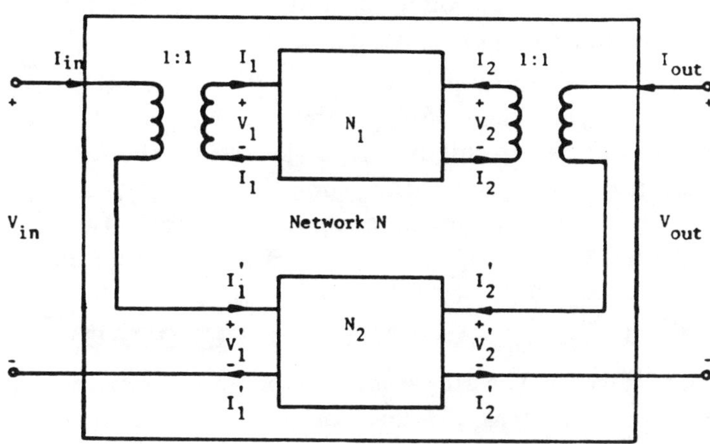

Fig. 4.

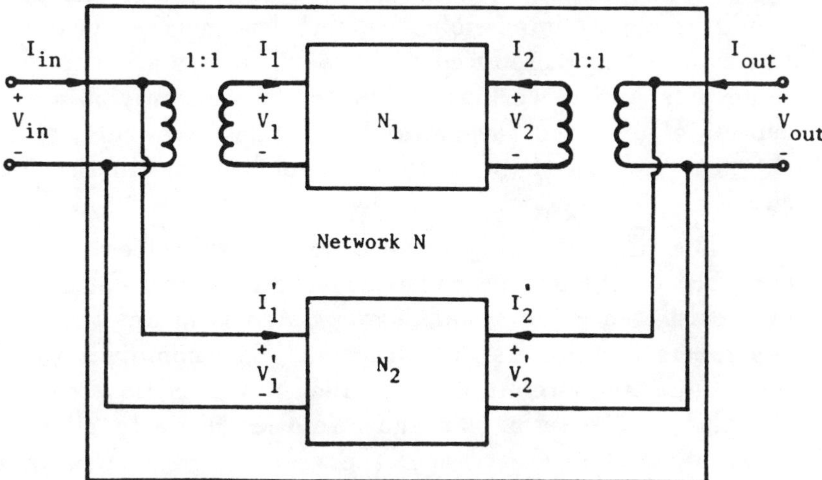

Fig. 5.

$$\begin{bmatrix} I_{in} \\ I_{out} \end{bmatrix} = \begin{pmatrix} Y_{11} + Y'_{11} & Y_{12} + Y'_{12} \\ Y_{21} + Y'_{21} & Y_{22} + Y'_{22} \end{pmatrix} \begin{bmatrix} V_{in} \\ V_{out} \end{bmatrix} .$$

The other three cases of interest occur when we have *hybrid* interconnections (series-parallel and parallel-series) and the cascaded interconnection under the same port currents isolation conditions. For series-parallel interconnections we use *H*-parameters, for parallel- series interconnection we use *G*-parameters, and for cascaded interconnection we use ABCD-parameters to describe the overall two-port network parameters.

These techniques to deal with interconnecting networks have many applications in *feedback* network amplifiers. Suppose circuit N_1 plays the role of the *feed-forward* or *open-loop* amplifier network and the circuit N_2 plays the role of the feedback network. Then using the above described circuit algebra with its corresponding conditions we can state the overall network parameters more simply than can be done using any conventional techniques. But before we express the relation between these two concepts let us review some basic aspects of feedback theory.

III-2.1. BASIC ASPECTS OF FEEDBACK THEORY

Perhaps the single most important contribution to engineering sciences-- in particular, electrical engineering-- was the introduction of the concept of feedback theory. The feedback or closed-loop system, as compared with the open-loop system, has many advantageous properties. To give a simple example of these open-loop and closed-loop situations, consider an oven and an inexperienced cook. Suppose this person sets the oven temperature selector at 400°F for about 20 minutes to prepare a dish. *Exactly 20 minutes later* this person opens the oven to see the result. On the other hand, an experienced cook who had tried this dish before may set the oven temperature at 300°F for about 12 minutes to get *the same* dish perfectly prepared. As one may guess the result of the first cook's efforts is not even close to satisfactory, let alone perfect. In both of the above situations no monitoring has been done by the cooks. Also there were no other mechanisms to monitor the development of the ultimate product until the job was completely done. The only difference between these two schemes is that in the second example some prior knowledge through the experience of the cook has been utilized. These are typical examples of the open-loop operations, i.e., the operations in which no mechanisms are used to monitor the development of the ultimate product. The result can either be satisfactory or otherwise but nothing can be done in between. The shortcoming of this situation to a large extent can be rectified by using the feedback or closed-loop system which is a mechanism that monitors the development of the ultimate product until it is completely finished. This closed-loop or feedback operation, of course, is much more involved than the open-loop operation. For example, developing a circuit to taste the preparation of a dish in an oven would be almost impossible. Nevertheless in many practical problems we are able to study the output as it is being produced and use this information to correct the system so that it remains within some prespecified working conditions. Based on this need we review some of the most elementary aspects of feedback theory in this section. We will see more applications of feedback amplifiers in the next part on the operational amplifiers and in the later parts on digital and analog integrated electronic circuits.

The schematic diagram of a closed-loop or feedback system is shown in Fig. 6. Here V_d is a disturbance to the system caused by the measurement devices and/or other plant uncertainties that may exist in a given system. By plant it is meant that part of dynamic system for which feedback controllers (or circuits) are designed. The combination of plant and feedback controllers and all interconnecting circuits is then called a dynamic system. In Fig. 6, the amplifier with gain $G_o(s)$ is the plant. The term plant uncertainties refers to all unknown parameters of the plant including those parameters that change (because of aging, environmental changes, *etc.*), with or without our knowledge. Furthermore, any time we insert a measuring device, say a voltmeter, into the system (or circuits) we inherently generate noise that will effect the ultimate response of our system. In the case of voltmeter, *per se*, that unit becomes a part of the overall network and therefore all the currents and/or voltages of the network are different than the case without this meter. Although these differences may not be noticeable they are there. Thus in Fig. 6, where we show an output by V_{out}, we mean we have inserted a voltmeter to read this output and as a result the output is *disturbed.* It is generally a common practice to model all such disturbances caused by measuring devices, plant uncertainties, *etc.* , with one external input; for example, V_d in Fig. 6. It is also a common practice to assume that this voltage is *added* to the system although in reality it may not be so.

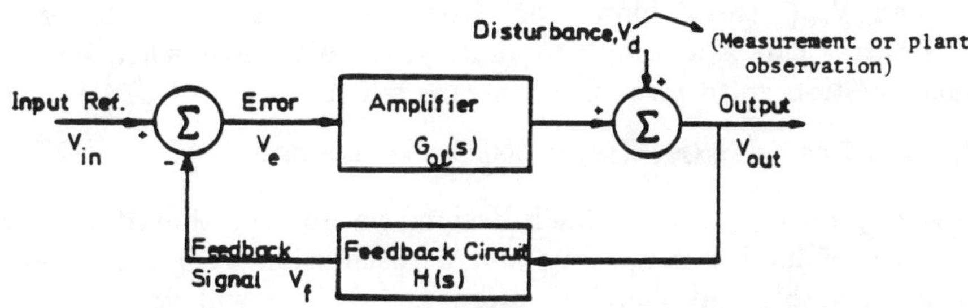

Fig. 6.   A Closed-Loop (or Feedback) System.

In this figure Σ is the comparator unit which simply compares the reference input signal V_{in} with the feedback signal V_f. Thus $V_e(s) = V_{in}(s) - V_f(s)$. This is called negative feedback. A simple signal analysis in the next comparator results in

$$V_{out}(s) = V_d(s) + G_{ol}(s)\, V_e(s) = V_d(s) + G_{ol}(s)\, [V_{in}(s) - V_f(s)].$$

Here $V_d(s)$ is the measurement and/or plant observation disturbance. Each signal that enters an *amplifier* circuit produces a signal that is proportional to the *gain* (or transfer function) of this amplifier as well as the *value* of the entering signal. The convention has always been to premultiply the signal value of this gain. Similarly, $V_f(s) = H(s)\, V_{out}(s)$, and if we substitute this $V_f(s)$ into the expression for $V_{out}(s)$ and rearrange the terms we have

$$V_{out}(s) = \frac{G_{ol}(s)}{1 + G_{ol}(s)H(s)}\, V_{in}(s) + \frac{1}{1 + G_{ol}(s)H(s)}\, V_d(s).$$

For now, assume that there is no disturbance in our system; that is , $V_d(s) = 0$. This choice will result in

$$V_{out}(s) = \frac{G_{ol}(s)}{1 + G_{ol}(s)H(s)}\, V_{in}(s) \triangleq G_{cl}(s)V_{in}(s) \ ,$$

where $G_{cl}(s)$ can be considered as a gain factor and correspondingly we call that the "closed-loop" gain or transfer function. $G_{ol}(s)$ and $H(s)$ can also be just constants.

From this expression we can conclude that if $G_{ol} \to \infty$, then $G_{cl} \to 1/H(s)$ which, in general, is finite. This result immediately tells us one of the basic advantages of the feedback configuration over that of an open-loop configuration; i.e., the feedback tends to stabilize the system at a cost of reducing its output amplitude. The term $G_{ol}(s)H(s)$ is called the *loop gain* and the term $[1 + G_{ol}(s)H(s)]$ is called the *return difference.*

Also recall Bode's sensitivity function which says if $x = f(y)$, then S_y^x (sensitivity of x with respect to changes in y) $\triangleq \dfrac{dx/x}{dy/y} = (\dfrac{dx}{dy})/(\dfrac{x}{y})$. Thus from G_{cl} we have $S_{G_{ol}}^{G_{cl}} = \dfrac{1}{1 + G_{ol}(s)H(s)}$. This tells us that if we want the output not to be susceptible to changes in the open-loop gain, $G_{ol}(s)$, we must increase the return difference. This is why the term return difference has also been called desensitivity factor. Also, we notice that if we increase the return difference then the effect of $V_d(s)$ in the output will decrease, since it is multiplied by the inverse of this quantity.

If, instead of negative feedback, we had positive feedback (which is mostly the case in electronic circuits, i.e., $V_e(s) = V_{in}(s) + V_f(s)$) , then

$$V_{out}(s) = \frac{G_o(s)}{1 - G_o(s)H(s)} \, V_{in}(s) + \frac{1}{1 - G_o(s)H(s)} \, V_d(s) \; .$$

Intuitively, and because of a minus sign in the denominator of the above expression, we guess that if $1 - G_o(s)H(s) \to 0$, then $V_{out}(s)$ may blow up. Now, the question is under what conditions will $1 - G_o(s)H(s) \to 0$? Because this is a complex quantity, the answer is when both the real and the imaginary parts of this equation go to zero. That means when

$$|G_o(j\omega)H(j\omega)| = 1 \quad \text{and} \quad \angle G_o(j\omega)H(j\omega) = 360^\circ \, k, \quad k=1,2,....$$

Therefore, as another property of a feedback design, we observe that if we are using positive feedback there is a chance of having instability and/or oscillation. On the other hand, negative feedback tends to be more stable and therefore more useful. We will do some design study regarding oscillator circuits in EXP. VI-3.

Unfortunately, given an electronic circuit to analyze, it is not always easy to recognize its open-loop amplifier circuit or its feedback circuit. Nevertheless, if one can recognize these units and separate them, then the above analysis can be easily used to discuss stability and/or any other pertinent question regarding the overall behavior of the circuit if the ports isolation property holds.

III-2.2. GENERAL DESCRIPTION OF FEEDBACK AMPLIFIERS

We now close this section with a general configuration of a two-port electrical network. Subsequently we will study this general representation in conjunction with a feedback amplifier network to obtain an *approximate* expression for the overall transfer function of this type of amplifier. The following derivations are closely gathered from the lecture notes of ECE 422 by Professor D.P. Brown at the University of Wisconsin-Madison in the Fall of 1972.

Consider the following two-port network with an input source (current θ_1 in parallel with source impedance Γ_s or voltage ϕ_1 in series with source impedance Γ_s) and an output load of Γ_l with its corresponding θ_2 or ϕ_2, Fig. 7.

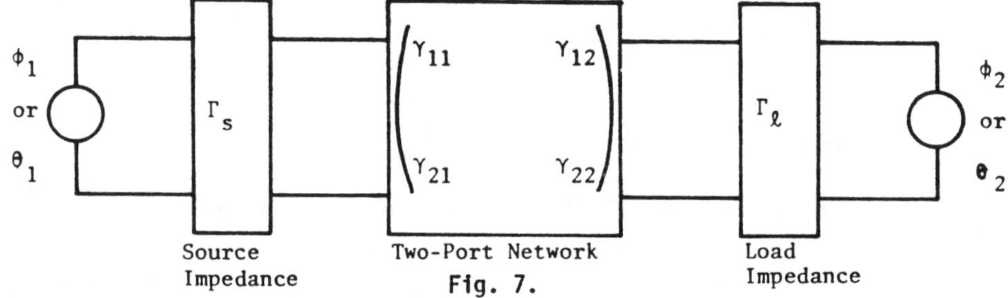

Fig. 7.

The γ_{ij}'s are the appropriate set of two-port network parameters. If the γ_{ij}'s are the Z-parameters, then both Γ_s and Γ_l are impedances and if γ_{ij}'s are the Y-parameters, then Γ_s and Γ_l are admittances, *etc.*

For the circuit of Fig. 7, we can easily write

$$\begin{bmatrix} \phi_1 \\ \phi_2 \end{bmatrix} = \begin{pmatrix} \Gamma_s + \gamma_{11} & \gamma_{12} \\ \gamma_{21} & \Gamma_l + \gamma_{22} \end{pmatrix} \begin{bmatrix} \theta_1 \\ \theta_2 \end{bmatrix} \; .$$

Suppose we want to compute $A_{11} \triangleq (\theta_1/\phi_1)$ (given $\phi_2 = 0$). Then from $\phi_2 = 0$ and the second equation in the above we have $\gamma_{21}\theta_1 + (\Gamma_l + \gamma_{22})\theta_2 = 0$. Solving for θ_2 and substituting the result in the first equation gives

$$A_{11} = (\Gamma_l + \gamma_{22})/[(\Gamma_s + \gamma_{11})(\Gamma_l + \gamma_{22}) - \gamma_{12}\gamma_{21}] \; .$$

Likewise we can compute $A_{21} \triangleq (\theta_2/\phi_1)$ (given $\phi_2 = 0$). This result is

$$A_{21} = -\gamma_{21}/[(\Gamma_s + \gamma_{11})(\Gamma_l + \gamma_{22}) - \gamma_{12}\gamma_{21}].$$

In general we can show that

$$\begin{cases} T_{in} = \left.\dfrac{\phi_1}{\theta_1}\right|_{\phi_2=0} = \Gamma_s + \gamma_{11} - \dfrac{\gamma_{12}\gamma_{21}}{\Gamma_l + \gamma_{22}} \;, \text{ and} \\[2em] T_{out} = \left.\dfrac{\phi_2}{\theta_2}\right|_{\phi_1=0} = \Gamma_l + \gamma_{22} - \dfrac{\gamma_{12}\gamma_{21}}{\Gamma_s + \gamma_{11}} \;. \end{cases}$$

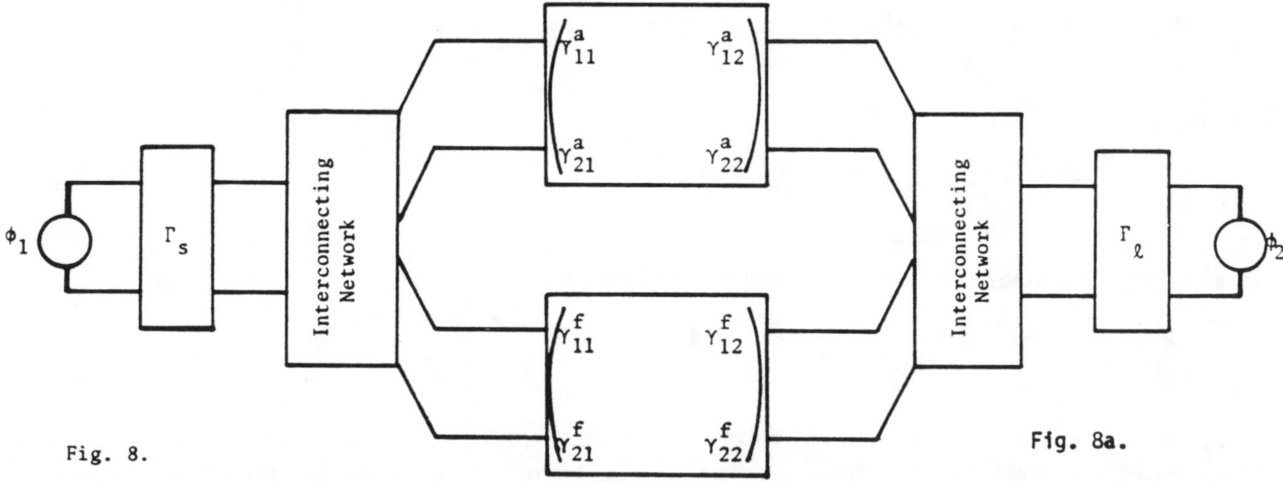

Fig. 8.

Fig. 8a.

In the terminology of feedback theory we can have a case such as Fig. 8a, which is the schematic circuit version of a general positive feedback design shown in Fig. 8b. In fact, in most circuit feedback amplifiers we have a positive feedback situation. Also, in the terminology of two-port networks, the interconnecting networks just tell us how the two networks of the feedforward amplifier [a(s) of the Fig. 8a] and the feedback amplifier [f(s) of the Fig. 8b] are connected to each other. Based on this interconnecting information and if the γ_{ij}' s are chosen correctly we can have the following overall parameters for this network:

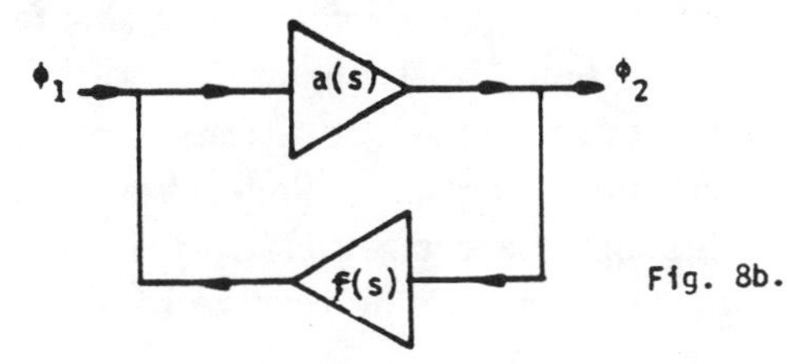

Fig. 8b.

$$\begin{pmatrix} \gamma_{11}^T & \gamma_{12}^T \\ \gamma_{21}^T & \gamma_{22}^T \end{pmatrix} \triangleq \begin{pmatrix} \Gamma_s + \gamma_{11}^a + \gamma_{11}^f & \gamma_{12}^a + \gamma_{12}^f \\ \gamma_{21}^a + \gamma_{21}^f & \Gamma_l + \gamma_{22}^a + \gamma_{22}^f \end{pmatrix}$$

$$\triangleq \begin{pmatrix} \Gamma_s + \gamma_{11}^T & \gamma_{21}^T \\ \gamma_{21}^T & \Gamma_l + \gamma_{22}^T \end{pmatrix} \;.$$

Note that the above algebra is correct if and only if we can separate various sub-networks such that the port currents isolation properties remain valid.

Under the above conditions we may present some practical considerations to simplify the overall parameters of this two-port network to yield some practical equations as follows.

(a)   In the ideal case no reverse transmission of the feedforward amplifier is expected, i.e., $\gamma_{12}^a \simeq 0$ or $\gamma_{12}^a \ll \gamma_{12}^f$.

(b)   Similarly, we expect $\gamma_{21}^a \gg \gamma_{21}^f$.

We also notice that

$$\frac{\phi_2}{\phi_1} = \frac{\Gamma_l\,\theta_2}{\phi_1} = \frac{-\Gamma_l\,\gamma_{21}^{\mathrm{T}}}{(\Gamma_s + \gamma_{11}^{\mathrm{T}})(\Gamma_l + \gamma_{22}^{\mathrm{T}}) - \gamma_{12}^{\mathrm{T}}\gamma_{21}^{\mathrm{T}}}\ .$$

From Fig. 8b, (and for positive feedback), we have

$$\frac{\phi_2}{\phi_1} = \frac{a(s)}{1 - a(s)f(s)}.$$

From the above two expressions we get

$$\frac{\phi_2}{\phi_1} = \frac{-\Gamma_l\,\gamma_{21}^{\mathrm{T}}/(\Gamma_s + \gamma_{11}^{\mathrm{T}})\,(\Gamma_l + \gamma_{22}^{\mathrm{T}})}{1 - \gamma_{12}^{\mathrm{T}}\gamma_{21}^{\mathrm{T}}/(\Gamma_s + \gamma_{11}^{\mathrm{T}})\,(\Gamma_l + \gamma_{22}^{\mathrm{T}})} \triangleq \frac{a(s)}{1 - a(s)f(s)}.$$

These relations result in

$$\begin{cases} a(s) \triangleq -\Gamma_l\,\gamma_{21}^{\mathrm{T}}/(\Gamma_s + \gamma_{11}^{\mathrm{T}})\,(\Gamma_l + \gamma_{22}^{\mathrm{T}})\ , \text{ and} \\ f(s) \triangleq -\gamma_{12}^{\mathrm{T}}/\Gamma_l\ . \end{cases}$$

Using the approximations stated in (a) and (b) yields

$$\begin{cases} a(s) = -\Gamma_l\,\gamma_{21}^{\mathrm{a}}/(\Gamma_s + \gamma_{11}^{\mathrm{T}})\,(\Gamma_l + \gamma_{22}^{\mathrm{T}})\ , \text{ and} \\ f(s) = -\gamma_{12}^{\mathrm{f}}/\Gamma_l\ . \end{cases}$$

The implication of this algebra is that we must divide the feedback problem into two separate sub-networks of feedforward and feedback circuits. Then we must calculate the open-loop transfer function of each block provided that the loading effects of the feedback network in both input and output ports of the feedforward network are appropriately incorporated (recall that $\gamma_{11}^{\mathrm{T}} = \gamma_{11}^{\mathrm{a}} + \gamma_{11}^{\mathrm{f}}$ and $\gamma_{22}^{\mathrm{T}} = \gamma_{22}^{\mathrm{a}} + \gamma_{22}^{\mathrm{f}}$). But, of course, the feedback network is not loaded by $a(s)$. Calculating both $a(s)$ and $f(s)$, then the transfer function between input and output becomes $\phi_2/\phi_1 = a(s)/[1-a(s)f(s)]$. In reality though these conditions are seldom met, what we are calculating is an approximation to the true transfer function. Nevertheless, the above method is much easier to use than the true exhaustive and complete circuit analysis of the overall network. A few subtle points regarding the above techniques must yet be resolved but we do not elaborate further.

III-3. COMMON ALGEBRA OF THREE-TERMINAL DEVICES (TTD)

III-3.0. INTRODUCTION

Three-terminal devices (TTD's) in general are non-linear elements in the sense that their corresponding port relations are nonlinear functions of their arguments. A typical representation of this device is given in Fig. 9.

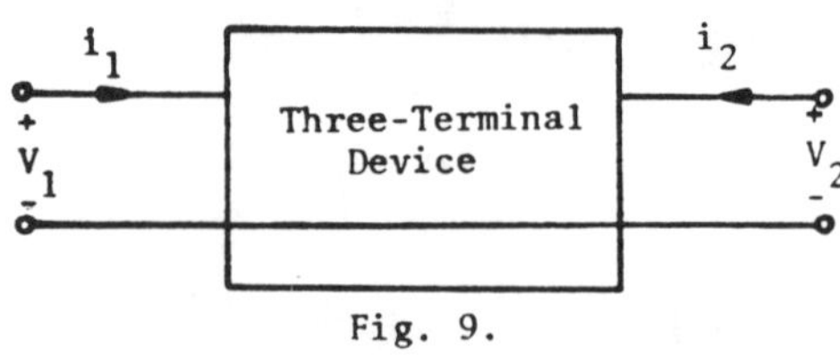

Fig. 9.

Here we have:

$$v_1 = v_1(i_1\,,\,i_2) \text{ or } v_1 = v_1'(i_1\,,\,v_2)\ ;$$

$$i_1 = i_1(v_1\,,\,i_2) \text{ or } i_1 = i_1'(v_1\,,\,v_2)\ ;$$

$$v_2 = v_2(i_2\,,\,i_1) \text{ or } v_2 = v_2'(i_2\,,\,v_1)\ ;$$

$$i_2 = i_2(v_2, i_1) \text{ or } i_1 = i_2'(v_2, v_1) \ .$$

Wherein all the above functionals are nonlinear with respect to their arguments. To analyze these devices, irrespective of their physical structure which will be partially reviewed later in this part, we must first choose one of the above models. After this choice is being made we have two different approaches to complete the analysis. These approaches are the *analytical* and the *graphical*.

To explain the above two methods we choose the following input and output relations which appear in the form of two nonlinear equations.

$$v_1(t) = v_1'[i_1(t), v_2(t)] \ , \quad \text{and}$$

$$i_2(t) = i_2[v_2(t), i_1(t)] \ .$$

The techniques we present in the following, with slight modifications, are applicable to a large body of relevant problems. We bear in mind that an engineering student may easily generalize this *algebra* to any particular case of future interest.

As is evident in the above model the $v_2(t)$ and $i_1(t)$ play the role of parameters for the input voltage equation and the output current equation, respectively. How do we analyze circuits involving these elements? Consider a general network N consisting of one TTD and three general subnetworks N_1, N_2 and N_3 as depicted in Fig. 10.

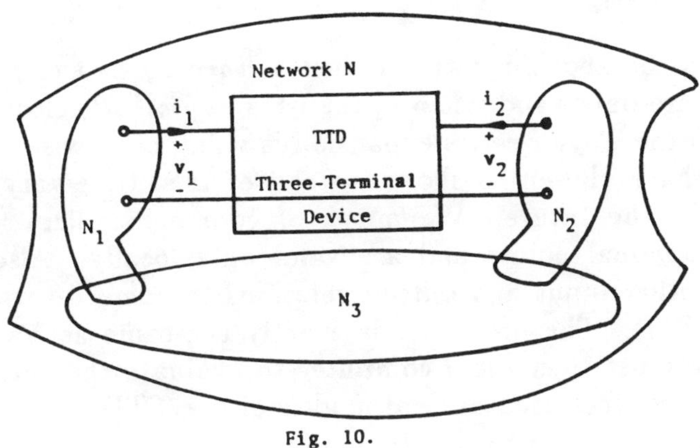

Fig. 10.

Our goal is to be able somehow to account for the effects of the TTD in network N. If we know the general nonlinear relations of, say, $v_1(t)$ and $i_2(t)$ of the device and the corresponding relations between (i_1, v_1) and (i_2, v_2) from the network N_1 and N_2, respectively, then we may be able to analyze the network N with computer help. But, in reality, we do not have definite analytical expressions for these equations. What we have at best are some nonlinear curves that are drawn in various coordinate systems depending on the nature of the experiments and the parameters involved. In the following we present two general approaches to handle the analysis of network N of Fig. 10 with an active TTD.

III-3.1. ANALYTICAL APPROACH TO STUDY NONLINEAR ACTIVE TTD

First we recall that an active device is one that requires external power to bring it into the active operational mode. This has not been the case for passive elements such as RLC's. Since we have already assumed that the active TTD is a non-linear element and we do not have the exact analytical expressions of either $v_1(t)$ or $i_2(t)$, we then consider perturbational analysis to look into this

- 194 -

problem. Our main *tool* for this analysis is, of course, the Taylor series expansion approach. We expand each of these input and output nonlinear equations around an *operating* point to derive a *local* or *small-signal* model of the TTD. The procedure is as follows (we omit t for convenience). Recall that v_1 (or i_2) is a multivariable function which can be expanded as

$$v_1 = \overline{v_1} + \left.\frac{\partial v_1}{\partial i_1}\right|_* \Delta i_1 + \left.\frac{\partial v_1}{\partial v_2}\right|_* \Delta v_2 + ..., \text{ and}$$

$$i_2 = \overline{i_2} + \left.\frac{\partial i_2}{\partial v_2}\right|_* \Delta v_2 + \left.\frac{\partial i_2}{\partial i_1}\right|_* \Delta i_1 + \cdots .$$

Here "-" stands for the *nominal* or the operating value, i.e., the value of a function at the point for which we are trying to find its Taylor series expansion, and $|_*$ means evaluated at this operating point. The ... means higher-order differentiation terms of v_1 and i_2 multiplied by higher-order terms of Δi_1, Δv_2 and $\Delta i_1 \Delta v_2$ as well as constants. *If* we assume that these Δi_1 and Δv_2 are *sufficiently* small, i.e., we can neglect Δi_1^2, Δv_2^2 and $\Delta i_1 \Delta v_2$, then we are left with the above three terms in each expression resulting in the incremental variations model of the given TTD as follows.

$$v_1 - \overline{v_1} \triangleq \Delta v_1 = \left.\frac{\partial v_1}{\partial i_1}\right|_* \Delta i_1 + \left.\frac{\partial v_1}{\partial v_2}\right|_* \Delta v_2 , \text{ and}$$

$$i_2 - \overline{i_2} \triangleq \Delta i_2 = \left.\frac{\partial i_2}{\partial v_2}\right|_* \Delta v_2 + \left.\frac{\partial i_2}{\partial i_1}\right|_* \Delta i_1 .$$

The implication of Δ in the above equation is just a *warning* sign to remind us that the right-hand side of these equations are only valid when v_1 (or i_2) is *sufficiently* close to v_1 (or i_2). As long as this requirement holds, then the Taylor series expansion is valid, otherwise this expansion loses its usefulness. Some textbooks have chosen to show, instead of Δ's, time-varying quantities. We may use this notation as well in the future. We must, of course, be alert that these quantities are all evaluated around the nominal points and are valid only locally. Also notice that we could have started with a host of other input and output relations to come up with several equally valid local representations of the TTD. The procedure is exactly the same as was done above. Having determined a local model we must then use a computer to evaluate the currents and the voltages of the network N for each of these incremental local models of the TTD.

The next question that may be raised is, what are the implications of these various terms in the above equations? Starting with the input equation and considering the physical dimensions of various terms, we propose to call $r_n \triangleq \left.\frac{\partial v_1}{\partial i_1}\right|_*$ and $\mu \triangleq \left.\frac{\partial v_1}{\partial i_1}\right|_*$ (reverse amplification). Similarly, $\frac{1}{r_o} \triangleq \left.\frac{\partial i_2}{\partial v_2}\right|_*$ and $\beta = \left.\frac{\partial i_2}{\partial i_1}\right|_*$

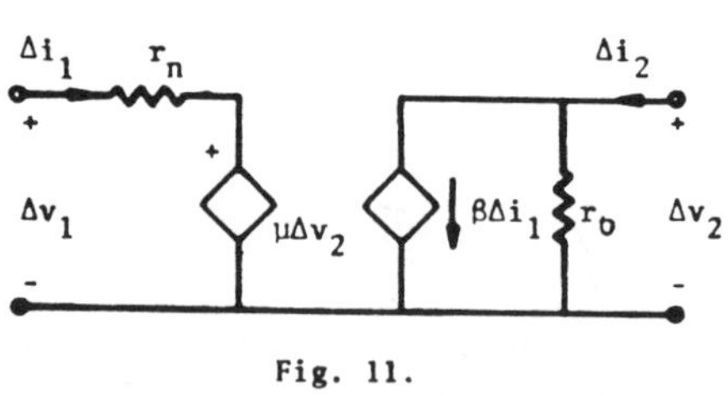

Fig. 11.

(current gain). Putting together these terms results in an equivalent small-signal model for the TTD as depicted by the circuit of Fig. 11. Substituting this equivalent model for the TTD in the network N of Fig. 10, (provided that all the biasing voltage sources associated with this TTD are shorted--the current sources opened), we can solve for various currents and voltages of the elements of this network. The result is, of course, local; i.e., it depends on the operating point for which this model is developed. If we want to have a result for a range wider than what we get from this model then we have to use a new operating point which is the old operating point plus or minus the maximum of Δi_2 or Δv_1 and repeat, generating another small-signal model. These are standard techniques given for the analysis of nonlinear networks.

III-3.2. GRAPHICAL APPROACH TO STUDY NONLINEAR TTD

Another common technique to analyze circuits involving nonlinear TTD's is the graphical approach. The procedure is rather simple and we describe its principal features as follows. Consider network N of Fig. 10. Suppose instead of analytical data regarding the input voltage $v_1(t)$ we had the graph of Fig. 12, and instead of the output current $i_2(t)$ we had the graph of Fig. 13. These are the TTD characteristic curves. We can now use the following techniques to find the operating current and voltage values of network N.

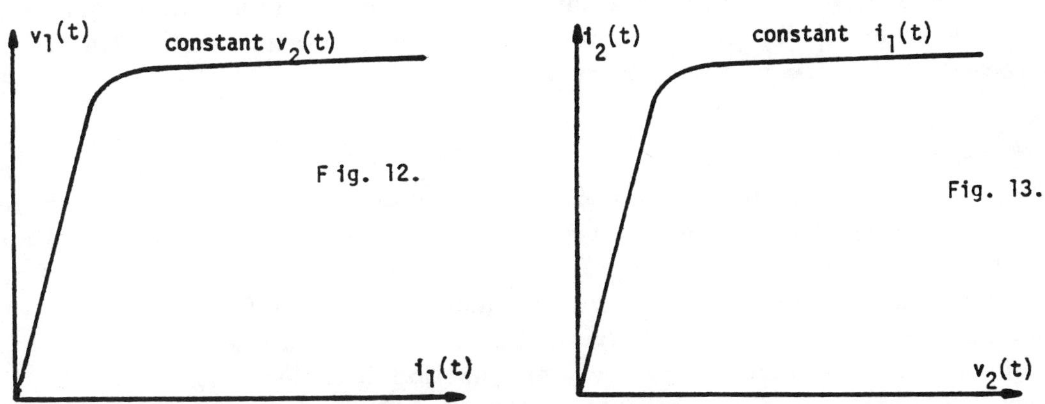

Notice: Here we are not concerned with any specific sign and/or polarities in these graphs. Our purpose is to demonstrate only the graphical method.

The next step is to develop a relation between (v_1, i_1) of network N_1 and (i_2, v_2) of network N_2. Suppose these relations become $v_1(t) = R_1[i_1(t)]$ and $i_2(t) = G_2[v_2(t)]$. These functions are called *load lines*. Obviously the input as well as the output operating points must satisfy both of these equations as well as the characteristic curves. Thus the problem becomes very manageable when we graph the function R_1 on the coordinates of Fig. 12, and graph the function G_2 on the coordinates of Fig. 13. Intersections of these two curves determine the input and output operating points for network N. Figs. 14 and 15 illustrate some typical examples of this method.

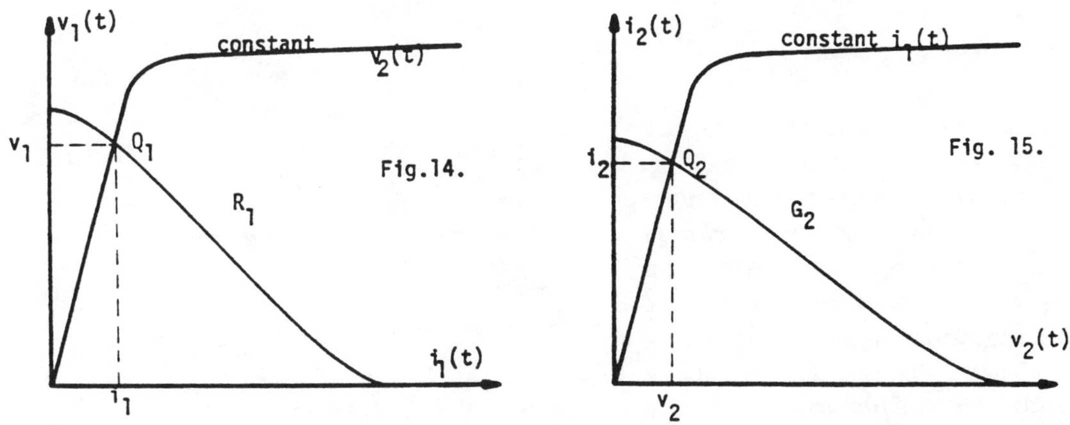

Although we have assumed that both R_1 and G_2 are nonlinear functions they will generally be linear functions. We do not elaborate further on these issues until we have a specific network and a specific TTD to work with. Nevertheless these graphical techniques will be quite useful for these types of problems as we will discover in the future experiments of this part.

III-4. PHYSICAL STRUCTURE OF ACTIVE THREE-TERMINAL DEVICES: AN OVERVIEW

To understand the behavior of an active three-terminal device requires a knowledge of the fundamentals of solid state electronics which is the subject of a separate course. We will attempt *to review* very briefly some basic aspects of solid state device theory although it is not our intention to teach this subject. This review will help us to understand the procedure behind each experiment of this part. In so doing we will be introducing a terminology that can not be addressed thoroughly due to space limitation and the fact that we are merely concerned with applications of these devices and not their physical structures. While there are many outstanding and comprehensive references on this subject either [1], [5], or [7] can be used for additional information on these matters. We closely gathered the following excerpts from [7].

III-4.1. p-n JUNCTIONS

A simple and an idealized diagram of a p-n junction is shown in Fig. 16. Here we have two pieces of the same semiconductor of the types n and p, that are uniformly doped and are brought close to each other to form the so-called *abrupt p-n junction*. In actuality this is, of course, a *graded* junction, i.e., the doping type does not abruptly change at the interface. But the abrupt assumption makes the analysis of the *built-in-electric field* at the interface much simpler than the graded assumption. The energy band diagram of this junction becomes that of Fig. 17; E_f is the Fermi energy level. Here the n-side is at a positive potential with respect to the p-side and the electric field is produced in the –x direction.

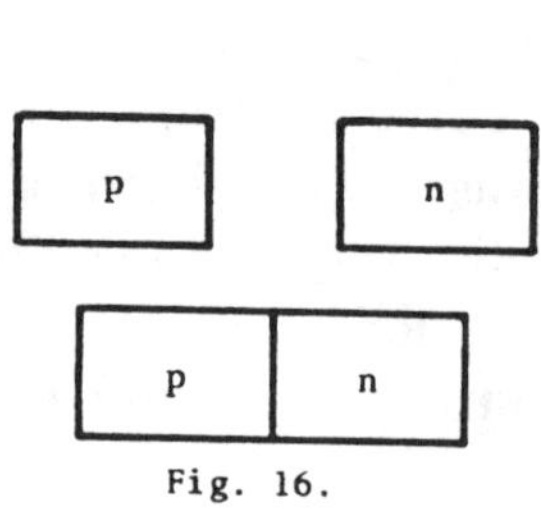

Fig. 16.

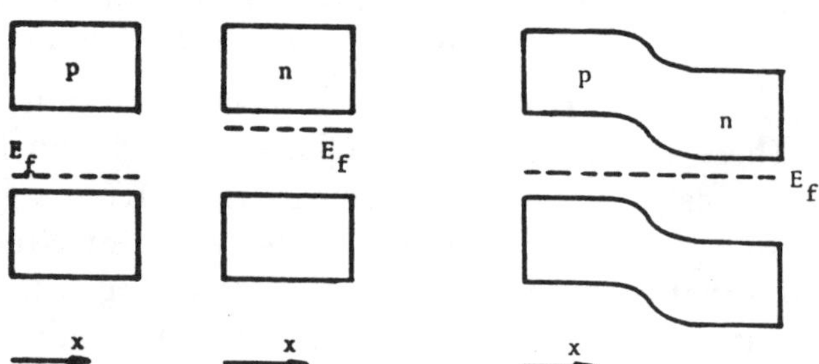

Fig. 17. Energy Diagram of p-n Junction Before and After Contact.

The electrostatic and the electric field as functions of x are given by Fig. 18. From a careful analysis of the *impurity density profile* and the *space charge density* of an abrupt p-n junction we can develop expressions for a *built-in-electric field* at the junction. Generally speaking this computation is done by using Gauss's equation $dE/dx = \rho/\epsilon$ or the Poisson equation $d^2\psi/dx^2 = -\rho/\epsilon$. The results can be found in [3], [5], and [7].

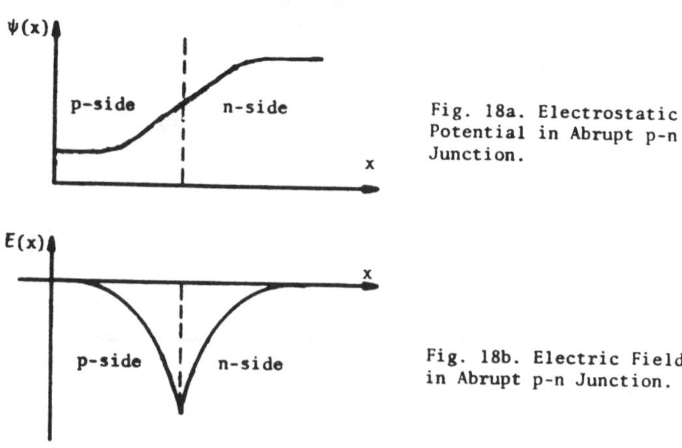

Fig. 18a. Electrostatic Potential in Abrupt p-n Junction.

Fig. 18b. Electric Field in Abrupt p-n Junction.

The analysis becomes more complicated when the junction is forward or reversed biased. We can also develop equations to determine the carrier concentration at the edges of the space charge layer. These relations are useful to show how a combination of these junctions works.

III-4.2. METAL-OXIDE-SEMICONDUCTOR (MOS) TRANSISTORS

This is the first example of a three-terminal active device. *A schematic* diagram of its cross section is given in Fig. 19. In the process of making this device, the metal contact touches both source and drain regions to form ohmic contacts. As evident from Fig. 19, the path from source to drain is actually two p-n junctions in a *back-to-back* configuration, such that for any external biasing one of these junctions become reverse-biased with a very low source to drain conductance. The gate acts as a modulator of this conductance to generate a conductive *channel* between source and drain. If the structure is as shown in Fig. 19, then this is called an

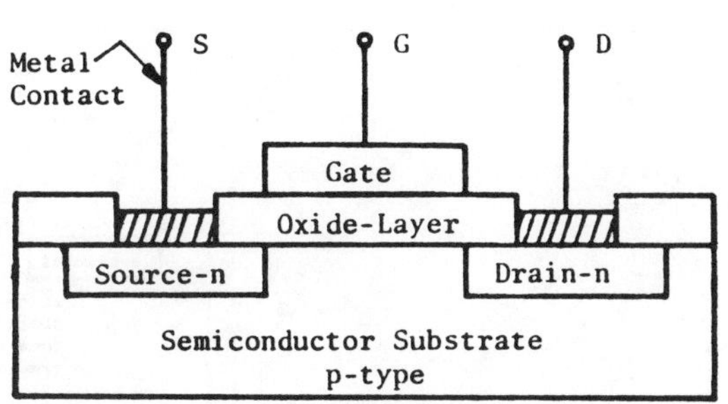

Fig. 19. Basic Structure of MOS

n-channel device. If the substrate is, instead, n-type and the source and the drain are p-types, then this is called a *p-channel* device. The conductance is controlled by the electric field in the gate and this is also why such a device is known as a *Field-Effect Transistor* or MOSFET for short. This device is also known as an IGFET (IG stands for insulated gate).

The main issue in understanding additional properties of MOS technology is the understanding of semiconductor surfaces and interfaces. As an oxide layer is grown on a semiconductor surface, a charge layer will be produced on this surface which will be *added* to the already existing charges in the oxide layer. These two charges will result in a *net charge* on the semiconductor surface layer giving rise to the concept of the MOS electrostatic structure. There are four possible conditions for this net surface charge.

Accumulation: In which the direction of the electric field is such that it attracts the *majority* carriers to the interface.

Flat-Band: As the name implies the bands at the interface become flat due to the lowering of the electric field of the accumulation state and resulting in the surface band-bending to become smaller and smaller. This is done by lowering the gate bias. Under this condition no electric field or space charge remains in the semiconductor.

Depletion: The direction of electric field at the semiconductor surface becomes such that the majority carriers are "repelled and the minority carriers are attracted."

Inversion: As the electric field direction changes from the accumulation condition to the depletion condition the oxide field is being reversed. Due also to the changes of this field strength, the Fermi energy level comes closer to the minority carrier band than to the majority carrier band. In this situation the conductivity type of the surface-layer also becomes reversed or *inverted.*

In the MOS transistor we want to have a controlled inversion. This is done by controlling the oxide field. There are different ways of controlling this inversion and as a result we have different types of MOS transistors.

Depletion-Mode Transistor: When a conducting channel (inversion layer) exists at the zero gate-bias. For this device, a finite *threshold* voltage v_T is needed to turn the source-drain channel *off.* (The threshold voltage corresponds to the distance between Fermi energy level and the bottom of conduction bands.)

Enhancement-Mode Transistor: When no conducting channel exists at the zero gate-bias. For this device, a finite threshold voltage v_T is needed to turn the source-drain channel *on.*

In short, since we may have n- or p-channel, depletion- or enhancement-mode devices we have four types of MOS transistors. These are summarized in Fig. 20. The symbol for this device is not unique in the literature; in fact, each of the following types is shown differently. In this part we will only use the symbol shown in Fig. 20, as this is consistent with reference [7]. In Section V-6.2, a set of slightly different symbols as used in the literature is shown.

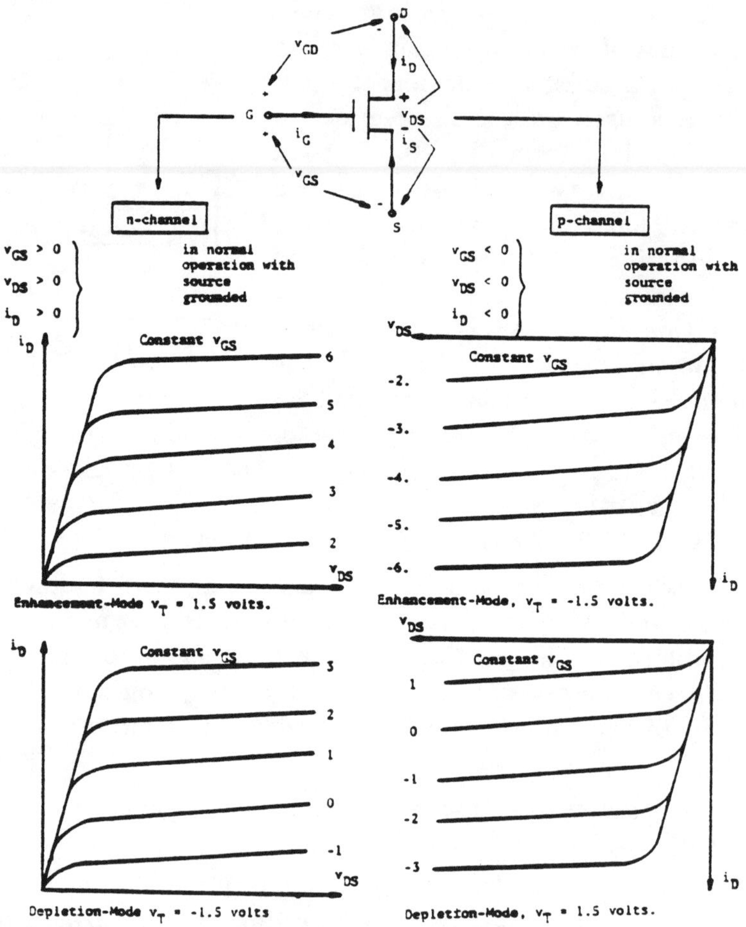

Fig. 20. Drain Characteristics of Four Types of MOS Transistors.

A careful look at this three-terminal device shows that because of the substrate this actually can be a four terminal device. A symbol for such a four-terminal device is given in Fig. 21a. In Fig. 21b, we show that the same device being used as three-terminal device. As a four-terminal device a bias of v_{BS} between the substrate (or *body*) and source is used. The effect of this v_{BS} is the same as changing the threshold voltage v_T. When $v_{BS} = 0$, this is the same as v_T, but when $v_{BS} \neq 0$, then v_T changes to v_T(effective) $= v_T(v_{BS}{=}0) + \Delta v_T$. This Δv_T is approximately proportional to $\sqrt{v_{BS}}$ and is usually given by the manufacturer. This new parameter becomes a useful tool for the circuit designer, since by changing it, the operating modes of the transistor, from the depletion mode to the enhancement mode, can be generally changed.

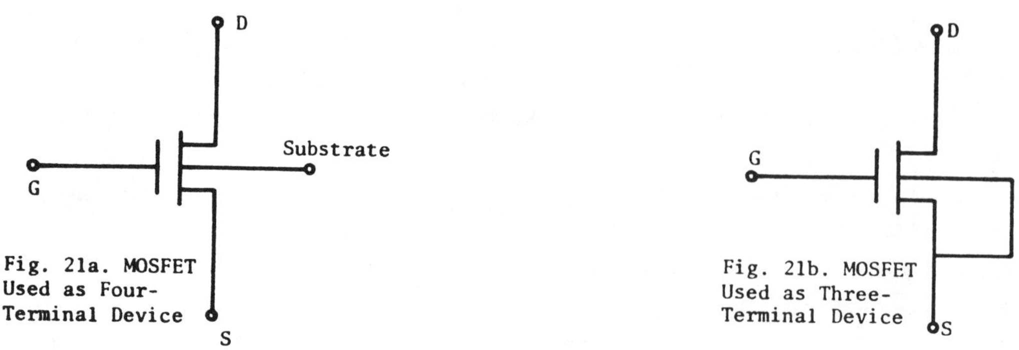

Fig. 21a. MOSFET Used as Four-Terminal Device

Fig. 21b. MOSFET Used as Three-Terminal Device

Finally, we close this sub-section with a few additional remarks on the general properties of field-effect transistors and their differences with the bipolar junction transistors to be explained next.

1. This is a unipolar (one type of carrier) device - its operation depends on the flow of majority carriers only.

2. It is much simpler to fabricate and is mainly used in integrated form.

3. It has very high input impedance at low frequencies. This also requires care, since very small amounts of charge on the gate will result in a large electric field and consequently oxide breakdown. Any attempt to protect the device will yield a degradation of the input impedance.

4. MOS transistors can be made to be symmetric in the sense that the source and drain can be interchanged without any difficulty (also used as bilateral switch). This property does not exist in the bipolar junction transistors (BJT).

5. If an n-channel and a p-channel device, with appropriate fabrication consideration, share the same substrate, then this new device is called the *complementary symmetry* MOS (CMOS or COSMOS). The CMOS has applications as a fast and low power-dissipated switching circuit. A schematic diagram of this device is shown in Fig. 22. This device works as follows. When the input signal switches the n-channel device on, it switches the p-channel device off, connecting V_B to the output. Likewise, when the p-channel device is on, the n-channel device is off and the output is grounded. Thus the output changes from V_B to 0 only because of changes in the input signal.

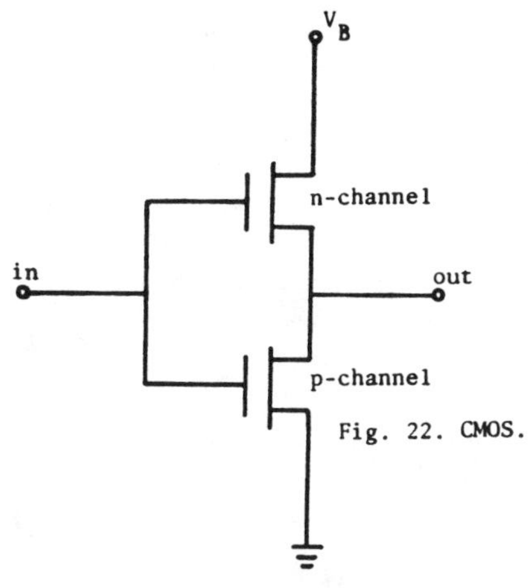

Fig. 22. CMOS.

6. MOS transistors are self-isolating in the sense that the active region of each device is completely surrounded by a depletion layer. This is not the case for the BJT.

7. MOS transistors act as voltage-variable resistors where the source to drain channel is the resistor. These changes are done by appropriate biasing of the gate. The channel resistance of one device is used as the load resistor of another device, eliminating the need for diffused resistors in MOS integrated circuits.

8. MOS transistors can act as memory devices because of the charge stored in the internal capacitances.

9. MOS transistors can also act as signal choppers because they have no offset voltage at zero drain current.

10. The disadvantage of the MOS transistor is that its output is fairly noisy. As a digital switch this does not cause a problem. But for low-level linear applications the junction FET (JFET) is preferred since it is less noisy than a bipolar transistor.

There is another type of Field-Effect Transistor in this family of devices called the Metal-Semiconductor FET (MESFET).

III-4.3. BIPOLAR JUNCTION TRANSISTORS (BJT)

The BJT is a three-terminal device made of three layers of p- and n-type as shown in Fig. 23. The current directions and voltage polarities are chosen as shown in Fig. 23. In a properly biased npn transistor, v_{CB} is positive; similarly for pnp transistors, v_{CB} is negative. The input impedance of the BJT is not infinite; it merely is that of a forward biased p-n diode. Thus we must be concerned with both input and output volt-ampere characteristics instead of only output characteristics. The BJT's are not structurally symmetric devices; neither are they doped symmetrically. Because most power dissipation in transistors takes place in the reverse biased collector junction, this asymmetry helps to reduce the effects of heat dissipation. The collector junction is designed for high breakdown voltage and the emitter junction is designed for high *emitter efficiency*. This is the ratio of minority carrier current injected into the base to the total emitter current.

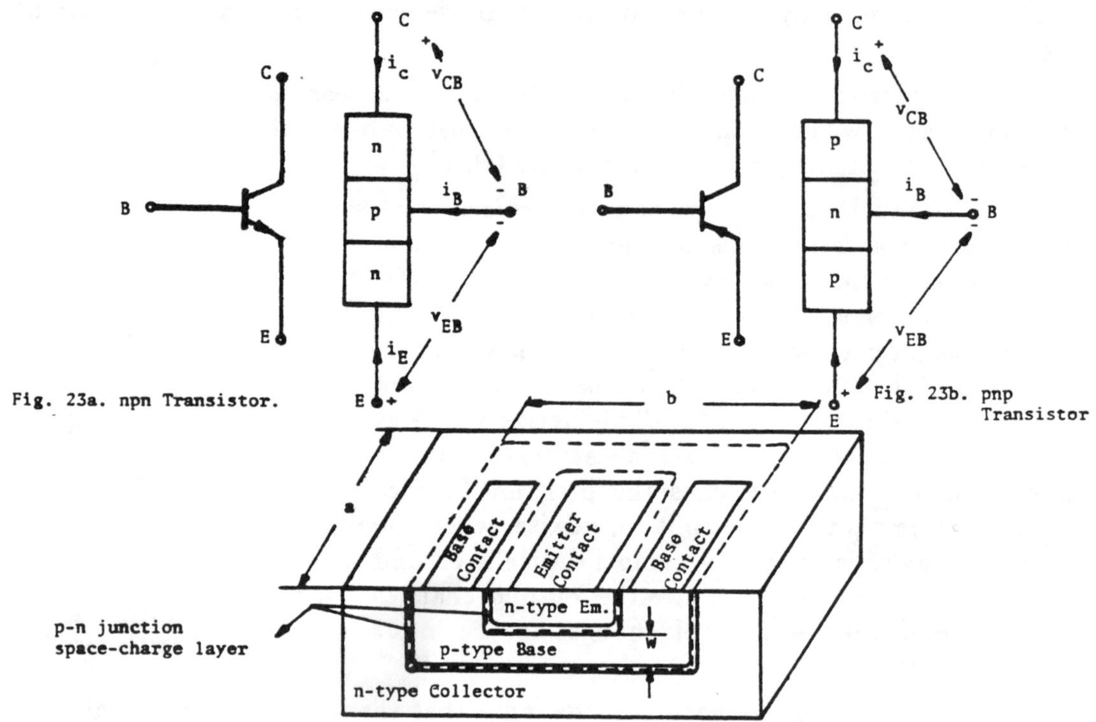

Fig. 23a. npn Transistor.

Fig. 23b. pnp Transistor

Fig. 23c. Structure of a Typical BJT.

    In most transistors the minority carriers diffuse from emitter to collector such that the recombination in the base is a minimum. To understand how a transistor works and what the current relationships and current gain definitions are, we must go back and learn how p-n junctions between the emitter and the base and between the base and the collector work. Actually, the transistor works because the minority carriers injected at one junction are *swept* through the other junction because of fields in the depletion range. This transport by diffusion can be described in the following way. We set the emitter junction forward biased, and $v_{CB}=0$, at first. Then we look into the *excess density profile* in the base. For this bias condition and referring to Fig. 24, we have the following equalities.

$$n'(0) = n_{oB}[\exp(qv_{BE}/kT) - 1] \triangleq n_{oB}\,\nu_{BE} \quad ,$$

$$p'(d_1) = p_{oE}[\exp(qv_{BE}/kT) - 1] \triangleq p_{oE}\,\nu_{BE} \quad ,$$

$$i_E = i_E(\text{hole}) + i_E(\text{electron}) \quad ,$$

$$= A[-qD_h dp(d_1)/dx + qD_e\,dn(0)/dx] \quad ,$$

$$\simeq -qA[D_h\,p'(d_1)/W_E + D_e\,n'(0)/W_B] \quad ,$$

$$\simeq -qA[D_h\,p_{oE}/W_E + D_e\,n_{oB}/W_B]\,\nu_{BE} \quad , \text{ and}$$

$$i_C = -qAD_e\,dn(d_2)/dx \simeq qAD_e n_{oB}\,\nu_{BE}/W_B \quad .$$

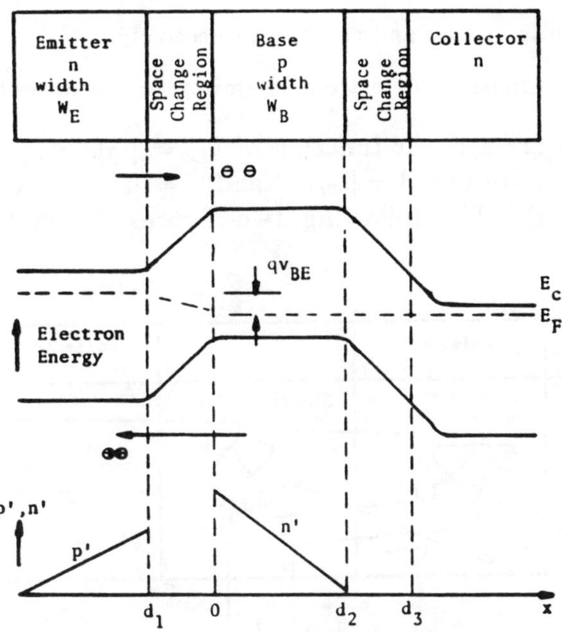

Fig. 24. npn Transistor with Forward Bias on Emitter
Junction and Zero Bias on Collector Junction.

The base current should not be computed from $i_B + i_C + i_E = 0$, since the two currents i_E and i_C are very close to each other resulting in $i_B = 0$. Instead we use $i_B = i_{B_1} + i_{B_2}$, where

$i_{B_1} \triangleq$ (current due to hole injected from base to emitter)

$\qquad = qAD_h\, p_{oE}\, \nu_{BE}/W_E$, and

$i_{B_2} \triangleq$ (current due to excess density profile)

$\qquad = q \displaystyle\int_{\text{volume of base}} Rd(\text{vol})$,

here R is the recombination rate in the base and is given by

$\qquad R = n'(x)/\tau$.

Thus,

$$i_{B_2} = qA \int_0^{d_2} \frac{n'(\tau)}{\tau}dx = qAW_B\, n'(0)/2\tau \ ,$$

resulting in

$\qquad i_B = qA(D_h\, p_{oE}/W_E + W_B\, n_{oB}/2\tau)\, \nu_{BE}$.

Using this i_B and the above i_C, we may now express the correction of i_E as

$\qquad i_E = -qA(D_h\, p_{oE}/W_E + D_e n_{oB}/W_B + W_B n_{oB}/2\tau)\, \nu_{BE}$.

In all the above equations we have used the following notations:

$\qquad n_{oB}\, , p_{oE} \triangleq$ Thermal equilibrium densities of electrons in the base and holes in the emitter, respectively.

$A \triangleq$ Cross sectional area .

W_B , $W_E \triangleq$ Width of bulk in base and emitter, respectively.

D_e , $D_h \triangleq$ Diffusion coefficients of electrons in emitter and holes in the base.

All these three currents are a linear function of $\nu_{BE} \triangleq [\exp(qv_{BE}/kT)-1]$, and thus these are controlled by this quantity. Here we call $\beta = |i_C/i_B|$ and $\alpha = |i_C/i_E|$. Using the above quantities it is easy to show that $\beta = \alpha/(1-\alpha)$. The following two figures summarize the essence of the above analysis.

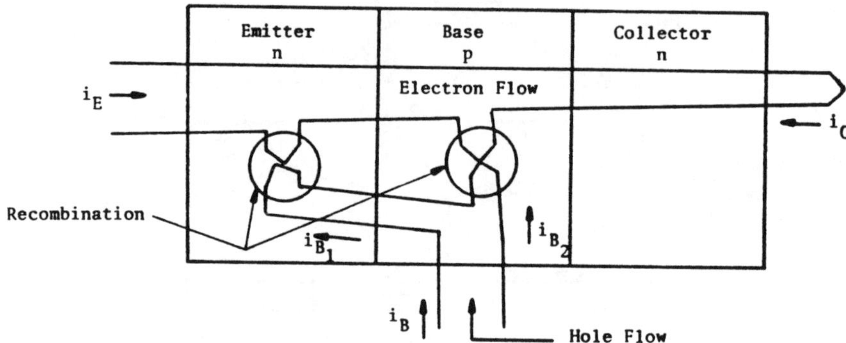

Fig. 25a. Currents in an npn Transistor. (Note that i_E must be negative.)

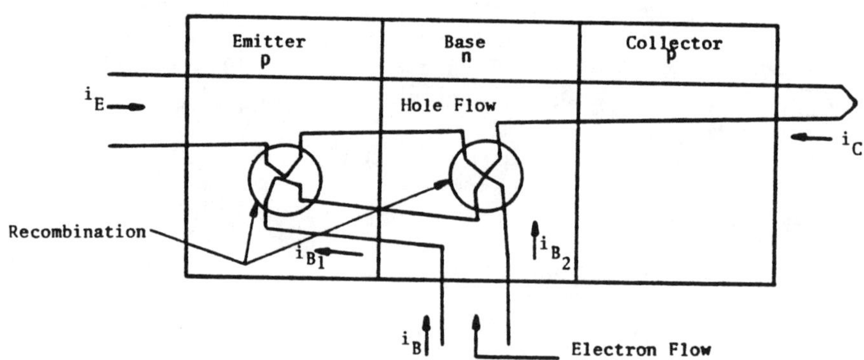

Fig. 25b. Currents in a pnp Transistor. (Note that i_B and i_C are both negative.)

The above analysis is carried out for $v_{CB} = 0$. This choice will cause problems in applications, since to control a short circuit current has little if no application. Thus a reverse bias is used for the collector junction to allow the control of a power flow in the output resistor. This bias changes the excess densities on both sides of the junction to the following values:

$$n'(d_2) = n_{oB}\,[\exp(qv_{BC}/kT)-1] \simeq -n_{oB} \ ,$$
$$p'(d_3) = p_{oC}\,[\exp(qv_{BC}/kT)-1] \simeq -p_{oC} \ .$$

The excess density profiles in the three regions and the new energy band diagram are now given by Fig. 26.

The effect on the base profile is essentially negligible because $n_{oB} \ll n'(0)$ under normal bias conditions. This reverse bias produces a small *leakage* current $I_C O$, which is independent of both junction voltages. Here we have

$$i_C = -\alpha i_E + I_{CO} \qquad , \text{ or}$$
$$i_C = \beta i_B + I_{CO}/(1-\alpha) \ .$$

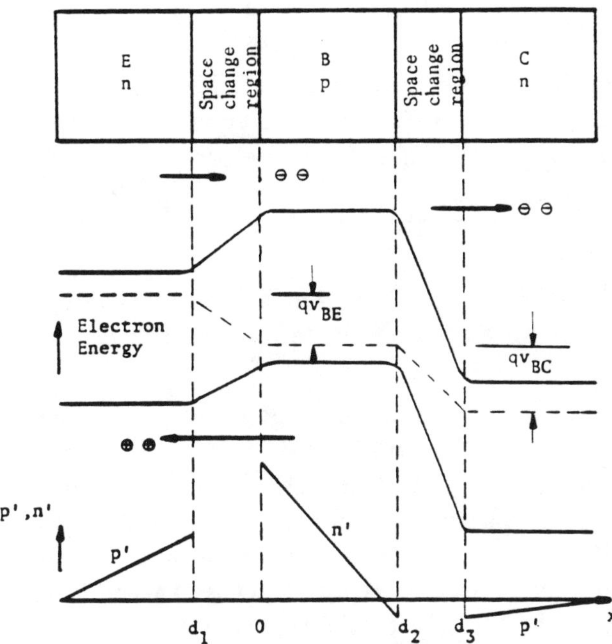

Fig. 26.   npn Transistor in Active Mode.

This I_{CO} is a very small quantity but it is important to understand how it develops. If we set $i_E = 0$, then $i_C = I_{CO}$. If $I_B = 0$ then $i_C = I_{CO}/(1-\alpha)$. To understand this basic difference between the open emitter and the open base is essential for knowing how a transistor works. The case $i_C = I_{CO}$ refers to a common base situation (i.e., the base is common between input and output as we will study later in EXP. III-1). This is due to the collector-base junction acting in reverse bias. No amplification is done since no injection is allowed at the emitter-base junction. The case common-emitter (which always produces much higher gains) works as follows. Because $i_B = 0$, the leakage current I_{CO} must now flow into the emitter. Then this current I_{CO} in the emitter produces the current αI_{CO} in the collector which is added to the previous collector current of I_{CO} resulting in $I_{CO} + \alpha I_{CO}$. As this amplification process continues we will have

$$i_C = I_{CO} + \alpha I_{CO} + \alpha^2 I_{CO} + ... = \sum_{n=0}^{\infty} \alpha^n I_{CO}. \text{ Since } \alpha < 1, \text{ then } i_C = I_{CO}/(1-\alpha).$$

Finally, and in short, how does a transistor work?

In the common base configuration the transistor works because the excess minority carriers which are drawn into the base by the bias on the base-emitter junction are swept into the collector circuit by a steep density gradient and a narrow base [7]. In this configuration the current gain is almost one.

In the common emitter configuration the current gain is much larger than one for a very small input base current.

Here we present a discussion (cf., reference [7]) of how a transistor works. Suppose the emitter of a transistor is *perfect*, thus the injection of minority carriers into the emitter is negligible compared with the injection into the base. For this case we have: $\beta = (qD_e\, n_{oB}/W_B)/(qW_B\, n_{oB}/2\tau) = 2D_e\tau/W_B^2$. This β corresponds to the ratio of collector current due to diffusion and base current due to recombination in the base. The average transit time for a minority carrier across the base becomes $\tau_r = \int_0^{W_B} dx/v_{\text{diff}}$. Here the diffusion velocity v_{diff} is computed from

$$V_{\text{diff}} = \frac{I_C}{Aqn'(x)} = \frac{n'(0)D_e/W_B}{n'(0)\,(1-x/W_B)} = D_e/(W_B - x) \quad .$$

Near the emitter side there are many slow traveling particles, while near the collector side there are only few fast traveling particles. Thus a diffusion velocity is specified according to the current flow. Using the above velocity results in $\tau_r = W_B^2/2D_e$ and therefore

$$\beta = \frac{\tau}{\tau_r} \triangleq \frac{\text{minority carrier life time}}{\text{minority carrier transit time}} \quad .$$

If we think of this β as the mechanism of current amplification or control, then it says as a hole enters into the base there must be an electron injected across the junction from the emitter in order to have charge neutrality.

After the lifetime τ the recombination is completed, but the steep density profile causes the electron to move toward the collector. The high field in the collector-base junction draws the electron across the base-emitter junction to maintain the charge neutrality. This electron falls off the cliff within another transit time and the process continues until eventually in a time of the order of lifetime the laws of probability catch up with the hole and it recombines. The gain factor $\beta = |i_C| \,/\, |i_E|$ is equal to the number of electrons lured across the collector base junction by this one hole and will then be the ratio of the lifetime to the transit time.

To design a high β transistor one needs high lifetime material and short base *width*. The first criterion causes problems if the transistor must be used for high speed circuits, because the excess density profile can not easily be changed in a time shorter than a lifetime. The second criterion can cause difficulty with the high voltage capability of the collector circuit. If the base is too narrow, the collector junction depletion region can *punch through* the base, i.e., the two depletion regions merge for high collector voltage. In modern transistors these problems are somewhat lessened through the use of nonuniformly doped base regions. We again emphasize that the above materials in Section III-4, on the physical structure of the active TTD's, are mainly gathered from [7].

III-5.   MANUFACTURER'S SPECIFICATION DATA SHEETS

On the following pages (cf., Fig. 27), we present the specification data sheets of one brand general purpose npn transistor for the sake of illustrations and review of its basic properties [11].

2N2218S, AS, 2N2219S, AS, 2N2221, A (SILICON)
2N2222, A, 2N5581, 2N5582

NPN SILICON ANNULAR HERMETIC TRANSISTORS

. . . widely used "Industry Standard" transistors for applications as medium-speed switches and as amplifiers from audio to VHF frequencies.

- DC Current Gain Specified — 1.0 to 500 mAdc
- Low Collector-Emitter Saturation Voltage —
  $V_{CE(sat)}$ @ I_C = 500 mAdc
  = 1.6 Vdc (Max) — Non-A Suffix
  = 1.0 Vdc (Max) — A-Suffix
- High Current-Gain—Bandwidth Product —
  f_T = 250 MHz (Min) @ I_C = 20 mAdc — All Types Except
  = 300 MHz (Min) @ I_C = 20 mAdc — 2N2219A, 2N2222A, 2N5582
- Complements to PNP 2N2904,A thru 2N2907,A
- JAN,JTX Available in all devices
- JTXV Available on 2N2222,A Series
- 2N2218 and 2N2219 available in TO-39 Package With 1/2" Leads (1)

NPN SILICON SWITCHING AND AMPLIFIER TRANSISTORS

SELECTION GUIDE

Device Type	BV$_{CEO}$ I_C = 10 mAdc Volts	h$_{FE}$ I_C = 150 mAdc Min/Max	h$_{FE}$ I_C = 500 mAdc Min	Package
2N2218 2N2219	30	40/120 100/300	20 30	TO-5
2N2221 2N2222	30	40/120 100/300	20 30	TO-18
2N5581 2N5582	40	40/120 100/300	25 40	TO-46
2N2218A 2N2219A	40	40/120 100/300	25 40	TO-5
2N2221A 2N2222A	40	40/120 100/300	25 40	TO-18

*MAXIMUM RATINGS

Rating	Symbol	2N2218 2N2219 2N2221 2N2222	2N2218A 2N2219A 2N2221A 2N2222A	2N5581 2N5582	Unit
Collector-Emitter Voltage	V_{CEO}	30	40	40	Vdc
Collector-Base Voltage	V_{CB}	60	75	75	Vdc
Emitter-Base Voltage	V_{EB}	5.0	6.0	6.0	Vdc
Collector Current — Continuous	I_C	800	800	800**	mAdc
		2N2218,A 2N2219,A	2N2221,A 2N2222,A	2N5581 2N5582	
Total Device Dissipation @ T_A = 25°C Derate above 25°C	P_D	0.8 5.33	0.5 3.33	0.5 3.33	Watt mW/°C
Total Device Dissipation @ T_C = 25°C Derate above 25°C	P_D	3.0 20	1.8 12	2.0 11.43	Watts mW/°C
Operating and Storage Junction Temperature Range	T_J, T_{stg}	←——— -65 to +200 ———→			°C

*Indicates JEDEC Registered Data.
**Motorola Guarantees this Data in Addition to JEDEC Registered Data.

Fig. 27. General Purpose npn Silicon Transistor (Courtesy of Motorola [11]).

2N2218S,AS, 2N2219S,AS, 2N2221,A, 2N2222,A, 2N5581, 2N5582 (continued)

***ELECTRICAL CHARACTERISTICS** ($T_A = 25^\circ C$ unless otherwise noted)

Characteristic		Symbol	Min	Max	Unit
OFF CHARACTERISTICS					
Collector-Emitter Breakdown Voltage		BV_{CEO}			Vdc
($I_C = 10$ mAdc, $I_B = 0$)	Non-A Suffix		30	–	
	A-Suffix, 2N5581,2N5582		40	–	
Collector-Base Breakdown Voltage		BV_{CBO}			Vdc
($I_C = 10$ μAdc, $I_E = 0$)	Non-A Suffix		60	–	
	A-Suffix, 2N5581,2N5582		75	–	
Emitter-Base Breakdown Voltage		BV_{EBO}			Vdc
($I_E = 10$ μAdc, $I_C = 0$)	Non-A Suffix		5.0	–	
	A-Suffix, 2N5581,2N5582		6.0	–	
Collector Cutoff Current		I_{CEX}			nAdc
($V_{CE} = 60$ Vdc, $V_{EB(off)} = 3.0$ Vdc)	A-Suffix, 2N5581,2N5582		–	10	
Collector Cutoff Current		I_{CBO}			μAdc
($V_{CB} = 50$ Vdc, $I_E = 0$)	Non-A Suffix		–	0.01	
($V_{CB} = 60$ Vdc, $I_E = 0$)	A-Suffix, 2N5581,2N5582		–	0.01	
($V_{CB} = 50$ Vdc, $I_E = 0$, $T_A = 150^\circ C$)	Non-A Suffix		–	10	
($V_{CB} = 60$ Vdc, $I_E = 0$, $T_A = 150^\circ C$)	A-Suffix, 2N5581,2N5582		–	10	
Emitter Cutoff Current		I_{EBO}			nAdc
($V_{EB} = 3.0$ Vdc, $I_C = 0$)	A-Suffix, 2N5581,2N5582		–	10	
Base Cutoff Current		I_{BL}			nAdc
($V_{CE} = 60$ Vdc, $V_{EB(off)} = 3.0$ Vdc)	A-Suffix		–	20	
ON CHARACTERISTICS					
DC Current Gain		h_{FE}			–
($I_C = 0.1$ mAdc, $V_{CE} = 10$ Vdc)	2N2218,A,2N2221,A,2N5581(1)		20	–	
	2N2219,A,2N2222,A,2N5582(1)		35	–	
($I_C = 1.0$ mAdc, $V_{CE} = 10$ Vdc)	2N2218,A,2N2221,A,2N5581		25	–	
	2N2219,A,2N2222,A,2N5582		50	–	
($I_C = 10$ mAdc, $V_{CE} = 10$ Vdc)	2N2218,A,2N2221,A,2N5581(1)		35	–	
	2N2219,A,2N2222,A,2N5582(1)		75	–	
($I_C = 10$ mAdc, $V_{CE} = 10$ Vdc, $T_A = -55^\circ C$)	2N2218A,2N2221A,2N5581		15	–	
	2N2219A,2N2222A,2N5582		35	–	
($I_C = 150$ mAdc, $V_{CE} = 10$ Vdc)(1)	2N2218,A,2N2221,A,2N5581		40	120	
	2N2219,A,2N2222,A,2N5582		100	300	
($I_C = 150$ mAdc, $V_{CE} = 1.0$ Vdc)(1)	2N2218A,2N2221A,2N5581		20	–	
	2N2219A,2N2222A,2N5582		50	–	
($I_C = 500$ mAdc, $V_{CE} = 10$ Vdc)(1)	2N2218,2N2221		20	–	
	2N2219,2N2222		30	–	
	2N2218A,2N2221A,2N5581		25	–	
	2N2219A,2N2222A,2N5582		40	–	
Collector-Emitter Saturation Voltage(1)		$V_{CE(sat)}$			Vdc
($I_C = 150$ mAdc, $I_B = 15$ mAdc)	Non-A Suffix		–	0.4	
	A-Suffix, 2N5581,2N5582		–	0.3	
($I_C = 500$ mAdc, $I_B = 50$ mAdc)	Non-A Suffix		–	1.6	
	A-Suffix, 2N5581,2N5582		–	1.0	
Base-Emitter Saturation Voltage(1)		$V_{BE(sat)}$			Vdc
($I_C = 150$ mAdc, $I_B = 15$ mAdc)	Non-A Suffix		0.6	2.0	
	A-Suffix, 2N5581,2N5582		0.6	1.2	
($I_C = 500$ mAdc, $I_B = 50$ mAdc)	Non-A Suffix		–	2.6	
	A-Suffix, 2N5581,2N5582		–	2.0	

* Indicates JEDEC Registered Data

Fig. 27. Continued.

2N2218S,AS, 2N2219S,AS, 2N2221,A, 2N2222,A, 2N5581, 2N5582 (continued)

***ELECTRICAL CHARACTERISTICS** (Continued)

Characteristic		Symbol	Min	Max	Unit
SMALL-SIGNAL CHARACTERISTICS					
Current-Gain—Bandwidth Product(2) (I_C = 20 mAdc, V_{CE} = 20 Vdc, f = 100 MHz)	All Types, Except 2N2219A,2N2222A,2N5582	f_T	250 300	— —	MHz
Output Capacitance(3) (V_{CB} = 10 Vdc, I_E = 0, f = 100 kHz)		C_{ob}	—	8.0	pF
Input Capacitance(3) (V_{EB} = 0.5 Vdc, I_C = 0, f = 100 kHz)	Non-A Suffix A-Suffix, 2N5581,2N5582	C_{ib}	— —	30 25	pF
Input Impedance (I_C = 1.0 mAdc, V_{CE} = 10 Vdc, f = 1.0 kHz) (I_C = 10 mAdc, V_{CE} = 10 Vdc, f = 1.0 kHz)	2N2218A,2N2221A,2N5581 2N2219A,2N2222A,2N5582 2N2218A,2N2221A,2N5581 2N2219A,2N2222A,2N5582	h_{ie}	1.0 2.0 0.2 0.25	3.5 8.0 1.0 1.25	k ohms
Voltage Feedback Ratio (I_C = 1.0 mAdc, V_{CE} = 10 Vdc, f = 1.0 kHz) (I_C = 10 mAdc, V_{CE} = 10 Vdc, f = 1.0 kHz)	2N2218A,2N2221A,2N5581 2N2219A,2N2222A,2N5582 2N2218A,2N2221A,2N5581 2N2219A,2N2222A,2N5582	h_{re}	— — — —	5.0 8.0 2.5 4.0	X 10^{-4}
Small-Signal Current Gain (I_C = 1.0 mAdc, V_{CE} = 10 Vdc, f = 1.0 kHz) (I_C = 10 mAdc, V_{CE} = 10 Vdc, f = 1.0 kHz)	2N2218A,2N2221A,2N5581 2N2219A,2N2222A,2N5582 2N2218A,2N2221A,2N5581 2N2219A,2N2222A,2N5582	h_{fe}	30 50 50 75	150 300 300 375	—
Output Admittance (I_C = 1.0 mAdc, V_{CE} = 10 Vdc, f = 1.0 kHz) (I_C = 10 mAdc, V_{CE} = 10 Vdc, f = 1.0 kHz)	2N2218A,2N2221A,2N5581 2N2219A,2N2222A,2N5582 2N2218A,2N2221A,2N5581 2N2219A,2N2222A,2N5582	h_{oe}	3.0 5.0 10 25	15 35 100 200	μmhos
Collector-Base Time Constant (I_E = 20 mAdc, V_{CB} = 20 Vdc, f = 31.8 MHz)	A-Suffix, 2N5581,2N5582	$r_b{'}C_c$	—	150	ps
Noise Figure (I_C = 100 μAdc, V_{CE} = 10 Vdc, R_S = 1.0 k ohm, f = 1.0 kHz)	2N2219A,2N2222A	NF	—	4.0	dB
SWITCHING CHARACTERISTICS (A-Suffix, 2N5581 and 2N5582)					
Delay Time	(V_{CC} = 30 Vdc, $V_{BE(off)}$ = 0.5 Vdc, I_C = 150 mAdc, I_{B1} = 15 mAdc) (Figure 14)	t_d	—	10	ns
Rise Time		t_r	—	25	ns
Storage Time	(V_{CC} = 30 Vdc, I_C = 150 mAdc, I_{B1} = I_{B2} = 15 mAdc) (Figure 15)	t_s	—	225	ns
Fall Time		t_f	—	60	ns
Active Region Time Constant** (I_C = 150 mAdc, V_{CE} = 30 Vdc)		T_A	—	2.5	ns.

*Indicates JEDEC Registered Data.
**Motorola Guarantees this Data in Addition to JEDEC Registered Data.
(1) Pulse Test: Pulse Width ≤ 300 μs, Duty Cycle ≤ 2.0%.
(2) f_T is defined as the frequency at which $|h_{fe}|$ extrapolates to unity.
(3) 2N5581 and 2N5582 are Listed C_{cb} and C_{eb} for these conditions and values.

Fig. 27. Continued.

2N2218S,AS, 2N2219S,AS, 2N2221,A, 2N2222,A, 2N5581, 2N5582 (continued)

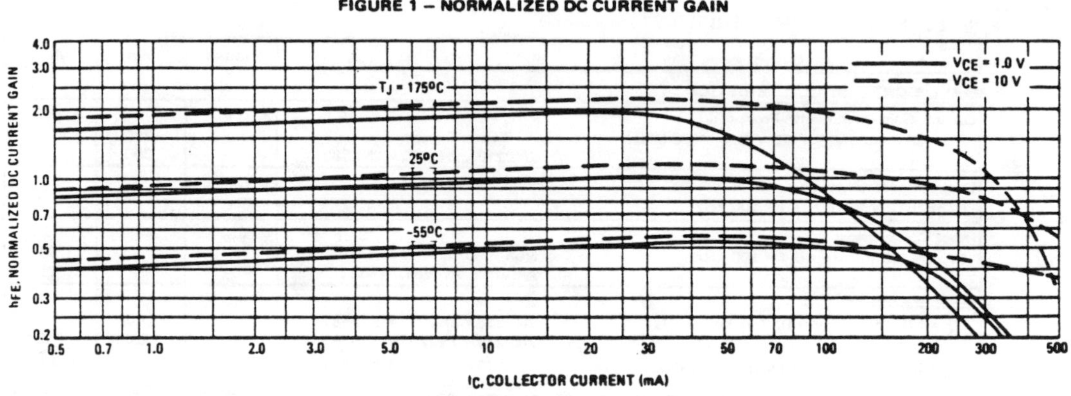

FIGURE 1 — NORMALIZED DC CURRENT GAIN

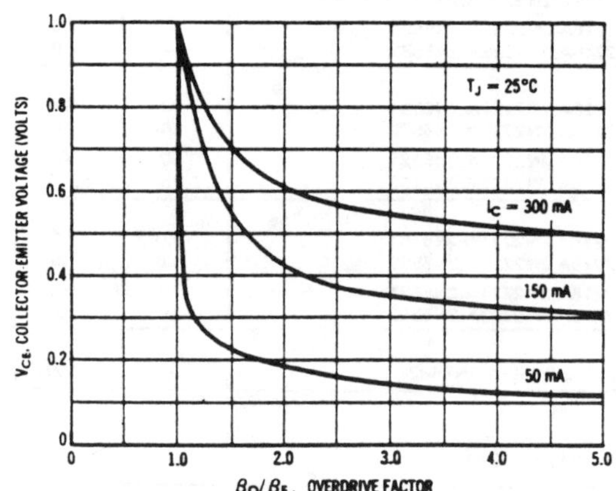

FIGURE 2 — COLLECTOR CHARACTERISTICS IN SATURATION REGION

This graph shows the effect of base current on collector current. β_o (current gain at the edge of saturation) is the current gain of the transistor at 1 volt, and $\beta_\flat$ (forced gain) is the ratio of $I_C/I_\flat$ in a circuit.

EXAMPLE: For type 2N2219, estimate a base current ($I_\flat$) to insure saturation at a temperature of 25°C and a collector current of 150 mA.

Observe that at I_C = 150 mA an overdrive factor of at least 2.5 is required to drive the transistor well into the saturation region. From Figure 1, it is seen that h_{FE} @ 1 volt is approximately 0.62 of h_{FE} @ 10 volts. Using the guaranteed minimum gain of 100 @ 150 mA and 10 V, β_o = 62 and substituting values in the overdrive equation, we find:

$$\frac{\beta_o}{\beta_\flat} = \frac{h_{FE} @ 1.0 V}{I_C/I_\flat} \qquad 2.5 = \frac{62}{150/I_\flat} \qquad I_\flat \approx 6.0 \ mA$$

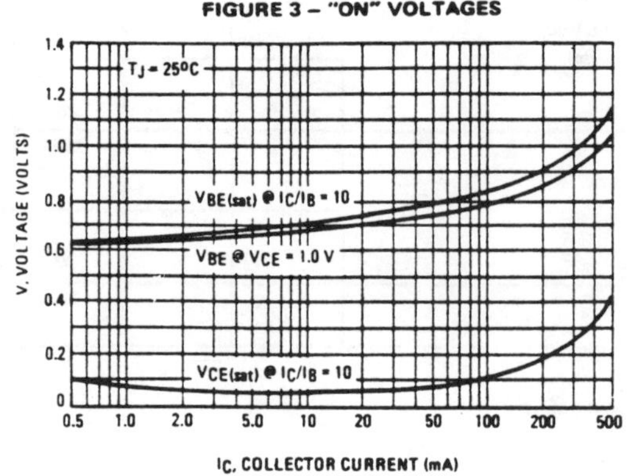

FIGURE 3 — "ON" VOLTAGES

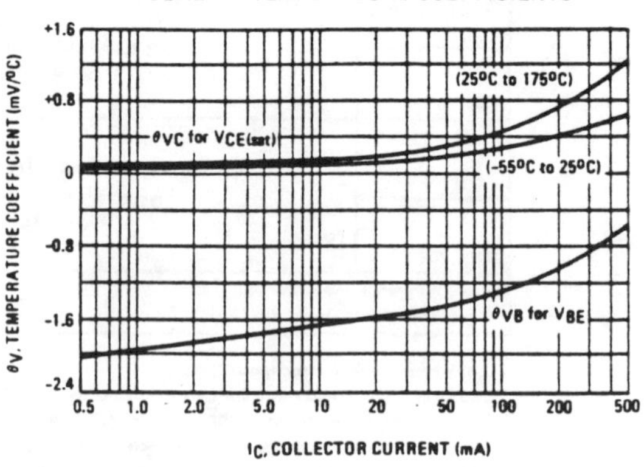

FIGURE 4 — TEMPERATURE COEFFICIENTS

Fig. 27. Continued.

2N2218S,AS, 2N2219S,AS, 2N2221,A, 2N2222,A, 2N5581, 2N5582 (continued)

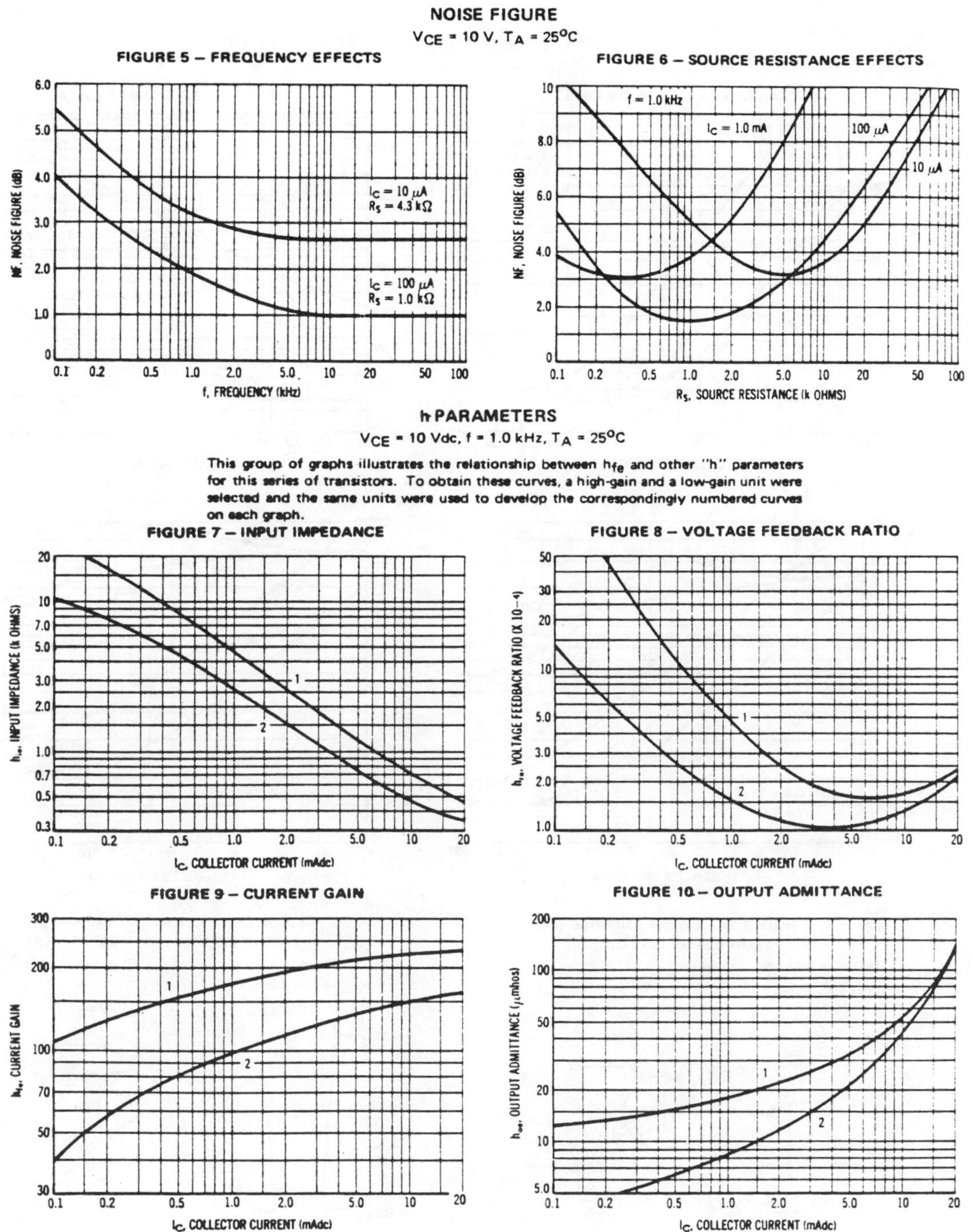

Fig. 27. Continued.

2N2218S,AS, 2N2219S,AS, 2N2221,A, 2N2222,A, 2N5581, 2N5582 (continued)

SWITCHING TIME CHARACTERISTICS

FIGURE 11 – TURN-ON TIME

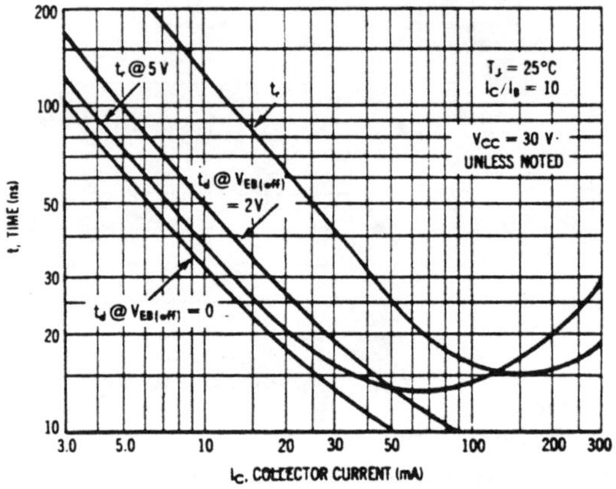

FIGURE 12 – CHARGE DATA

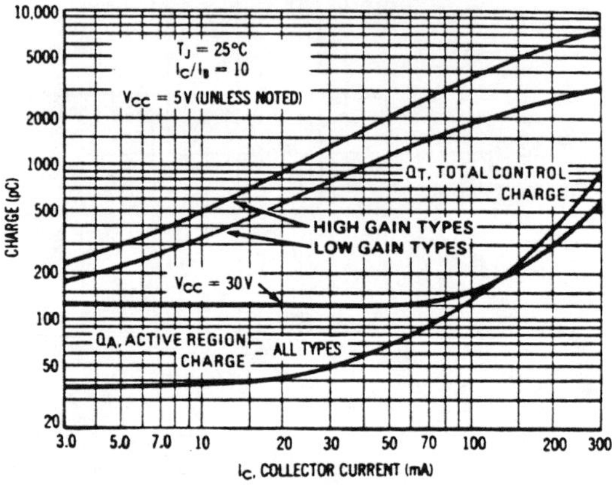

FIGURE 13 – TURN OFF BEHAVIOR

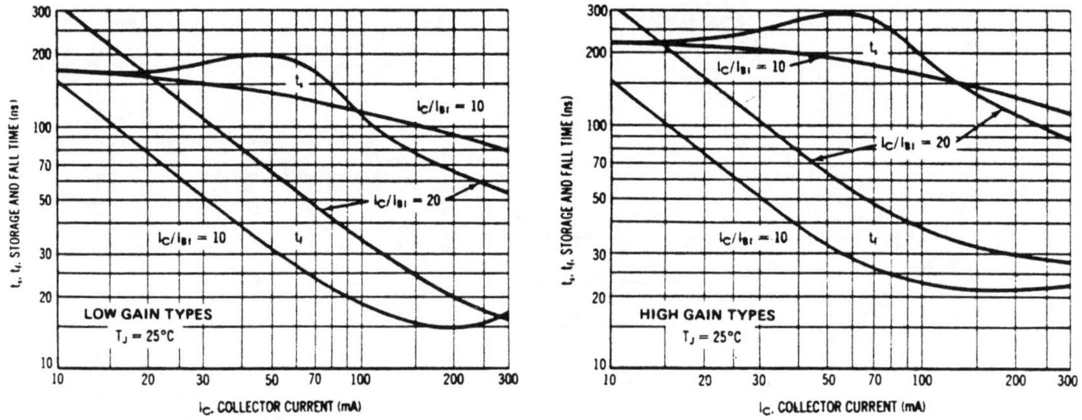

FIGURE 14 – DELAY AND RISE TIME EQUIVALENT TEST CIRCUIT

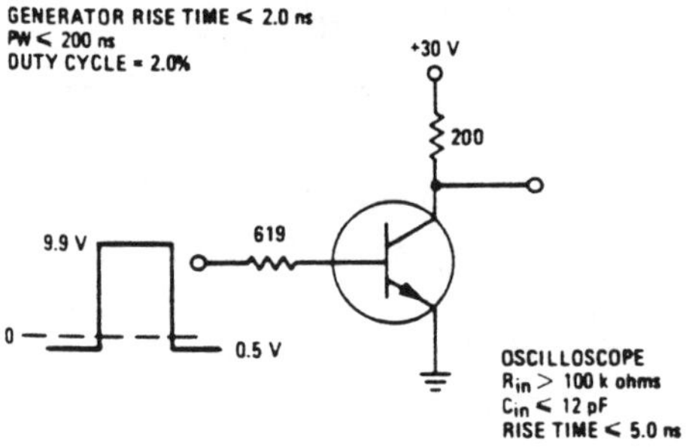

FIGURE 15 – STORAGE TIME AND FALL TIME EQUIVALENT TEST CIRCUIT

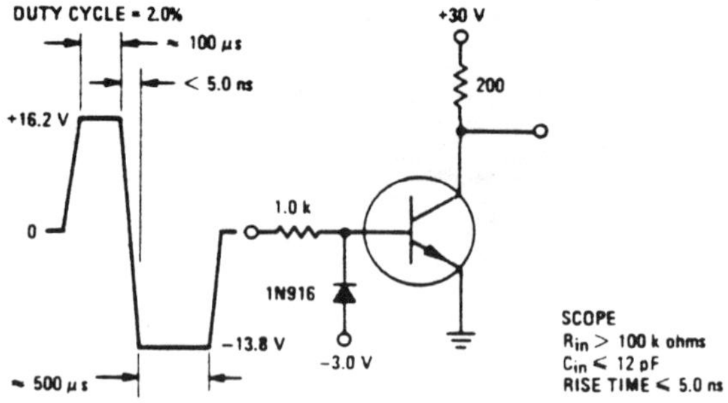

Fig. 27. Continued.

2N2218S,AS, 2N2219S,AS, 2N2221,A, 2N2222,A, 2N5581, 2N5582 (continued)

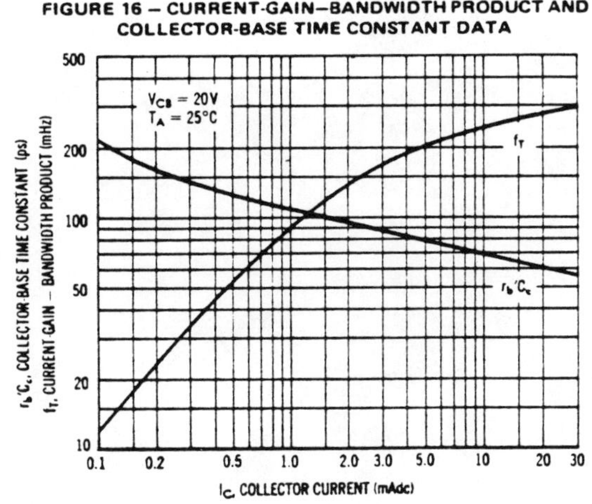

FIGURE 16 — CURRENT-GAIN—BANDWIDTH PRODUCT AND COLLECTOR-BASE TIME CONSTANT DATA

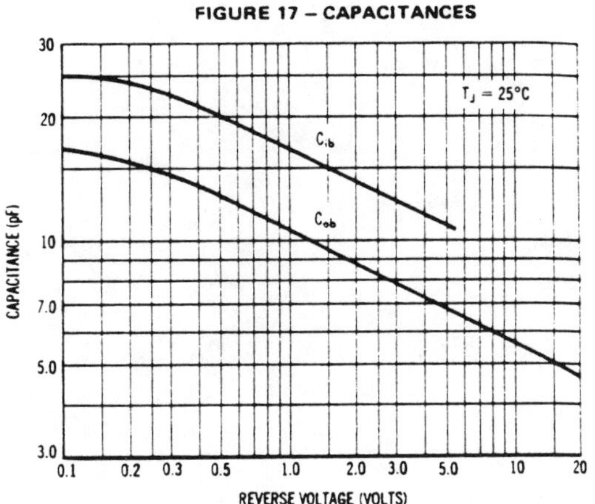

FIGURE 17 — CAPACITANCES

FIGURE 18 — ACTIVE-REGION SAFE OPERATING AREAS

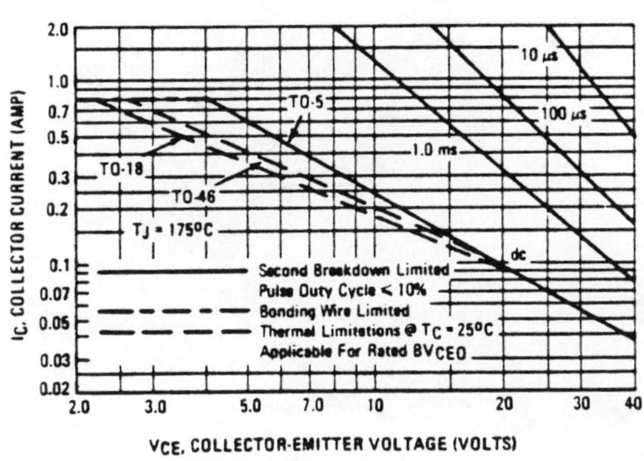

This graph shows the maximum I_C-V_{CE} limits of the device both from the standpoint of thermal dissipation (at 25°C case temperature), and secondary breakdown. For case temperatures other than 25°C, the thermal dissipation curve must be modified in accordance with the derating factor in the Maximum Ratings table.

To avoid possible device failure, the collector load line must fall below the limits indicated by the applicable curve. Thus, for certain operating conditions the device is thermally limited, and for others it is limited by secondary breakdown.

For pulse applications, the maximum I_C-V_{CE} product indicated by the dc thermal limits can be exceeded. Pulse thermal limits may be calculated by using the transient thermal resistance curve of Figure 19.

FIGURE 19 — THERMAL RESPONSE

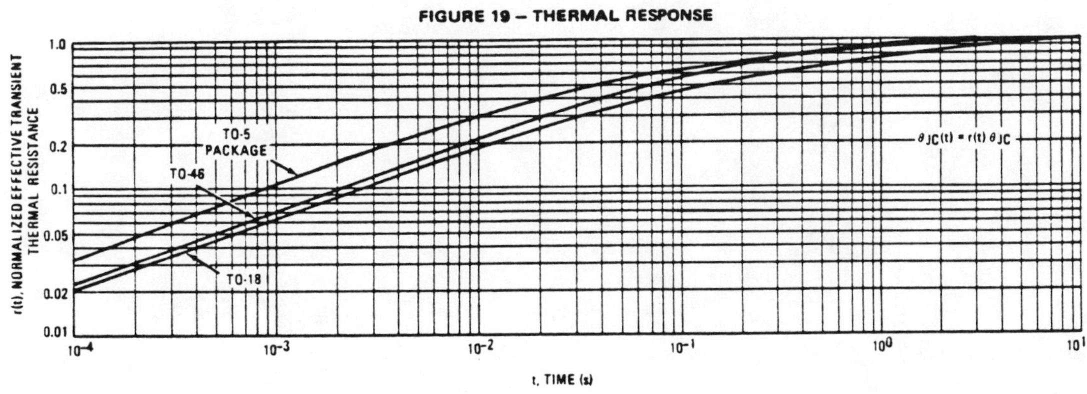

Fig. 27. Continued.

WORKSHEET

WORKSHEET

WORKSHEET

EXPERIMENT III-1

**BIPOLAR JUNCTION TRANSISTORS
CHARACTERISTICS AND SMALL-SIGNAL MODEL**

EQUIPMENT:

EEP	CURVE TRACER
BJTs (3)	GR-DECADE RESISTANCE BOX
VOM	PRECISION RESISTOR (1 MΩ)
VTVM	RESISTORS (47 Ω, 1 kΩ/2, 100 kΩ)

Table of Contents   Page No.

1. PURPOSE

The purpose of this experiment is to obtain the DC volt-ampere characteristics curves of a bipolar junction transistor (BJT). Subsequently, to construct a single stage transistor amplifier whose basic operation is studied with direct measurements. The results of this measurement must then be compared with those predicted from the characteristic curves developed for this BJT. This experiment shows how we may have a small current in one circuit that controls a large current in another circuit *via* the BJT. This is the reason transistors are sometimes considered to be current sources. Finally, from this experimental work the student is asked to develop a linear model for the BJT as it is used in its *active* mode.

Comment: It is the author's opinion that it is most instructive to have students perform this experiment with graphical analysis first. Then, upon gaining some familiarity with this device, they may be asked to develop the actual numerical values for the small-signal model of the transistor. For this reason this experiment is written in a slightly different fashion than the standard procedure used in the earlier experiments.

Note: Save your transistor(s) for the next three experiments.

2. INTRODUCTION

In general transistors and/or the related semiconductor devices are made either *discrete* or *integrated*. Here discrete means individual, of course, and integrated means a complete circuit in which several of these semiconductor components with the necessary circuitries are packed densely on one small semiconductor *wafer*. For our purpose of learning the fundamentals of these active devices it is important to study exhaustively the basic properties of a single transistor. In future experiments we will be looking into advanced analog and digital electronic designs that use circuits with integrated components.

We recall that there are two types of transistors, namely npn and pnp. From a circuit point of view these are complementary to each other with respect to biasing and the current directions are of opposite sign. Each transistor has three terminals called the Emitter, the Base and the Collector. For the sake of simplicity we will concentrate our studies on the properties of npn transistors only. Depending on which of these three terminals becomes common between the input and output ports we will have one of the following three basic configurations:

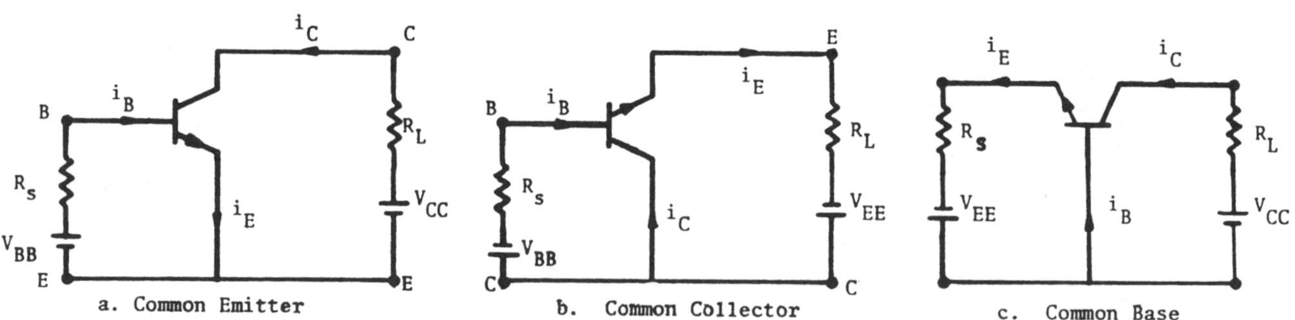

Fig. 1. Symbols for an npn Transistors.

We recall that transistors are active devices require some external power before they go into any operational modes. This is called *biasing* and it is responsible for the *proper* functioning of the device. We will study a self biasing technique in EXP. III-2. We recall from Section III-4, that we must be careful not to exceed the forward bias and, of course, the reverse bias beyond certain maximum values to avoid junction breakdown. The maximum breakdown voltages in forward or reverse bias for each junction are generally given to us. The resistors in Fig. 1, are included in order to provide paths for leakage currents of junction diodes as well as for device protection. Also recall that transistors are not electrically symmetric, i.e., they do not operate the same in the reverse direction. Getting back again to Fig. 1, we have the following general observations about each of these three basic configurations.

(a) The common-emitter has large current and voltage gains, resulting in a power gain much larger than unity. The input resistance is of order of kilo-ohms, the output resistance is on the order of 10^5 ohms with $180°$ phase difference in the output voltage.

(b) The common-collector has large current gain but the voltage gain does not exceed unity. The power gain is much larger than one. The input resistance is of order of 10^5 ohms, the output resistance is of order of a few ohms with no phase difference in the output voltage.

(c) The common-base has a current gain of approximately one and a voltage gain larger than one, resulting a power gain greater than unity. The input resistance is on the order of a few ohms; the output resistance is in the megaohms range with no phase difference in output voltage.

2.1. COMMON EMITTER CONFIGURATION

From the three choices mentioned earlier we choose the common emitter configuration for the purpose of testing as well as learning the fundamental properties of these types of active devices. Clearly, Fig. 2 shows a two-port three-terminal device, and from our earlier studies in Section III-2, we can represent its input and output relations as follows:

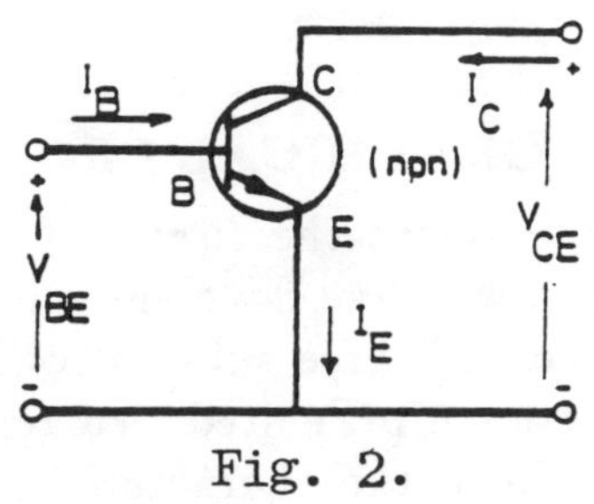

Fig. 2.

$$v_{BE} = v_{BE}(i_B, v_{CE}), \text{ or } i_B = i_B(v_{BE}, v_{CE}), \text{ and}$$

$$i_C = i_C(i_B, v_{CE}).$$

We could, of course, use other choices for representing the input and the output relations as described in general form in Section III-2. But we are mainly concerned with the amplification aspect of this device now, and that is why we choose to study i_C with i_B as its running parameter. We could have chosen v_{BE} instead as the parameter for i_C, but experience has shown that this choice exhibits severe nonlinearity. If, for a typical npn transistor, we draw the above two functions (we can actually do this graphically as it will be shown in EXP. III-4), we then get the following typical characteristic curves.

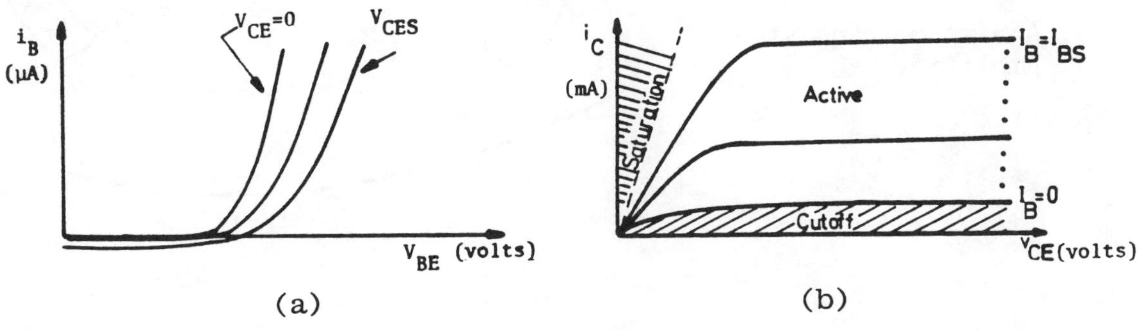

Fig. 3. Typical Input and Output Characteristic Curves for an npn Transistor.

In most of our present applications we use the transistor in its active region of operation as shown in Fig. 3b. The other two regions are used when the transistor acts as a digital switch. In this case the transistor becomes a highly nonlinear device. Generally speaking, for an npn transistor we have following operational modes:

Active Mode:　　　　　　When the Emitter-Base (EB) junction is forward biased and the Collector-Base (CB) junction is reverse biased.

Reverse Active Mode:　　When the CB junction is forward biased and the EB junction is reverse biased.

Cut Off Mode:　　　　　When both EB and CB junctions are reverse biased.

Saturation Mode:　　　　When both EB and CB junctions are forward biased.

As mentioned before we presently use the transistor when it operates in its active region. This mode is perhaps the closest that a transistor can get to be a linear device. Even in this region, if we are not careful to choose a correct *operating point*, we may run into difficulties. The right choice of operating or *quiescent* point is generally made with the help of proper biasing of the transistor. We will not consider the details of biasing until the next experiment.

We thus summarize our thoughts so far:

Transistors are three-terminal nonlinear devices which have a set of input and output characteristic curves similar to those given in Fig. 3. How we use these curves to analyze a circuit involving these elements is the subject of our present studies.

2.2. GRAPHICAL PROCEDURE TO STUDY A TRANSISTOR

Suppose the circuit of Fig. 4 has the characteristic curves of Fig. 3. Then, by the general procedure we have developed in Section III-3, we may redraw this circuit in terms of a network N which consists of three sub-networks; namely N_1, N_2 and a three-terminal device-transistor. In section III-3, we have presented two general procedures to handle this problem. We have shown in this section that given the characteristic curves of the TTD we must then look into the input and output relations between this device and the networks N_1 and N_2.

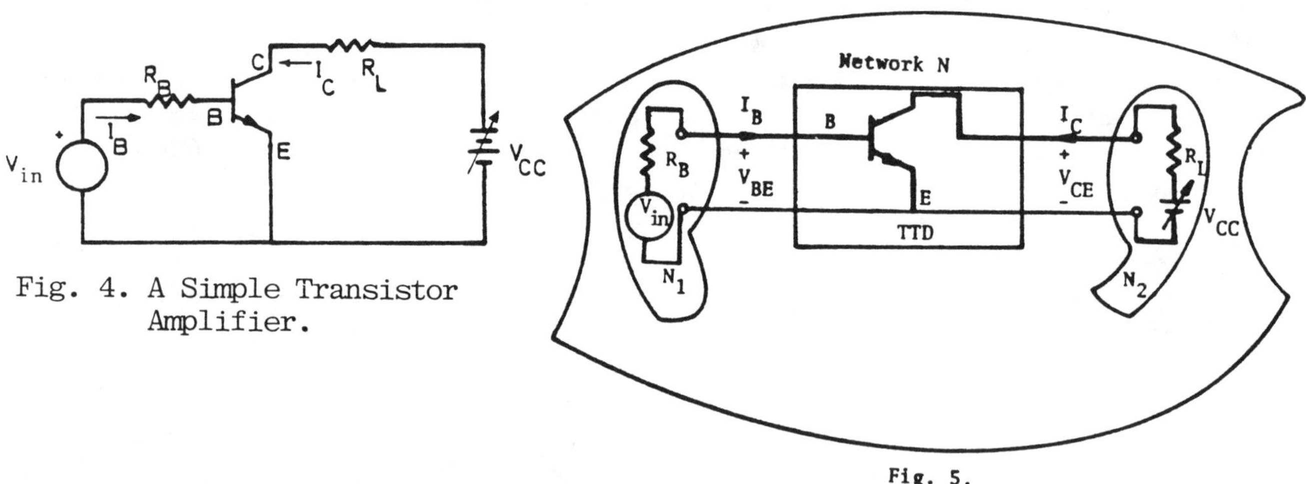

Fig. 4. A Simple Transistor Amplifier.

Fig. 5.

The input port relation, by inspection, is $V_{BE} = V_{in} - R_B I_B$, and the output port relation, by inspection again, is $V_{CE} = V_{CC} - R_L I_C$. Each of these equations are linear in their unknowns which are taken to be (I_B, V_{BE}) and (I_C, V_{CE}).

The operating point of the transistor not only must satisfy these equations coming from networks N_1 and N_2, but must satisfy the input and output characteristic curves of the TTD as well. As we discussed in Section III-3.2, the only way this can happen is when both the appropriate characteristic curve and the corresponding port equations cross each other. This intersection point is the appropriate operating point of the transistor. For the sake of discussion suppose the input unknowns are determined first. Then given the output characteristic curve of the TTD, we can draw the output port equation ($V_{CE} = V_{CC} - R_L I_C$) in the same characteristic coordinate plane. The intersection of this equation with the characteristic curve corresponding to the input current I_B, determined first, is the operating point of the transistor. The procedure is shown in Fig. 6. The port equations are called *load lines*. Since in this experiment we deal only with direct current, we call these *DC-load lines*. For the sake of example we have chosen $V_{CC} = 10$ volts, $R_L = 333\ \Omega$ and $I_B = 40\ \mu A$ to draw Fig. 6. The DC operating point becomes $V_{CE} = 2.9$ volts and $I_C = 21$ mA.

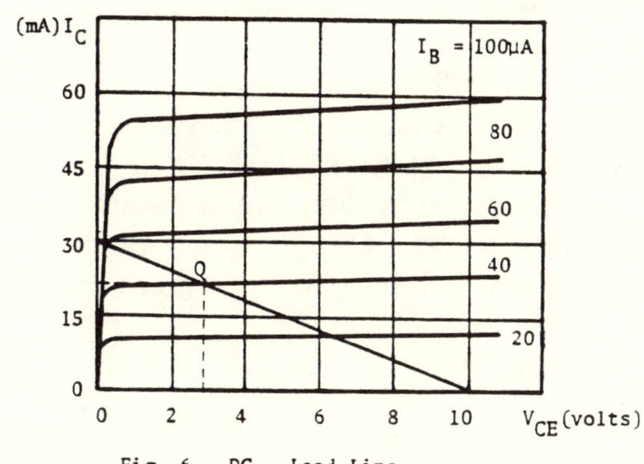

Fig. 6. DC - Load Line.

Here some readers may wonder how we determined I_B. Graphically this is not an easy problem. We do, however, realize that in spite of the mathematical complication of analyzing a set of nonlinear equations (this is exactly what we do when we use graphical techniques), we are dealing with a real engineering problem that always has a unique solution. If someone wants to know the graphical details about finding the DC operating point of a transistor then we suggest the following. First, draw the input load line in the (i_B, v_{BE}) plane. This gives an upper and a lower value for the operating current I_B as well the output parameter such as v_{CE}. Then draw the output load line in the (i_C, v_{CE}) plane and choose this range of I_B that was found earlier to shrink the range of I_B. Repeating this process several times will converge our efforts into a final solution for the DC operating point of the transistor. This is why we avoid the problem by just giving a value for I_B which incidently may only be an approximation to its true operating value.

Before we close this section we may raise the following question which is related to the input current I_B. Suppose I_B changes, say, as a function of time, then how do we account for these changes in the output operating point?

There are, of course, several reasons for changes in I_B. Perhaps the most convincing reason is when we have a sinusoidal input instead of the DC input voltage V_{in}; then, I_B changes as a function of time. Another reason may be that we are not sure about the exact value of I_B, thus we want to see both extremes of the output operating point as though I_B is changing between its two extreme values. We show the answer to this question in Fig. 7. But first, recall that the DC output load line remains the same as before. Suppose that I_B changes from I_{B_2} to two values of upper I_{B_3} and lower I_{B_1}, yielding Q^U and Q_L, respectively. Here clearly the I_C changes from I_C^L to I_C^U and V_{CE} changes from V_{CE}^L to V_{CE}^U. If these changes of I_B were static then we consider the operating point as a single point in the range between Q_L to Q^U. But suppose I_B purposely changes as, say, a sinusoidal function of time and its maximum becomes I_{B_3} and its minimum becomes I_{B_1}.

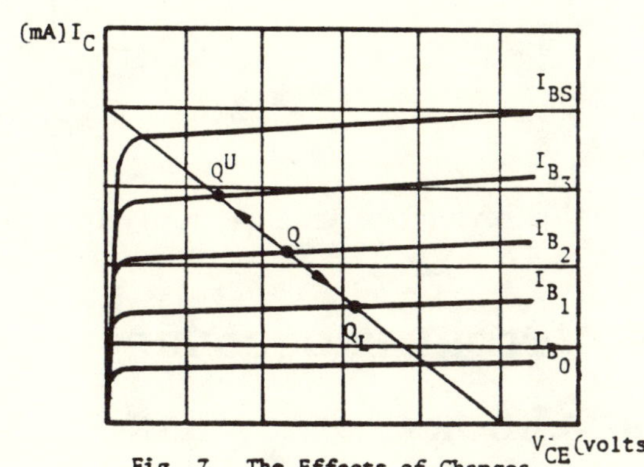

Fig. 7. The Effects of Changes of I_B in the Operating Point Q.

Then the I_C and V_{CE} also change as sinusoidal functions of time with the same frequency as I_B.

To provide a better understanding of the variations of the Q point in Fig. 7, we assume this operating point is changing as a function of time in the plane perpendicular to the output characteristic plane and passes through the DC-load line. The time axis is perpendicular to the characteristic plane, of course. We then rotate this I_B plane 90° around the DC-load line to become parallel with the characteristic plane. This is how we get Fig. 8, which shows that if the initial Q_o (or I_B) were chosen cleverly then the transistor operates linearly and its output current and voltage are without distortions. On the other hand distortion can be caused if the initial Q_o (or I_B) is chosen close to the edges of the active region of the transistor (Q'), as is shown in Fig. 8.

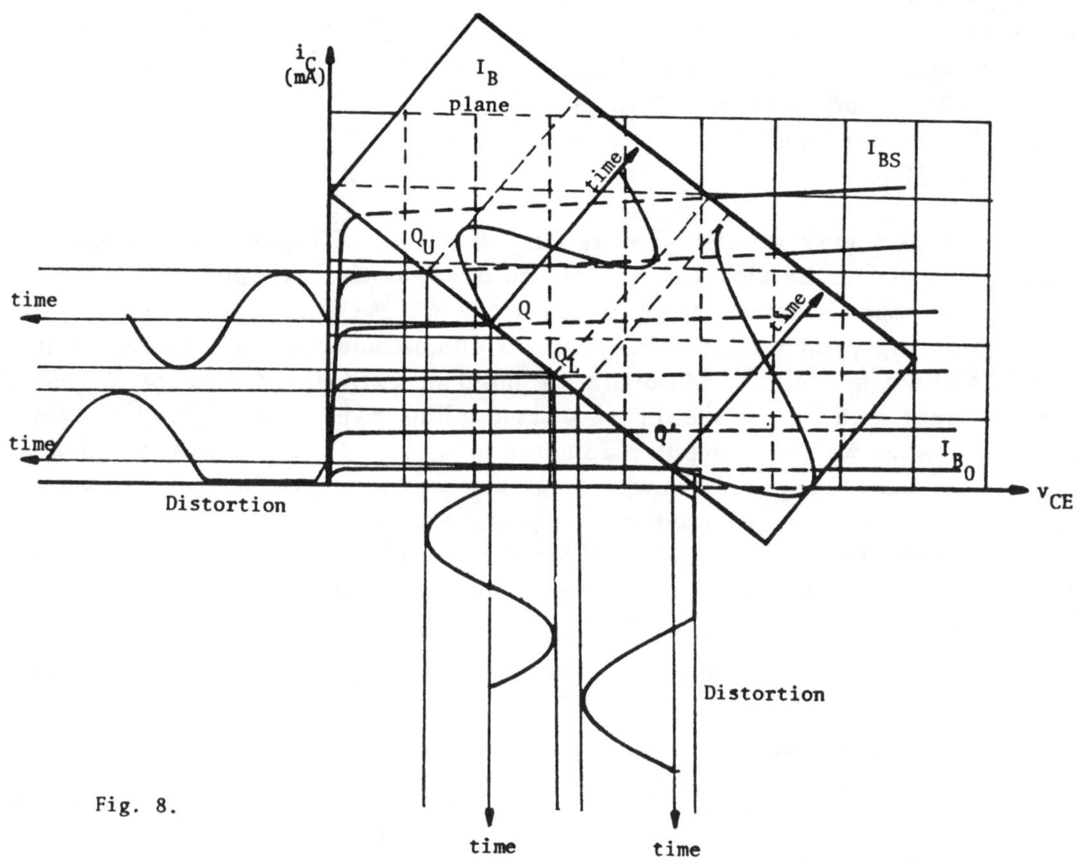

Fig. 8.

With this review we are now ready to perform the following laboratory work.

3. LABORATORY PROCEDURE

3.1. PRE-LABORATORY WORK

Carefully review the preceding materials as well as the fundamentals of two-port networks, *etc.*, that have been presented in the earlier sections of this part. If this is your first experiment in a new laboratory course, make sure that you have also read completely the materials presented in Part Zero of this textbook.

3.2. LABORATORY WORK

WARNING: To avoid damaging your transistor, before making any changes from one test to another, TURN OFF your power supply.

3.2.1. COLLECTOR CHARACTERISTIC CURVES OF THE BJT

Construct the circuit of Fig. 9. *Set* the voltage reference source (VRS) to provide $I_B \simeq 0$ μA on the current meter (full scale 80 μA). *Vary* V_{CE} from 0 to 7 volts in 1 volt increments by changing V_{CC} (the transistor power supply). *Use* the DMM to measure I_B, the oscilloscope to measure V_{CE}, and the laboratory current meter (VOM) to measure I_C. If using the VTVM be sure to select the VTVM scale which provides the maximum deflection when making the measurement. If the VOM is used, then be sure to provide an appropriate shunt in order to obtain a 5 mA full scale meter. *Plot* I_C versus V_{CE} with $I_B = 0$ μA for at least six different V_{CE} and the corresponding I_C each time. *Repeat* this entire procedure with values of I_B up to 35 μA in 5 μA increments. *Do not exceed* these voltage and current values. If there is one current meter to be used for both input and output loops, *do not short* C to E; just *turn off* the power supply first, then change I_B to reach its new value. Otherwise you may smell the transistor burning! *Plot your data on the chart of Fig. 10, with proper labeling and scales, of course!*

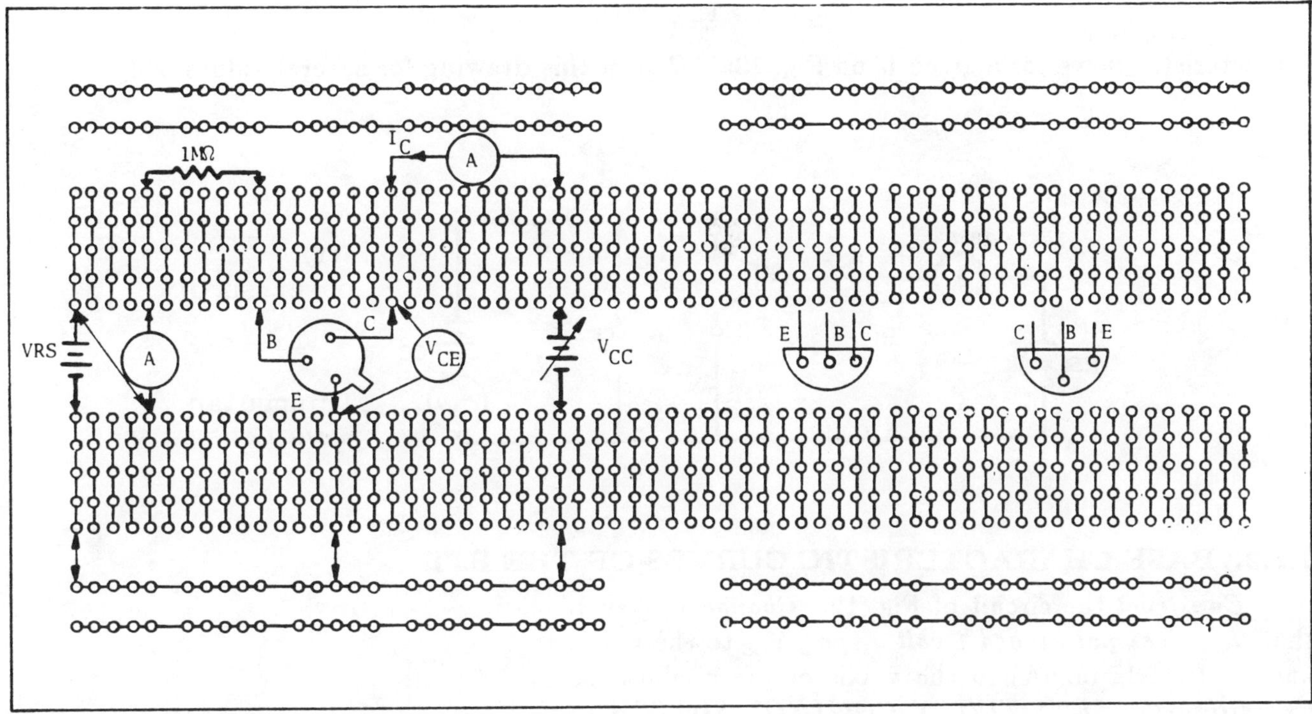

Fig. 9. Simple BJT Circuit with Three Typical Bottom Views of BJT.

The above procedure, though correct, is perhaps a little tedious. Instead we may construct the following circuit to get a similar result (cf., EXP. I-4).

Construct the circuit of Fig. 11. *Use* a low frequency triangular (or rectangular) wave generator with peak-to-peak amplitude that together with V_{CC} covers the entire range of the desired V_{CE}. *Set the SENSITIVITY CONTROL SWITCH* to 1 V/DIV on both channels. *Apply* V_{CE} to the x-axis and V_Y (i.e., I_C in mA) to the y-axis of the oscilloscope to see one characteristic curve for a given I_B. To change I_B repeat the same procedure as described earlier in the first part of this section. Notice that the triangular or rectangular waveform provides the necessary swept values for V_{CE}. *Plot* each

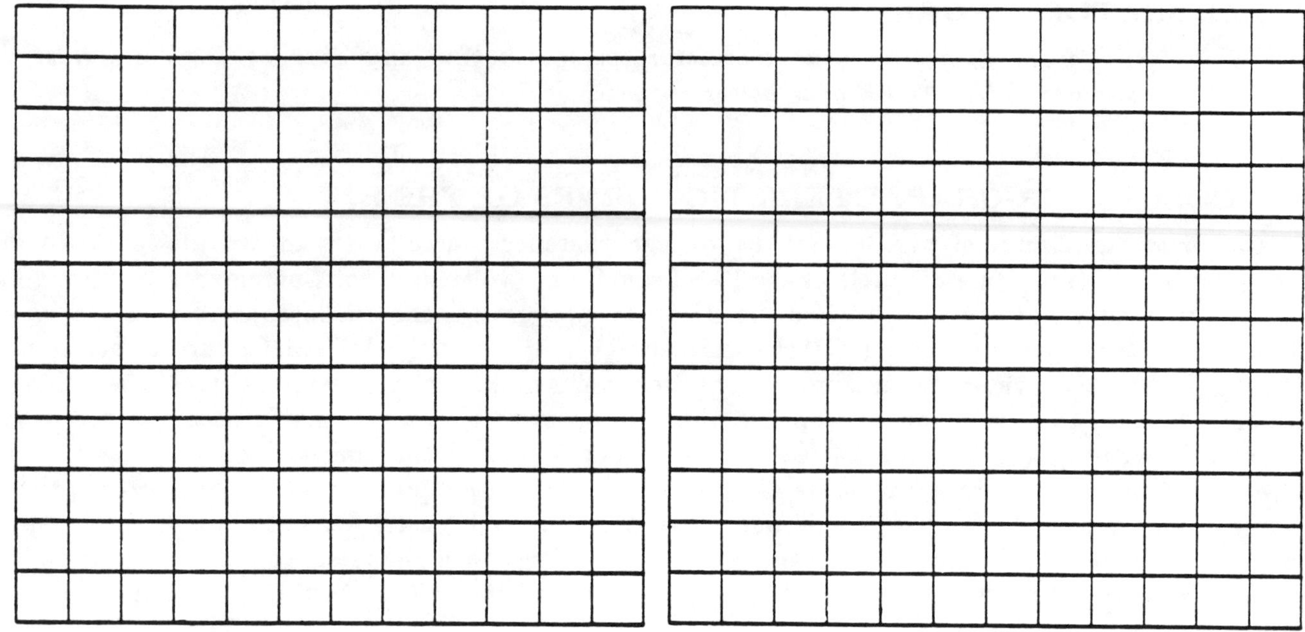

Fig. 10.

characteristic curve for a given I_B on Fig. 13a. *Repeat* this drawing for several values of I_B.

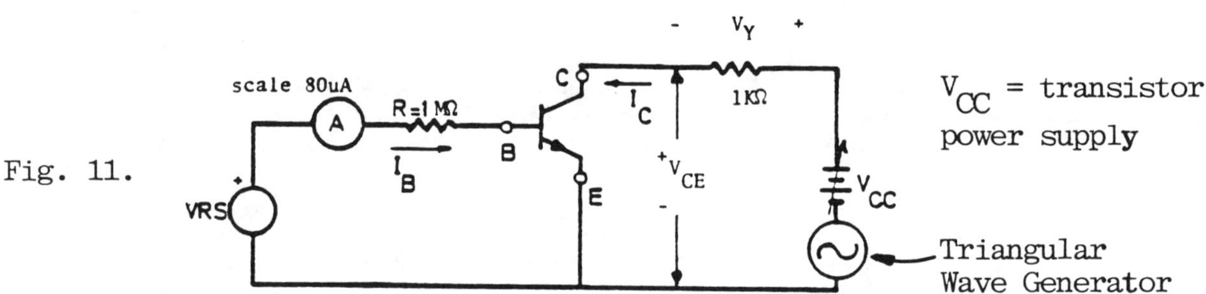

Fig. 11.

3.2.2. BASE CHARACTERISTIC CURVES OF THE BJT

Construct the circuit of Fig. 12. *Change* V_{CC} such that V_{CE} *does not exceed 1 volt.* *Apply* V_{BE} to the x-axis and V_Y (i.e., I_B in μA) to the y-axis of the oscilloscope by *setting the SENSITIVITY CONTROL SWITCH* to 0.1 V/DIV on both channels. *Repeat* this experiment for several values of $V_{CE} < 1$ volt. *Plot* these curves with proper labeling on Fig. 13b.

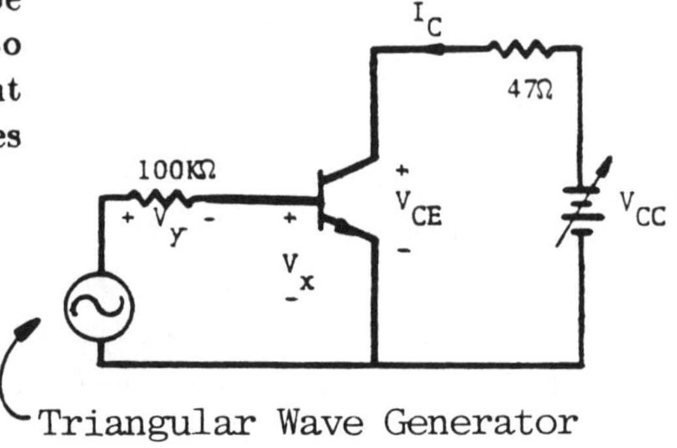

Fig. 12.

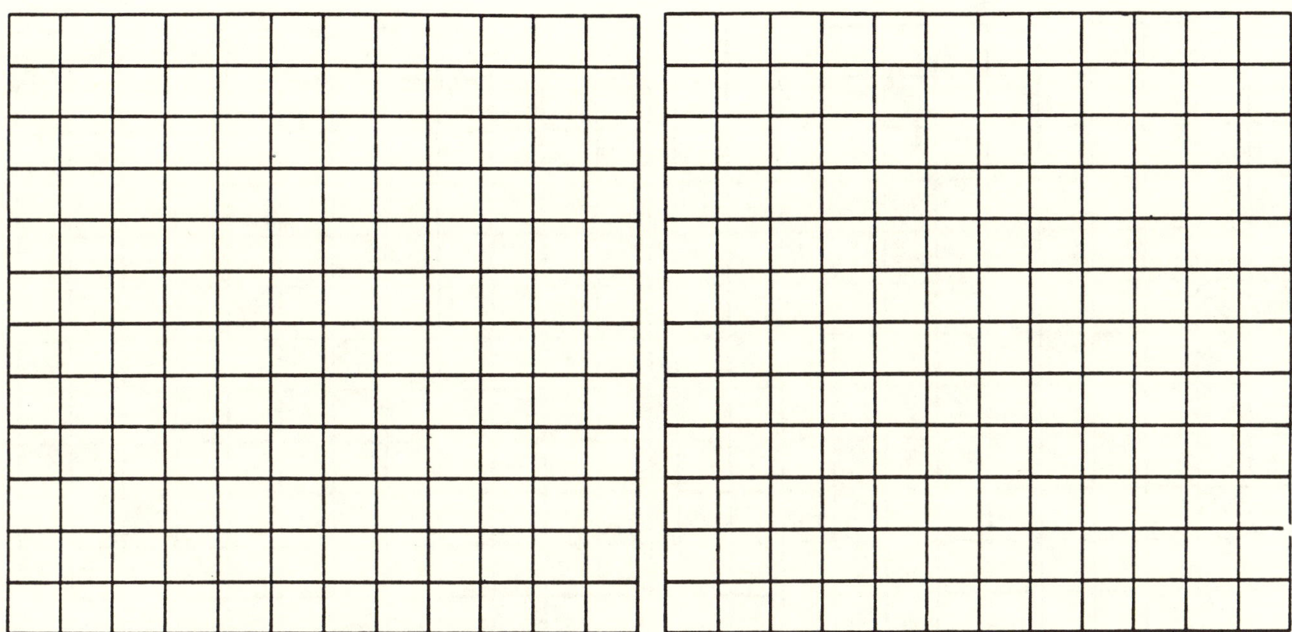

Fig. 13a.   Fig. 13b.

3.2.3. CHARACTERISTIC CURVES BY CURVE TRACER

Using the electronic curve tracer, *draw* the input-output characteristic curves of your BJT's. *Plot* these curves with proper labeling on the chart of Fig. 14. *Compare* your previous experimental data with these curves obtained from the curve tracer. *Discuss* any discrepancies.

Also generate these characteristic curves for the other two transistors using the curve tracer in Fig. 14c, and Fig. 14d. Save all of these characteristic curves for future experiments.

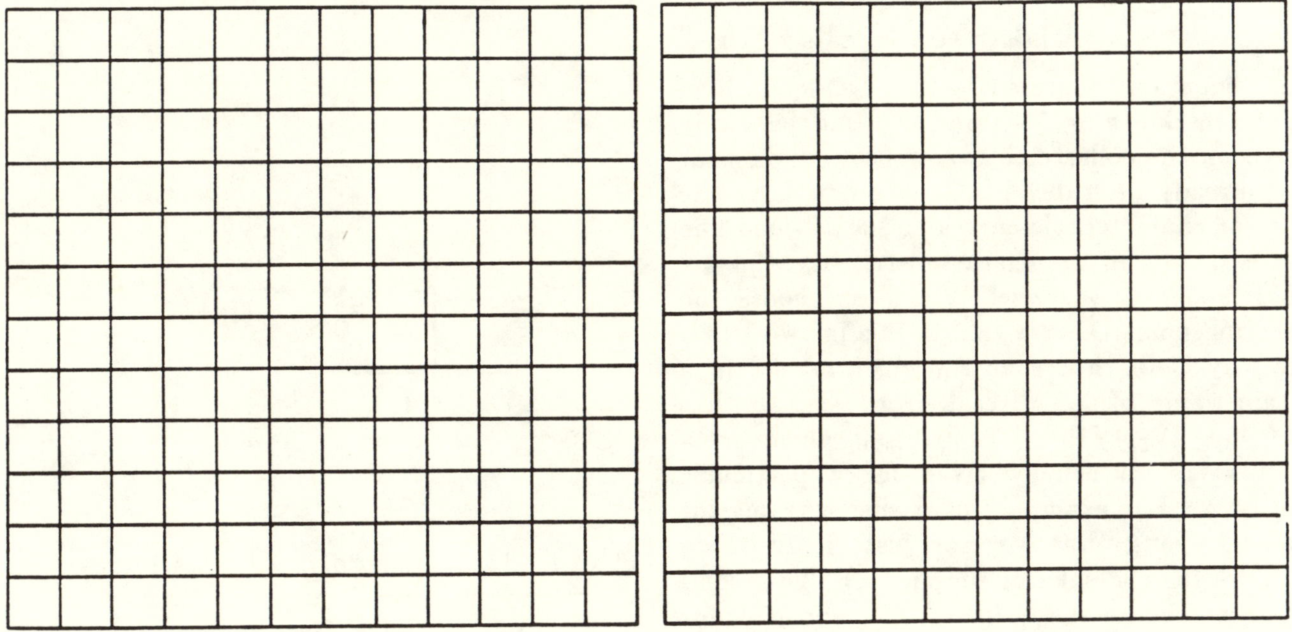

Fig. 14a.   Fig. 14b.

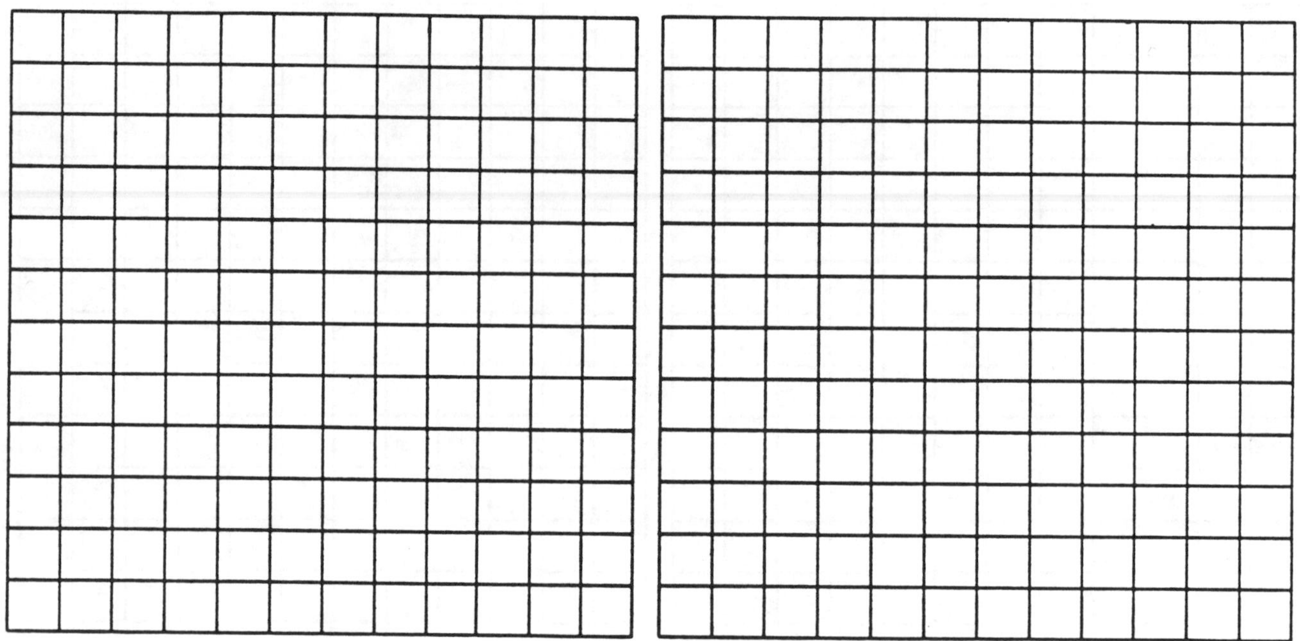

Fig. 14c.   Fig. 14d.

3.2.4. THE BJT AMPLIFIER

To show some basic properties of the BJT whose characteristic curves have been found in this experiment, we study the following simple one stage amplifier.

Construct the circuit shown in Fig. 15, using the same transistor whose characteristic curves have been found earlier. The resistor R_L is called the load resistor and the purpose of the *feedback* resistor R_B is for self biasing of the base (this concept will be studied in the next experiment). Because in most cases $I_C \gg I_B$, the Kirchhoff's voltage law results in

$$V_{CC} = (I_B + I_C) R_L + V_{CE} \simeq I_C R_L + V_{CE}.$$

This equation represents the DC-load line, which, by knowing V_{CC} and R_L, can be easily drawn on the collector characteristic curves that have already been developed. We now have the collector characteristic curves and a DC-load line that is drawn on the characteristic curves plane. *What is left to be done?* We must choose the operating point Q. We can do this in two ways. One way is to choose an R_B which results in a certain value of I_B. (Recall from Fig. 15 that $I_B = (V_{CE} - V_{BE}) / R_B$. This V_{BE}, which is a small value, is usually given for a particular transistor. For example, for a silicon transistor $V_{BE} = 0.65$ volts, thus we may just choose $I_B \simeq V_{CE} / R_B$ for the sake of simplicity.) The second way is to choose the Q (or I_B) first and look for the corresponding R_B to give such an operating point.

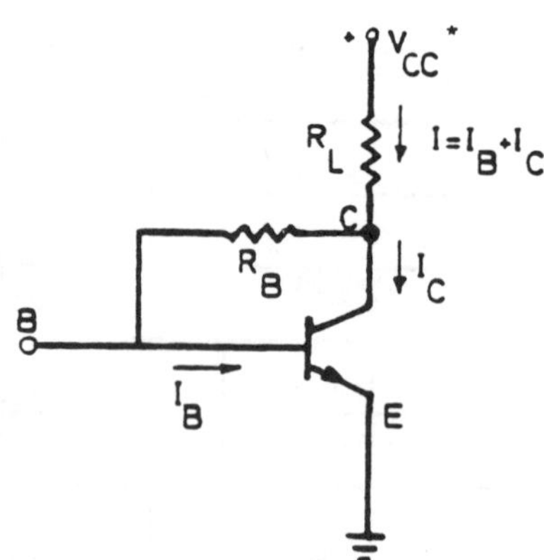

Fig. 15. A Simple BJT Amplifier.

---

* Usually the + is understood and is not written anymore.

For numerical values *set* $V_{CC} = 10$ volts, $R_L = 1$ kΩ. *Draw* the DC-load line on the collector characteristic plane of the BJT that you found earlier. *Choose* an operating point Q in the middle of the active region of this transistor. This choice allows the maximum possible changes in input current. Then *find* the value for R_B that gives such an I_B. *Use* the GR-Decade Resistance Box for R_B. What is the value of R_B? *Check* again to make sure this R_B will give the same Q. If not, *find out* how much the difference is. Do any drastic changes occur in the first operating point you selected? *Explain* your observations in your report.

3.2.5. EFFECTS OF INPUT SIGNAL

Add an input signal current to the circuit of Fig. 15 by constructing a current source (voltage source plus a large series resistance). *Connect* this source as shown in Fig. 16, where R_s is chosen to be a 1 MΩ precision resistor.

The output current of the current source is approximately the output voltage indicated on the VRS divided by 1 MΩ. *Using* the current source to provide base signal currents from 0 to $+40$ μA in 5 μA steps, *determine* the corresponding values of I_C and V_{CE} for each value of signal current. This *base current must be measured by the current meter in the base circuit.* *Repeat* this process for values of base current from 0 to -40 μA in 5 μA steps. *Tabulate* and *plot* on Fig. 17 these measured values of I_C and V_{CE}. *Compare* these values with those expected from the earlier characteristic curves. *Note* that the appropriate value of the base current for determination of the expected values of I_C and V_{CE} (for a particular signal current) is the sum of the base signal current and the base bias current. Does the collector current increase with an increase in the base current? Does the collector voltage increase with an increase in the base current?

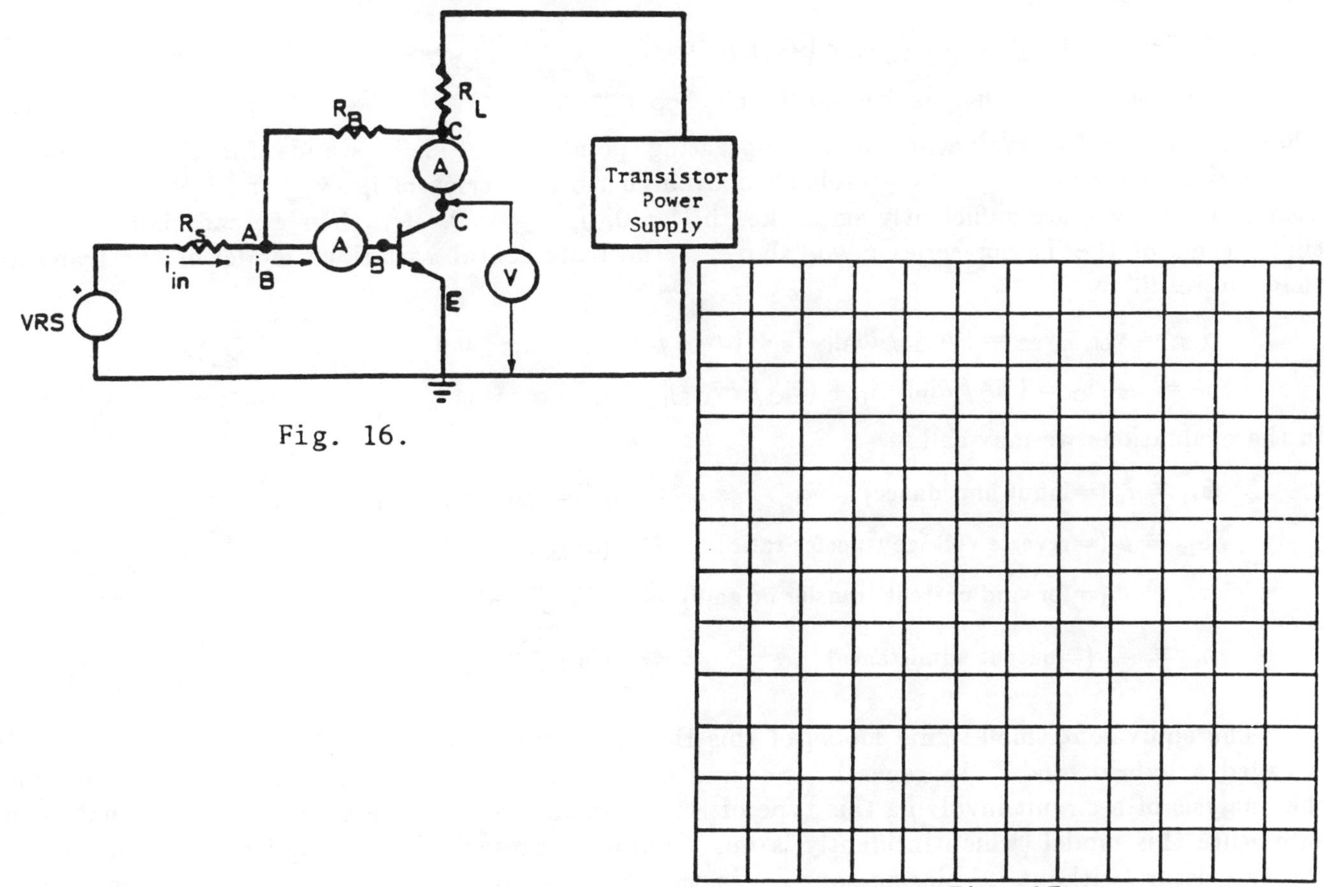

Fig. 16.

Fig. 17.

3.2.6. TIME VARYING INPUT SIGNAL

Replace the VRS with a sinusoidal wave from the function generator (FG). *Use* the oscilloscope to measure both the input signal voltage V_{in} and the output voltage V_{CE} between the point C and ground. *Set* the output of FG to 1 kHz. *Increase* the input voltage until the output waveform just begins to distort. *Record* this value of input signal current. *Determine* the voltage gain G_v of the circuit where $G_v = V_{out} / V_{in}$. *Generate* a dynamic operating point curve and the corresponding I_C and V_{CE} as described by Fig. 8.

4. SMALL-SIGNAL MODEL OF THE BJT

From our experimental work and the resulting characteristic curves we are absolutely convinced that the BJT is a nonlinear device that under proper conditions can be approximated by a linear equivalent small-signal model. Here we reiterate the analytical procedure developed in Section III-3.1 to find this model for the BJT.

We recall that the following input and output relations were chosen for this BJT:

$$v_{BE} = v_{BE}(i_B, v_{CE}), \text{ or } i_B = i_B(v_{BE}, v_{CE}), \text{ and}$$

$$i_C = i_C(i_B, v_{CE}).$$

Suppose a proper biasing has been made for this transistor to bring its operation into its active region. If we show this operating point by Q and the corresponding quantities by $\bar{i}_C$, $\bar{i}_B$, $\bar{v}_{CE}$, and $\bar{v}_{BE}$, and the incremental variations around this operating point by $\Delta i_C \triangleq \hat{i}_C = i_C - \bar{i}_C$, $\Delta i_B \triangleq \hat{i}_B = i_B - \bar{i}_B$, $\Delta v_{CE} \triangleq \hat{v}_{CE} = v_{CE} - \bar{v}_{CE}$, and $\Delta v_{BE} \triangleq \hat{v}_{BE} = v_{BE} - \bar{v}_{BE}$, then we can use a Taylor series expansion of v_{BE} and i_C as follows:

$$v_{BE} = \bar{v}_{BE} + (\partial v_{BE} / \partial i_B)|_* \hat{i}_B + (\partial v_{BE} / \partial v_{CE})|_* \hat{v}_{CE} + \cdots, \text{ and}$$

$$i_C = \bar{i}_C + (\partial i_C / \partial i_B)|_* \hat{i}_B + (\partial i_C / \partial v_{CE})|_* \hat{v}_{CE} + \cdots,$$

where $|_*$ stands for evaluating at the operating point and + ... , stands for the higher-order differentiation terms of i_C and v_{BE} multiplied by higher-order terms of $\hat{i}_B$, $\hat{v}_{CE}$ and $\hat{i}_B \hat{v}_{CE}$, *etc.* If we assume $\hat{i}_B$ and $\hat{v}_{CE}$ are sufficiently small; i.e., $|\hat{i}_B|^2 \simeq 0$, $|\hat{i}_B \hat{v}_{CE}| \simeq 0$, *etc.*, then we can choose the first three terms of the Taylor series given above as the incremental variations model of the transistor that will result in

$$\hat{v}_{BE} = v_{BE} - \bar{v}_{BE} = (\partial v_{BE} / \partial i_B)|_* \hat{i}_B + (\partial v_{BE} / \partial v_{CE})|_* \hat{v}_{CE}, \text{ and}$$

$$\hat{i}_C = i_C - \bar{i}_C = (\partial i_C / \partial i_B)|_* \hat{i}_B + (\partial i_C / \partial v_{CE})|_* \hat{v}_{CE}.$$

In these equations we may call

$$h_{11} \triangleq r_n \ (=\text{input impedance}) \qquad \triangleq (\partial v_{BE} / \partial i_B)|_* \ ;$$

$$h_{12} \triangleq \mu \ (=\text{reverse voltage transfer ratio}) \qquad \triangleq (\partial v_{BE} / \partial v_{CE})|_* \ ;$$

$$h_{21} \triangleq \beta \ (=\text{forward current transfer or gain}) \qquad \triangleq (\partial i_C / \partial i_B)|_* \ ;$$

$$h_{22} \triangleq \frac{1}{r_o} \ (=\text{output admittance}) \qquad \triangleq (\partial i_C / \partial v_{CE})|_* \ .$$

The equivalent small-signal model of this BJT becomes that of Fig. 18. This particular model is called a hybrid model. In general, $\mu \simeq 0$ and $r_o \simeq \infty \ \Omega$. Therefore, this choice further simplifies the analysis of a circuit involving this type of BJT. At any rate for the purpose of local analysis we substitute this model (which incidently is not a unique representation of the transistor) in place of the transistor (without its biasing sources) in the network consisting of this TTD. The remaining work becomes that of simple circuit analysis. To generate a *global* analysis for the active region we

need to repeat the above process for several operating
points. This analysis requires more sophisticated techniques such as computer-aided circuit design.

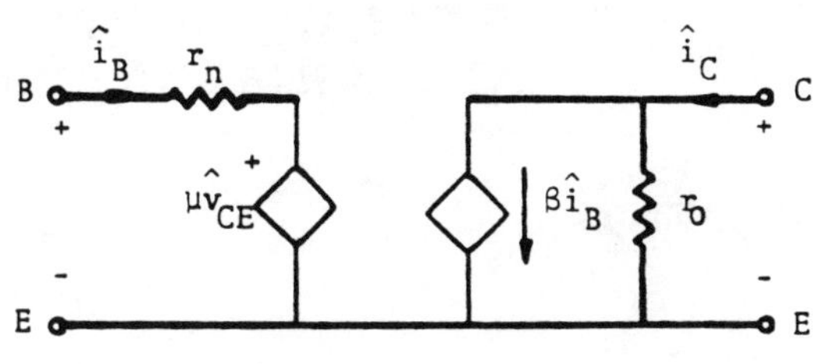

Fig. 18.   Small-Signal Model of a BJT.

The real issue, however, is how do we measure these parameters for a given BJT and at a particular operating condition?

Think about it until the next experiment.

5. FOR YOUR REPORT

To document your data, follow the procedures of preparing our standard report as described in
Part Zero. Make sure that the characteristic graphs that you submit with your report are properly
labeled. Save copies of these graphs which are to be used in the next several experiments. Answer
all the relevant questions including those in the text of this experiment.

EXPERIMENT III-2

BIPOLAR JUNCTION TRANSISTORS BIASING AND SMALL–SIGNAL MODEL PARAMETERS MEASUREMENTS

EQUIPMENT:

EEP	PRECISION RESISTOR (1 MΩ)
VOM	POTS (10 kΩ/2, 100 kΩ/2)
BJTs (3)	GR-DECADE RESISTANCE BOX
HAIR DRYER	CAPACITORS (0.1 μF, 100 μF)
RESISTORS (47 Ω, 1 kΩ, 4.7 kΩ/2, 10 kΩ/2, 47 kΩ/2, 100 kΩ)	

Table of Contents

Page No.

1. PURPOSE

The purpose of this experiment is to review thoroughly the implications of self biasing of, say, a single stage transistor amplifier that has been introduced in the last experiment and to introduce measurement procedures to find the small-signal model of a transistor, in particular that of the transistor being used in this experiment.

2. INTRODUCTION

The circuit of Fig. III-1-15 shows the simplest self-biasing method that exists. This so-called inverter transistor amplifier is indeed the fundamental block of all amplifier circuits. The main reason for having any biasing circuits in the first place is to avoid the use of several power supplies as would be otherwise needed for the various active elements of a given circuit. Thus we must devise some auxiliary circuits that are going to generate various voltage levels for intermediate applications of the active devices of the circuit. All of these biasing circuits share the same main power supply of the EEP as their only DC source. Naturally, the main power supply must provide sufficient current for all of these circuits. *To reduce the parasitic effects of these auxiliary circuits on the main power supply, it is advisable to put a large capacitor between each of these circuits and the main line.*

Proper design and functioning of these auxiliary circuits is the essence of the *self-biasing* technique. Any wrong choice of elements or any other design variation may cause drastic changes in the transistor's operating point. Recalling Fig. III-1-8, we realize that changes in I_B, *per se*, can shift the transistor's operating point greatly, thus moving the operation of the transistor into any one of three different modes described in EXP. III-1.

Another source of variation of the transistor's operating point (or any active device, in general) is the circuit temperature which can rapidly increase for a number of reasons thus causing severe problems. This effect is very critical and generally affects the entire performance of the circuit. Most manufacturers provide specification data sheets which state the temperature operating range of the device. It is thus advisable to review these data before using any particular active element in a circuit (cf., Fig. III-27).

Finally, we may say that by a proper or stable biasing of a transistor we can keep its operation within any reasonable range of temperature variations as we please.

2.1. STABLE BIASING OF THE BJT OPERATING POINT

Recall the circuit in Fig. III-1-1a, wherein V_{CC} and V_{BB} are needed for basic operation of that simple common-emitter BJT. In the circuit of Fig. III-1-15, we used only V_{CC} and we generated from the same power supply V_{BB} *via* R_B. Here we give a more complete biasing circuit with better flexibilities than that of the circuit in Fig. III-1-15. Consider the circuit in Fig. 1, which is equivalent to that in Fig. 2, since the functions of resistors R_1 and R_2 are the same as a voltage divider which generates $V_{BB} = R_2 V_{CC} / (R_1 + R_2)$, with $R_B = R_1 \| R_2$. Thus the circuit in Fig. 2 is the same (except for R_E) as that in Fig. III-1-2. Clearly we see that there is no need to have two power supplies to generate V_{CC} and V_{BB}. The choices of R_1 and R_2 must be made such that I_B (corresponding to the operating point) falls in the center of the active region of the transistor and remains in that vicinity for the maximum swing of the operating point. This is called stable biasing of the BJT.

Referring to Fig. 2, an expression for I_B and that for I_C can be developed from the following equations:

$$V_{BB} = I_B R_B + V_{BE} + (I_B + I_C) R_E \,,$$

$$V_{CC} = I_C R_C + V_{CE} + (I_C + I_B) R_E \,.$$

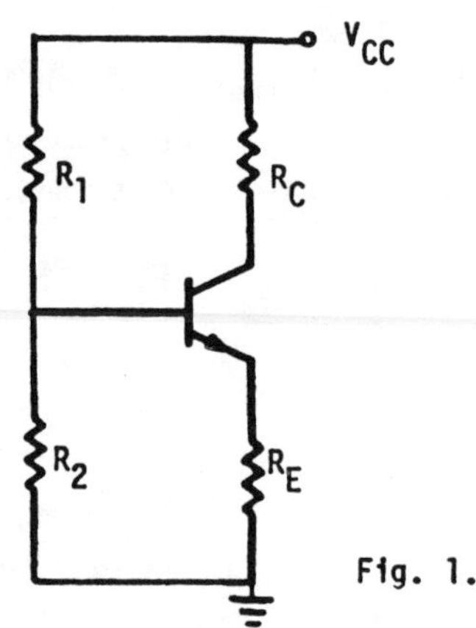

Fig. 1.

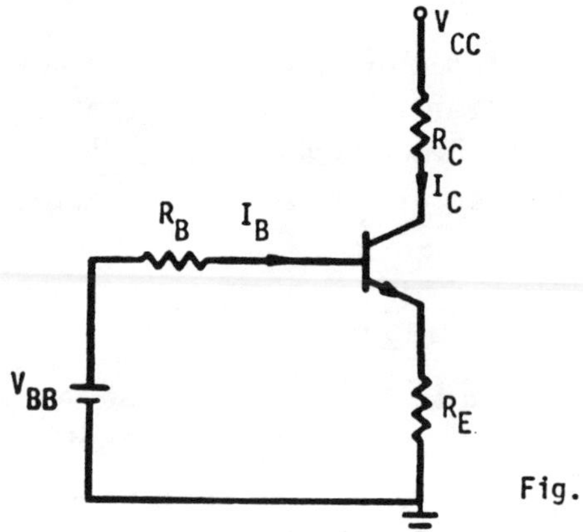

Fig. 2.

Here we conclude that (with $I_B = I_C / \beta$) we have

$$I_B = (V_{BB} - V_{BE}) / [R_B + (1 + \beta) R_E] , \text{ and}$$

$$I_C = (V_{CC} - V_{CE}) / [R_C + (1 + 1/\beta) R_E] .$$

If we choose β sufficiently large such that $\beta R_E \gg R_B$ we get

$$I_B \simeq (V_{BB} - V_{BE}) / \beta R_E , \text{ and}$$

$$I_C \simeq (V_{BB} - V_{BE}) / R_E .$$

The above analysis actually implies that this I_C is approximately independent of V_{CE} when the transistor is in its active mode. But in reality this I_C is a function of β, V_{BE}, and the collector leakage current I_{CO} that has been introduced in Section III-4.3. Any changes in these quantities will result in an ultimate change in the I_C. To complete our discussions on this matter we suggest a relation for I_C as a function of β, V_{BE} and I_{CO} be generated. Using the basic concepts of error analysis as discussed in Section 0.9-4, a relation for $\Delta I_C / I_C$ as a function of $\Delta \beta$, ΔV_{BE} and ΔI_{CO} is then to be developed. Study the effects of these changes on the operating conditions of the transistor.

2.2. PARAMETER DETERMINATION OF A SMALL-SIGNAL MODEL OF THE BJT

In Section 4 of EXP. III-1, we have developed a small-signal model for the BJT, that has been represented by Fig. III-1-18 and is repeated here for convenience in Fig. 3. Here we have

$$h_{11} \triangleq r_n \triangleq (\partial v_{BE} / \partial i_B)|_*$$

$$h_{12} \triangleq \mu \triangleq (\partial v_{BE} / \partial v_{CE})|_*$$

$$h_{21} \triangleq \beta \triangleq (\partial i_C / \partial i_B)|_*$$

$$h_{22} \triangleq \frac{1}{r_o} \triangleq (\partial i_C / \partial v_{CE})|_* .$$

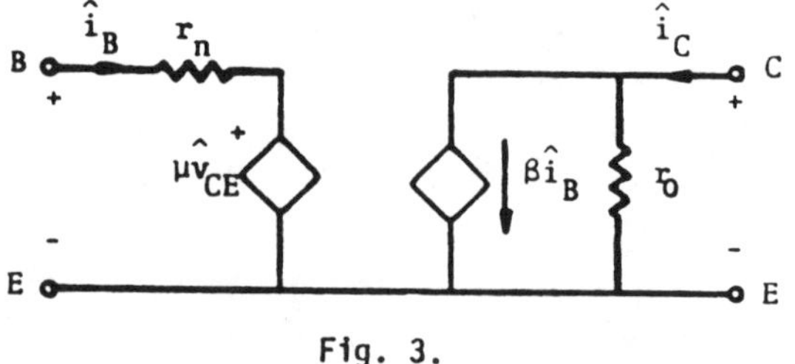

Fig. 3.

Also recall that the only available information about this transistor, so far, is its input and output characteristic curves developed earlier in EXP. III-1. Thus it is natural to ask if a relationship exists between these parameters and those characteristic curves. Actually, the relationship is clear. First, we have to select an operating point; then we have to compute these various derivative functions at this point. We may approximate these values by $\Delta x / \Delta y$ instead of $\partial x / \partial y$. We may also select $\mu \simeq 0$ and $r_o = \infty$ for simplicity of analysis. Then we compute the other two parameters from these characteristic curves.

An alternative to this approach is to measure these various quantities and/or parameters directly from the special circuits such as those given in Fig. 4 and Fig. 5.

Consider the circuit of Fig. 4a, with its equivalent small-signal model in Fig. 4b. Here we have made the necessary adjustment to the biasing sources in order to have a particular desired operating point of I_C and V_{CE}.

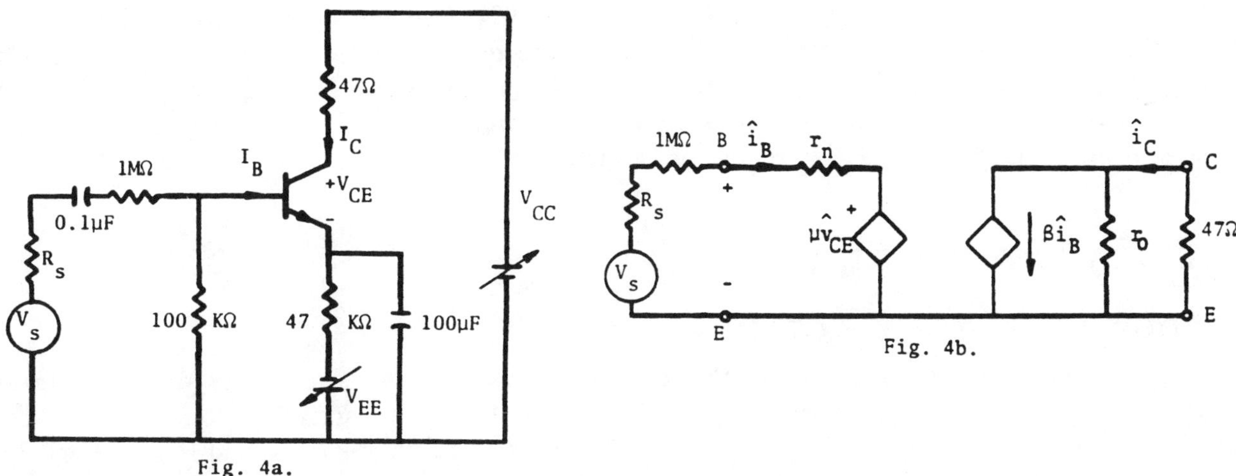

Fig. 4a.

Fig. 4b.

These capacitors make the measurement less noisy. We choose an input frequency of 1 kHz in order that the effects of these capacitors be minimum. We also dropped the 100 kΩ resistor because it is much larger than r_n. Similarly r_o can be dropped because it is much larger than 47 Ω. If we use an AC voltmeter with very high input impedance to read the voltage across a 1 MΩ precision resistor, then this reading is proportional to the I_B measured in μA. Similarly the voltage across a 1000 Ω resistor is proportional to I_C in mA. Then we define $\beta = I_C / I_B$. Also, to measure r_n we measure the input resistance across BE by measuring V_{BE} and I_B as described before.

To measure the other two parameters we have to excite the output circuit as shown in Fig. 5a, with the small-signal equivalent model of Fig. 5b.

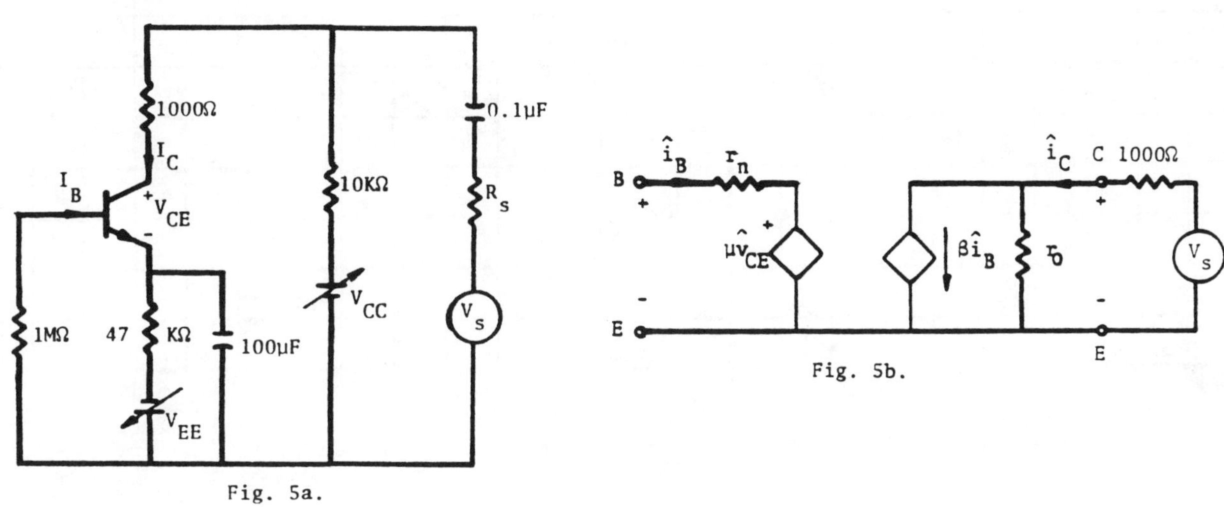

Fig. 5a.

Fig. 5b.

Here we keep the same operating point and the same frequency of the 1 kHz to reduce the effect of coupling capacitors. Since $i_B \simeq 0$ (because of 1 MΩ input resistance that makes the input acts like an open circuit), we have $r_o = V_{CE} / I_C$ and $\mu = V_{BE} / V_{CE}$. Directly measuring these quantities will yield the corresponding other two parameters.

2.3. REMARKS AND CONCLUSIONS

In this and the last experiment we have tried to study the transistor in a most elementary manner. We have not yet mentioned problems associated with the frequency range over which these models are valid. Our small-signal model is valid for low frequency operation only. As the frequency increases this model loses its accuracy due to the presence of many stray capacitors between various junctions that have been left out in our small-signal model. The high frequency model of a transistor is much different than the above small-signal model. Finally we must point out that our small-signal model is valid for active region of the transistor only.

3. LABORATORY PROCEDURE

3.1. PRE-LABORATORY WORK

From the best and in your opinion the most accurate characteristic curves you have developed for your transistor choose an operating point in the center of its active region. Decide the base current that should be chosen for this operating point and the values of R_1 and R_2 for a self-biasing operation as depicted by Fig. 1. From these characteristic curves, develop an approximate small-signal model as was suggested in Section 2.2, for this transistor. But remain flexible about these values since you may have made mistakes in your measurement. Thus check your values against those found by your partner and settle all the differences before choosing any element values for actual construction of your circuit. Use the chart of Fig. 6 to draw your characteristic curves.

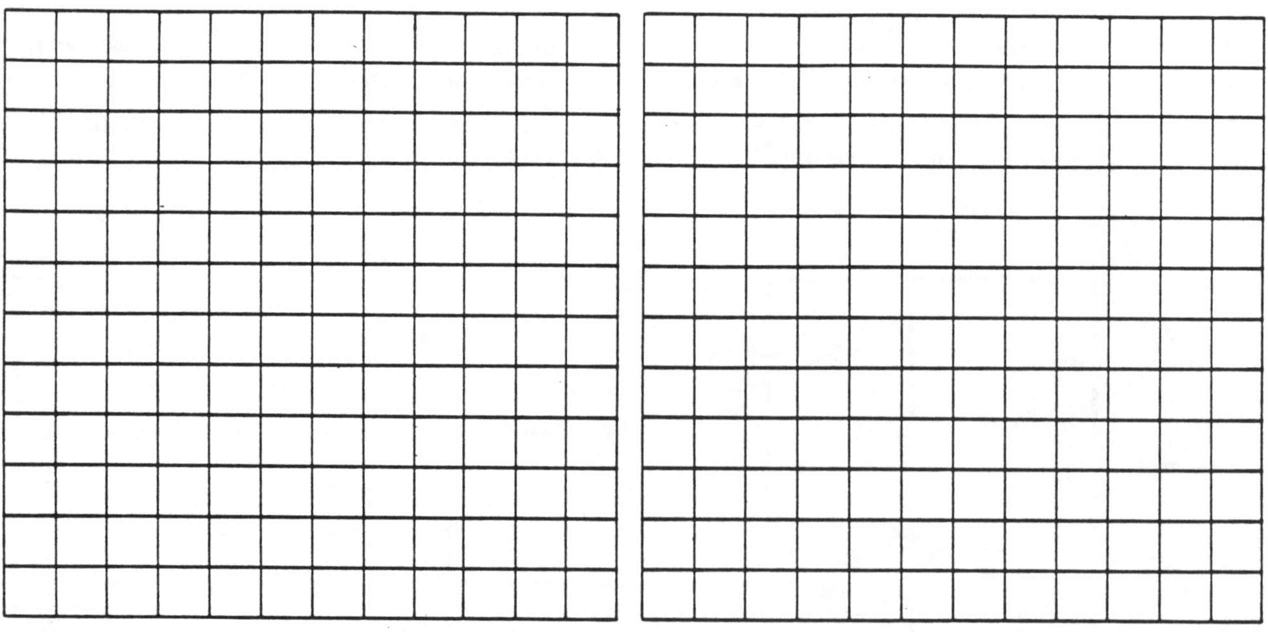

Fig. 6.

3.2. LABORATORY WORK

Construct the circuit of Fig. 7. With the help of two pots and the GR-Decade Resistance Box, *adjust all* the biasing parameters to achieve the desired transistor operating point. *Record* V_o as sinusoidal input with the frequency of 1 kHz is being applied. *Compare* these values to that predicted from the small-signal model calculated before. *Discuss* any drastic discrepancies you have observed. *Increase* V_s in amplitude and *detect* the value for which output begins to distort. *Turn off* the power supply and V_s and *remove* the emitter by-pass RC network. Then *repeat* the above testing for the same input function you have applied earlier. *Discuss*

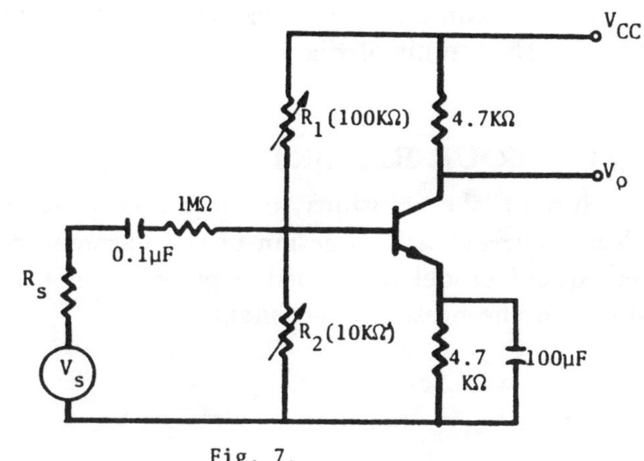

Fig. 7.

your observations. If it is feasible to increase the transistor temperature, for instance, by blowing hot air (from a hair dryer) at it, *study* the effects of heat change on its operating point.

Then *turn* the power supply and input source *off* and *construct* the circuit in Fig. 4a, and subsequently that in Fig. 5a, *to measure* the small-signal model of the transistor as described in Section 2.2, and *for the same operating point as before*. Now, *use* these parameters to *verify* and/or *contradict* your earlier small-signal model of the transistor. By the end of this experiment you must be able to identify the best or the more accurate small-signal model of your BJT.

4. PROBLEM SET

The common emitter amplifier with an emitter resistor is also called an emitter follower amplifier (cf., Fig. 8).

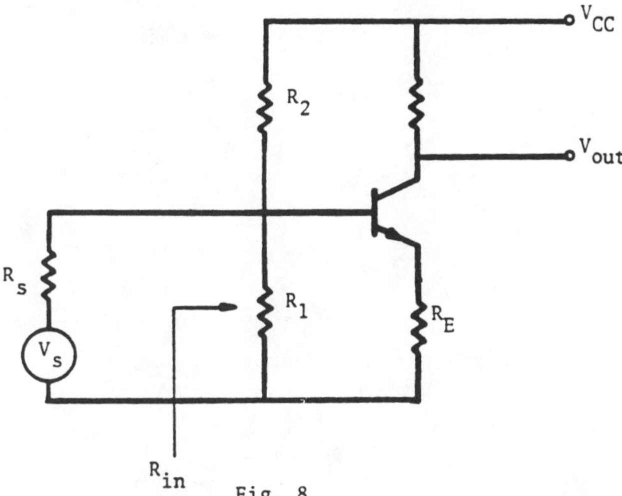

Fig. 8.

1. Determine the input resistance of this circuit.

2. Using an equivalent small-signal model of the transistor, determine the value of the input resistance and voltage gain V_{out} / V_s.

3. From this gain what are your conclusions about this new circuit?

4. Can you find an approximate expression for the above gain?

5. What is the physical meaning of this gain?

6.   Can we make this gain expression independent of certain parameters?

7.   Obtain an expression for the output resistance R_{out} in terms of the remaining parameters of the circuit of Fig. 8.

5. FOR YOUR REPORT

Submit the solutions to the above problem set and your computations of the changes in the collector current as a function of the parameters asked in Section 2.1. Your report must contain the small-signal model developed *a priori* from the characteristic curves of EXP. III-1, as well as your data from the present experiment.

EXPERIMENT III-3

**BIPOLAR JUNCTION TRANSISTOR CAPACITIVE
COUPLED LOAD AND MULTI-STAGE FEEDBACK AMPLIFIERS**

EQUIPMENT:

EEP	GR-DECADE RESISTANCE BOX
BJTs (3)	PRECISION RESISTOR (1 MΩ)
POTS (10 kΩ, 100 kΩ/2)	CAPACITORS (0.1 μF, 100 μF)

RESISTORS (100 Ω, 1 kΩ, 4.7 kΩ/2, 10 kΩ/2, 47 kΩ/2, 100 kΩ/2)
pnp SILICON POWER DARLINGTON BD896 (OPTIONAL)

Table of Contents .... Page No.

1. PURPOSE

The main purpose of this experiment is to familiarize the student with a method of blocking the DC component of the output signal voltage of an amplifier stage and thus allowing only the AC component of this signal to pass to the next stage or load. The concept of the AC-load line will be studied. Also the frequency response of an amplifier and its operation outside its active region will be investigated. We will construct a simple feedback amplifier circuit that consists of a two-stage feedforward transistor amplifier and a resistive feedback circuit to study its basic applications and/or properties in order to verify our theoretical expectations from this circuit. Finally a Darlington transistor pair is constructed to investigate some of its basic properties.

2. INTRODUCTION

A transistor has many applications in both digital and analog electronic circuits. These applications stem from the operational regions that we choose for a transistor. In the analog electronics that we are now mainly concerned with, the transistors work in their active region, basically as amplifiers. In the last two experiments we have learned the basics, and we have collected data regarding the fundamental properties of transistors. We now have a reasonably accurate small-signal model as well as an accurate set of characteristic curves for our transistors. We need this information to complete our experiment.

2.1. AC-LOAD LINE

Consider the circuit of Fig. III-1-15, that is properly biased, resulting in an operating point Q (i.e., I_C and V_{CE} as well as I_B (or R_B) are all determined) that is in the center of the active region of the transistor. This operating point is also known as the zero-excitation operating point. The DC-load line of this amplifier goes through point Q and can easily be drawn on the characteristic curves plane. If a sinusoidal input is now applied to the base of this amplifier we will have a sinusoidal wave with nonzero average at the collector of the transistor. This experiment suggests that V_{CE} will be changed to a new value which is the former V_{CE} plus or minus the contribution of this sinusoidal input to the output voltage. Further, if we want to have a pure zero average AC waveform at the output we must add an RC filter to this circuit to block the DC component of the output signal (cf., Fig. 1). The effect of capacitor C is that it blocks the output DC level and its impedance in the frequency range of the applied input signal is very small (or is effectively a short circuit).

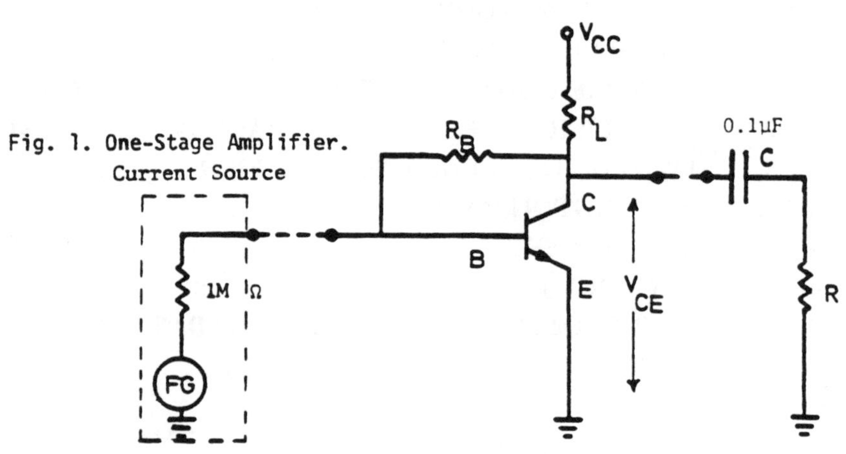

Now, the output that is a pure sinusoidal wave can be seen across a new load resistor which is $R_L || R$ (since the impedance of C is negligible). This new effective load resistance, and the fact that the operating point changes as a result of a sinusoidal input will yield a new load line that is called the AC-load line, shown in Fig. 2. In short, the AC-load lines are lines having a slope of $-(R + R_L) / RR_L$, which corresponds to the effective output resistance of circuits with only AC component output signals. These lines are drawn in the same output characteristic plane and at the same DC operating point. If this operating point does not provide enough space for input current variations and brings the transistor outside of its active region, then we may change the operating point accordingly and the new AC-load line can be drawn crossing this new point with a slope of $-(R + R_L)/RR_L$. The intersection of the AC-load line with V_{CE} can be at V_{CC} plus or minus the contribution of the maximum sinusoidal effect of the input source on V_{CE}. Additional comments regarding the DC blocking in multi-stage amplifier circuits are given in EXP. VI-4.

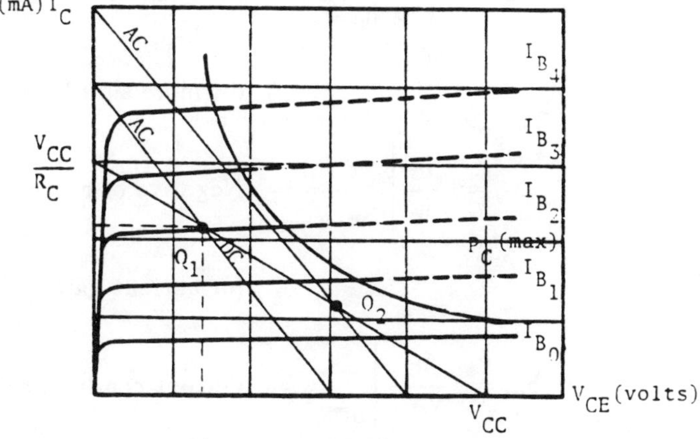

Fig. 2.   Load-Lines.

There is also a maximum power limitation associated with each transistor that is the loci of $V_{CE}I_{CE} =$ constant. This is shown in Fig. 2, by the parabola.

2.2. MULTI-STAGE AMPLIFIERS

Seldom does a single amplifier stage (due to its low output amplitude) find an application. Thus by appropriate cascading of several (transistor) amplifier stages, Fig. 3, we may produce a signal that is at a measurable or usable level for some practical purposes. In the terminol-

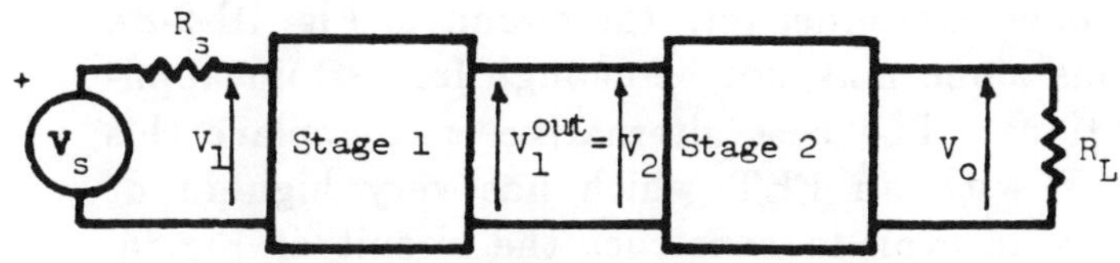

Fig. 3. Two–Stage Amplifier.

ogy of feedback theory, this is called an open-loop amplifier. To analyze this amplifier we notice that

$$G_v = \frac{V_o}{V_1} = \frac{V_o}{V_2} \cdot \frac{V_2}{V_1} = G_{v_2} G_{v_1} .$$

Here G_{v_1} and G_{v_2} are two voltage gains corresponding to stage one and stage two of the cascaded amplifier. The overall gain is

$$G_{v_s} = \frac{V_o}{V_s} = \frac{V_o}{V_1} \cdot \frac{V_1}{V_s} = G_v \frac{R_{in}^1}{R_{in}^1 + R_s} ,$$

where R_{in}^1 is the effective input resistance of stage one of this amplifier. The current gain, of course, is a different story and we can not, in general, multiply the current gain of each stage as above to get the overall current gain. (Further discussions on multi-stage amplification are given in EXP. VI-4.) As this derivation suggests, in order to analyze several stages of the transistor amplifier circuits, we need the so-called *H*-parameters for each amplifier. The equivalent small-signal model for a transistor with this parameter set is shown in Fig. 4. Here the *H*-parameters are as follows:

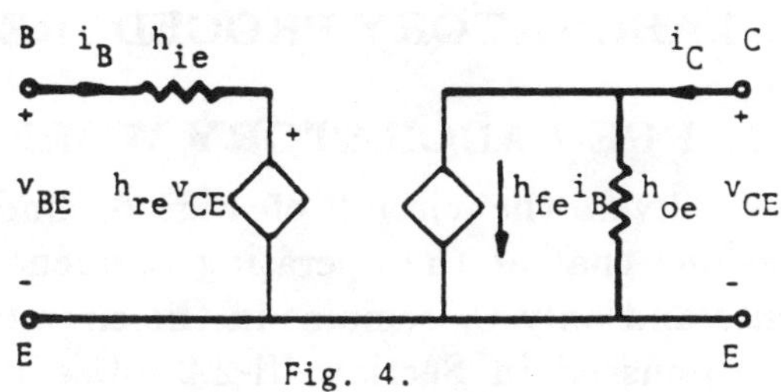

Fig. 4.

$$h_{1e} = \frac{v_{BE}}{i_B}\bigg|_{v_{CE}=0} = \text{Input resistance } (\Omega) \text{ with output short–circuited.}$$

$$h_{re} = \left.\frac{v_{BE}}{v_{CE}}\right|_{i_B=0} = \text{Fraction of output voltage at input with input open-circuited ,}$$

or, more simply, reverse-open-circuit voltage amplification gain .

$$h_{fe} = \left.\frac{i_C}{i_B}\right|_{v_{CE}=0} = \text{Negative of current transfer ratio ,}$$

(or current gain) with output short-circuited .

$$h_{oe} = \left.\frac{i_C}{v_{CE}}\right|_{i_B=0} = \text{Output conductance } (\Omega)^{-1} \text{ with input open-circuited.}$$

2.3. DARLINGTON PAIR

The input resistance of a BJT is not large for many applications. One approach to improve this input resistance is to use an emitter follower amplifier (cf., the circuit of Fig. III-2-8). This alone may not be enough for certain applications. The best alternative is to replace this BJT with an FET which has very high input resistance or to construct the circuit of Fig. 5, which is known as a Darlington pair (though it is mostly built as one unit). The input resistance of T_2 is the effective load resistance of T_1 and the input resistance of T_1 can be in the hundred megaohms range. The overall voltage gain is less than one. Due to the amplification of the leakage currents we can not cascade more of these pairs for higher input resistance. Furthermore, the self-biasing circuits used to bias T_1 and T_2, will effectively decrease the input resistance as determined in the above. There is a way to get around this problem and that is by the *bootstrap principle* of biasing the transistors [3].

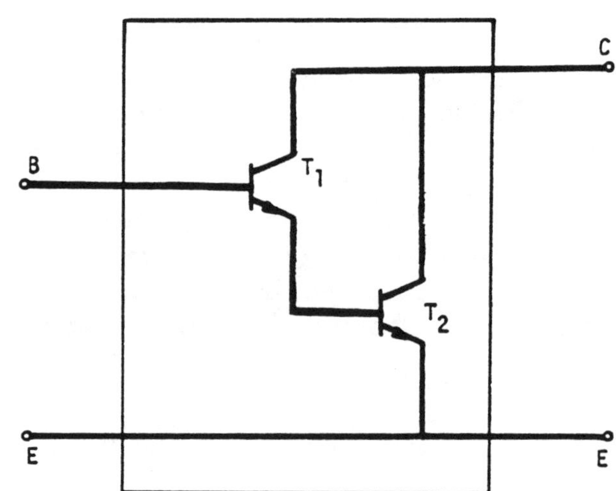

Fig. 5. Darlington Pair.

3. LABORATORY PROCEDURE

3.1. PRE-LABORATORY WORK

Given the circuit of Fig. 6, find an expression for $V_o(s)/V_{in}(s)$ directly from the circuit. Assume that at the operating frequency of the input signal all the internal capacitors are short circuits and only C_L remains in the circuit. Try this problem using the appropriate set of Γ-parameters as discussed in Section III-2.2. Use in either of the above two cases the best or more accurate equivalent small-signal model of your transistor in order to determine its operating point as well as the value of biasing circuit elements. You will need these results for your experimental work. Draw the best (or the most accurate) output characteristic curves of your transistor on the chart of Fig. 7a.

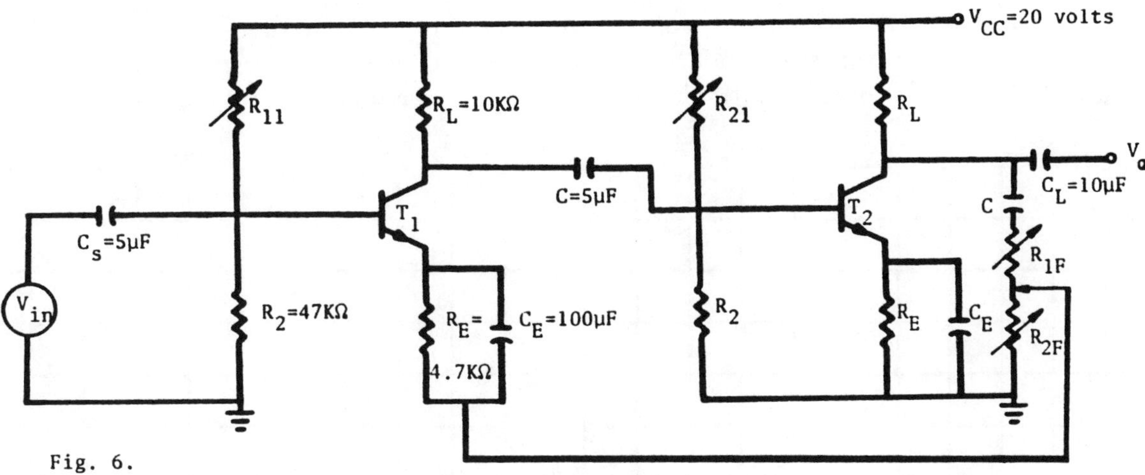

Fig. 6.

Also decide what and how much work should be done in the laboratory and what or how much work must be deferred to the end of class. You must learn to budget your time. Otherwise you will not be able to finish this experiment.

3.2. LABORATORY WORK

3.2.1. AMPLIFIER OUTPUT WITHOUT CAPACITIVE COUPLING

Construct the circuit of Fig. 1, without R, C, and the input source being connected. *Draw* the best output characteristic curves of your transistor on the chart of Fig. 7a. Then *find* the DC-load line and the Q point for your circuit using $V_{CE} = 6$ volts by calculating R_B and using the GR-Decade Resistance Box to set this R_B. *Apply* a sinusoidal input signal with a frequency of approximately 2 kHz to the base of this transistor. *Observe* the output as the current source output is gradually increased. *Increase* the

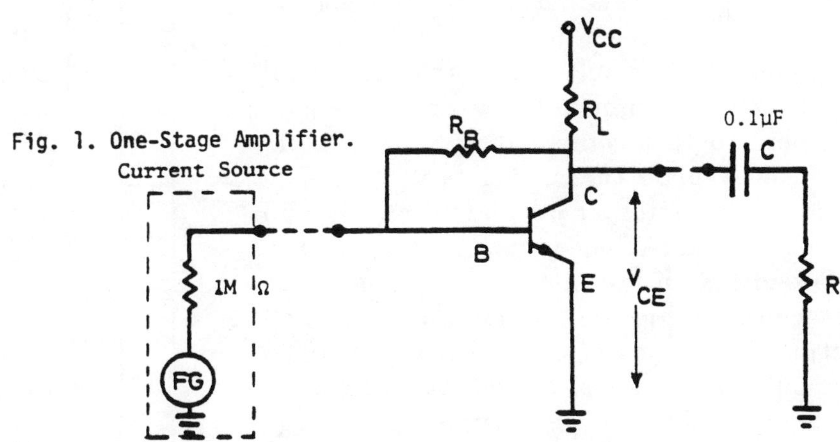

Fig. 1. One-Stage Amplifier. Current Source

amplitude of the current source output until the time-varying sinusoidal output V_{CE} begins to distort. *Record* the peak-to-peak value of the output voltage when the distortion first begins. What is the cause of this distortion? *Explain* in terms of the DC-load line and the collector characteristic curves of the transistor. For the maximum input current amplitude without the output distortion, *measure* the base to emitter voltage of the transistor with the oscilloscope. *Determine* the voltage gain of the circuit for the sinusoidal signal.

3.2.2. AMPLIFIER OUTPUT WITH CAPACITIVE COUPLING

Connect the RC network to the circuit of Fig. 1. *Consider* the voltage across R as the output voltage of the circuit. *Draw* the AC-load line in Fig. 7a, crossing at point Q. With the current source *set* to zero, *measure* the output voltage of the circuit. With the frequency of the current source at 4 kHz, *increase* the amplitude of this source and *sketch* the circuit output on Fig. 7b. *Increase* the current source output until the output voltage signal (across R) begins to distort.

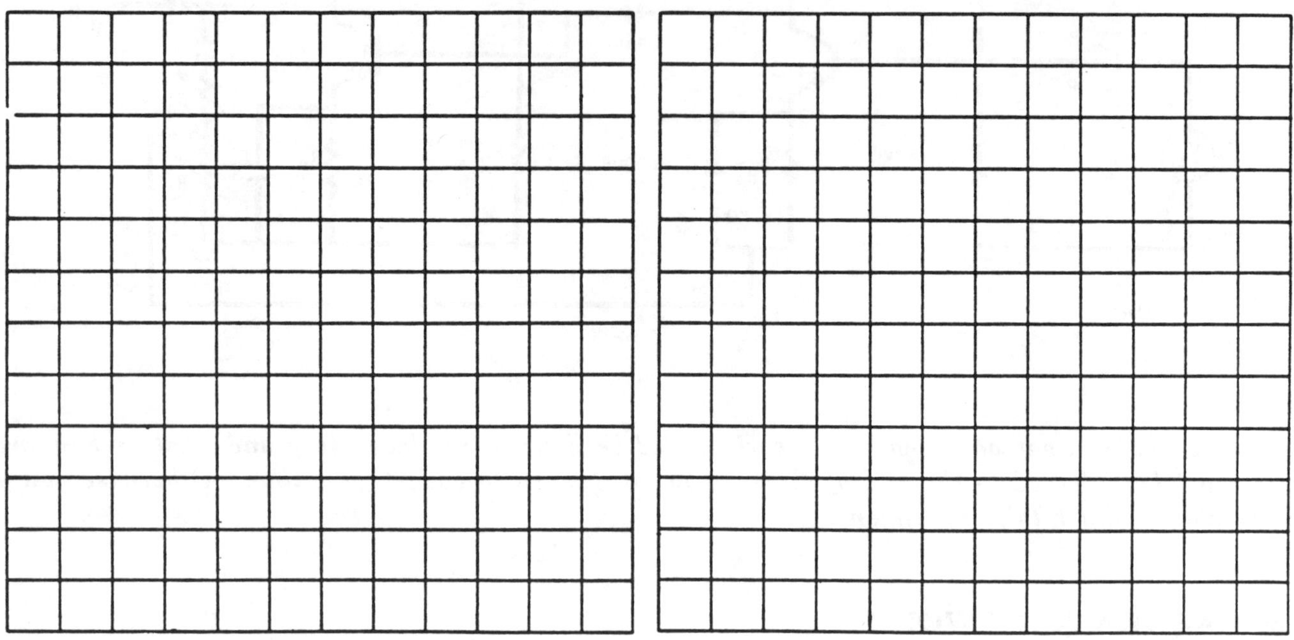

Fig. 7a.   Fig. 7b.

Measure this output voltage and the input voltage v_{BE} and *determine* the voltage gain of the circuit for sinusoidal signals. *Compare* this result with the analytical result for this circuit. How does this gain compare to that computed in Section 3.2.1? Can you explain these differences, if any, in terms of collector characteristic curves and the DC- and AC-load lines? *Draw* a new AC-load line, if you notice changes in Q, on the chart of Fig. 7a. *Draw* the input and output waveforms of this capacitive coupled circuit on the chart of Fig. 8. *Set* the output voltage of the FG just below the voltage that causes the circuit output to distort. Now, starting at low frequencies, *increase* the frequency and *record* the amplitude of the output voltage as a function of frequency. *Adjust* the output voltage of the FG so it remains constant as the frequency changes. *Plot* the output voltage versus frequency in

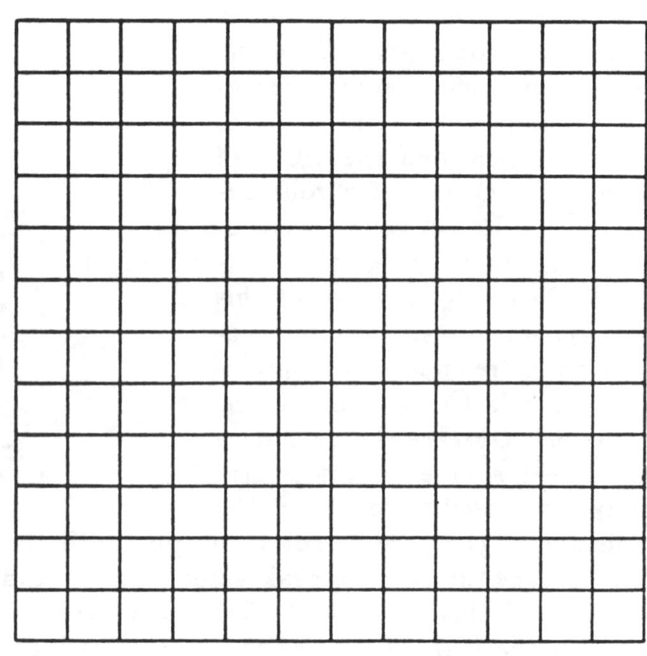

Fig. 8.

Fig. 9. At what frequency is the output 0.707 of its maximum? If you cannot maintain output of the FG as is, then *plot* instead the transfer function

$$T(s) = V_o(s) / V_{in}(s)$$

as a function of frequency on the chart of Fig. 9.

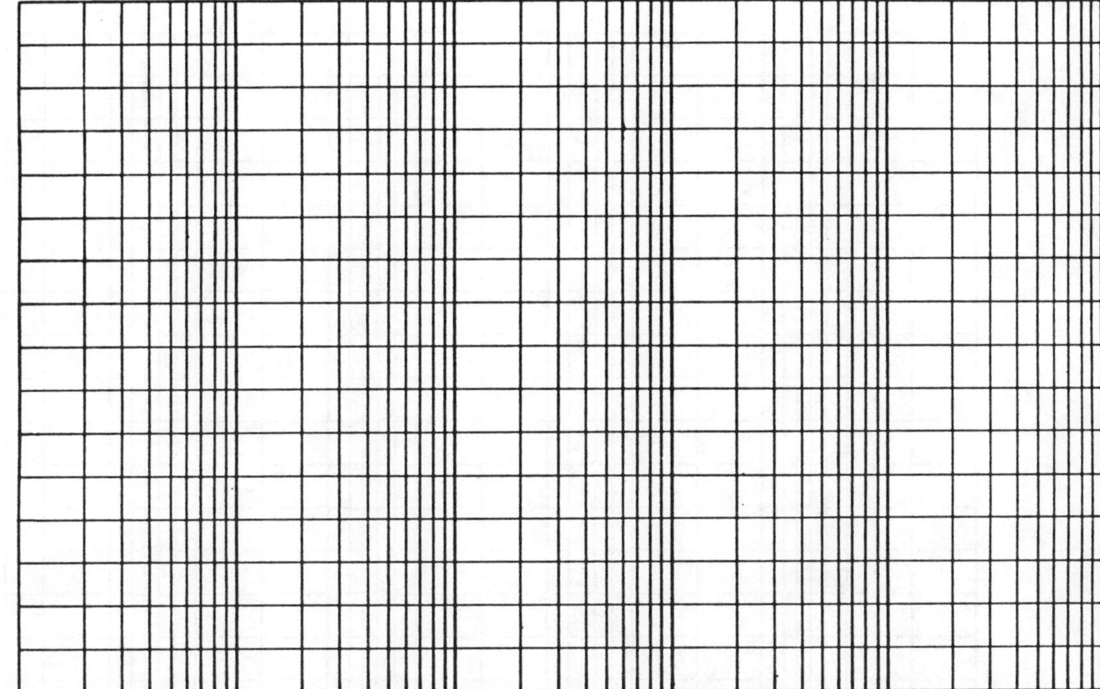

Fig. 9.

3.2.3. TWO-STAGE FEEDBACK AMPLIFIER

WARNING: Care must be exercised to avoid possible oscillation.

Construct the feedforward amplifier of the circuit in Fig. 6, which is redrawn below for convenience in Fig. 10. *Use* the element values that you have calculated in the pre-laboratory homework of this experiment.

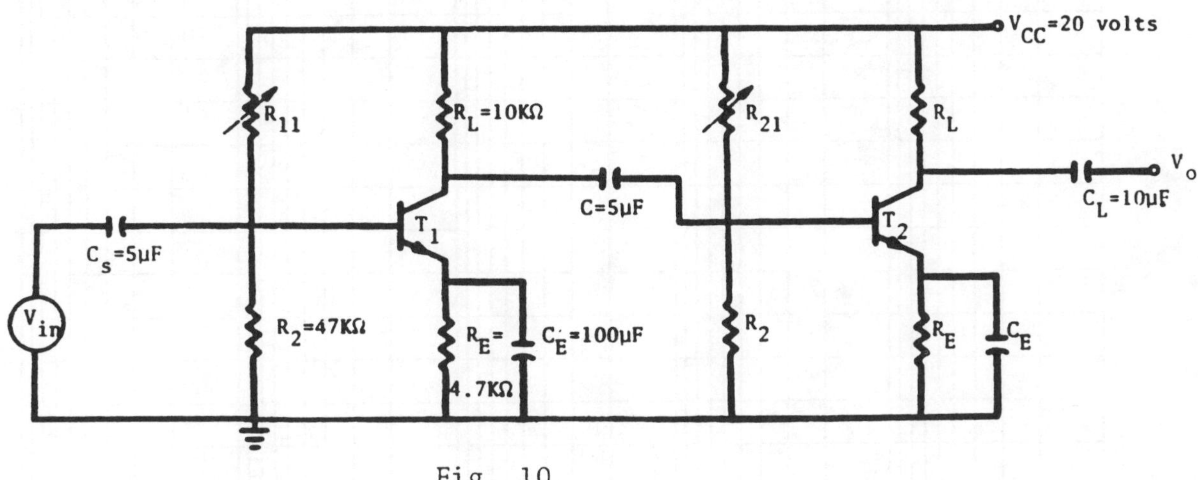

Fig. 10.

Measure the overall sinusoidal voltage gain of this open-loop amplifier at the frequency range of 10 Hz to 2 kHz. Then *turn off* the power supply and put the feedback R_F C network into the circuit. *Make* use of a pot for R_{1F} and R_{2F} such that $R_{1F} + R_{2F} = 10$ kΩ. Then gradually *increase* R_{2F} from a few ohms to almost 10 kΩ for at least four trials. For each setting of R_{2F}, *measure* the overall voltage gain of now the closed-loop or feedback amplifier at the same frequency values of the open-loop amplifier. *Plot* all of these sets of data on the same semilog chart of Fig. 11 using different colors or proper labelings. Then *draw* the *analytical* expression that you have developed in the pre-laboratory work for only one value of R_{2F} on the same figure. What are your conclusions? *Discuss* your observations?

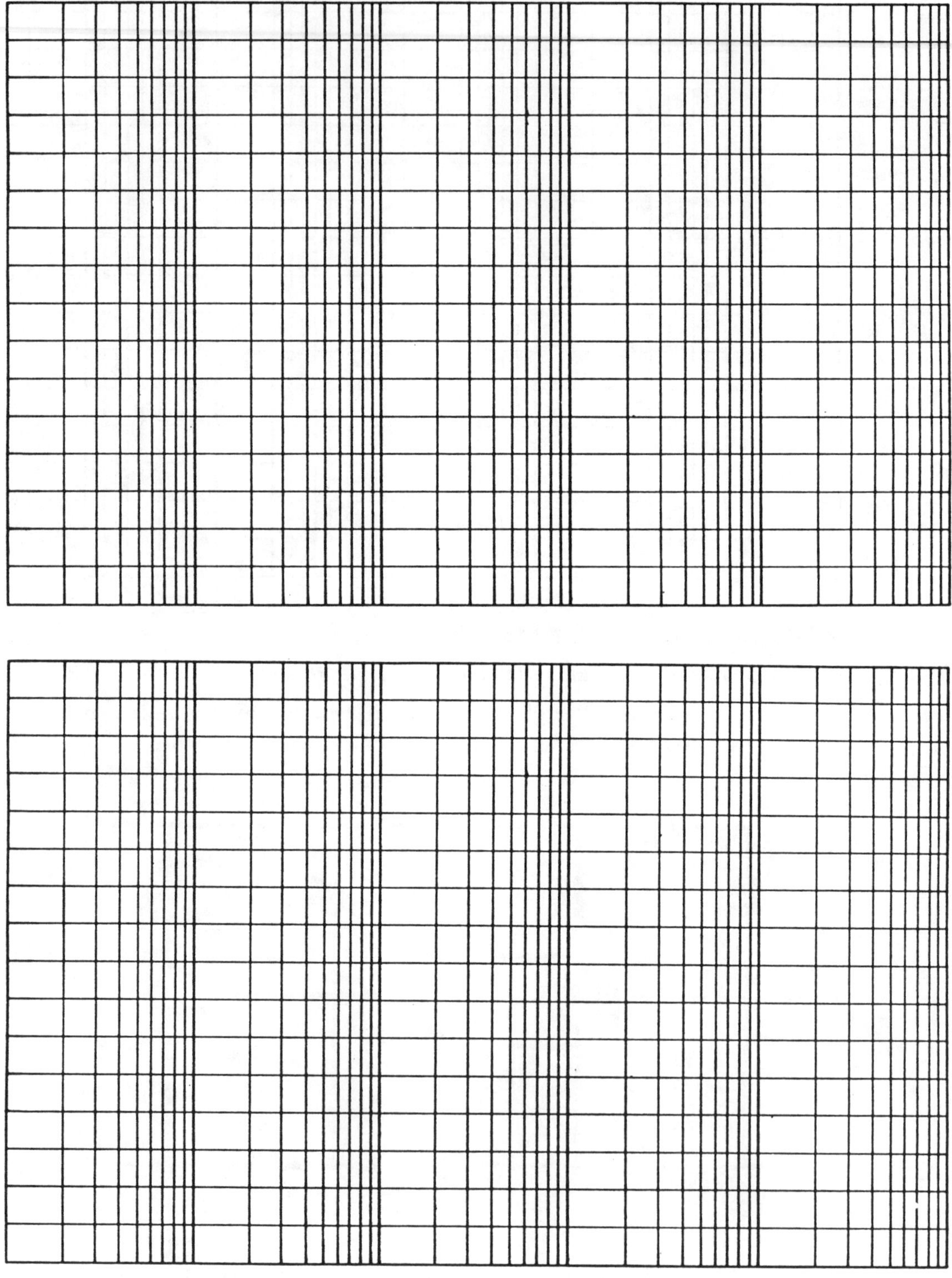

Fig. 11.

3.2.4. DARLINGTON PAIR

Construct the circuit of Fig. 5, with the standard self-biasing circuits as described in EXP. III-2, and an output resistance of 5 kΩ. *Measure* its input and output resistance as though this circuit were just one unit. Then *use* the *bootstrap principle* for biasing the transistor in order to increase its output resistance. *Recall* the simple biasing circuit, using the bootstrap principle, that is shown in the circuit of Fig. 12, [3]. *Choose* $C_b =$ 100 μF and $R_8 =$ 100 kΩ. If you have difficulty constructing the circuit of Fig. 5, you may use a pnp silicon power Darlington pair as shown in Fig. 14 to answer the above questions.

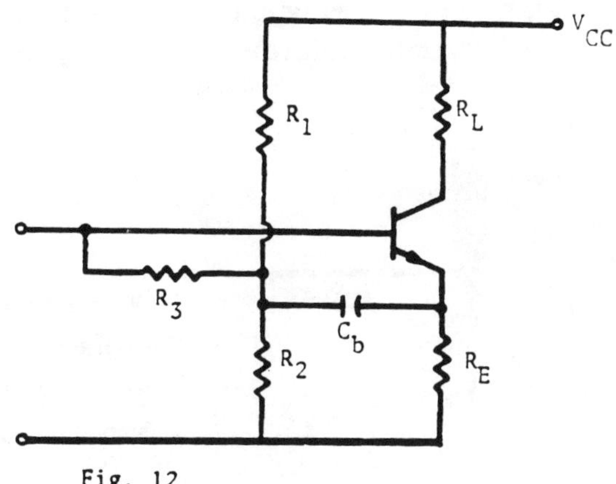

Fig. 12.

4. PROBLEM SET

Find the overall transfer function of the network of Fig. 13. Use your transistor equivalent small signal-model for T_1 and T_2.

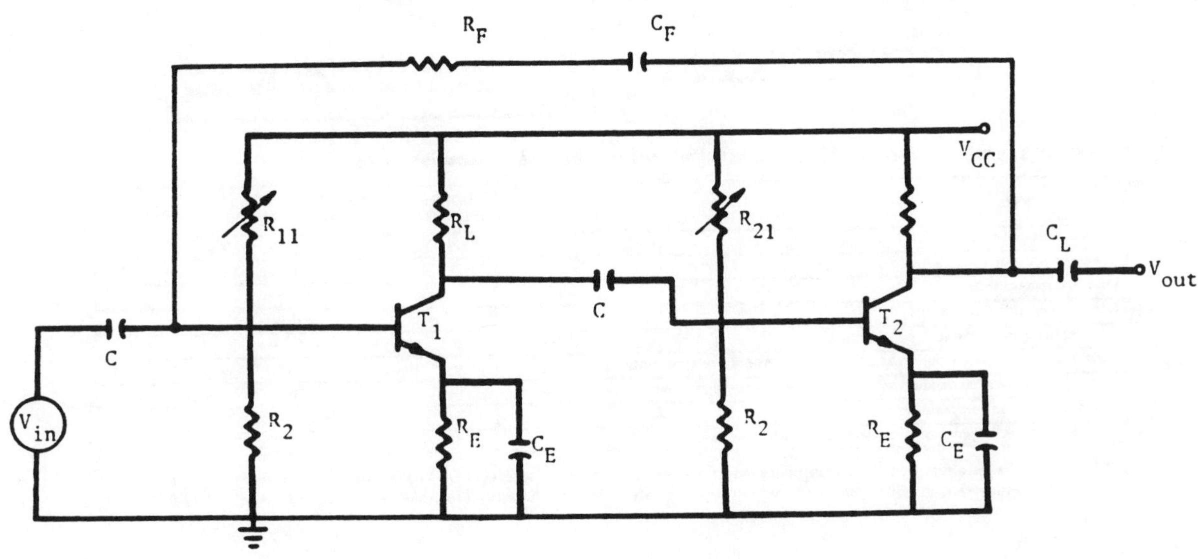

Fig. 13.

5. FOR YOUR REPORT

Draw the DC- and AC-load lines and the input and output waveforms of your experiment on the characteristic curves plane of your transistor. Answer the detailed analytical derivations of your experiment and compare your experimental results with those of the analytical as requested in the laboratory procedure. Submit your completed report written in our standard form, including the solution to the above problem.

6. MANUFACTURER'S SPECIFICATION DATA SHEETS

On the following pages (cf., Fig. 14), we present the specification data sheets of one type pnp Silicon Power Darlington for the purpose of illustrations and review of its basic properties [13].

**BD896, BD896A, BD898, BD898A,
BD900, BD900A, BD902
P-N-P SILICON POWER DARLINGTONS**

REVISED OCTOBER 1984

- **70 W at 25°C Case Temperature**
- **8 A Continuous Collector Current**
- **Min h$_{FE}$ of 750 at 3 A or 4 A**
- **Designed for Power Amplifier and High-Speed Switching Applications**

device schematic

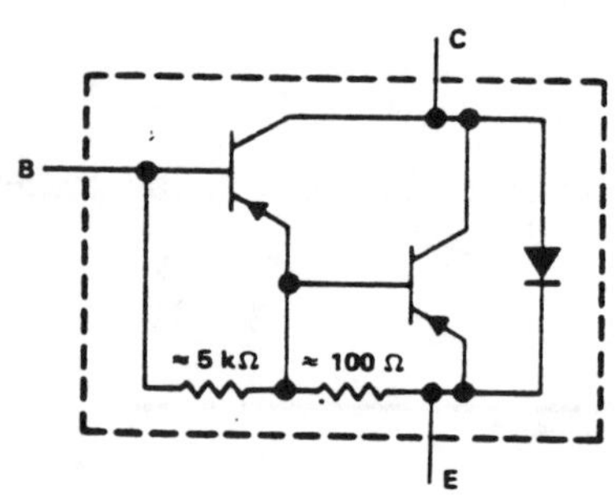

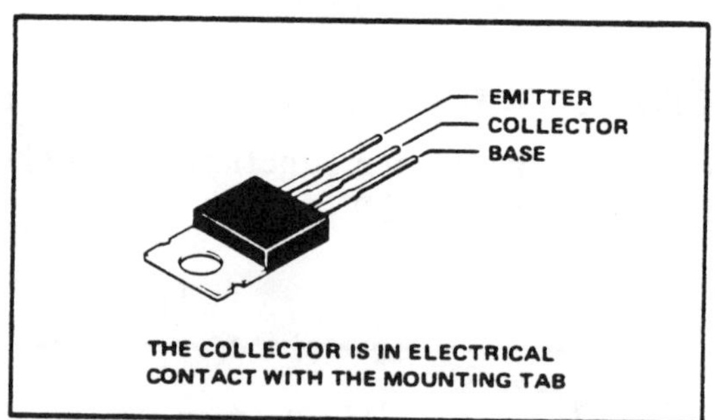

absolute maximum ratings at 25°C case temperature (unless otherwise noted)

	BD896 BD896A	BD898 BD898A	BD900 BD900A	BD902
Collector-base voltage	− 45 V	− 60 V	− 80 V	− 100 V
Collector-emitter voltage (I$_B$ = 0)	− 45 V	− 60 V	− 80 V	− 100 V
Emitter-base voltage		− 5 V		
Continuous collector current		− 8 A		
Continuous base current		− 0.3 A		
Continuous device dissipation at 25°C case temperature (see Note 1)		70 W		
Continuous device dissipation at 25°C free-air temperature (see Note 2)		2 W		
Operating free-air temperature range		− 65°C to 150°C		
Operating collector junction and storage temperature range		− 65°C to 150°C		

NOTES: 1. Derate linearly to 150°C case temperature of the rate of 0.56 W/°C or refer to Dissipation Derating Curve, Figure 7.
2. Derate linearly to 150°C free-air temperature of the rate of 16 mW/°C or refer to Dissipation Derating Curve, Figure 8.

Fig. 14. pnp Silicon Power Darlingtons (Courtesy of Texas Instruments [13]).

BD896, BD896A, BD898, BD898A, BD900, BD900A, BD902
P-N-P SILICON POWER DARLINGTONS

electrical characteristics at 25°C case temperature (unless otherwise noted)

PARAMETER	TEST CONDITIONS		BD896, BD896A			BD898, BD898A			UNIT
			MIN	TYP	MAX	MIN	TYP	MAX	
$V_{(BR)CEO}$	$I_C = -100\,mA$, $I_B = 0$, See Note 3		-45			-60			V
I_{CEO}	$V_{CE} = -30\,V$, $I_B = 0$				-500			-500	µA
I_{CBO}	$V_{CB} = -45\,V$, $I_E = 0$				-200				µA
	$V_{CB} = -60\,V$, $I_E = 0$							-200	µA
	$V_{CB} = -45\,V$, $I_E = 0$, $T_C = 100°C$				-2				mA
	$V_{CB} = -60\,V$, $I_E = 0$, $T_C = 100°C$							-2	mA
I_{EBO}	$V_{EB} = -5\,V$, $I_C = 0$				-2			-2	mA
h_{FE}	$V_{CE} = -3\,V$, $I_C = -3\,A$, See Notes 3 and 4	BD896, BD898	750			750			
	$V_{CE} = -3\,V$, $I_C = -4\,A$, See Notes 3 and 4	BD896A, BD898A	750			750			
$V_{BE(on)}$	$V_{CE} = -3\,V$, $I_C = -3\,A$, See Notes 3 and 4	BD896, BD898			-2.5			-2.5	V
	$V_{CE} = -3\,V$, $I_C = -4\,A$, See Notes 3 and 4	BD896A, BD898A			-2.5			-2.5	
$V_{CE(sat)}$	$I_C = -3\,A$, $I_B = -12\,mA$, See Notes 3 and 4	BD896, BD898			-2.5			-2.5	V
	$I_C = -4\,A$, $I_B = -16\,mA$, See Notes 3 and 4	BD896A, BD898A			-2.8			-2.8	
V_F	$I_F = 8\,A$				3.5			3.5	V

electrical characteristics at 25°C case temperature (unless otherwise noted)

PARAMETER	TEST CONDITIONS		BD900, BD900A			BD902			UNIT
			MIN	TYP	MAX	MIN	TYP	MAX	
$V_{(BR)CEO}$	$I_C = -100\,mA$, $I_B = 0$, See Note 3		-80			-100			V
I_{CEO}	$V_{CE} = -40\,V$, $I_B = 0$				-500				µA
	$V_{CE} = -50\,V$, $I_B = 0$							-500	
I_{CBO}	$V_{CB} = -80\,V$, $I_E = 0$				-200				µA
	$V_{CB} = -100\,V$, $I_E = 0$							-200	µA
	$V_{CB} = -80\,V$, $I_E = 0$, $T_C = 100°C$				-2				mA
	$V_{CB} = -100\,V$, $I_E = 0$, $T_C = 100°C$							-2	mA
I_{EBO}	$V_{EB} = -5\,V$, $I_C = 0$				-2			-2	mA
h_{FE}	$V_{CE} = -3\,V$, $I_C = -3\,A$, See Notes 3 and 4	BD900, BD902	750			750			
	$V_{CE} = -3\,V$, $I_C = -4\,A$, See Notes 3 and 4	BD900A	750						
$V_{BE(on)}$	$V_{CE} = -3\,V$, $I_C = -3\,A$, See Notes 3 and 4	BD900, BD902			-2.5			-2.5	V
	$V_{CE} = -3\,V$, $I_C = -4\,A$, See Notes 3 and 4	BD900A			-2.5				
$V_{CE(sat)}$	$I_C = -3\,A$, $I_B = -12\,mA$, See Notes 3 and 4	BD900, BD902			-2.5			-2.5	V
	$I_C = -4\,A$, $I_B = -16\,mA$, See Notes 3 and 4	BD900A			-2.8				
V_F	$I_F = 8\,A$				3.5			3.5	V

NOTES: 3.   These parameters must be measured using pulse techniques. $t_W = 300\,\mu s$, duty cycle $\leqslant 2\%$.
        4.   These parameters are measured with voltage-sensing contacts separate from the current-carrying contacts and located within 3,2 mm (0.125 inch) from the device body.

Fig. 14.   Continued.

BD896, BD896A, BD898, BD898A,
BD900, BD900A, BD902
P-N-P SILICON POWER DARLINGTONS

thermal characteristics

PARAMETER	MIN	TYP	MAX	UNIT
$R_{\theta JC}$			1.79	°C/W
$R_{\theta JA}$			62.5	

resistive-load switching characteristics

PARAMETER	TEST CONDITIONS			MIN	TYP	MAX	UNIT
t_{on}	$I_C = -3\,A$,	$I_{B1} = -12\,mA$,	$I_{B2} = 12\,mA$,			1	µs
t_{off}	$V_{BE(off)} = 3.5\,V$,	$R_L = 10\,\Omega$,	See Figure 1			5	

PARAMETER MEASUREMENT INFORMATION

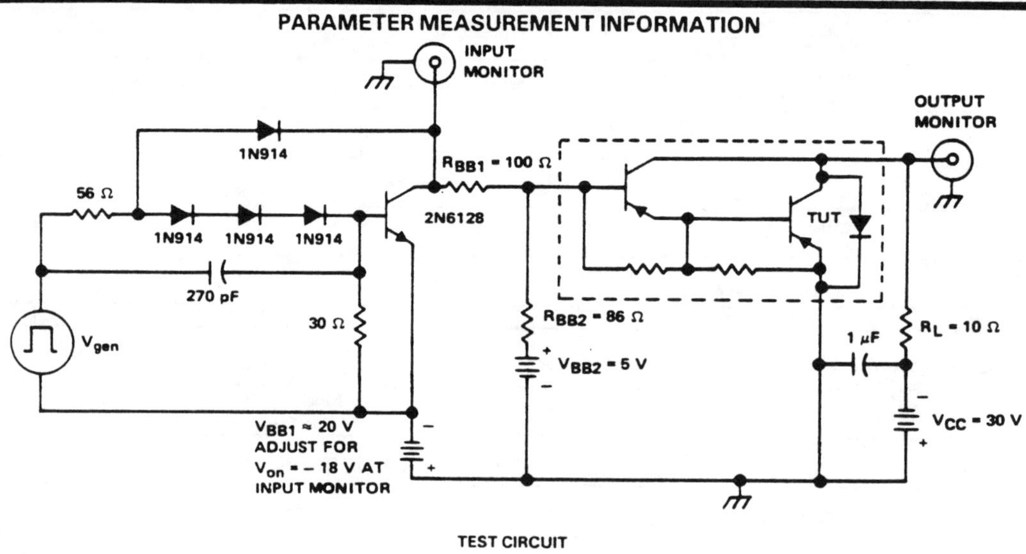

TEST CIRCUIT

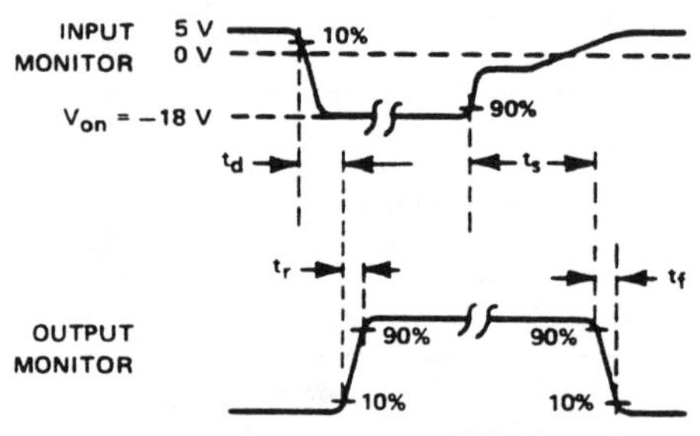

VOLTAGE WAVEFORMS

NOTES: A. V_{gen} is a 30-V pulse into a 50 Ω termination.
B. The V_{gen} waveform is supplied by the following characteristics: $t_r \leqslant 15\,ns$, $t_f \leqslant 15\,ns$, $Z_{out} = 50\,\Omega$, $t_w = 20\,µs$, duty cycle $\leqslant 2\%$.
C. Waveforms are monitored on an oscilloscope with the following characteristics: $t_r \leqslant 15\,ns$, $R_{in} \geqslant 10\,M\Omega$, $C_{in} \leqslant 11.5\,pF$.
D. Resistors must be noninductive types.
E. The d-c power supplies may require additional bypassing in order to minimize ringing.

FIGURE 1.   RESISTIVE-LOAD SWITCHING

TEXAS
INSTRUMENTS
POST OFFICE BOX 225012 ● DALLAS TEXAS 75265

Fig. 14.   Continued.

BD896, BD896A, BD898, BD898A,
BD900, BD900A, BD902
P-N-P SILICON POWER DARLINGTONS

TYPICAL CHARACTERISTICS

STATIC FORWARD CURRENT TRANSFER RATIO
vs
COLLECTOR CURRENT

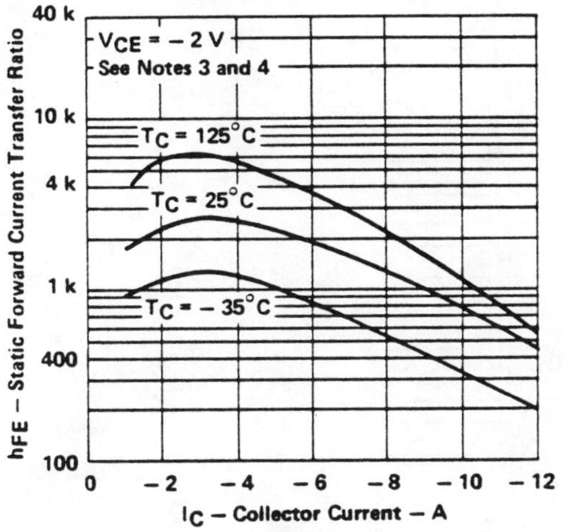

FIGURE 2

STATIC FORWARD CURRENT TRANSFER RATIO
vs
COLLECTOR CURRENT

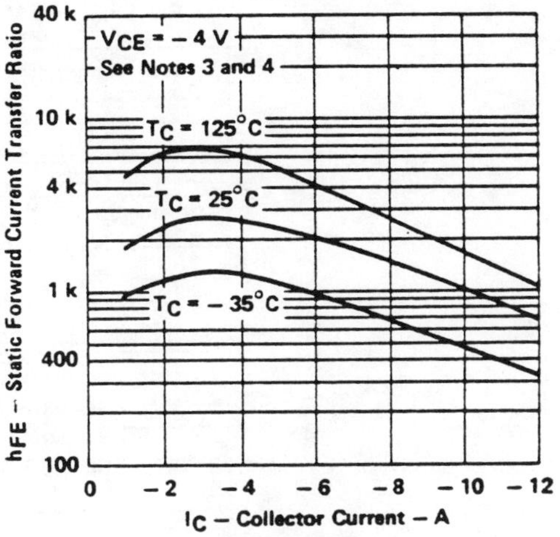

FIGURE 3

COLLECTOR-EMITTER SATURATION VOLTAGE
vs
COLLECTOR CURRENT

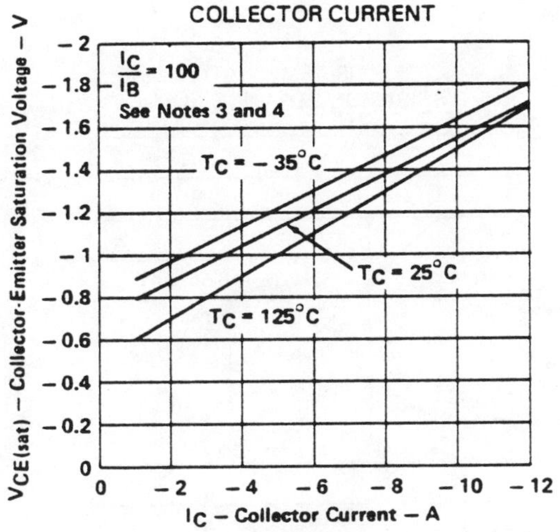

FIGURE 4

BASE-EMITTER VOLTAGE
vs
COLLECTOR CURRENT

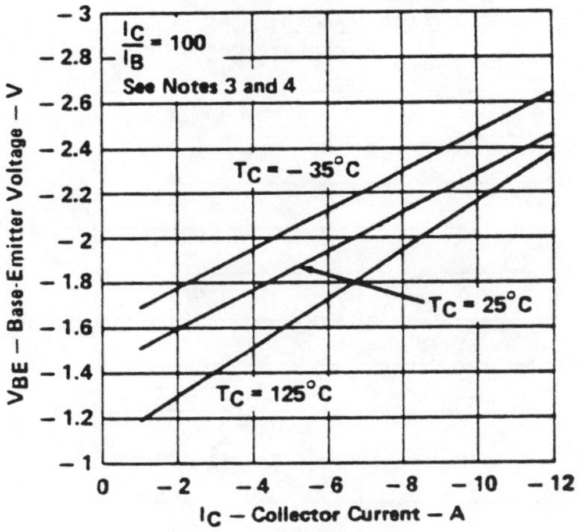

FIGURE 5

NOTES: 3. These parameters must be measured using pulse techniques, t_w = 300 µs, duty cycle ≤ 2%.
4. These parameters are measured with voltage-sensing contacts separate from the current-carrying contacts and located within 3,2 mm (0.125 inch) from the device body.

Fig. 14. Continued.

BD896, BD896A, BD898, BD898A,
BD900, BD900A, BD902
P-N-P SILICON POWER DARLINGTONS

MAXIMUM SAFE OPERATING AREA

MAXIMUM COLLECTOR CURRENT
vs
COLLECTOR-EMITTER VOLTAGE

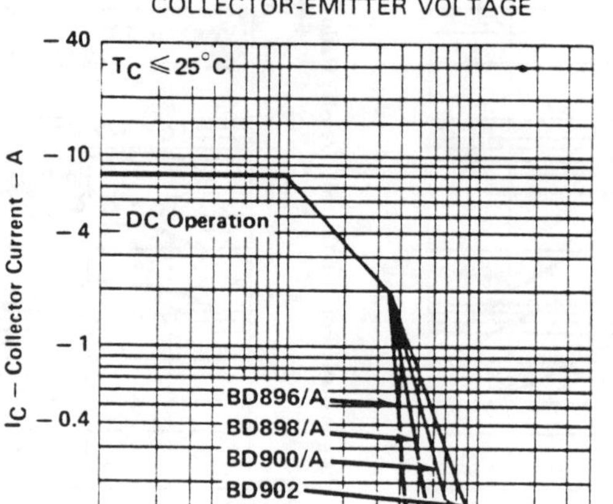

FIGURE 6

THERMAL INFORMATION

CASE TEMPERATURE
DISSIPATION DERATING CURVE

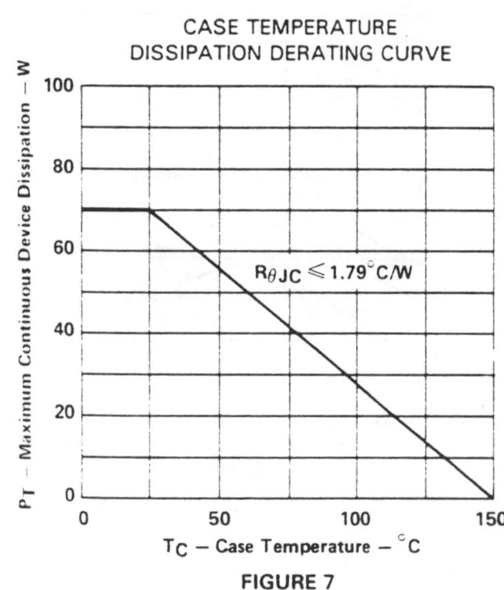

FIGURE 7

FREE-AIR TEMPERATURE
DISSIPATION DERATING CURVE

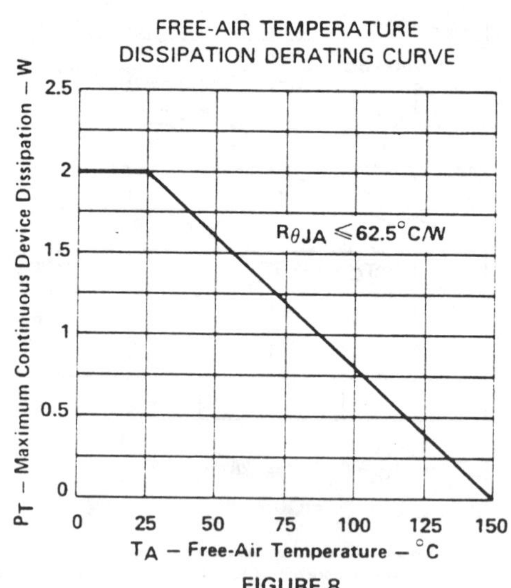

FIGURE 8

Fig. 14. Continued.

EXPERIMENT III-4

BIPOLAR JUNCTION TRANSISTORS
LARGE–SIGNAL BEHAVIOR

EQUIPMENT:

    EEP
    CURVE TRACER
    BJTs (3)
    RESISTORS (10 Ω, 1 kΩ, 10 kΩ)

Table of Contents   Page No.

1. PURPOSE

The purpose of this experiment is to acquaint the student with the large-signal and/or switching behavior of a BJT. This problem is actually the same as studying the conditions under which the operating modes of a transistor change. We are, of course, interested in changes that take place between the two modes (or states) of *cut off* and *saturation* of a transistor, although there is always a transition mode in between. We will make our analysis and/or presentation based on the original work of Ebers and Moll [1].

2. LARGE-SIGNAL BEHAVIOR OF THE BJT

2.1. INTRODUCTION

For many years the way that a transistor switches from one mode to another and its behavior at these boundaries have been of great concern and extensive research. One prominent source of knowledge on this subject is the original work of Ebers and Moll [1]. It is well known that the main tool for this variation between the operational modes is the proper biasing of the transistor. Correspondingly, under the so-called generalized biasing in Section 7.5 of [5], the ways that one may generate a large-signal model of the transistor that is useful for all three operational regions of this device is studied. In this reference it is shown how, by proper biasing, one may accomplish this goal. This ultimate large-signal model involves four parameters for which only three are independent and all four are related to the geometry and the substance of the device. Here we directly review this topic, using the same notation as found in the original work of Ebers and Moll [1].

In the terminology of Ebers and Moll [1], there are three regions in the collector characteristic plane as follows:

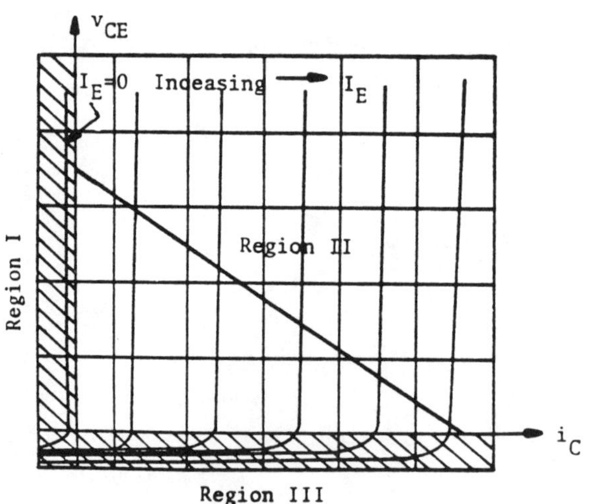

Fig. 1. Transistor Collector Characteristics, Common-Base Configuration.

Region I (RI):	Collector *current cut off*, or collector *voltage saturation*. Here both the emitter junction and the collector junction are reverse biased.
Region II (RII):	*Active* region. Here the emitter junction is forward biased and the collector junction is reverse biased.

Region III (RIII):   Collector *current saturation*, or collector *voltage cut off*.
Here both the emitter junction and the collector junction are forward biased.

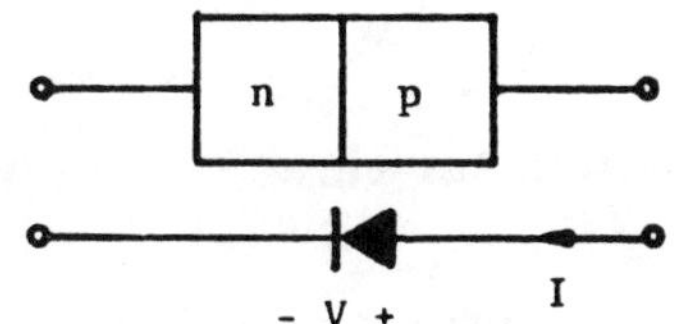

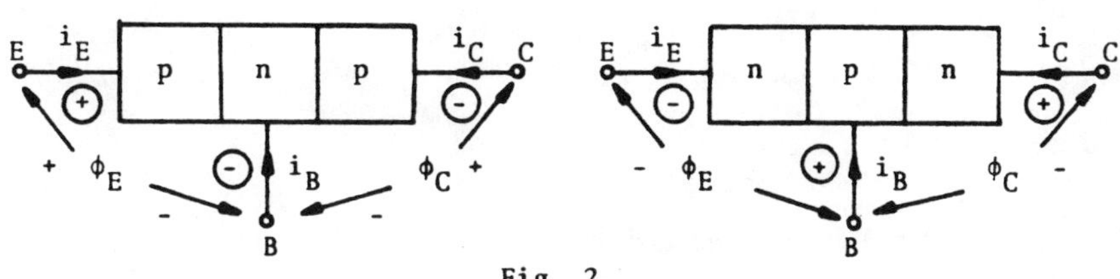

Fig. 2.

Consider the p-n diode and transistors of Fig. 2. Here we show the correct signs of ϕ_E and ϕ_C (the junction voltages) as being the case in each transistor. But we show the same directions for i_E and i_C in both transistors. The actual (or physical) directions of each current are shown by ($\pm$, in circles) under the above signs. The plus sign means the actual sign coincides with the figure and the minus sign means it is reversed. Our goal is to find expression for these i_E and i_C in terms of various transistor parameters. Since these transistors are made of several p-n junctions and for each such junction the expression for the current I through, versus the voltage V across, is

$$I = I_S \left[\exp(qv/kT) - 1\right] ,$$

where I_S is the reverse saturation current and $q/kT \simeq 40$ V^{-1} at 25 °C. Thus we propose the following expressions for the currents in the generalized common-base configuration of Fig. 2.

$$\begin{bmatrix} i_E \\ i_C \end{bmatrix} = \begin{pmatrix} a_{11} & a_{12} \\ a_{21} & a_{22} \end{pmatrix} \begin{bmatrix} \left[\exp(q\phi_E / kT) - 1\right] \\ \left[\exp(q\phi_C / kT) - 1\right] \end{bmatrix} .$$

(Here we used i instead of I that is used in [1]). The quantities of ϕ_E and ϕ_C are junction voltages with respect to the references (base) and with the positive end at the p-side of each diode. These equations clearly show that each current is a function of two junction voltages. Also for the most general case we have $a_{12} = a_{21}$. These equations are mainly written with the assumptions that the resistivities of the semiconductor region and the injected current densities are both low.

Here we must now relate the four a_{ij} parameters to the following transistor parameters. (These are mainly gathered from [1], with some additional comments).

$I_{EO} = I_{EBO} =$ saturation current of emitter junction with zero collector current (emitter-base junction is reverse biased): *emitter leakage current.*

$I_{CO} = I_{CBO} =$ saturation current of collector junction with zero emitter current (collector-base junction is reverse biased): *collector leakage current.*

In the sense of Fig. 2, and compared to i_E and i_C, both I_{EO} and I_{CO} are: in the same direction for an npn transistor; and in the opposite direction for a pnp transistor.

$\alpha_N =$ transistor current gain with the emitter functioning as an emitter and the collector functioning as a collector (normal alpha or simply our previous α).

α_I = transistor current gain with the collector functioning as an emitter and the emitter functioning as a collector (inverted or reversed alpha, or α_R).

The above alphas may be considered as DC alphas since these correspond to the fraction of collected current at the collector. It can be easily shown that

$$\alpha_N I_{EO} = \alpha_I I_{CO}.$$

Our goal has been to relate these four (or three) above quantities to the a_{ij} coefficients of i_E and i_C equations.

In the active or linear region and with the collector being reverse biased, we have (recall $\exp(q\phi_C / kT) \simeq \exp(40\,\phi_C) \simeq 0$, for $\phi_C < 0$):

$$I_E = a_{11}[\exp(q\phi_E / kT) - 1] - a_{12} \ , \ \text{and}$$

$$I_C = a_{21}[\exp(q\phi_E / kT) - 1] - a_{22} \ .$$

From the above two equations we have

$$i_C = \frac{a_{21}}{a_{11}} \, i_E + \left(\frac{a_{12}\, a_{21}}{a_{11}} - a_{22} \right) \triangleq - \alpha_N i_E + I_{CO}.$$

For low current values this α_N is constant. *Note that this equation suggests that* $i_C = I_{CO}$ *for* $i_E = 0$.

On the other hand, if emitter is reverse biased and the collector is used as the emitter, then we have

$$i_E = \frac{a_{12}}{a_{22}} \, i_C + \left(\frac{a_{12}\, a_{21}}{a_{22}} - a_{11} \right) \triangleq - \alpha_I i_C + I_{EO}.$$

Also we notice that $i_E = I_{EO}$ *for* $i_C = 0.$

Based on the above terminologies, we can easily show that

$$\begin{pmatrix} a_{11} & a_{12} \\ a_{21} & a_{22} \end{pmatrix} = \frac{1}{1 - \alpha_N \alpha_I} \begin{pmatrix} -I_{EO} & \alpha_I I_{CO} \\ \alpha_N I_{EO} & -I_{CO} \end{pmatrix} \triangleq \begin{pmatrix} -I_{ES} & \alpha_I I_{CS} \\ \alpha_N I_{ES} & -I_{CS} \end{pmatrix} \ .$$

Where $a_{12} = a_{21}$ implies that $\alpha_N I_{EO} = \alpha_I I_{CO}$ and/or $\alpha_N I_{ES} = \alpha_I I_{CS}$. The above two new parameters are defined as follows:

$I_{ES} = I_{EBS}$ = emitter current that flows when collector and base are shorted and the emitter-base junction is reverse biased: *emitter saturation current.*

$I_{CS} = I_{CBS}$ = collector current that flows when emitter and base are shorted and the collector-base junction is reverse biased: *collector saturation current.*

Putting back these results into the original expressions for i_E and i_C yields

$$\begin{bmatrix} i_E \\ i_C \end{bmatrix} = \begin{pmatrix} -I_{ES} & \alpha_I I_{CS} \\ \alpha_N I_{ES} & -I_{CS} \end{pmatrix} \begin{bmatrix} [\exp(q\phi_E / kT) - 1] \\ [\exp(q\phi_C / kT) - 1] \end{bmatrix} \ .$$

From the above two equations we conclude that

$$i_E = -\alpha_I i_C - (1 - \alpha_N \alpha_I)\, I_{ES} [\exp(q\phi_E / kT) - 1] \quad , \ \text{or}$$

$$\ = -\alpha_I i_C - I_{EO} [\exp(q\phi_E / kT) - 1] \qquad\qquad , \ \text{and}$$

$$i_C = -\alpha_N i_E - (1 - \alpha_N \alpha_E)\, I_{CS} [\exp(q\phi_C / kT) - 1] \quad , \ \text{or}$$

$$\ = -\alpha_N i_E - I_{CO} [\exp(q\phi_C / kT) - 1] \qquad\qquad .$$

The above two input and output equations in either of their two forms can be easily synthesized to result in a large-signal model of the transistor. These equations express that each of the input or output circuit is a simple dependent current source plus a junction diode. The only problem that must be studied carefully is choosing the correct sign and/or direction of the current and voltage values for either cases of npn or pnp transistors. This problem is already addressed in Fig. 2, with the additional comments in the line below the definitions of I_{EO} and I_{CO}.

A similar result is the following. Since both in RI and RII, the collector junction is reverse biased, thus we may have (note that $\exp(q\phi_C / kT) \simeq \exp(40\phi_C) \simeq 0$ for $\phi_C < 0$):

$$i_E = -I_{ES}\,[\exp(q\phi_E / kT) - 1] - \alpha_I\,I_{CS}, \text{ and}$$

$$i_C = \alpha_N\,I_{ES}\,[\exp(q\phi_E / kT) - 1] + I_{CS}.$$

From the above two equations we can easily conclude that

$$i_C = -\alpha_N\,i_E + I_{CO} \qquad ,$$

a result which we have already seen before.

To synthesize for an equivalent circuit of a transistor in RI we need to choose i_E and $i_C = -\alpha_N\,i_E + I_{CO}$; similarly to synthesize for such circuit in RII we need to choose i_C and $i_E = (-i_C + I_{CO}) / \alpha_N$.

The boundaries of these large-signal regions are of particular interest, since the transistor's properties change very rapidly as its operating point gets close to these boundaries. Mathematically the problem boils down to how to treat these exponential functions in the chosen characteristics plane as well as in the chosen boundaries of the regions. For example the boundary between RI and RII may be set at $i_E = 0$ and that between RII and RIII may be set at $v_C = 0$, in the (i_C, v_C) plane. For the sake of completing this analysis we may now continue to study RIII in which both emitter and collector junctions are forward biased.

For this study, we start with

$$i_E + \alpha_I\,i_C = -\,I_{EO}\,[\exp(q\phi_E / kT) - 1] \,, \text{ and}$$

$$i_C + \alpha_N\,i_E = -\,I_{CO}\,[\exp(q\phi_C / kT) - 1].$$

Solving for ϕ_E and ϕ_C will result in

$$\phi_E = (kT / q) \ln\,[1 - (i_E + \alpha_I\,i_C) / I_{EO}] \simeq (kT / q) \ln\,[-(i_E + \alpha_I\,i_C) / I_{EO}] \,,$$

$$\phi_C = (kT / q) \ln\,[1 - (i_C + \alpha_N\,i_E) / I_{CO}] \simeq (kT / q) \ln\,[-(i_C + \alpha_N\,i_E) / I_{CO}] \,.$$

The equivalent model of the above two equations may be simply found by noting that each ϕ_E and ϕ_C (though very small) represents the corresponding voltage source between each junction and reference (base). It is also shown in [1] that we need a base spreading resistance that is measured in the current saturation region for this model.

2.1.1. COMMON EMITTER CONFIGURATION

In most of our studies and/or experiments we will be looking into the common emitter configuration. Therefore, we now study this case with its equivalent model. Recall that we had the following expressions for currents which must be slightly changed to accommodate this new configuration.

$$i_E + \alpha_I\,i_C = -\,I_{EO}\,[\exp(q\phi_E / kT) - 1],$$

$$i_C + \alpha_N\,i_E = -\,I_{CO}\,[\exp(q\phi_C / kT) - 1] \,, \text{ and}$$

$$i_B + i_C + i_E = 0 \,\cdot$$

It is easy to show that

$$i_C = \frac{\alpha_N}{1 - \alpha_N}\, i_B - \frac{I_{CO}}{1 - \alpha_N}\, [\exp(q\phi_C / kT) - 1]\,,$$

which is valid in all three regions. We may now define $\beta_N \triangleq \alpha_N / (1 - \alpha_N)$, resulting in

$$i_C = \beta_N\, i_B - \frac{I_{CO}}{1 - \alpha_N}\, [\exp(q\phi_C / kT) - 1]\cdot$$

We must recall that this ϕ_C, in the sense of Fig. 2, is the same as $\phi_{CB} = \phi_{CE} + \phi_{EB} = \phi_{CE} - \phi_{BE}$. Thus the actual i_C becomes

$$i_C = \beta_N\, i_B - \frac{I_{CO}}{1 - \alpha_N}\, [\exp(q\phi_{CE} / kT)\, \exp(-q\phi_{BE} / kT) - 1]\cdot$$

From the above equation and with the help of various relevant parameters we may generate i_C versus ϕ_{CE} for some given i_B and/or ϕ_{BE}.

We may also use a similar algebra to generate i_B as follows:

$$i_B = (1 - \alpha_N)\, I_{ES}\, [\exp(q\phi_{EB} / kT) - 1] + (1 - \alpha_I)\, I_{CS}\, [\exp(q\phi_{CB} / kT) - 1]\,,$$

$$= (1 - \alpha_N)\, I_{ES}\, [\exp(-q\phi_{BE} / kT) - 1] + (1 - \alpha_I)\, I_{CS}\, [\exp(-q\phi_{BE} / kT)\, \exp(q\phi_{CE} / kT) - 1]\cdot$$

This, of course, can be used to generate the input characteristic curves provided that transistor parameters are found experimentally.

2.2. EQUIVALENT LARGE-SIGNAL MODEL OF THE BJT

To make our analysis clear we present the following input and output large-signal models of the BJT. These models are not unique. As an example we choose an npn transistor.

2.2A.   Common-Base

We choose the following equations to synthesize this device in its active region:

$$i_E = -\alpha_I\, i_C - I_{EO}\, [\exp(q\phi_E / kT) - 1]\,,\ \text{and}$$

$$i_C = -\alpha_N\, i_E + I_{CO}\cdot$$

Let $I_{EO}\, [\exp(q\phi_E / kT) - 1]$ be represented by a diode D_E. Also note that ϕ_E is positive at the base (p-side). Then clearly the circuit that represents the above two equations with its linearized model is as shown in Fig. 3.

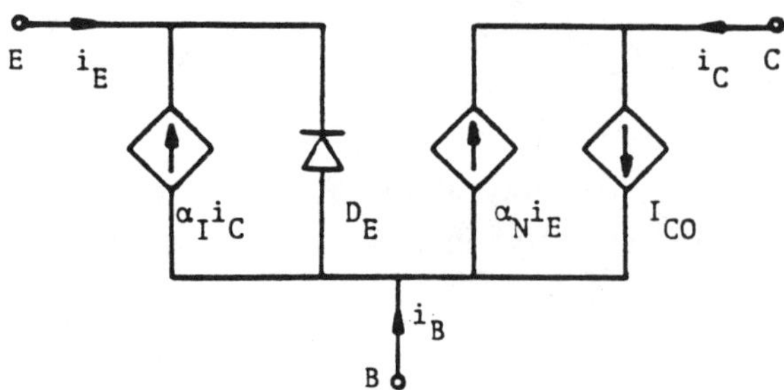

Fig. 3a.   A Large-Scale Model of BJT.

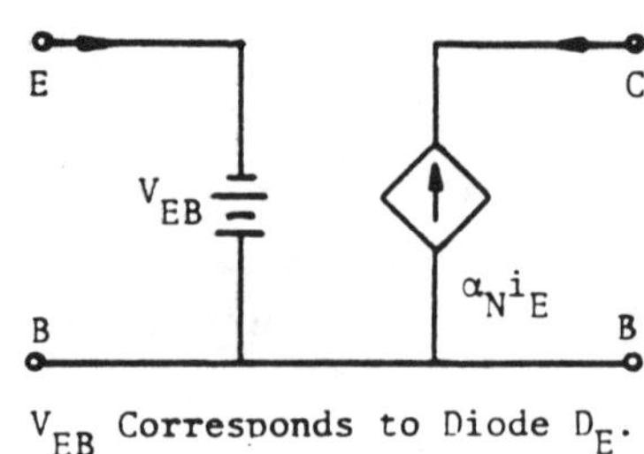

Fig. 3b.   An Approximate of the
Circuit of Fig. 3a.

2.2B.   Common-Emitter

We choose the following equations to synthesize this device in its active region:

$$i_B = (1 - \alpha_N) \, I_{ES} \, [\exp(q\phi_E / kT) - 1] - (1 - \alpha_I) \, I_{CS} \text{, and}$$

$$i_C = \beta_N \, i_B + I_{CO} / (1 - \alpha_N).$$

Let $(1 - \alpha_N) \, I_{ES} \, [\exp(q\phi_E / kT) - 1]$ be represented by a diode D_B. Then the circuit that represents these equations with its linearized model is as shown in Fig. 4.

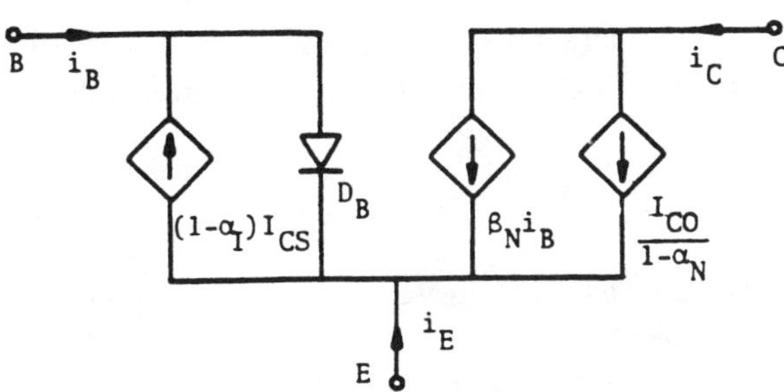

Fig. 4a.   A Large-Signal Model of BJT.

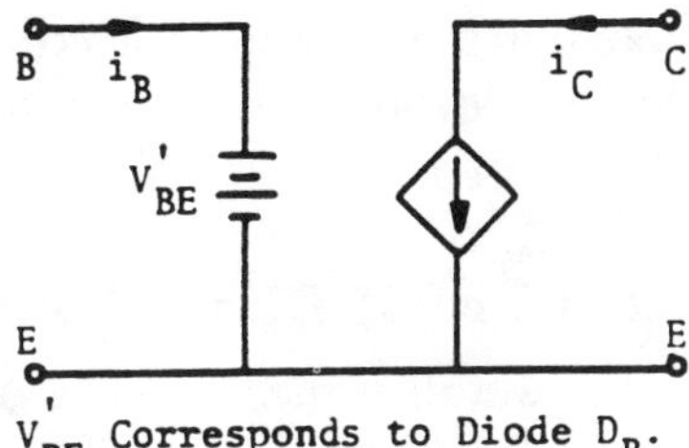

V'_{BE} Corresponds to Diode D_B.

Fig. 4b.   An Approximate of the Circuit of Fig. 4a.

In the common-emitter case we have two more approximate equivalent models for the cut off and saturation modes of the transistor as follows:

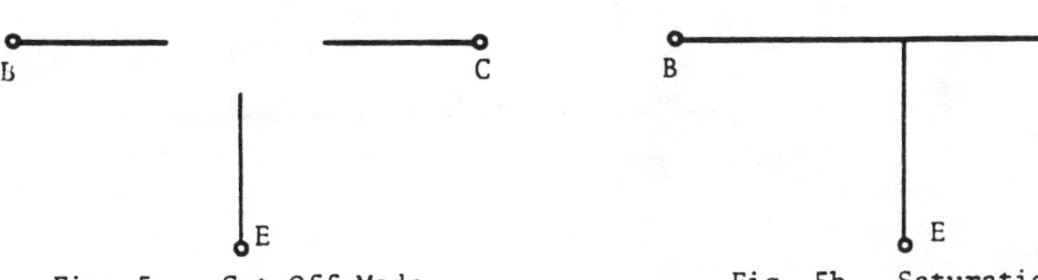

Fig. 5a.   Cut Off Mode.   Fig. 5b.   Saturation Mode.

2.3. COMMENTS ON NOTATIONS

In the last three experiments we have been using various notations representing a certain partial differential of a function at certain points or along a certain particular path. For example, we saw $h_{11} \triangleq (\partial v_{BE} / \partial i_B) \,|_* $, (similarly h_{12}, h_{21}, and h_{22}); $h_{1e} \triangleq (v_{BE} / i_B) \,\big|_{v_{CE}=0}$, (similarly h_{re}, h_{fe}, h_{oe}); and $\beta = i_C / i_B$, $\beta = I_C / I_B$, $\beta \triangleq (\partial i_C / \partial i_B)|_*$, or $\beta = (\Delta I_C / \Delta I_B) \big|_{v_{CE}=constant}$. These different notations for perhaps the same quantity may look very confusing and inconsistent. In general, one must expect that, for example, $\beta \triangleq (I_C / I_B) \neq (\partial i_C / \partial i_B)|_*$. This indeed is the case. But working within the linear active region of the transistor enables us, with a good degree of accuracy, to choose any one of these parameters for *almost* the same purpose. In other words, as long as we know what we are doing we will be safe! We must, of course, be alert to possible errors that may be developed by the wrong choices of these parameter values or evaluating the wrong partial differentials.

Perhaps we must emphasize more accurately one particular parameter that is frequently used to describe the transistor. That is the β or h_{fe} parameter, which is also called the transistor current gain.

For the small-signal operation of the transistor we choose the AC beta parameter which is defined as follows:

$$\beta = h_{fe} = \frac{\hat{i}_C}{\hat{i}_B}\bigg|_{v_{CE}=constant} = \frac{\Delta i_C}{\Delta i_B}\bigg|_*.$$

For the large-signal operation of transistor we choose the DC beta parameter that is defined as follows:

$$\beta_0 = h_{FE} = \frac{I_C}{I_B}\bigg|_{V_{CE}=constant}.$$

If we have both DC and AC components, then the corresponding beta becomes the combination of the above two betas. The same distinction can be made for the α parameter.

2.4. HYBRID-PI MODEL OF THE BJT

The various small-signal models of a transistor that we have seen before (based on the Taylor series expansion of a nonlinear function) must be modified (based on the physical consideration of the device) to accommodate a few additional features of this device. The so-called hybrid-pi model of Fig. 6 is just an alternative to the small-signal model that we have looked at earlier. This circuit is mainly used for low-frequency operation.

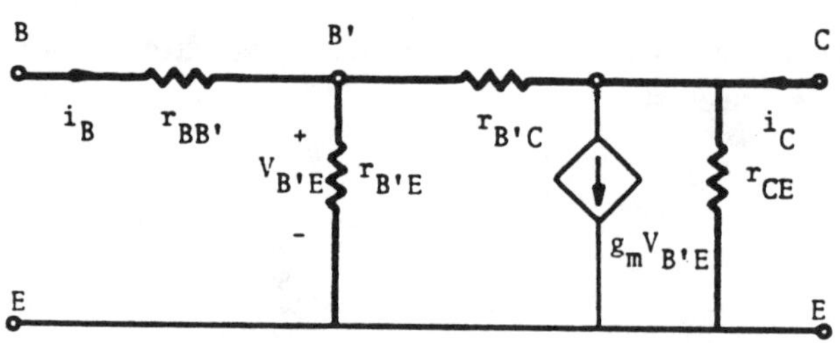

Fig. 6.   Hybrid-Pi Model of BJT.

The various parameters of the circuit in Fig. 6 are calculated as follows.

Given a collector current I_C and the circuit temperature T, we have

$$g_m \simeq \frac{|I_C|}{V_T} \quad (\triangleq \text{transconductance: } V_T = kT/q \simeq 26 \text{ mV at } 25°C);$$

$$r_{B'E} = (1/g_{B'E}) = h_{fe}/g_m \quad \text{and} \quad r_{BB'} = h_{ie} - r_{B'E};$$

$$r_{B'C} = (1/g_{B'C}) = (1/\text{feedback conductance : very large}) = r_{B'E}/h_{re};$$

$$r_{CE} = (1/g_{CE}) = h_{oe} - (1+h_{fe})g_{B'C}.$$

The circuit in Fig. 6, has to be modified for high frequency (the frequency greater than $1/\tau$, where τ is the minority carrier life time) operation of the transistor. The new hybrid-pi model becomes that of the circuit in Fig. 7.

Here C_C and C_E are given by the manufacturer. Actually we have

$$C_E = C_{DE} \text{ (diffusion capacitance)} + C_{TE} \text{ (emitter-junction capacitance)}$$

$$\simeq C_{DE} \simeq \tau/r_{B'E}.$$

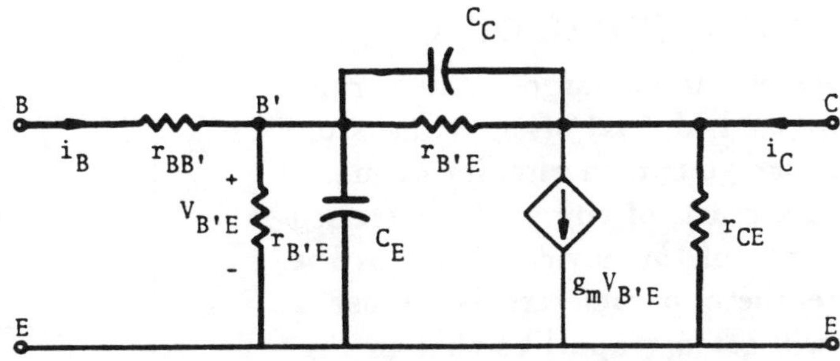

Fig. 7.   High Frequency Hybrid-Pi Model of BJT.

2.5. LARGE-SIGNAL OPERATION OF THE BJT

In this section we briefly study the conditions under which the large-signal operations of the BJT takes place. We choose to study such an operation for only two modes of cut-off and saturation. The main application of this choice is in the switching of digital circuits. To analyze a circuit in these modes requires the application of an equivalent circuit model of the transistor in these modes. Here we recall such models as those given by Fig. 8. As an example, consider the transistor inverter of Fig. 9. To develop the condition under which this circuit acts in either cut-off or saturation mode, we must substitute the appropriate model from Fig. 8 in place of the transistor in Fig. 9. Then, if $V_{BB} < V_{BE}$, $V_{out} = V_{CC}$ and we have the cut off mode; if $V_{out} = 0$, then we have the saturation mode.   Here $V_{out} = V_{CC} - R_C I_C = V_{CC} - \beta R_C (V_{BB} - V_{BE}) / R_B = 0$.   This yields $V_{BB} \geq V_{BE} + R_B V_{CC} / \beta R_C \triangleq V_{BBS}$. Thus if $V_{BE} < V_{BB} < V_{BBS}$ the transistor remains in its active mode.

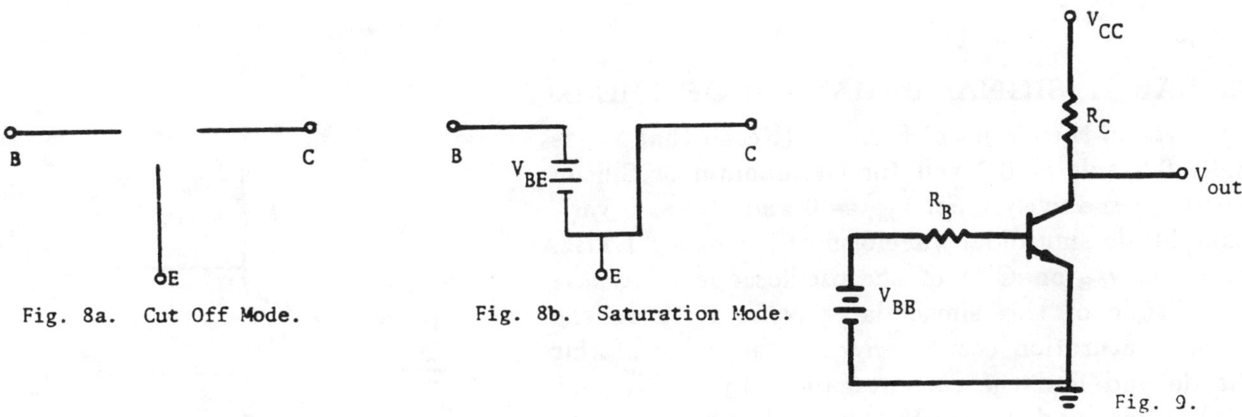

Fig. 8a.   Cut Off Mode.   Fig. 8b.   Saturation Mode.   Fig. 9.

3. LABORATORY PROCEDURE

3.1. PRE-LABORATORY WORK

Carefully study the analysis developed in the text of this experiment. Devise your own circuits to measure the various parameters that were introduced in this experiment. Complete the solution to problems given in Section 4, and submit the solution with your report due this session.

3.2. LABORATORY WORK

3.2.1. THE EBERS-MOLL MODEL OF THE BJT

From the definitions of various large-signal parameters (I_{EO}, I_{CO}, I_{ES}, I_{CS} and alphas) of the transistors studied in Section 2, *device* your own circuits to measure these quantities. Care must, of course, be exercised to avoid permanent burnout of the device. Perhaps the easiest way to generate these parameters is to use a curve tracer. Here we will get a graph like that in Fig. 10. The vertical axis corresponds to i_C or v_B, and the horizontal axis corresponds to v_{CE} or v_{CB}, with the constant being input current or input voltage. These various large-signal parameters are actually generated by increasing the variable on the horizontal axis. When it reaches the breakdown we have to be careful not to exceed the device limitation. The maximum reverse

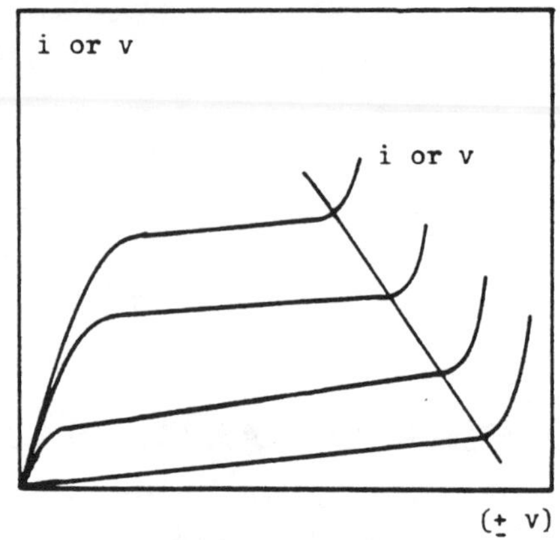

Fig. 10. A Schematic of Breakdown Points.

breakdown voltages are labeled as BV_{BE} (or BV_{BEO} meaning that the missing terminal, here C, is open), BV_{BC} and BV_{CE} that can be easily calculated or be found by using a curve tracer with proper terminals applied as junction diodes. Having determined these parameters, *use* an approximate Ebers-Moll equation to generate by computer a set of input and output characteristic curves of a common-emitter configuration. These characteristic curves may not show the effects of some additional device nonlinearities such as the base-width modulation that causes breakdown or upward tilt in the collector characteristic curves to occur. *This simulation should be completed after the laboratory session.*

3.2.2. LARGE-SIGNAL BEHAVIOR OF THE BJT

Construct the circuit of Fig. 11. (Recall that V_{BE} is typically 0.3 volt or 0.7 volt for Germanium or Silicon transistors, respectively.) *Set* $V_{BB} = 0$ and v(t) at a variable amplitude sinusoidal waveform of frequency 1 kHz. *Observe* the v_{CE} on CH1 of the oscilloscope. *Increase* the amplitude of this sinusoidal waveform up to the point that saturation occurs. *Record* the value of this amplitude and find the current gain β by reading i_E. *Set* v(t) at zero and *change* V_{BB} (variable power supply) to observe both cut-off and saturation modes with respect to i_C and v_{CE}. *Select* V_{BB} such that $V_{CE} = V_{CC}/2$. *Apply* this V_{BB} and the variable sinusoidal waveform v(t) of frequency 1 kHz. *Increase* the amplitude of v(t) to see both saturation and cut-off clipping

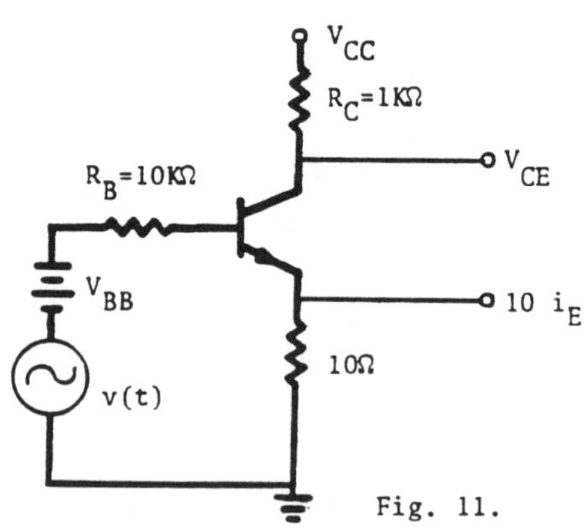

of this sinusoidal waveform. *Adjust* V_{BB} if necessary to maintain a symmetric clipped waveform. For values of v(t) just before any clipping takes place *measure* the voltage gain of $v_{CE}(t)/v(t)$ corresponding to the peak-to-peak value. *Find* the collector efficiency of this transistor which is the ratio of average AC power dissipated in R_C to that supplied by the input sources.

4. PROBLEM SET

1. Develop limits for V_{BB} of the circuit in Fig. 12, so that the transistor acts in either cut-off or saturation mode. What is the purpose of R_B? How do you choose R_B?

2. Develop the corresponding large-signal Ebers-Moll equivalent circuit of a pnp transistor with all its detailed currents direction and/or voltage polarities.

3. For a pnp transistor with $\alpha_N = \alpha_I = 0.9$, $I_{CO} = I_{EO} = 10\ \mu A$ and at $25\,^\circ C$ temperature, find an expression for i_C and discuss its values as a function of various voltages, *etc.*, at the three regions we have introduced earlier. Choose some typical values for junction voltages and calculate the corresponding values for i_C.

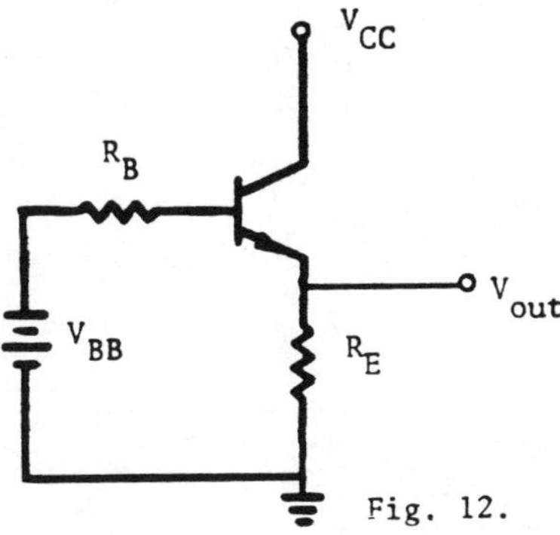

5. FOR YOUR REPORT

Submit the results of your simulation with a detailed report of your observation of the present experiment in our standard format.

EXPERIMENT III-5

**FIELD-EFFECT TRANSISTORS
CHARACTERISTICS AND APPLICATIONS**

EQUIPMENT:

    EEP
    CURVE TRACER
    FET (MOSFET OR JFET)
    RESISTORS (100 Ω, 1 kΩ, 10 kΩ, 1 MΩ)
    CAPACITOR (0.1 μF)

Table of Contents   Page No.

1. PURPOSE

The purpose of this experiment is to acquaint the student with the basic characteristics as well as the applications of a typical field-effect transistor. The further applications and/or properties of this family of devices which are comparable to those of BJT's are not included in this experiment. We must expect that many common features between these two families of devices, such as load line analysis, small- and large-signal models, amplifications, *etc.*, do exist. But here we will only review a few properties associated with a typical field-effect transistor; namely, the JFET (Junction FET).

2. INTRODUCTION

In Section III-4.2, we have studied some basic physical aspects of field-effect transistors, in particular, their differences with the BJT's. In the following we review a few definitions of various essential parameters or operational modes of a typical field-effect transistor, the JFET. Recall that the three terminals of the JFET are the source (S), the drain (D), and the gate (G). As a TTD, one may apply the general procedures that have been developed in Section III-3, for its analysis. There are, of course, some basic differences between this TTD and those we studied earlier in EXPs. III-1 to 4.

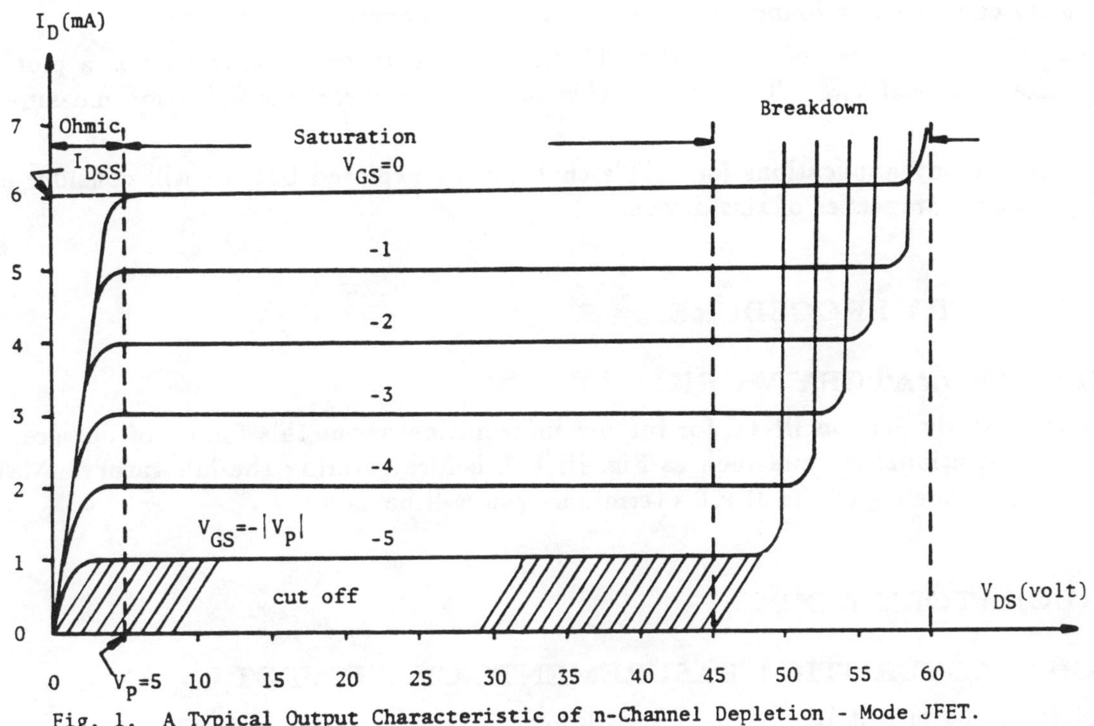

Fig. 1. A Typical Output Characteristic of n-Channel Depletion - Mode JFET.

A typical set of volt-ampere characteristic curves for, say, an n-channel depletion mode JFET is shown in Fig. 1. In this figure, the various operational regions of the JFET are described. The V_P is a voltage (pinch off) that corresponds to the beginning of the saturation region. Before V_{DS} increases to V_P the device is linear and acts as a voltage-variable resistor. The cut-off takes place when $|V_{GS}| > |V_P|$. The ratio of V_{DS}/I_D at the origin is called the *ON* drain resistance. The JFET's are perhaps the most widely used transistors with both digital and analog applications.

To develop its small-signal model for low-frequency application we may use the standard Taylor series expansion. We also recall that since the input impedances of these devices are infinite, this small-signal model in the ohmic region simplifies to the circuit of Fig. 2. Here we can easily show that

$$i_D = g_m v_{GS} + v_{DS} / r_D ,$$

with $g_m \triangleq$ mutual conductance or transconductance, we have

$$g_m \triangleq \left. \frac{i_D}{v_{GS}} \right|_{v_{DS}=0} = \left. \frac{\partial i_D}{\partial v_{GS}} \right|_{v_{DS}} .$$

Also we have $r_D (= 1/g_D) \triangleq \left. \dfrac{v_{DS}}{i_D} \right|_{v_{GS}=0} = \left. \dfrac{\partial v_{DS}}{\partial i_D} \right|_{v_{GS}} .$

The amplification factor μ is defined as

$$\mu \triangleq - \left. \frac{v_{DS}}{v_{GS}} \right|_{i_D=0} = - \left. \frac{\partial v_{DS}}{\partial v_{GS}} \right|_{i_D} = r_D g_m .$$

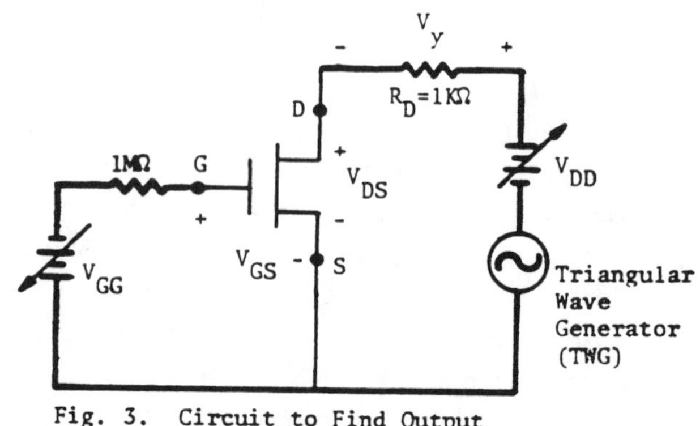

Fig. 2. The Low-Frequency Small-Signal Model of JFET.

Since $I_{DS} = I_{DSS}(1 - V_{GS} / V_P)^2$, here V_P is the pinch off for the depletion-mode JFET (or instead the threshold voltage V_T for an enhancement-mode JFET), and I_{DSS} is the value of saturation current for $V_{GS} = 0$, [3], thus

$$g_m = \left. (\partial I_{DS} / \partial V_{GS}) \right|_{V_{DS}} = 2 (I_{DSS} I_{DS})^{1/2} / |V_P| .$$

This quantity can be easily found from the characteristic curves.

Another relevant piece of information is the so-called transfer curve that is a plot of I_D versus V_{GS} for a fixed value of V_{DS}. The slope of this curve is the direct amplification measure of the device.

There are many applications for FET's that can be explored but we will consider only the following few sample properties of this device.

3. LABORATORY PROCEDURE

3.1. PRE-LABORATORY WORK

Carefully study Section III-4.2 for further information about this family of devices. Draw your own various operational circuits such as Fig. III-1-9, before entering the laboratory. Make sure you know the proper labeling of the JFET's terminals you will be using.

3.2. LABORATORY WORK

3.2.1. CHARACTERISTIC MEASUREMENTS OF THE JFET

Construct the circuit in Fig. 3. *Use a low* frequency (1 kHz) triangular (or rectangular) wave generator with peak-to-peak amplitude that together with V_{DD} covers the entire range of desired V_{DS}. This usually is decided based on some prior information about the device. *Set the SENSITIVITY CONTROL SWITCH to 1 V/DIV on both channels. Apply* V_{DS} to the x-axis (CH1) and V_Y (i.e., I_D in mA) to the y-axis (CH2) of the oscilloscope to see one characteristic curve for a given V_{GS}. *Change* V_{GS} *and repeat* the above procedure. *Plot* each characteristic curve for a given V_{GS} in Fig. 4a. *Repeat* this drawing for several values of V_{GS}.

Fig. 3. Circuit to Find Output Characteristic of JFET.

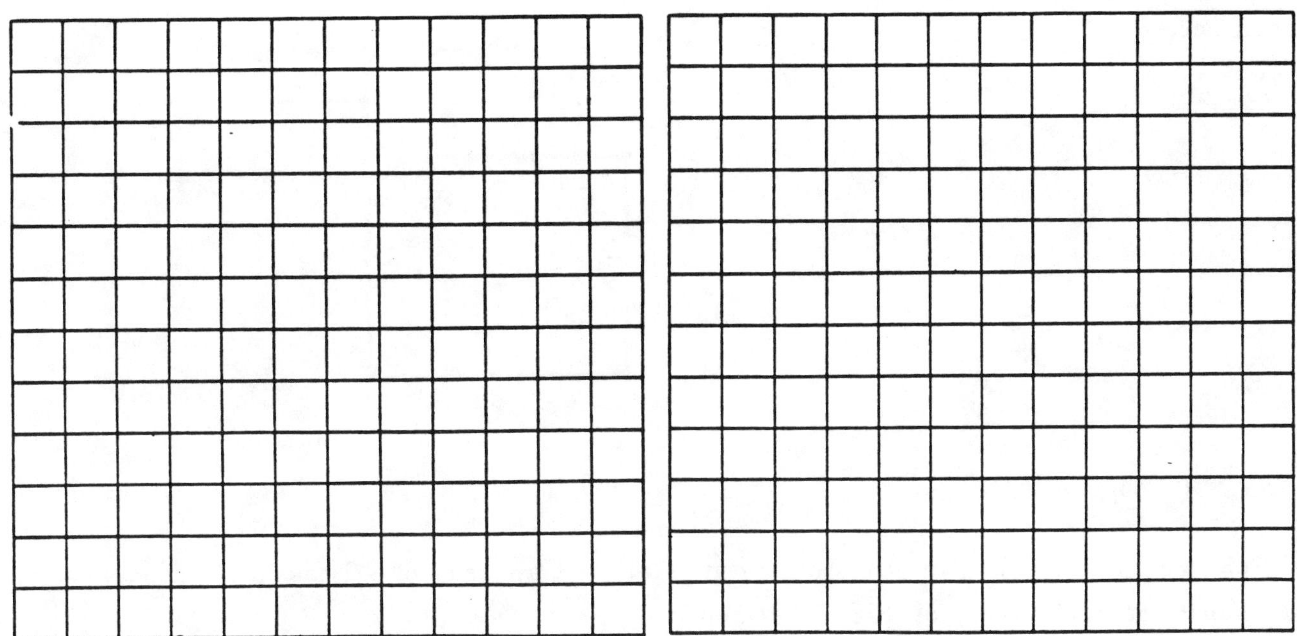

<table>
<tr><td>Fig. 4a.</td><td>Fig. 4b.</td></tr>
</table>

3.2.2. TRANSFER CHARACTERISTIC MEASUREMENTS

Construct the circuit in Fig. 5. *Use* a low frequency (less than 100 Hz) triangular (or rectangular) wave generator with peak-to-peak amplitude that together with V_{GG} covers the entire range of the characteristic curves that you generated earlier for this device. *Set SENSITIVITY CONTROL SWITCH* to 1 V/DIV on CH1 and 0.1 V/DIV on CH2 of the oscilloscope. *Apply* V_{GS} to CH1 and V_y to CH2. *Adjust* the input source to see one transfer characteristic curve. *Draw* this curve on Fig. 4b. *Change* V_{DS} by adjusting V_{DD}. Perhaps the most linear results can be developed when the JFET is in the saturation region. If you are experiencing difficulty tracing this curve, then change the setting of the *VOLTS/DIV* switches as required.

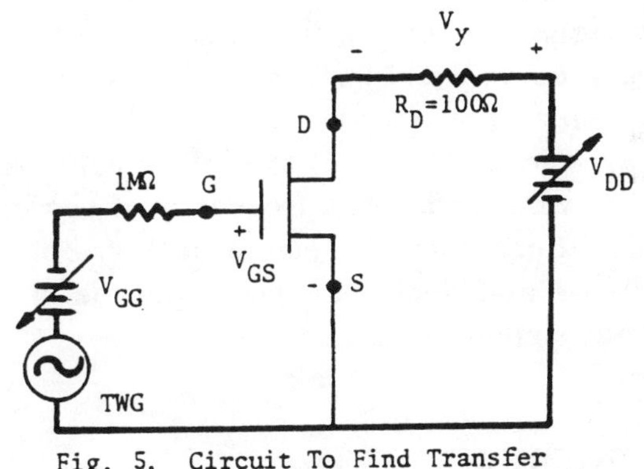

Fig. 5. Circuit To Find Transfer Characteristic of JFET.

Note: The above two measurements can also be done using a curve tracer.

3.2.3. SOME APPLICATIONS OF THE JFET

3.2.3A. An Amplifier JFET

Construct a simple JFET amplifier as shown in Fig. 6. *Apply* a small amplitude and low frequency sinusoidal input $v_{in}(t)$ and *observe* the $v_{out}(t)$ on the oscilloscope. Gradually *increase* the input amplitude and *monitor* the $v_{out}(t)$ until distortion (or clipping) at one or both ends occurs. Some adjustment may be needed to have a symmetric clipped $v_{out}(t)$. For values of $v_{in}(t)$ just before this clipping happens *find* the amplifier gain. Also *find* the transfer characteristic of this circuit when clipping takes place.

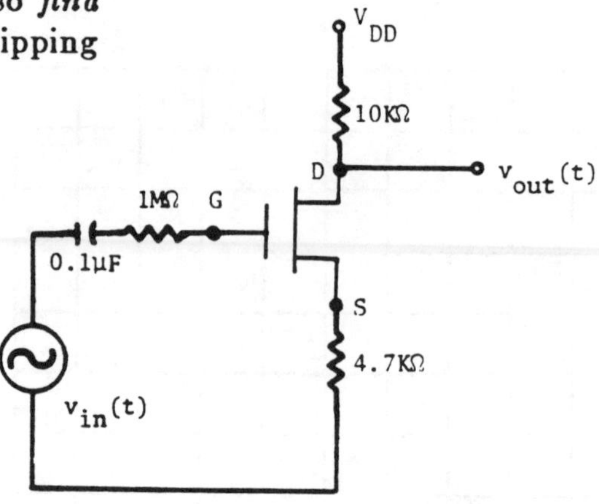

Fig. 6.   A Simple JFET Amplifier.

3.2.3B.   A Chopper JFET

This circuit is credited to professor Scidmore of the University of Wisconsin-Madison.

Construct the circuit of Fig. 7. *Choose* $v_{in}(t)$ to be a low frequency (less than 100 Hz) small amplitude (less than 2 volts) sinusoidal waveform and $v_m(t)$ (modulation signal) a square waveform of 1 kHz frequency and 10 volts amplitude. *Trigger* the oscilloscope trace on input signal and *sketch* the $v_{out}(t)$ on Fig. 8a. This is called a chopper. *Change* the triggering from input signal to modulation signal. What are its effects on $v_{out}(t)$? *Change* the amplitude and frequency of $v_m(t)$ and *observe* the $v_{out}(t)$. *Increase* the frequency of the input signal to much

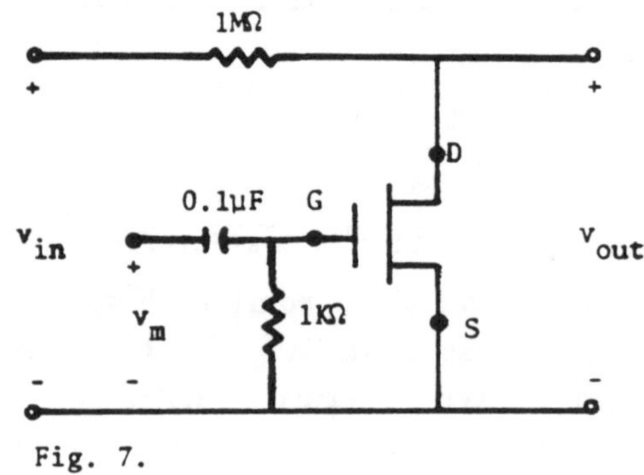

Fig. 7.

higher values than the frequency of the modulation signal (30 kHz) and observe the $v_{out}(t)$. *Draw* a typical output of this signal on Fig. 8b. This result is called gating rather than chopping. *Change* (increase and decrease) the amplitude of $v_m(t)$ and *study* its effects on the output signal. *Draw* this output signal on Fig. 8c.

4. PROBLEM SET

1.   From the definition of the I_{DSS}, devise a circuit that measures I_{DSS} for $V_{DS} = 5$ volts.

2.   From the definition of the pinch-off voltage, devise a circuit that measures V_P.

3.   From the definition of the *ON* drain resistance, devise a circuit that measures this parameter.

5. FOR YOUR REPORT

Submit your results with solutions to the problem set of this experiment in our standard report form.

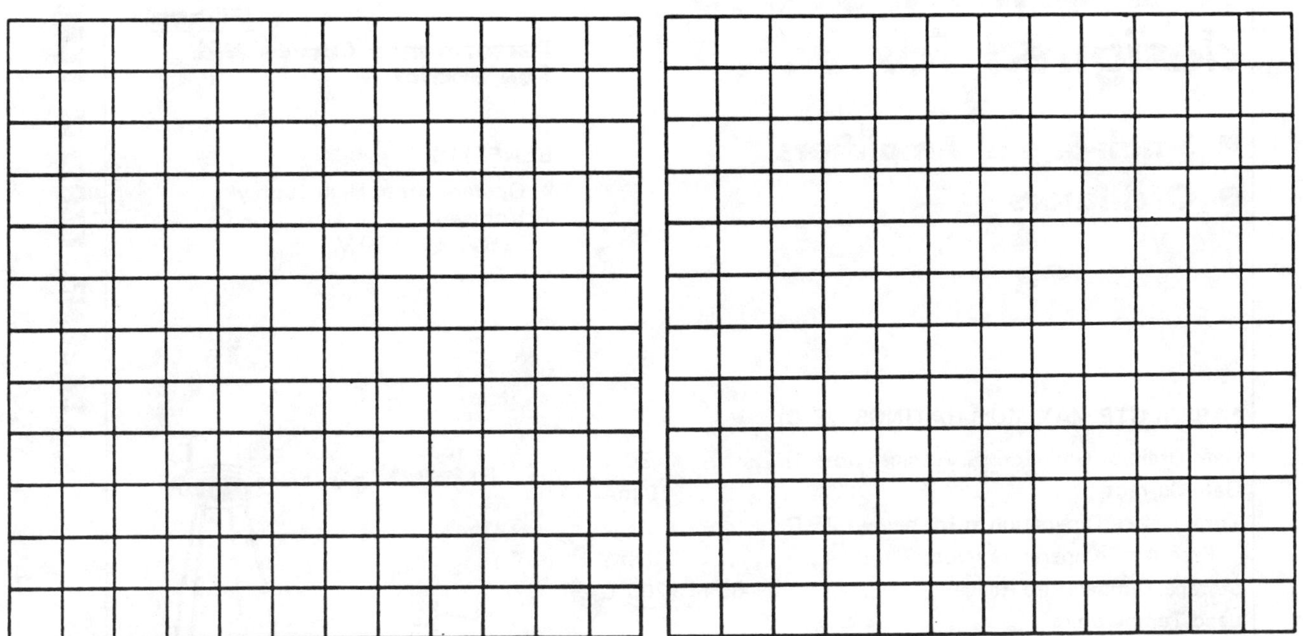

Fig. 8a.   Fig. 8b.

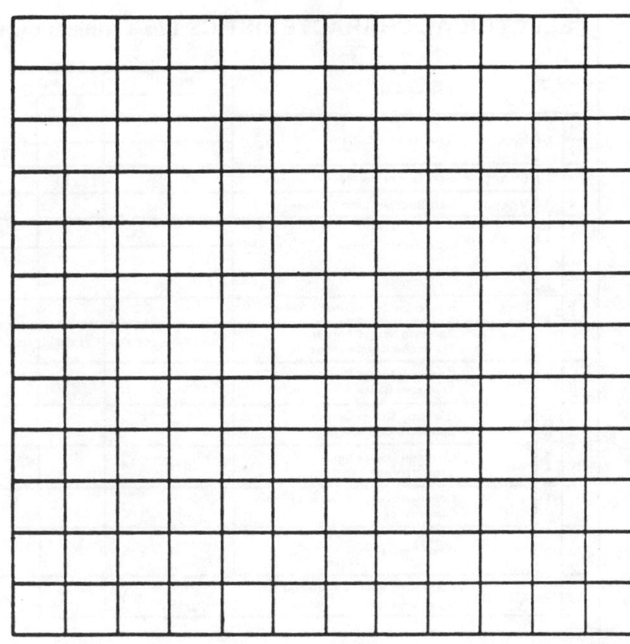

Fig. 8c.

6. MANUFACTURER'S SPECIFICATION DATA SHEETS

On the following pages (cf., Fig. 9), we present the specification data sheets of one brand of general purpose n-channel JFET for the sake of illustration and review of its basic properties [12].

n-channel JFETs
designed for . . .

Siconix — Siliconix

**Performance Curves NRL
See Section 4**

- ■ **Small-Signal Amplifiers**
- ■ **Oscillators**

BENEFITS

- ● Operates from High Supply Voltages $BV_{GSS} > 50$ V

*ABSOLUTE MAXIMUM RATINGS (25°C)

Gate-Drain or Gate-Source Voltage (Note 1) . . . –50 V
Gate Current . . . . 10 mA
Total Device Dissipation at (or below) 25°C
   Free-Air Temperature (Note 2) . . . . . . 300 mW
Storage Temperature Range . . . . –65 to +200°C
Lead Temperature
   (1/16" from case for 10 seconds) . . . . . . 300°C

TO-72
See Section 6

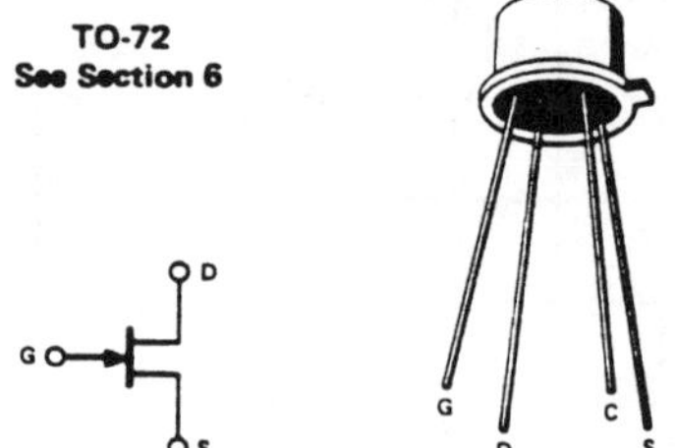

*ELECTRICAL CHARACTERISTICS (25°C unless otherwise noted)

#		Characteristic	2N3821		2N3822		2N3823		Unit	Test Conditions	
			Min	Max	Min	Max	Min	Max			
1	S T A T I C	I_{GSS} — Gate Reverse Current		-0.1		-0.1		-0.5	nA	V_{GS} = -30 V, V_{DS} = 0	
2				-0.1		-0.1		-0.5	µA		150°C
3		BV_{GSS} — Gate-Source Breakdown Voltage	-50		-50		-30		V	I_G = -1 µA, V_{DS} = 0	
4		$V_{GS(off)}$ — Gate-Source Cutoff Voltage		-4		-6		-8		V_{DS} = 15 V, I_D = 0.5 nA	
5		V_{GS} — Gate Source Voltage	-0.5	-2						V_{DS} = 15 V, I_D = 50 µA	
					-1	-4				V_{DS} = 15 V, I_D = 200 µA	
							-1.0	-7.5		V_{DS} = 15 V, I_D = 400 µA	
6		I_{DSS} — Saturation Drain Current (Note 3)	0.5	2.5	2	10	4	20	mA	V_{DS} = 15 V, V_{GS} = 0	
7	D Y N A M I C	g_{fs} — Common-Source Forward Transconductance (Note 3)	1500	4500	3000	6500	3.500	6.500	µmho	V_{DS} = 15 V, V_{GS} = 0	f = 1 kHz
8		y_{fs} — Common-Source Forward Transadmittance	1500		3000		3.200				f = 100 MHz
9		g_{os} — Common-Source Output Conductance (Note 3)		10		20		35			f = 1 kHz
10		C_{iss} — Common-Source Input Capacitance		6		6		6	pF		f = 1 MHz
11		C_{rss} — Common-Source Reverse Transfer Capacitance		2		2		2			
12		NF — Noise Figure		5		5		6	dB	V_{DS} = 15 V, V_{GS} = 0, R_{gen} = 1 meg, BW = 5 Hz	f = 10 Hz
13		$\bar{e}_n$ — Equivalent Short-Circuit Input Noise Voltage		200		200		200	$\frac{nV}{\sqrt{Hz}}$	V_{DS} = 15 V, V_{GS} = 0, BW = 5 Hz	

*JEDEC Registered Data.

NRL

NOTES:
1. Due to symmetrical geometry, these units may be operated with source and drain leads interchanged.
2. Derate linearly to 175°C free-air temperature at rate of 2 mW/°C.
3. These parameters are measured during a 2 msec interval 100 msec after d–c power is applied.

Siliconix

Fig. 9. n-channel JFET (Courtesy of Siliconix [12]).

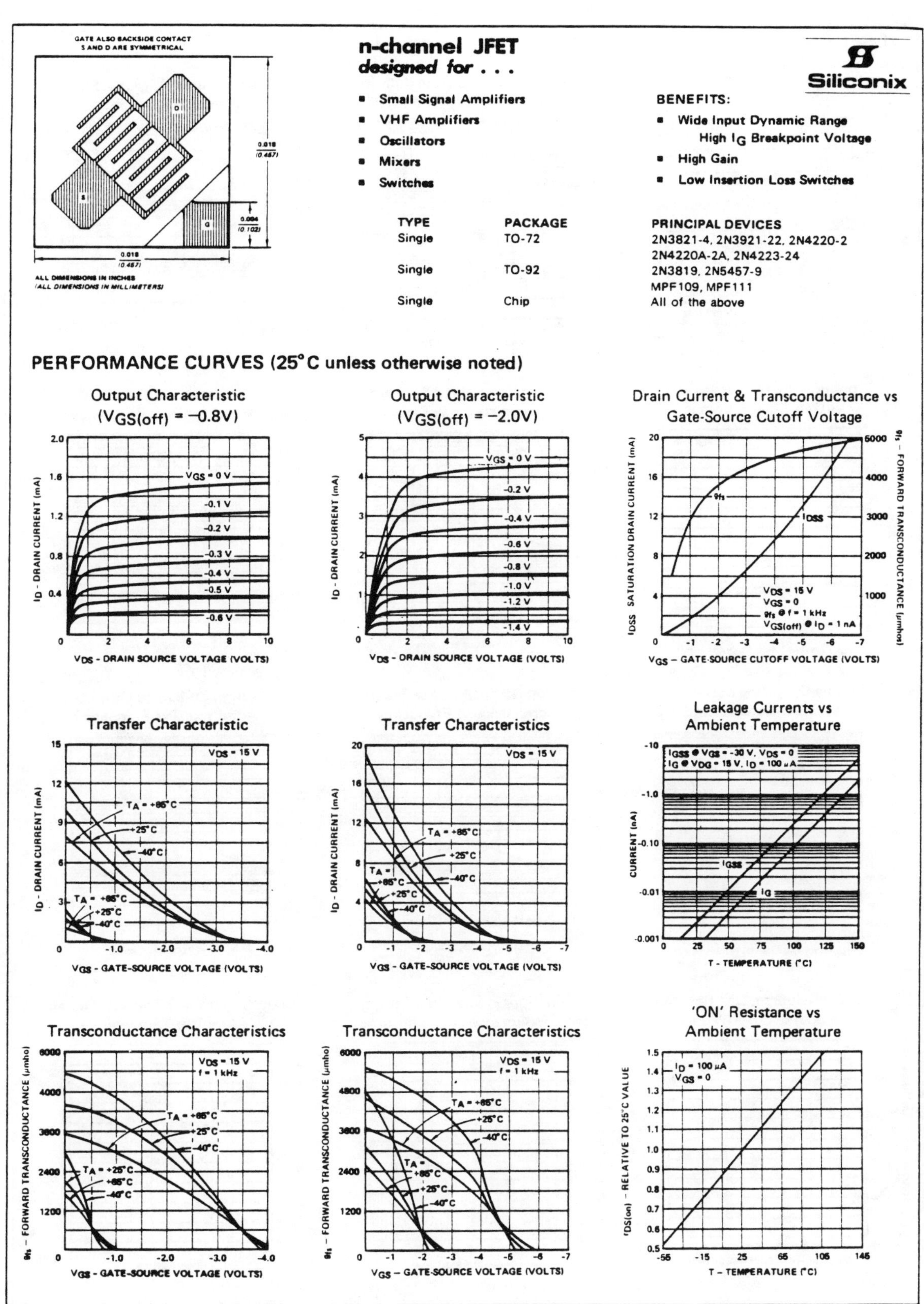

Fig. 9. Continued.

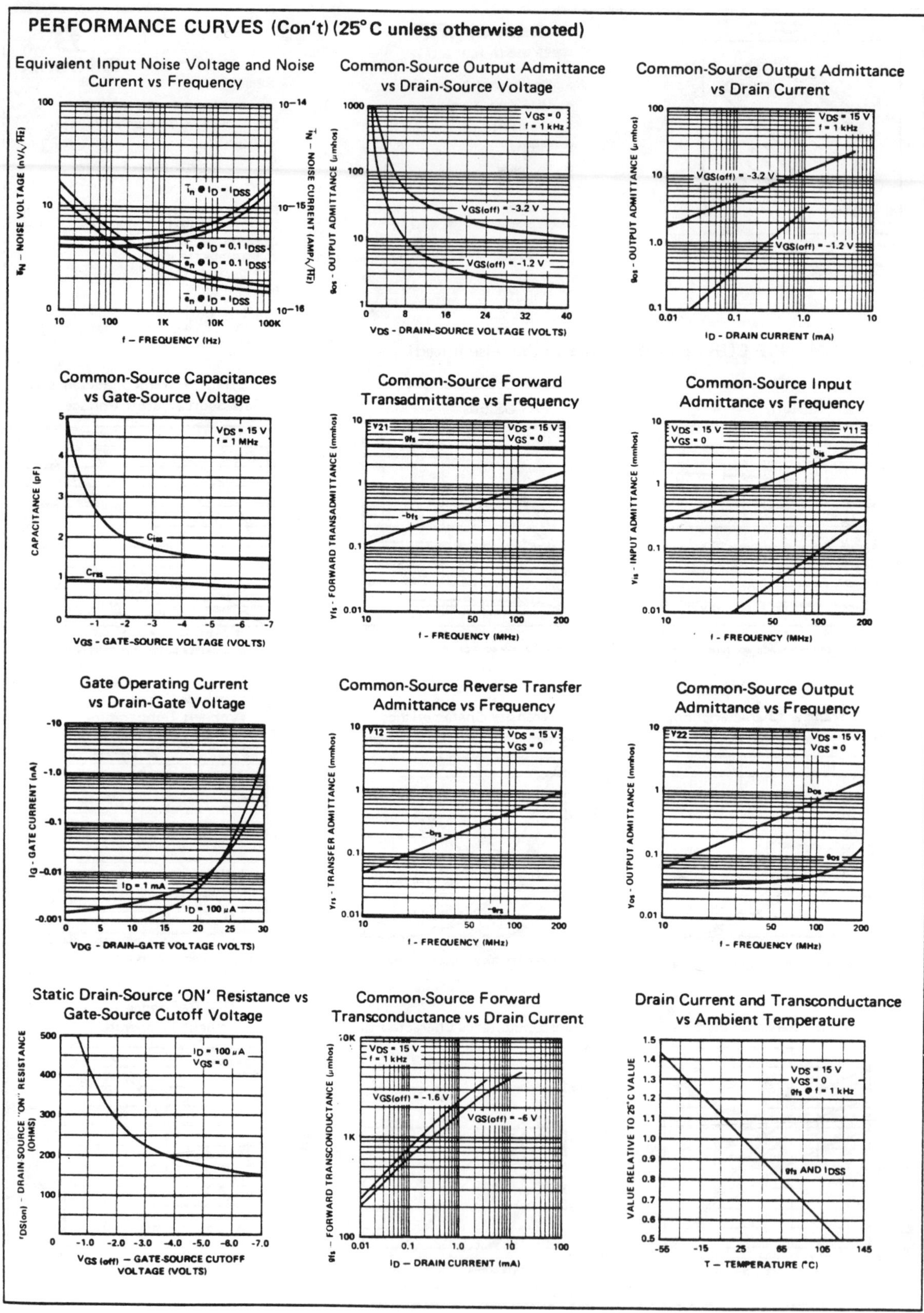

Fig. 9. Continued.

EXPERIMENT III-6

SILICON CONTROLLED RECTIFIERS

EQUIPMENT:

EEP	RESISTORS (100 Ω-5W, 1 kΩ, 2kΩ, 15 kΩ, 33 kΩ)
POTS (20 kΩ, 150 kΩ)	ISOLATION TRANSFORMER
SCR; TRIAC; DIAC	LIGHT BULB (25 W)
RECTIFIER DIODES (4)	CAPACITORS (0.1 μF, 10 μF, 90 μF)

Table of Contents   Page No.

1. PURPOSE

The purpose of this experiment is to familiarize the students with the operation and applications of the silicon controlled rectifier (SRC). This solid state device is used both as a switching device and as a high power-gain device.

2. INTRODUCTION

Reliable electronic switching devices have many applications in industry, especially in the power electronic industry. These switches are extremely close to the ideal *ON* and *OFF* switches needed in many applications. Actually these devices have very high impedances in their off states under forward bias and very low impedances in their on states. This transition can be made by external sources as we wish and in an extremely well-controlled manner as needed in a particular application. We have already seen some of these switches such as transistors and diodes. But there are, of course, some more with many different names. For example, we have *semiconductor* (mainly made of silicon) *controlled rectifiers* (SCR); *triacs*; and *Shockley four layer diodes* that are members of the *thyristor* family [7]. Also we have *unijunction transistors* (UJT). All can act as electronic switching devices.

The mechanism under which these devices work and many of their interesting relevant properties are discussed in one form or another in the references of this part. Perhaps one may consider Chapter 11 of [5] as a starting point to learn more about this subject. The following experiment, credited to Professor Stremler, is, however, selected from [6], that is based on [0], [2], [4], [8], [10], and [14]. Certain information reported in [6] is mainly gathered "with permission" from [14]. Originally the SCR used in [6] was a Westinghouse 209M. This is now changed to TIC116E SCR [13], 2N6072 TRIAC, and D3202U DIAC, made by Texas Instruments Inc.

2.1. SOME THEORETICAL BACKGROUND

The silicon controlled rectifier (SRC) is a three-terminal, three-junction semiconductor device which is capable of very high current gains. The three terminals are called the cathode, anode, and gate, as shown in Fig. 1. A device allowing access to the fourth region of the pnpn arrangement is called a silicon controlled switch. Another device, with only a cathode and an anode lead, is called the Shockley diode. All devices with this pnpn arrangement belong to the thyristor family.

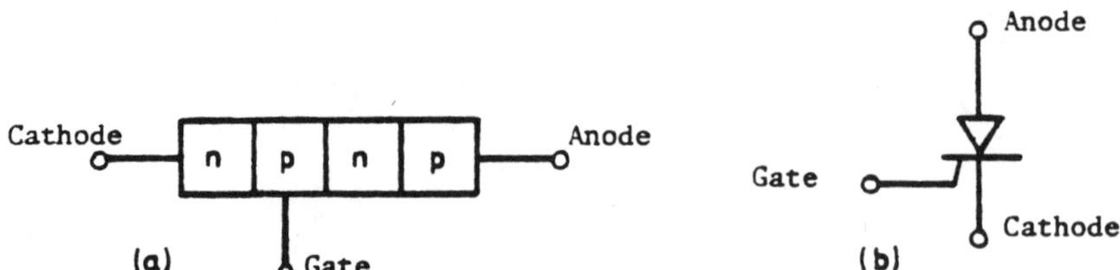

Fig. 1. The Physical Layout and Schematic Symbol for the SCR.

It is evident from Fig. 1a that sufficient p-n junctions exist in the SCR to block current flow in either direction. The operation of the SCR can be described by considering the device to consist of an interconnected arrangement of a pnp- and an npn-transistor, as shown in Fig. 2. If we forward-bias the device (cf., Fig. 2) and apply a gate signal, we note that: (1) the collector current for the npn transistor is the base drive for the npn transistor; (2) the collector current for the pnp transistor is the base drive for the pnp unit; and (3) when circuit current is sufficient to maintain transistor saturation, gate drive can be removed and the SCR will conduct until current is markedly reduced or the device is reverse biased. *From the preceding, it is obvious that this will conduct, as a diode, upon initiation of gate control. This makes this device extremely useful in power control circuits.*

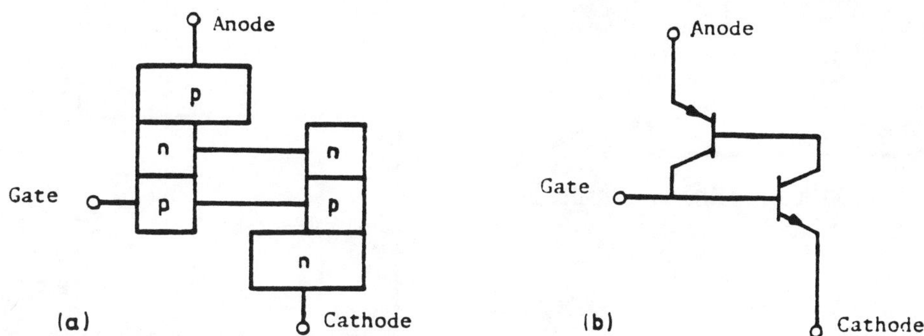

Fig. 2. SCR Equivalent Consisting of Two Transistors: (a) Physical Construction; (b) Equivalent Circuit.

The characteristics of the SCR which make it important are: (1) when the gate is not positive with respect to the cathode, a large voltage can be applied to the anode with no accompanying current conduction; (2) raising the gate voltage a small amount to the triggering level causes a very high current flow with very little voltage drop from anode to cathode; and (3) the gate cannot be used to cause the anode-to-cathode current to stop flowing; only the combination of a gate voltage below the trigger level and the temporary interruption of anode current will restore the SCR to the nonconducting state.

2.2. SCR OPERATION

A silicon controlled rectifier (SCR) is a solid state device that performs two major functions in present day applications; as a switching device, and as a high-power-gain device. As a switching device, SCR's are available that switch up to 1600 V with a maximum current of 470 amps RMS. As a high-power-gain device, a control signal of a few volts and less than an ampere can activate an SCR, resulting in a power gain of 100,000. Application of SCR's results in solid state reliability limited only by electrical, thermal, and mechanical considerations. In its normal state, the SCR blocks applied voltage in either direction, but when an appropriate gate-to-cathode voltage and current pulse is applied to the gate, the anode-to-cathode impedance approaches zero and the power is transferred to the load. When the polarity of the applied voltage is reversed, the device reverts to the blocking state.

The operation of an SCR in controlling the power in a load from an AC source is shown in Fig. 3. The load does not see any voltage during time A, since there is no gate signal. During time B, a gate pulse applied at the start of a positive voltage cycle results in the transfer of power to the load over the entire positive half of the voltage cycle. Once the SCR has been brought into the conducting state, the gate circuit no longer has control. Time C illustrates the control feature of an SCR. Conduction is started at a predetermined time during the cycle, thereby supplying only a portion of the input power to the load.

The (V-I) characteristics of SCR operation are shown in Fig. 4. (Appendix A lists some of the commonly used symbols and definitions for the SCR.) The SCR basically blocks the anode-to-

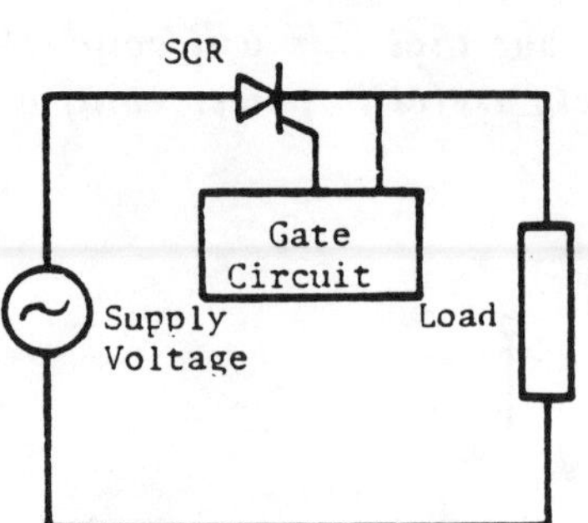

Fig. 3. Basic Voltage Relationship
of the SCR Circuit.

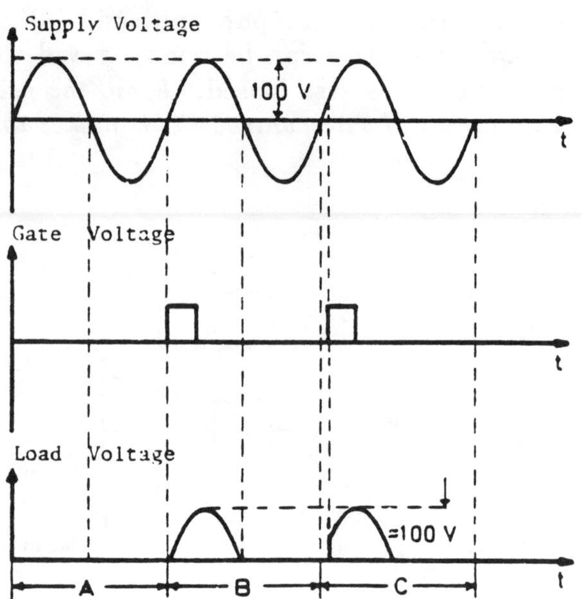

cathode voltage in both directions. When voltage blocking occurs in the forward direction, a small amount of leakage current, I_{FB}, exists. The device continues to block in the forward direction until the two-terminal (anode-to-cathode) rating is exceeded. This two-terminal rating is designated as the maximum voltage breakover, V_{BO}. Because normal SCR operation is to block voltage until the device is switched on by a control signal, most manufacturers limit their device to a lower value called the forward blocking voltage V_{FB}. The forward blocking voltage capability of a device is determined with a small positive voltage applied to a gate, typically 0.15 V. This 0.15 V bias is applied during testing to protect the user from false firing, which could occur from noise pickup or a small amount of leakage current in the gate circuit.

Once the SCR has been triggered to the conducting state by a gate signal, the anode-to-cathode voltage drops to the forward voltage drop value, V_F, and varies with the current through the device. The SCR remains in a con-

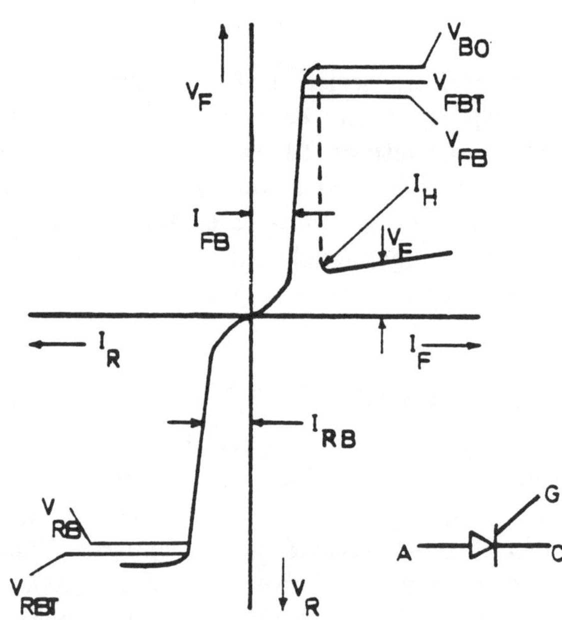

Fig. 4. (V-I) Characteristics
of SCR Operation.

ducting state until the forward current through the device drops below the holding current, I_H. As the supply voltage approaches zero, the forward current in a resistive network instantaneously follows this voltage to the value of I_H, at which time the SCR reverts to its voltage blocking mode. A highly capacitive or inductive load allows the SCR to turn off at that instant of time when the holding-current value is no longer maintained regardless of the applied voltage.

The peak reverse breakover voltage, V_{RB}, is the maximum repetitive voltage the SCR can block without damage to the device in its reverse direction. The transient peak reverse voltage, V_{RBT} is the maximum transient voltage-blocking capability of the SCR in the reverse direction. This allows the SCR to be applied up to its full reverse blocking voltage and still be protected in case transient voltages should appear across the device within the limit of the transient rating, V_{RBT}. As a solid-state switch, the SCR has a reverse impedance of several hundred-thousand ohms. Consequently, a reverse voltage results in a reverse leakage current, I_R, which is normally on the order of a few milli-amperes in magnitude.

Example 1: The SCR shown in the circuit of Fig. 5 requires a holding current of 100 mA and a gate trigger voltage of 0.75 V and gate trigger current of 10 mA. Calculate: (a) the value of e_{in} that will cause the SCR to begin conducting; (b) the SCR conduction current; and (c) the value to which V_{AA} must be reduced to turn the SCR off. Assume that the internal anode-to-cathode SCR resistance is 0.10 Ω.

Solution: (a) $e_{in} = 0.75 + (10 \times 10^{-3})(500) = 5.75$ V.
        (b) $I_A = [300 - (0.10) I_A] / 20$,
           $I_A = 15 / 1.005 = 14.93$ A $\simeq 15$ A.
        (c) To drop I_A below the 100 mA holding current, the voltage across R_L must be less than $0.1 \times 20 = 2$ V.

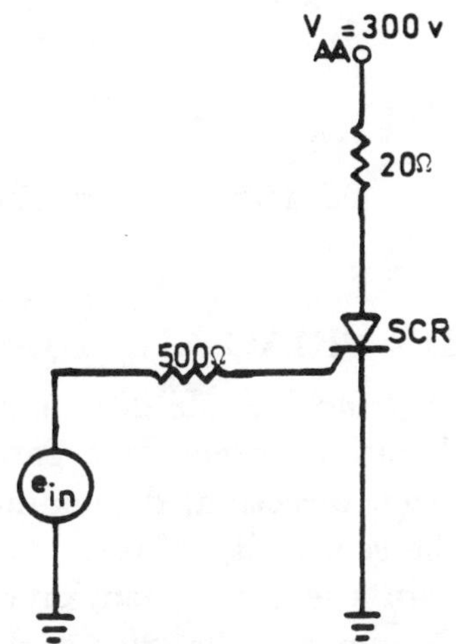

Fig. 5. Example of a Simple SCR Circuit.

We have seen that the gate voltage must be less than the minimum turn-on voltage and the anode current must be temporarily interrupted to turn the SCR off. Interrupting or diverting the anode current can present a serious problem. For an AC power supply voltage, the anode current stops flowing when the voltage becomes negative. For a DC supply, additional circuitry must be used to turn the SCR off.

One method of diverting the anode current is to use a shunt transistor, as shown in Fig. 6a. During the on time of the SCR, the transistor must be reverse-biased. When the SCR is to be turned off, a pulse is applied to the base of the transistor to turn it on (into saturation), diverting the anode current through the shunt path. The transistor must, of course, have a maximum collector-to-emitter breakdown voltage equal to or exceeding the power supply voltage. It also must have a current rating which is high enough such that the entire anode current can be diverted through the transistor.

In the capacitor commutation scheme of Fig. 6b, a second SCR is used to turn off the first. Suppose that SCR(1) is on and SCR(2) is off. The voltage at the anode of SCR(1) is *LOW* while the voltage at the anode of SCR(2) is *HIGH*. Application of a pulse voltage to the gate of SCR(2) causes it to turn on. The voltage at the node of SCR(2) drops rapidly and this voltage is transferred to the anode of SCR(1). This causes the voltage at the anode of SCR(1) to go slightly negative, turning SCR(1) off. Note that because one SCR is always on, the net current drain from the source remains *HIGH*. We shall investigate some of these methods in this experiment. Discussion on SCR selection is given in Appendix B.

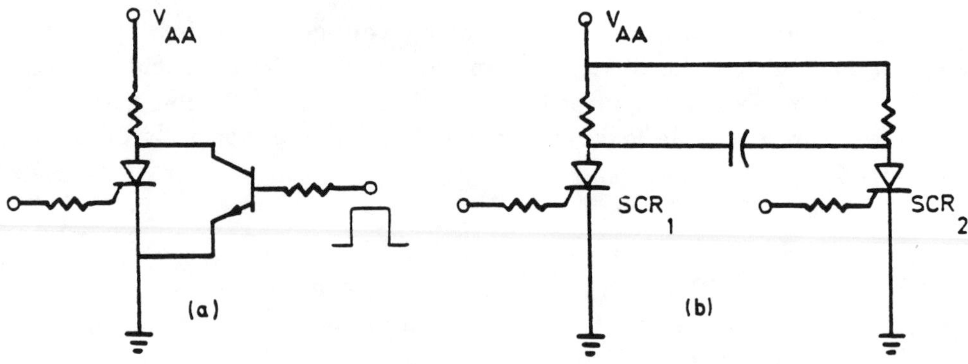

Fig. 6. DC turn-off methods: (a) Shunt Transistor; (b) Capacitor Commutation

2.3. THE TRIAC AND DIAC

The *triac* is a bi-directional switch with one gate electrode. Positive or negative gate voltages will initiate conduction in each direction, depending on the anode-to-cathode voltage. The sensitivity of response to the gate signal, however, is different in the four quadrants of the current/voltage plane, Fig. 7. In general, the triac performs the same circuit functions as an inverse-parallel SCR pair, with the advantages of compactness and simplicity of trigger circuitry. It is commonly used for the brightness control of incandescent lamps.

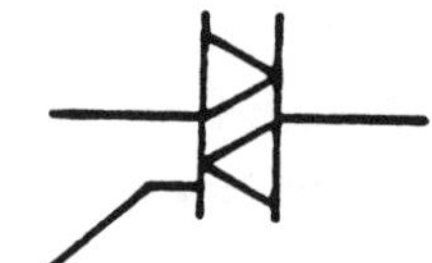

Fig. 7. Circuit Symbol for the Triac.

The *diac* is a two-terminal version of the triac without a gate electrode. It is designed to be operated in the voltage breakover mode and is intended for thyristor phase-control circuits such as are used for light-dimming applications. For such applications, it is used in the gate circuit of a triac to hold off the initiation of conduction until a given point (phase angle) is reached.

In this experiment, we shall have an opportunity to observe the circuit of the SCR, the triac and the diac.

3. LABORATORY PROCEDURE

The silicon controlled rectifier and triac are very reliable devices providing their ratings are never exceeded. *It is expected that the specification data sheets for the devices that are being used will be furnished in the laboratory. Consult these ratings before using your devices. Also be aware that the thyristors are quite unforgiving if mistakes ae made which cause their ratings to be exceeded.* The most common failure mode is a short circuit between terminals.

It is also expected that all the necessary components needed to complete this experiment have already been mounted on special boards with sufficient binding posts. Thus it is just a matter of proper wiring of these elements to perform the experiment.

3.1. DIAC CHARACTERISTICS

The objective of the diac is to provide a *snap action* to triggering pulses and similar applications. The diac has a nonconducting and a conducting state, with a steep negative resistance slope separating the two.

Construct the circuit of Fig. 8a. With the voltmeter disconnected, *set* both adjustable power supplies at zero. *Increase* the voltage at about 2 volt increments. *Read* the values of voltage and current across the diac. As soon as the diac breaks over, *change* scales in the ammeter, *increase* the voltage to 40 volts and *connect* the voltmeter across the diac itself. *Start* reducing the voltage at about 2 volt increments, while recording V and I until the diac breaks back into non-conduction. *Reverse* the diac and *repeat* the above. *Plot* V versus I for the diac on the chart of Fig. 8b.

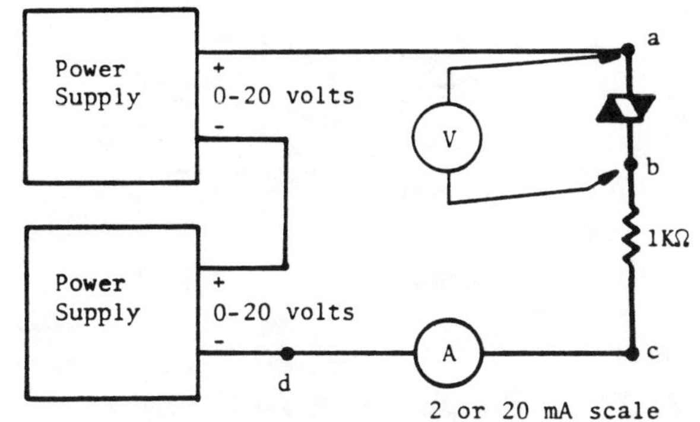

Fig. 8a.

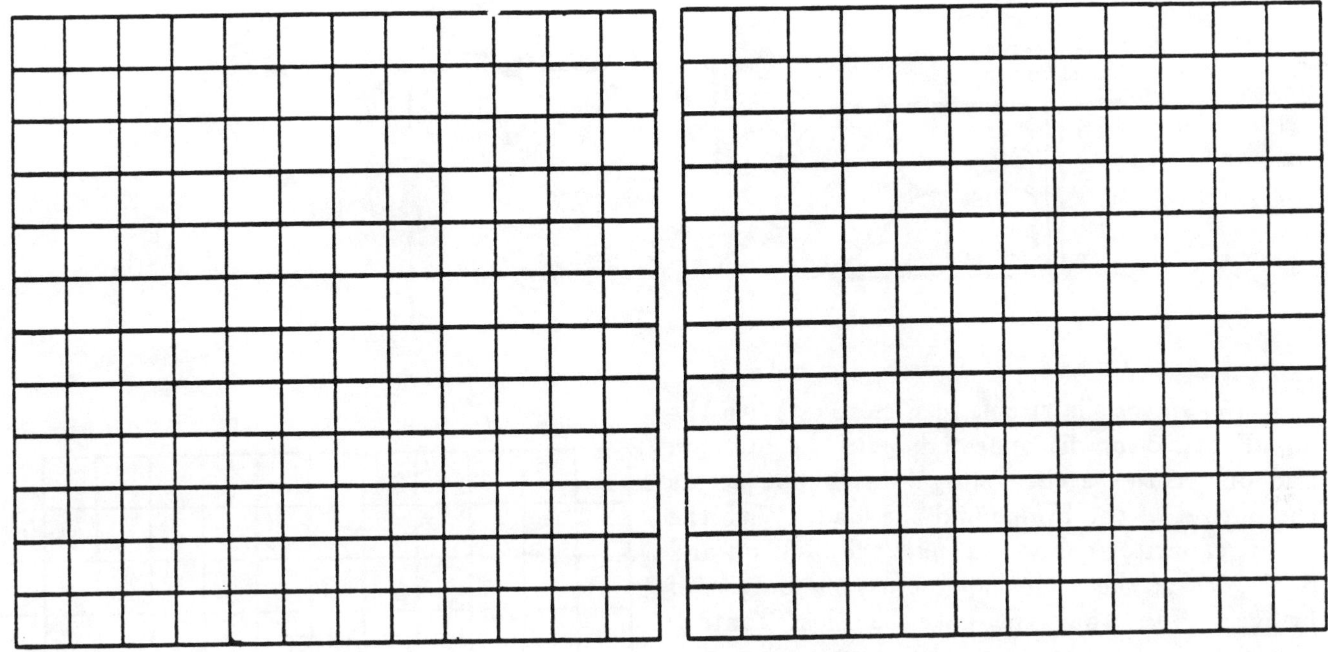

Fig. 8b.

3.2. SCR CHARACTERISTICS

Construct the circuit in Fig. 9. *Measure* the gate current before the SCR will fire. *Compare* this value with its rated value. *Cause* the SCR to fire; then *open* the gate circuit. *Reduce* the power supply voltage and *observe* the holding current. This is the minimum current just before the SCR turns off. *Compare* this value with its rated value. After causing the SCR to fire, *open* the gate circuit. *Use* the following turn-off techniques:

Interruption: *Short* circuit the cathode to the anode of the SCR using a jumper lead. *Remove* the short circuit.

Capacitor Commutation: *Take* a 10 μF capacitor and charge it from the 20 volt supply. *Place* the capacitor across the SCR cathode-to-anode to back-bias the SCR. *Repeat* the above two steps with the capacitor initially discharged.

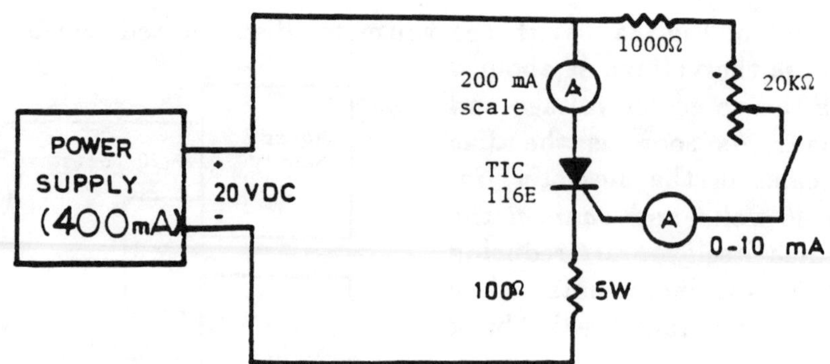

Fig. 9.

3.3. CONTROL OF A DC LOAD

Warning: Do not short circuit the bridge.

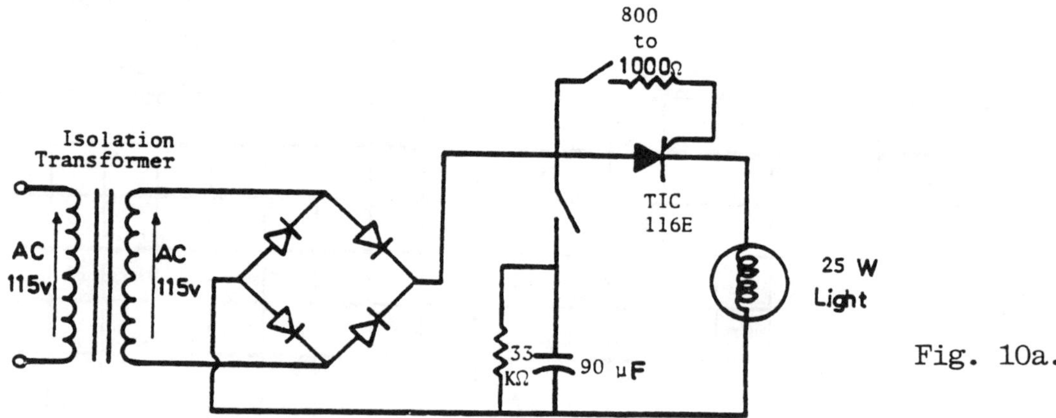

Fig. 10a.

Construct the circuit in Fig. 10a. With the 90 μF capacitor disconnected, *gate* the SCR on and off. *Observe* the voltage waveshape across the output of the bridge and the load. With the 90 μF capacitor connected, *gate* the SCR on and off. *Observe* the waveshape across the output of bridge. *Note* that the gate has lost control. *Draw* these waveshapes on the chart of Fig. 10b.

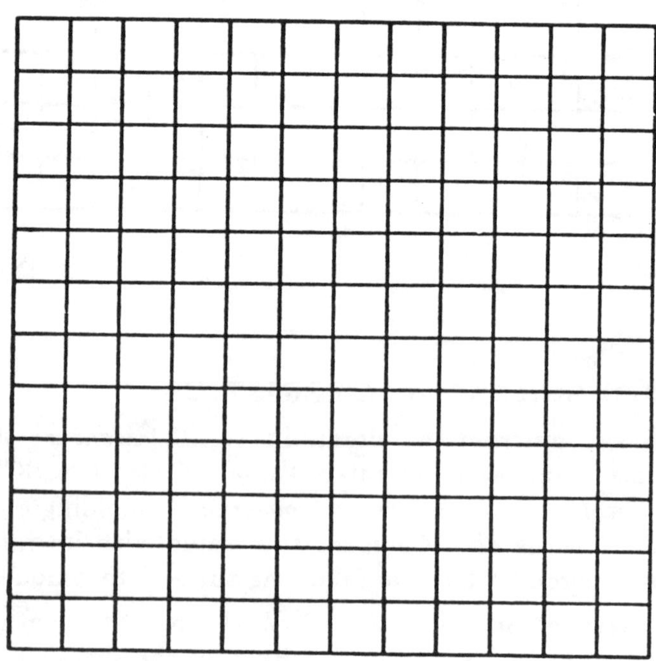

Fig. 10b.

Construct the circuit in Fig. 11. *Operate* the control switch and observe how the light bulb is controlled. *Explain* the behavior of the lamp as the 90 μF capacitor is connected or disconnected in this circuit. What is the purpose of the 33 kΩ resistor? What changes in the circuit will be needed to control a 300 W load? Why is an isolation transformer used?

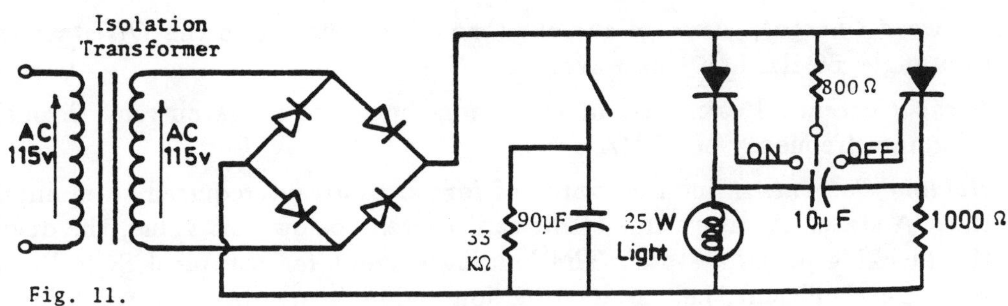

Fig. 11.

3.4. THE LIGHT DIMMER

Construct the circuit in Fig. 12. *Observe* the brightness of the 25 W lamp as the potentiometer control is moved. Also *observe* the waveform of the voltage across the lamp (use the higher scale on the *SENSITIVITY CONTROL SWITCH* of the oscilloscope). *Explain* the operation of this circuit. For a pot setting of 20 kΩ, *calculate* the time at which the SCR will fire, knowing the characteristics of the diac. What is the RMS voltage seen by the lamp? Use as many *reasonable* assumptions as necessary.

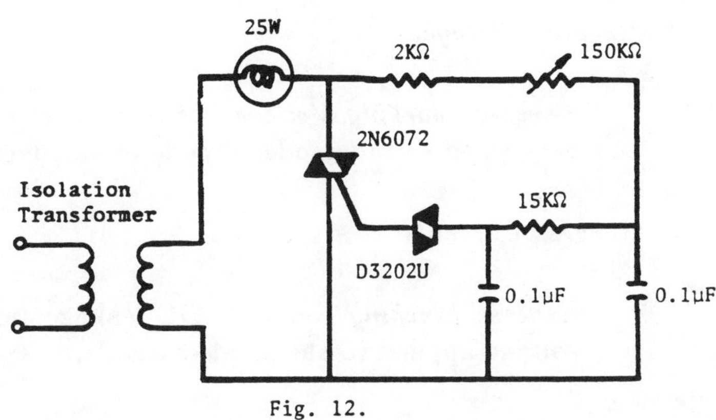

Fig. 12.

4. APPENDICES

4.1. APPENDIX A: SYMBOLS AND DEFINITIONS

4.1.1. Voltages

V_{AA} *Anode Supply Voltage.*

V_{FB} *Forward Blocking Voltage:* Minimum forward DC voltage the device is capable of blocking in the forward direction.

P_{FV} *Repetitive Peak Forward Blocking Voltage:* Maximum peak forward voltage that may be applied to the anode-cathode in the forward direction. SCR should not be fired as a two-terminal device above this voltage.

V_F *On Voltage:* DC voltage drop across the anode-cathode with the specified value of DC current flowing through the device.

4.1.2. Currents

I_{FB} *Forward Blocking Current:* DC leakage current through the device in the *OFF*-state.

I_F *Forward Current:* RMS current through the device in the *ON*-state.

$I_{F(AV)}$ *Forward Current:* Average current through the device in the *ON*-state.

I_O *Forward Current:* Average current through the device in the *ON*-state for 180° conduction angle, 60 Hz, half sine wave.

$I_{FM(surge)}$ *Surge Current:* Peak one-half cycle non-repetitive surge current through the device at minimum frequency of 60 Hz.

I_H *Holding Current:* Minimum value of forward current required to maintain the device in the *ON*-state. At any value of forward current below this value, the device will revert to the blocking or *OFF*-state. The holding current for a typical SCR has a negative temperature coefficient; i.e., as its junction temperature drops, its holding current requirements increase.

$I^2 t$ This defines the surge current capability for pulse widths less than one-half cycle (8.3 milliseconds or less). I is RMS amperes, t is pulse width in milliseconds.

4.1.3. Reverse Voltages

PRV   *Reverse Blocking Voltage:* Maximum repetitive peak reverse blocking voltage that may be applied to the anode-cathode of the device with the gate open.

4.1.4. Reverse Currents

I_R *Reverse Blocking Current:* DC leakage current through the device with reverse blocking voltage applied to the anode-cathode.

4.1.5. Gate Voltages

V_{GT} *Gate Trigger Voltage:* Maximum DC gate voltage required to trigger the device under the specified conditions.

V_{GNT} *Gate Non-Trigger Voltage:* Maximum DC gate voltage that can be applied to the device and it will still block rated forward blocking voltage.

4.1.6. Gate Currents

i_{GF} *Forward Gate Current:* Maximum peak gate current that can be applied to the device.

i_{GR} *Reverse Gate Current:* Maximum peak reverse gate current that can be applied to the device.

I_{GT} *Gate Trigger Current:* Maximum gate current required to trigger the device under the conditions specified.

I_{GNT} *Gate Non-Trigger Current:* Maximum DC gate current that can be applied to the device and it will still block rated forward voltage.

4.1.7. Power

P_{GM}	*Gate Power Dissipation:* Maximum allowable peak gate power dissipation.
$P_{G(AV)}$	*Gate Power Dissipation:* Maximum allowable average gate power dissipation.
P_{GR}	*Gate Power Dissipation:* Maximum allowable reverse peak gate power dissipation.
$P_{GR(AV)}$	*Gate Power Dissipation:* Maximum allowable reverse average gate power dissipation.

4.1.8. Switching

t_{off} *Turn-Off Time:* The Turn-off time is the sum of the device reverse recovery time plus the device gate recovery time at a specified rate of rise of reapplied forward voltage.

t_{on} *Turn-On Time:* The Turn-on time is the time interval between the time the gate current pulse reaches 50% of the final value and the time when the resulting forward current reaches 90% of its maximum value during switching from the *OFF*-state to the *ON*-state under specified conditions.

t_d *Delay Time:* The delay time is the time interval between the time the gate current pulse reaches 50% of the final value and the time when the resulting forward current reaches 10% of its maximum value during switching from the *OFF*-state to the *ON*-state under specified conditions.

t_r *Rise Time:* The rise time is the time between the time the forward current reaches 10% of its final value and the time the forward current reaches 90% of its final value during switching from the *OFF*-state to the *ON*-state under specified conditions.

t_f *Fall Time:* The fall time is the time interval between the time the current decreases from 90% to 10% values during switching from the *ON*-state to the *OFF*-state under specified conditions.

t_{gr} *Gate Recovery Time:* The gate recovery time is the minimum time interval between the reverse recovery point and the time the device voltage crosses zero at a specified rate of voltage rise.

t_{rr} *Reverse Recovery Time:* The reverse recovery time is the interval between the time the device current becomes negative and the time the reverse current falls to a specified value.

t_p *Pulse Width:* The pulse width is the time interval between the 50% magnitude points on the leading and trailing edge of the gate current pulse.

4.1.9. Thermal Impedances

θ *Effective Thermal Resistance:* The effective thermal resistance is the temperature rise per unit power dissipation of a designated junction above the temperature of a stated external reference point under conditions of thermal equilibrium.

$\theta_{(t)}$ *Transient Thermal Impedance:* The transient thermal impedance is the junction temperature change per unit change of junction power dissipation, measured at the end of a specified period of time during which time there was a step change in junction power dissipation with the device case or ambient temperature held constant.

θ_{JC} *Effective Thermal Resistance:* Junction to case.

θ_{CS} *Effective Thermal Resistance:* Case to sink.

θ_{JA} *Effective Thermal Resistance:* Junction to ambient.

$\theta_{JC(t)}$ *Transient Thermal Impedance:* Junction to case.

$\theta_{JA(t)}$ *Transient Thermal Impedance:* Junction to ambient.

4.1.10. Miscellaneous

R_{GK} *Gate-to-Cathode Resistance:* The two transistor analogy shows that a low external resistance between gate and cathode bypasses some current around the gate junction, thus requiring a higher anode current to initiate and maintain conduction. Low-current high-sensitivity SCR's are triggered by such a low current through the gate junction that a specified external gate-to-cathode resistance is required to prevent triggering by thermally generated leakage current.

4.2. APPENDIX B: SCR SELECTION

Selection of an SCR involves knowledge of such considerations as component limitation, voltage requirement, current at a conduction angle, cooling, firing circuits, protective networks, and atmospheric conditions.

The basic limitation of the SCR device is junction temperature. The junction formed by alloying a p-type material to an n-type material results in a eutectic compound which has a normal melting point in excess of 300°C. Exceeding this junction limitation of 300°C degrades and eventually destroys a device. Diffused or expitaxial-growth processes are thermally limited due to small imperfections in the junctions.

Since the SCR is required to block voltage in both the forward and reverse directions, a further limitation of 125°C is usually placed on the allowable junction temperature, based on the characteristic of semiconductor carrier mobility versus temperature. Thus, normal operating conditions allow for a maximum junction temperature of 125°C and transient current surges that heat the junction to approximately 300°C (typical of a motor starting application-high current, no voltage blocking capability).

Due to the variety of waveshapes and average currents available in SCR applications, manufacturers supply a family of curves which illustrate the relationship of the waveform and the average current to the maximum allowable stud temperature of the device, Fig. 13-3. For any fixed stud temperature, allowable average current decreases with conduction angle due to the RMS current limitations. The maximum power dissipation in watts can be determined from a family of curves, such as those shown in Fig. 13-2. These curves are used in conjunction with the heat-sink characteristics for the proper transfer of heat away from the stud of the device.

The system steady-state and transient voltages dictate minimum comparable V_{FB}, V_{RB} and transient ratings. Most manufacturers supply SCR devices rated in steps of 100 V through 1,600 V, V_{FB}. The transient peak voltage is given as some value in excess of the V_{FB} and V_{RB} ratings. Typical system designs use a 2.5:1 safety ratio of V_{FB} and V_{RB} to operating line voltage, unless a transient peak voltage is known. If a non-repetitive transient rating is known, the V_{FBT} and V_{RBT} value can be safely used. Both cases assume the use of additional transient protection such as a selenium surge protector and an RC network across the SCR.

The required heat sink is determined by the thermal conductivity of the stud to heat sink, and of the heat sink to ambient. The SCR manufacturer normally supplies the thermal conductivity from the device stud to the heat sink in the form of a typical thermal impedance value; the heatsink manufacturer, however, supplies the typical thermal impedance value of the heat sink to ambient. The sum of these thermal impedances multiplied by the power dissipation determined from Fig. 13-2, is the temperature differential from the stud to ambient. Subtracting this tempera- ture differential from the maximum stud temperature, as determined from Fig. 13-3, results in the maximum ambient temperature for reliable operation.

5. FOR YOUR REPORT

Prepare your report as usual and answer the questions which have been raised in Section 3.1 to 3.4. Write your answers in the same order as that of Sections 3.1 to 3.4.

6. MANUFACTURERS' SPECIFICATION DATA SHEETS

On the following pages (cf., Figs. 13 and 14), we present the specification data sheets of two brands of SCR's for the purpose of illustration and review of its basic properties [9] and [13].

SCR

2N5204-07

The 2N5204-07 series of silicon controlled rectifiers are reverse blocking triode thyristor semiconductor devices for use in medium power switching and phase control applications requiring blocking voltage up to 1200 volts, and average load current (single-phase, 180° conduction angle) up to 22 amperes.

General Electric's C137 SCR is recommended where a higher level of performance is required for a device of this size.

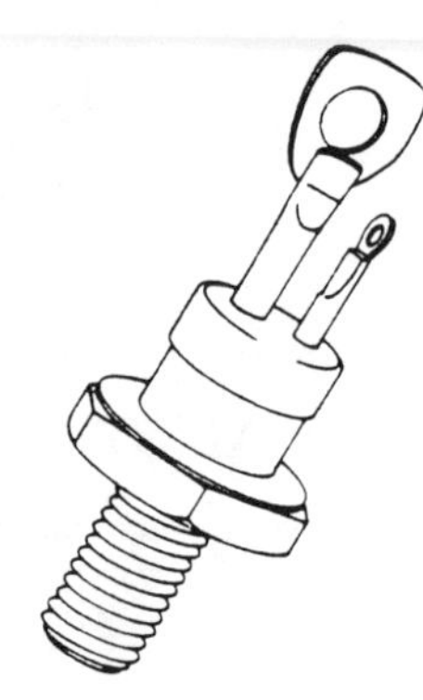

MAXIMUM ALLOWABLE RATINGS

Type	Repetitive Peak Off-State Voltage, V_{DRM} [1] [2] $T_C = -40°C$ to $+125°C$	Repetitive Peak Reverse Voltage V_{RRM} [1] [2] $T_C = -40°C$ to $+125°C$	Non-repetitive Peak Reverse Voltage V_{RSM} [1] [3] $T_C = -40°C$ to $+125°C$
2N5204	600 Volts†	600 Volts†	720 Volts†
2N5205	800 Volts†	800 Volts†	960 Volts†
2N5206	1000 Volts†	1000 Volts†	1200 Volts†
2N5207	1200 Volts†	1200 Volts†	1440 Volts†

(1) Values apply for gate terminal open-circuited. (Negative gate bias is permissible.)
(2) Maximum case-to-ambient thermal resistance for which maximum V_{DRM} and V_{RRM} ratings apply equals 5.0°C per watt for full sine wave or full-wave rectified sinusoidal volatge waveform. (3.0°C per watt is maximum case-to-ambient thermal resistance for pure dc voltage waveform.)
(3) Half sine wave voltage pulse, 10 millisecond maximum duration.
(4) di/dt rating is established in accordance with EIA Standard RS-397, Section 5.2.2.6. Off-state (blocking) voltage capability may be temporarily lost immediately after each current pulse for duration less than the period of the applied pulse repetition rate. The pulse repetition rate for this test is 400 Hz. The duration of the JEDEC di/dt test condition is 5.0 seconds (minimum).

RMS On-State Current, $I_{T(RMS)}$ . . . . .35 Amperes (all conduction angles)
Average On-State Current, $I_{T(AV)}$ . . . . .Depends on conduction angle (See Charts 3 and 5)
Critical Rate-of-Rise of On-State Current, di/dt: [4]
   Gate triggered operation . . . . (See Chart 6)
     Switching from 1200 volts . . . . 75 Amperes per microsecond†
               1000 volts . . . . 80 Amperes per microsecond†
               800 volts . . . . 90 Amperes per microsecond†
               600 volts . . . . .100 Amperes per microsecond†
   Breakover voltage triggered operation . . . . 10 Amperes per microsecond
Peak One Cycle Surge (non-rep) On-State Current, I_{TSM} . . . . .300 Amperes†
I^2t (for fusing), for time = 1.0 milliseconds (See Chart 9) . . . . .200 Ampere² seconds
            for time = 8.3 milliseconds (See Chart 9) . . . . .375 Ampere² seconds
Peak Gate Power Dissipation, P_{GM} . . . . .60 Watts for 500 microseconds†
Average Gate Power Dissipation, $P_{G(AV)}$ . . . . .10 Watts†
Peak Negative Gate Voltage, V_{GM} . . . . .5 Volts†
Storage Temperature, T_{STG} . . . .—40°C to +150°C†
Operating Temperature, T_J . . . .—40°C to +125°C†
Maximum Stud Torque . . . . .30 Lb-in (35 Kg-cm)

†Indicates data included on JEDEC Type Number Registration.

Fig. 13. SCR (Courtesy of General Electric Company [9]).

CHARACTERISTICS

Test	Symbol	Min.	Max.	Units	Test Conditions
Peak Off-State or Reverse Current [1] [2]	I_{DRM} or I_{RRM}			mA	$T_C = -40°$ to $+125°C$
2N5204		—	3.3†		$V_{DRM} = V_{RRM} =$ 600 Volts Peak
2N5205		—	2.5†		800
2N5206		—	2.0†		1000
2N5207		—	1.7†		1200
D.C. Gate Trigger Current	I_T	—	40	mAdc	$T_C = +25°C$, $V_D = 12$ Vdc, $R_L = 12$ ohms
			80†		$T_C = -40°C$, $V_D = 12$ Vdc, $R_L = 12$ ohms
D.C. Gate Trigger Voltage	V_{GT}	—	3.0	Vdc	$T_C = +25°C$, $V_D = 12$ Vdc, $R_L = 12$ ohms
		—	3.0†.		$T_C = -40°C$, $V_D = 12$ Vdc, $R_L = 12$ ohms
		0.25†	—		$T_C = +125°C$, Rated V_{DRM}, $R_L = 1000$ ohms
Peak On-State Voltage	V_{TM}	—	2.3†	Volts	$T_C = +25°C$, $I_{TM} = 70$ A peak, 1 msec wide pulse. Duty cycle $\leq 2\%$.
Holding Current	I_H			mAdc	Anode supply $= 24$ Vdc, Gate supply $= 10$ V, 20 ohms. Initial Forward Current Pulse $= 0.5$ A, 0.1 to 10.0 msec. wide.
		—	100		$T_C = +25°C$
		—	200†		$T_C = -40°C$
Critical Rate of Rise of Forward Blocking Voltage. (Higher values may cause device switching.)	dv/dt	100†	—	Volts/μsec	$T_C = +125°C$, Rated V_{DRM}, Gate open circuited.
Thermal Resistance	Θ_{J-C}	—	1.5†	°C/watt	Junction-to-case, dc

(1) Values apply for gate terminal open-circuited. (Negative gate bias is permissible.)
(2) Maximum case-to-ambient thermal resistance for which maximum V_{DRM} and V_{RRM} ratings apply equals 5.0°C per watt for full sine wave or full-wave rectified sinusoidal voltage waveform. (3.0°C per watt is maximum case-to-ambient thermal resistance for pure dc voltage waveform.)
†Indicates data included on JEDEC Type Number Registration.

OUTLINE DRAWING

(COMPLIES WITH JEDEC TO-48)

NOTES:
1. Complete threads to extend within 2½ threads of seating plane. Diameter of unthreaded portion. 249" (6.32MM) Maximum, .220" (5.59MM) Minimum.
2. Angular orientation of these terminals is undefined.
3. ¼-28 UNF-2A. Maximum pitch diameter of plated threads shall be basic pitch diameter .2268" (5.76MM), minimum pitch diameter .2225" (5.66MM), reference: screw thread standards for Federal Service 1957, Handbook H28, 1957, P1.
4. A chamfer (or undercut) on one or both ends of hexagonal portions is optional.
5. Case is anode connection.
6. Large terminal is cathode connection.
7. Small terminal is gate connection.
8. Insulating kit available upon request.
A. ¼-28 steel nut, Ni. plated, .178 min. thk.
B. Ext. tooth lockwasher, steel, Ni. plated, .023 min. thk.

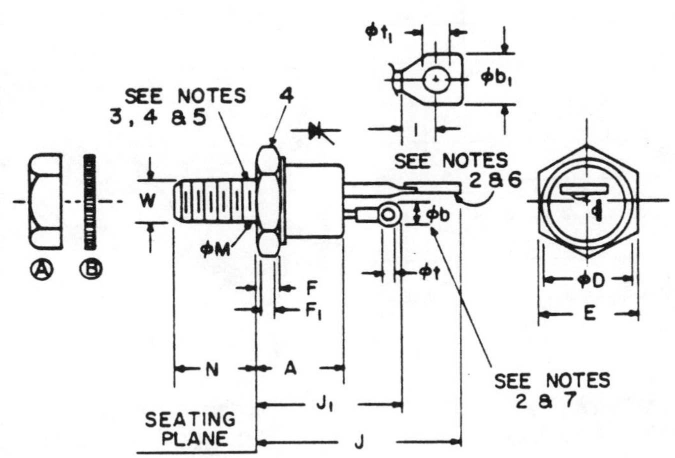

Fig. 13. Continued.

SYMBOL	INCHES		MILLIMETERS		NOTES
	MIN.	MAX.	MIN.	MAX.	
A	.330	.505	8.38	12.83	
ϕb	.115	.140	2.92	3.56	2
ϕb_1	.210	.300	5.33	7.62	2
ϕD		.544		13.82	
E	.544	.562	13.82	14.27	
F	.113	.200	2.87	5.08	4
F_1	.060		1.52		
J		1.193		30.30	
J_1		.875		22.23	
I	.120		3.05		
ϕM					1
N	.422	.453	10.72	11.51	
ϕt	.060	.075	1.52	1.91	
ϕt_1	.125	.165	3.18	4.19	
W					3

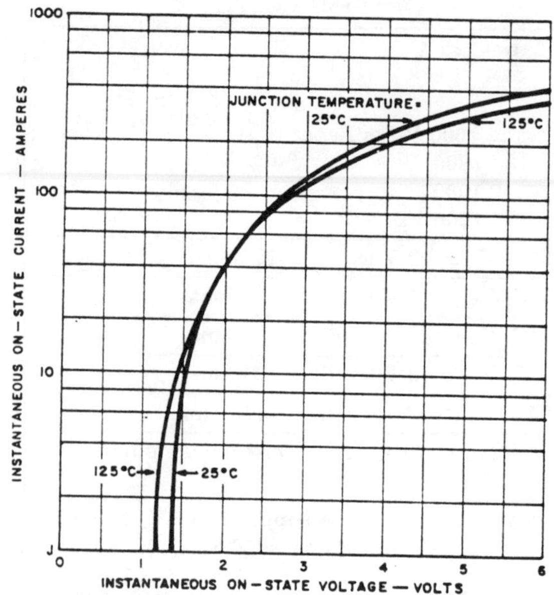

1. MAXIMUM ON-STATE CHARACTERISTICS

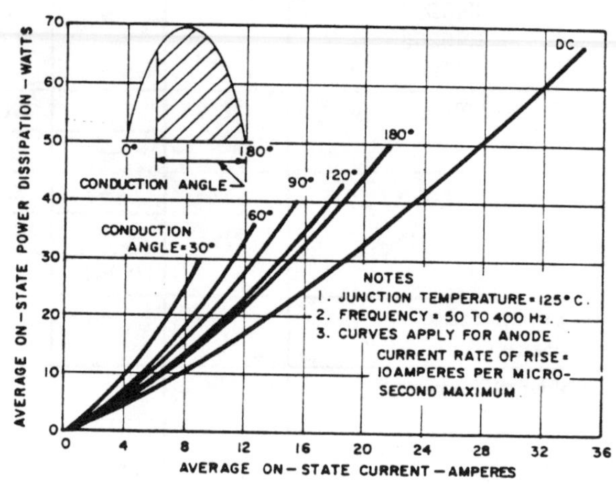

2. MAXIMUM ON-STATE POWER DISSIPATION FOR HALF-WAVE RECTIFIED SINE WAVE OF CURRENT

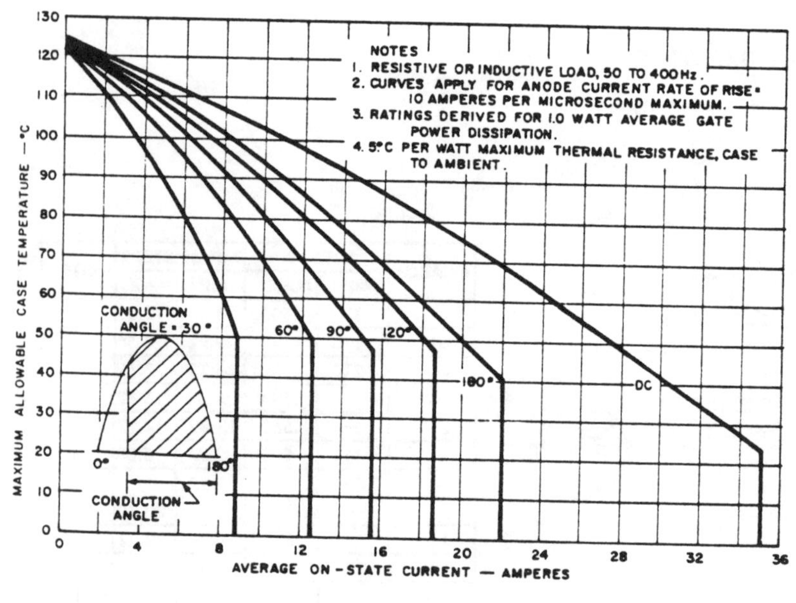

3. MAXIMUM ALLOWABLE CASE TEMPERATURE FOR HALF-WAVE RECTIFIED SINE WAVE OF CURRENT

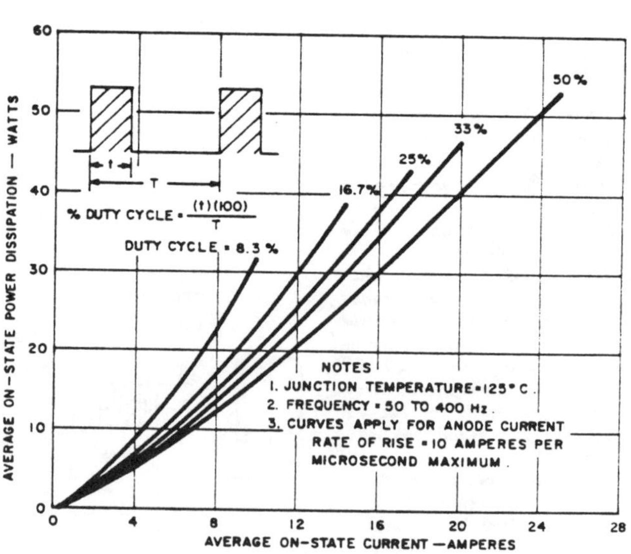

4. MAXIMUM ON-STATE POWER DISSIPATION FOR RECTANGULAR CURRENT WAVEFORM

Fig. 13. Continued.

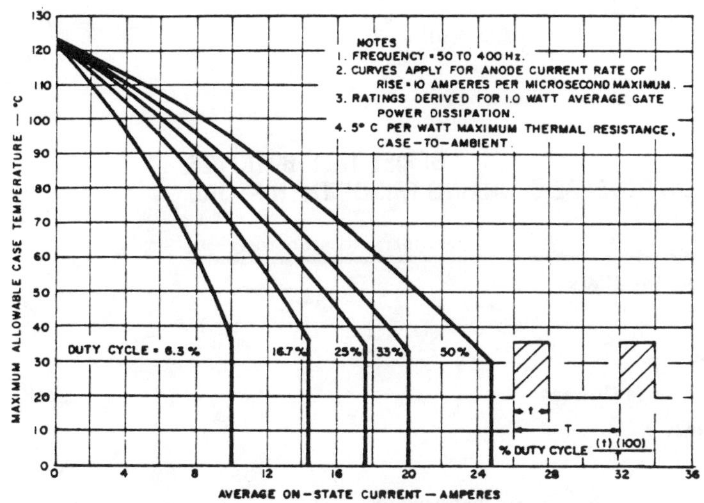

5. MAXIMUM ALLOWABLE CASE TEMPERATURE FOR RECTANGULAR CURRENT WAVEFORM

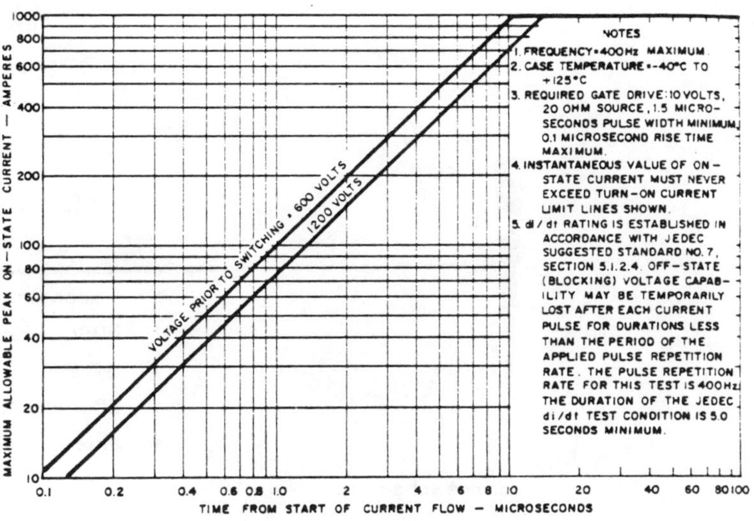

6. TURN-ON CURRENT LIMIT

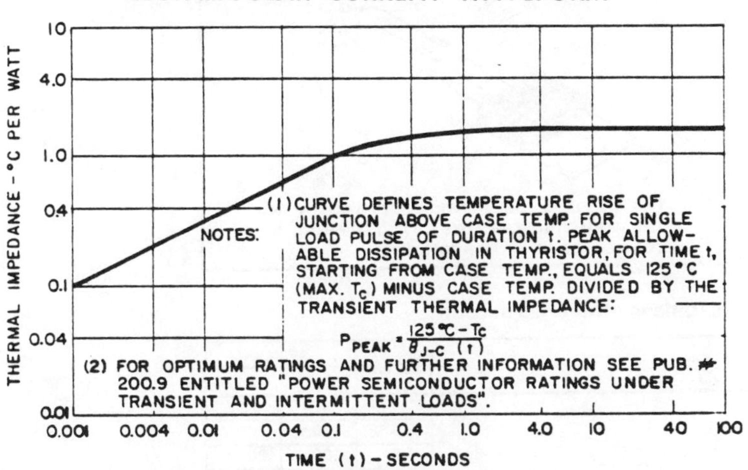

7. MAXIMUM TRANSIENT THERMAL IMPEDANCE, JUNCTION-TO-CASE

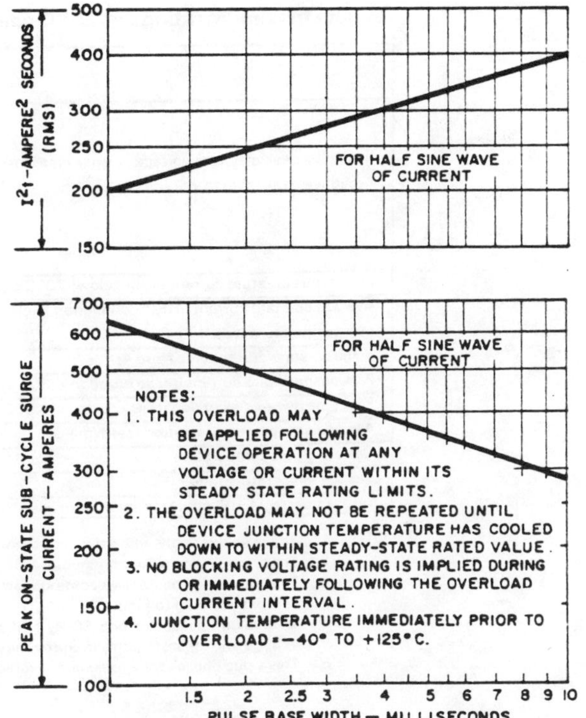

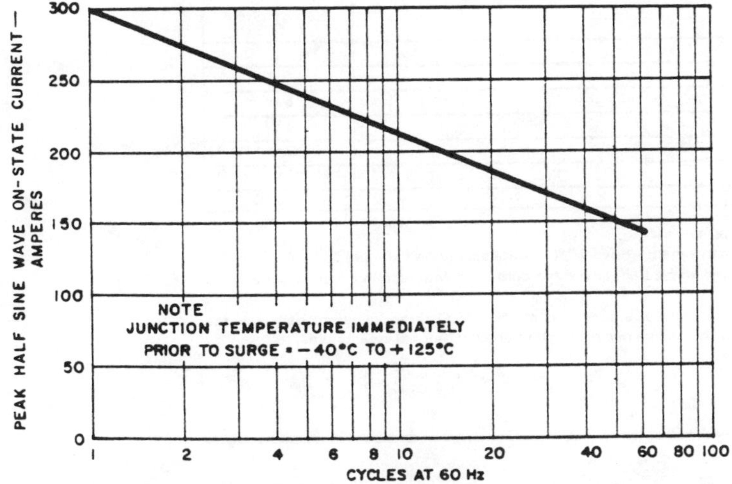

8. MAXIMUM ALLOWABLE SURGE (NON-REPETITIVE) ON-STATE CURRENT

9. MAXIMUM ALLOWABLE SUB-CYCLE SURGE (NON-REPETITIVE) ON-STATE CURRENT AND I²T RATING

Fig. 13. Continued.

SERIES TIC116, TIC126
P-N-P-N SILICON REVERSE-BLOCKING TRIODE THYRISTORS

APRIL 1971 – REVISED OCTOBER 1984

- **Silicon Controlled Rectifiers**
- **50 V to 600 V**
- **8 A and 12 A DC**
- **80 A and 100 A Surge Current**
- **Max I$_{GT}$ of 20 mA**

device schematic

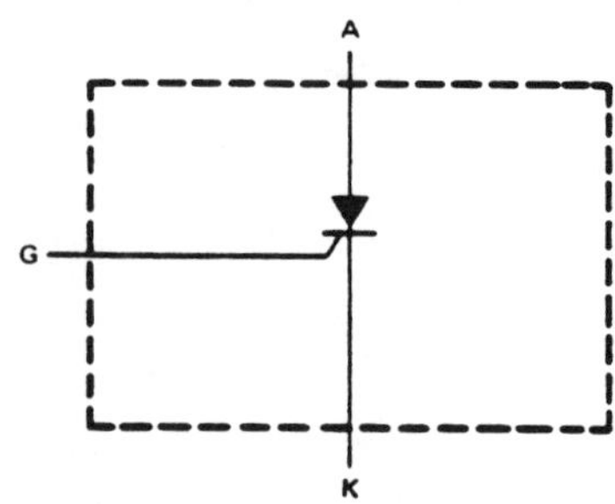

TO-220AB PACKAGE

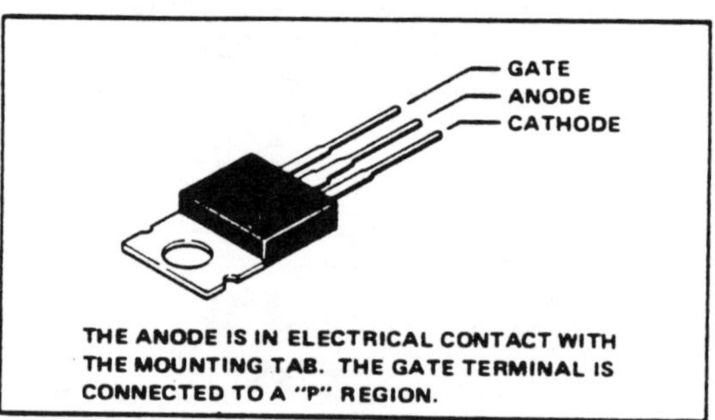

absolute maximum ratings at 25°C case temperature (unless otherwise noted)

		SUFFIX	SERIES	
			TIC116	TIC126
Repetitive peak off-state voltage, V$_{DRM}$ (see Note 1) Repetitive peak reverse voltage, V$_{RRM}$		F	50 V	50 V
		A	100 V	100 V
		B	200 V	200 V
		C	300 V	300 V
		D	400 V	400 V
		E	500 V	500 V
		M	600 V	600 V
Continuous on-state current at (or below) 70°C case temperature (see Note 2)			8 A	12 A
Average on-state current (180° conduction angle) at (or below) 70°C case temperature (see Note 3)			5 A	7.5 A
Surge on-state current (see Note 4)			80 A	100 A
Peak positive gate current (pulse duration ≤ 300 µs)			3 A	
Peak gate power dissipation (pulse duration ≤ 300 µs)			5 W	
Average gate power dissipation (see Note 5)			1 W	
Operating case temperature range			– 40 °C to 110 °C	
Storage temperature range			– 40 °C to 125 °C	
Lead temperature 1.6 mm (1/16 inch) from case for 10 seconds			230 °C	

NOTES:
1. These values apply when the gate-cathode resistance R$_{GK}$ = 1 kΩ.
2. These values apply for continuous d-c operation with resistive load. Above 70°C derate according to Figure 3.
3. This value may be applied continuously under single-phase 50-Hz half-sine-wave operation with resistive load. Above 70°C derate according to Figure 9.
4. This value applies for one 50-Hz half-sine-wave when the device is operating at (or below) rated values of peak reverse voltage and on-state current. Surge may be repeated after the device has returned to original thermal equilibrium.
5. This value applies for a maximum averaging time of 20 ms.

TEXAS INSTRUMENTS
POST OFFICE BOX 225012 ● DALLAS, TEXAS 75265

Fig. 14. SCR (Courtesy of Texas Instruments [13]).

SERIES TIC116, TIC126
P-N-P-N SILICON REVERSE-BLOCKING TRIODE THYRISTORS

electrical characteristics at 25°C case temperature (unless otherwise noted)

	PARAMETER	TEST CONDITIONS			MIN	TYP	MAX	UNIT
I_{DRM}	Repetitive Peak Off-State Current	V_D = Rated V_{DRM},	R_{GK} = 1 kΩ,	T_C = 100°C			2	mA
I_{RRM}	Repetitive Peak Reverse Current	V_R = Rated V_{RRM},	I_G = 0,	T_C = 110°C			2	
I_{GT}	Gate Trigger Current	V_{AA} = 6 V,	R_L = 100 Ω,	$t_{w(g)} \geqslant 20\,\mu s$		5	20	mA
V_{GT}	Gate Trigger Voltage	V_{AA} = 6 V, R_{GK} = 1 kΩ,	R_L = 100 Ω, T_C = 40°C	$t_{w(g)} \geqslant 20\,\mu s$,		.	2.5	V
		V_{AA} = 6 V, R_{GK} = 1 kΩ	R_L = 100 Ω,	$t_{w(g)} \geqslant 20\,\mu s$,		0.8	1.5	
		V_{AA} = 6 V, R_{GK} = 1 kΩ,	R_L = 100 Ω, T_C = 110°C	$t_{w(g)} \geqslant 20\,\mu s$,	0.2			
I_H	Holding Current	V_{AA} = 6 V, R_{GK} = 1 kΩ,	Initiating I_T = 100 mA, T_C = −40°C				70	mA
		V_{AA} = 6 V, R_{GK} = 1 kΩ	Initiating I_T = 100 mA,				40	
V_{TM}	Peak On-State Voltage	I_{TM} = 8 A,	See Note 6	SERIES TIC116			1.7	
		I_{TM} = 12 A,	See Note 6	SERIES TIC126			1.4	
dv/dt	Critical Rate of Rise of Off-State Voltage	V_D = Rated V_D,	I_G = 0,	T_C = 110°C		100		V/μs

NOTE 6: These parameters must be measured using pulse techniques, t_w = 300 μs, duty cycle ≤ 2 %. Voltage-sensing contacts, separate from the current-carrying contacts, are located within 3,2 mm (1/8 inch) from the device body.

thermal characteristics

PARAMETER	SERIES TIC116			SERIES TIC126			UNIT
	MIN	TYP	MAX	MIN	TYP	MAX	
$R_{\theta JC}$			3			2.4	°C/W
$R_{\theta JA}$			62.5			62.5	

resistive-load switching characteristics at 25°C case temperature

	PARAMETER	TEST CONDITIONS			MIN	TYP	MAX	UNIT
t_{gt}	Gate-Controlled Turn-On Time	V_{AA} = 30 V, V_{in} = 20 V,	R_L = 6 Ω, See Figure 1	$R_{GK(off)}$ = 100 Ω,		0.8		μs
t_q	Circuit-Commutated Turn-Off Time	V_{AA} = 30 V, See Figure 2	R_L = 6 Ω,	I_{RM} = 10 A,		11		

Fig. 14. Continued.

PARAMETER MEASUREMENT INFORMATION

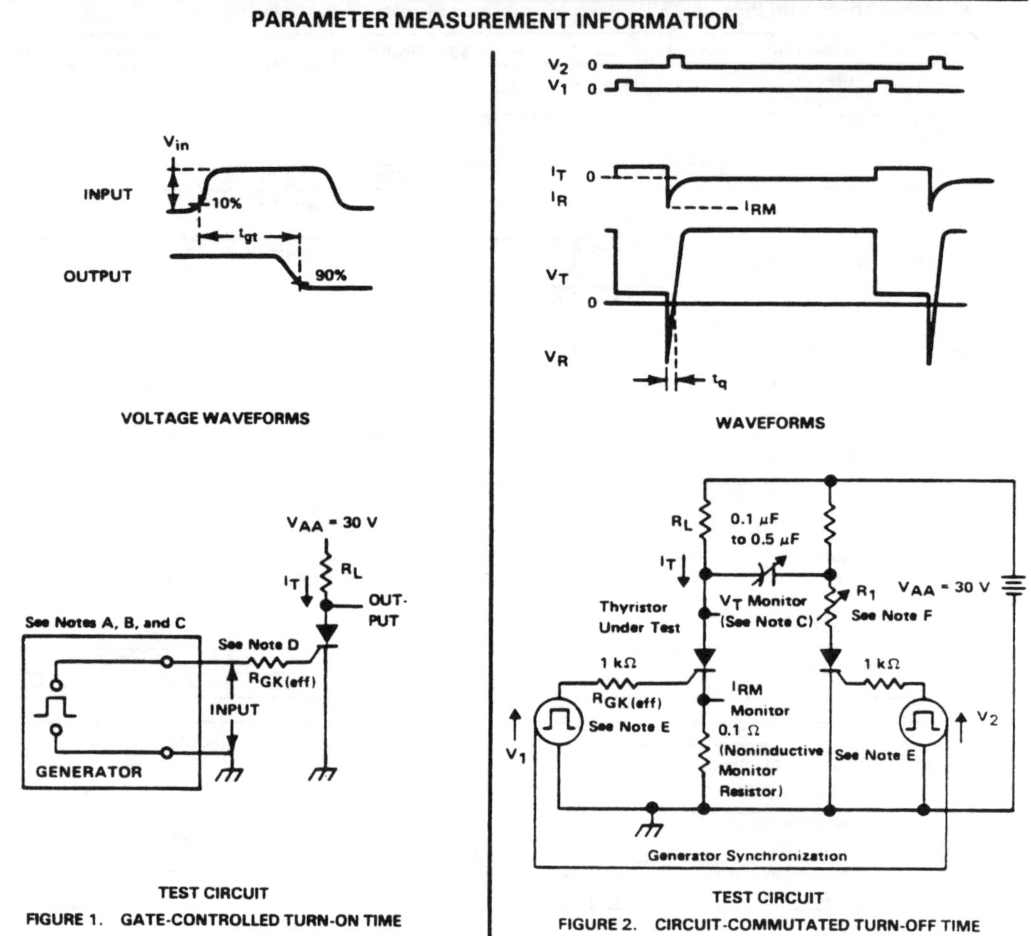

FIGURE 1.   GATE-CONTROLLED TURN-ON TIME

FIGURE 2.   CIRCUIT-COMMUTATED TURN-OFF TIME

NOTES:
A. V_{in} is measured with gate and cathode terminals open.
B. The input waveform of Figure 1 has the following characteristics: $t_r \leqslant 40$ ns, $t_w \geqslant 20$ μs.
C. Waveforms are monitored on an oscilloscope with the following characteristics: $t_r \leqslant 14$ ns, $R_{in} \geqslant 10$ MΩ, $C_{in} \leqslant 12$ pF.
D. $R_{GK(eff)}$ includes the total resistance of the generator and the external resistor.
E. Pulse generators for V_1 and V_2 are synchronized to provide an anode current waveform with the following characteristics: $t_w = 50$ to 300 μs, duty cycle $= 1$ %. The pulse duration of V_1 and V_2 are $\geqslant 10$ μs.
F. Resistor R_1 is adjusted for $I_{RM} \approx 10$ A.

Fig. 14. Continued.

SERIES TIC116, TIC126
P-N-P-N SILICON REVERSE-BLOCKING TRIODE THYRISTORS

THERMAL INFORMATION

AVERAGE ON-STATE CURRENT DERATING CURVE

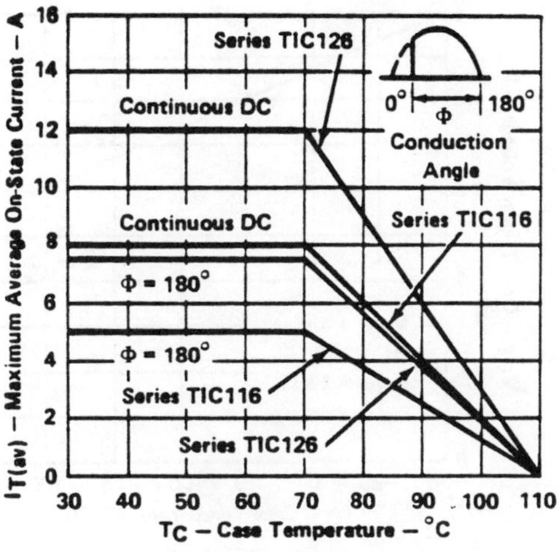

FIGURE 3

MAXIMUM CONTINUOUS ANODE POWER DISSIPATED vs CONTINUOUS ON-STATE CURRENT

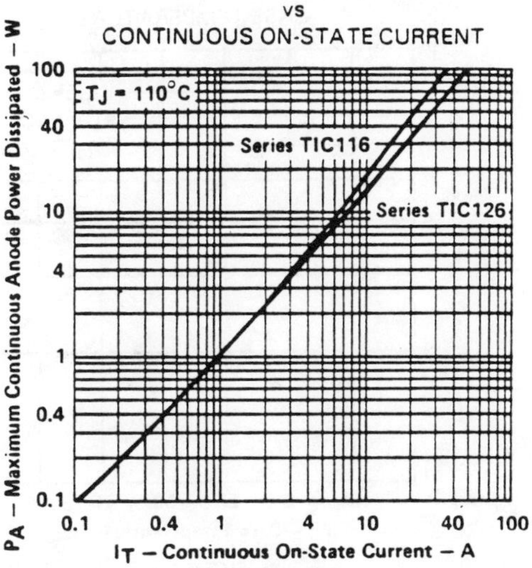

FIGURE 4

SURGE ON-STATE CURRENT vs CYCLES OF CURRENT DURATION

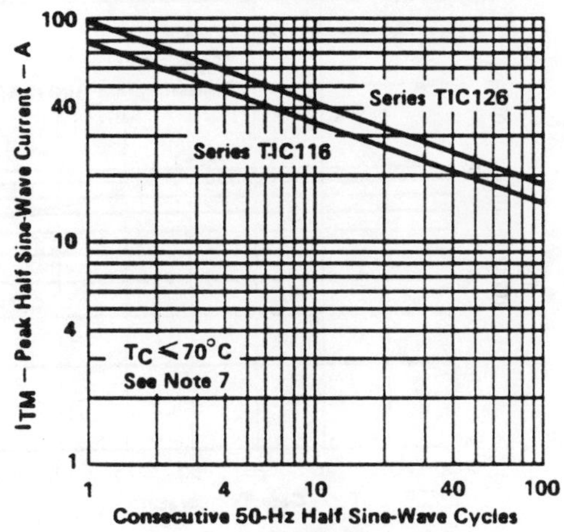

FIGURE 5

TRANSIENT THERMAL RESISTANCE vs CYCLES OF CURRENT DURATION

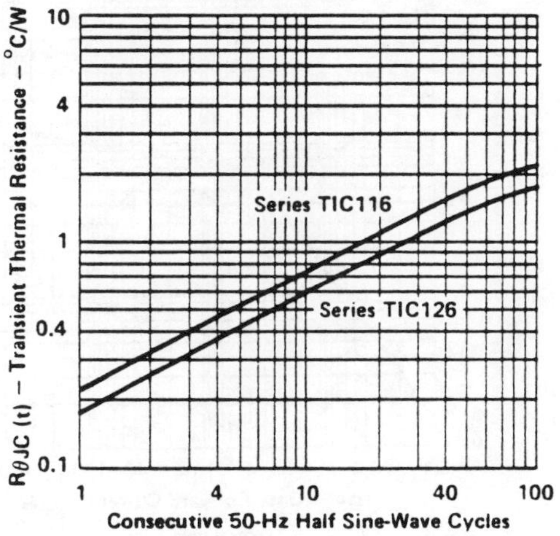

FIGURE 6

NOTE 7: This curve shows the maximum number of cycles of surge current for which gate control is guaranteed provided the device is initially at nonoperating thermal equilibrium.

TEXAS INSTRUMENTS
POST OFFICE BOX 225012 ● DALLAS, TEXAS 75265

Fig. 14. Continued.

TYPICAL CHARACTERISTICS

GATE TRIGGER CURRENT
vs
CASE TEMPERATURE

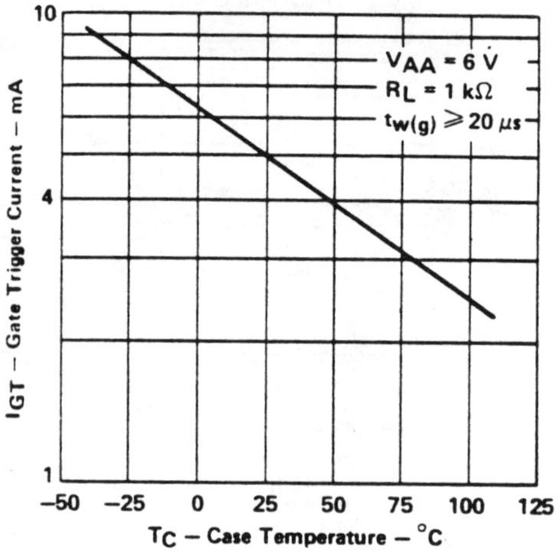

FIGURE 7

GATE TRIGGER VOLTAGE
vs
CASE TEMPERATURE

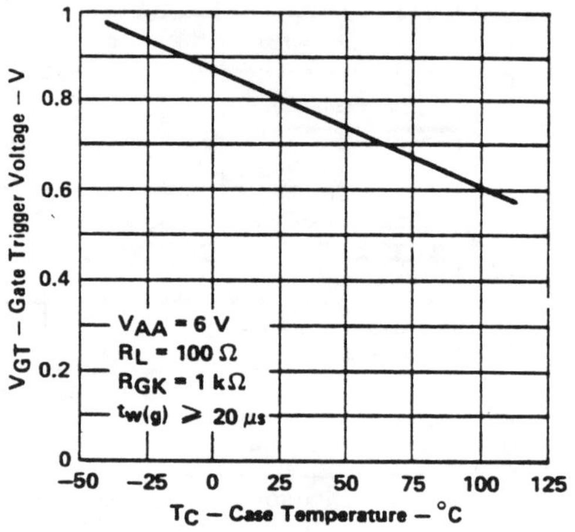

FIGURE 8

GATE FORWARD VOLTAGE
vs
GATE FORWARD CURRENT

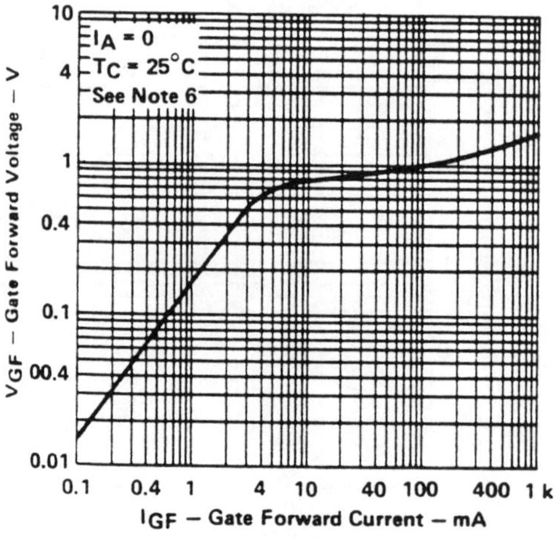

FIGURE 9

HOLDING CURRENT
vs
CASE TEMPERATURE

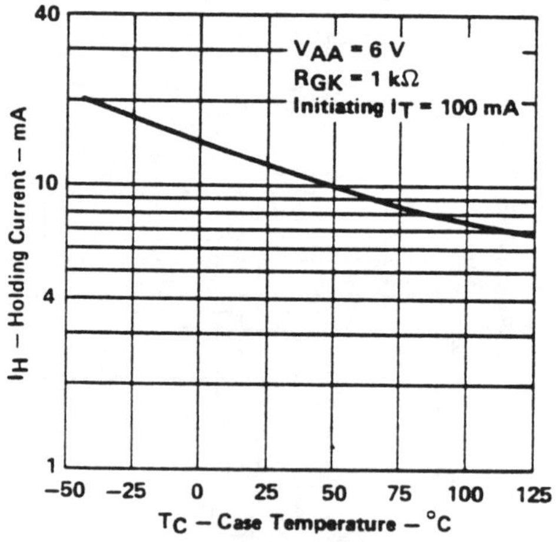

FIGURE 10

NOTE 6: These parameters must be measured using pulse techniques, t_w = 300 μs, duty cycle $\leqslant$ 2 %. Voltage-sensing contacts, separate from the current-carrying contacts, are located within 3,2 mm (1/8 inch) from the device body.

Fig. 14. Continued.

SERIES TIC116, TIC126
P-N-P-N SILICON REVERSE-BLOCKING TRIODE THYRISTORS

TYPICAL CHARACTERISTICS

PEAK ON-STATE VOLTAGE
vs
PEAK ON-STATE CURRENT

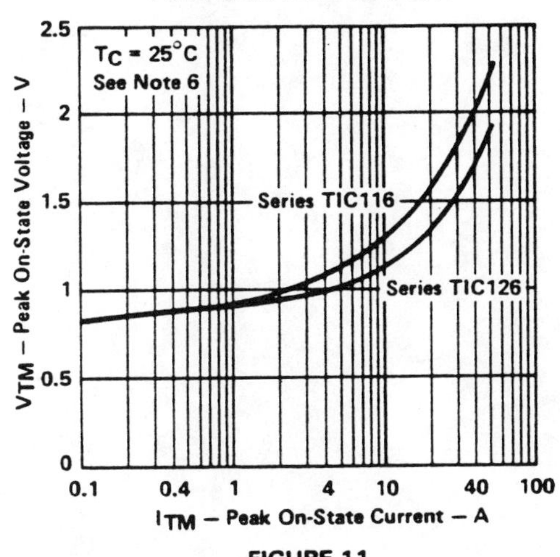

FIGURE 11

GATE-CONTROLLED TURN-ON TIME
vs
GATE CURRENT

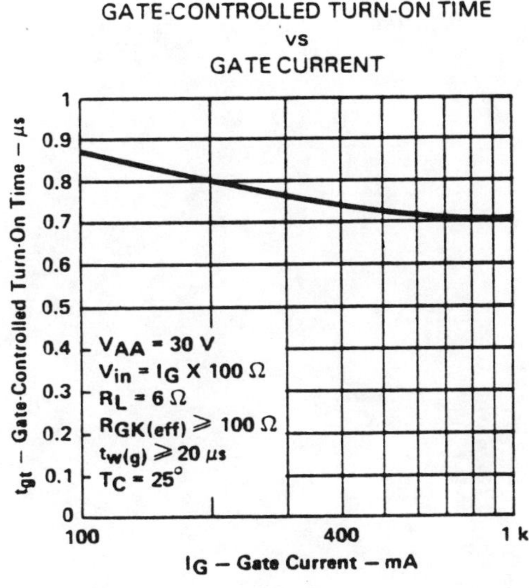

FIGURE 12

CIRCUIT-COMMUTATED TURN-OFF TIME
vs
CASE TEMPERATURE

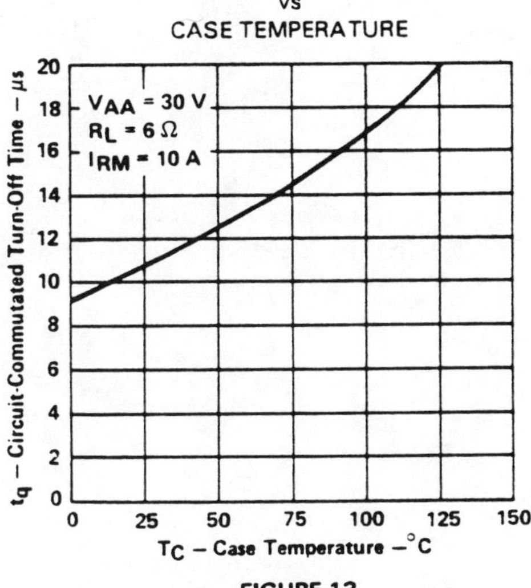

FIGURE 13

NOTE 6: These parameters must be measured using pulse techniques, t_w = 300 μs, duty cycle ≤ 2 %. Voltage-sensing contacts, separate from the current-carrying contacts, are located within 3,2 mm (1/8 inch) from the device body.

Fig. 14. Continued.

WORKSHEET

WORKSHEET

WORKSHEET

PART
FOUR

OPERATIONAL AMPLIFIERS

PART FOUR

OPERATIONAL AMPLIFIERS

IV-0. INTRODUCTION

In this part the basic properties of an operational amplifier are reviewed. This family of two-port active devices has an extremely important role in overall basic electronic circuits design. As an active two-port network it has all the relevant properties of such networks. In particular, this device presents the best example of applications of the concept of feedback amplifier, as these have been described in Section III-2. In this part we will study the fundamental properties of this family of devices from a circuit theory point of view. We present our case by first introducing a purely mathematical model (the ideal case) for this device. Then we present a device that is different from this mathematical (or ideal) model by demonstrating its behavior and/or its deviations from the perfect model.

The table of contents of the remaining part of this review is as follows.

The list of the experiments in this part is as follows.

EXPERIMENT IV-5: ACTIVE FIRST- AND SECOND-ORDER NETWORKS
EXPERIMENT IV-6: NEGATIVE RESISTANCE CIRCUITS

In future experiments we will be studying additional applications of this type of integrated device in conjunction with more advanced circuit design problems.

The material in this part are well known. Similar information on these matters can be found in the sources cited in Section IV-1.

Please review page 41 of this textbook for basic laboratory procedures.

IV-1.   REFERENCES

[A] Textbooks:

[0]   W.G. Jung, IC Op-Amp Cookbook. Indianapolis, IN: Howard W. Sams & Co., Inc., 1974.

[1]   K. Ogata, Modern Control Engineering. Englewood Cliffs, NJ: Prentice-Hall, 1970.

[2]   J.K. Roberge, Operational Amplifiers: Theory and Practice. New York: Wiley, 1975.

[3]   A.K. Scidmore, ECE 341 Lecture Notes. Madison, WI: The University of Wisconsin-Madison, 1977.

[4]   G.E. Tobey, J.G. Graeme and L.P. Huelsman, Eds., Operational Amplifiers Design and Applications. New York: McGraw-Hill, 1971.

[5]   J.V. Wait, L.P. Huelsman and G.A. Korn, Introduction to Operational Amplifier Theory and Applications. New York: McGraw-Hill, 1975.

[B] Manufacturers' Manuals

[6]   National Semiconductor, Linear Data Book. Santa Clara, CA: National Semi. Corp., 1980.

[7]   Signetics, Analog Data Manual. Sunnyvale, CA: Signetics Corp., 1983.

[8]   Texas Instruments, Linear Circuits Data Book. Dallas: Texas Instruments, 1984.

IV-2.   GENERAL REMARKS

In the earlier experiments of Part Three we studied some basic properties of transistors in conjunction with two-port active networks. These circuits with discrete elements have fewer applications today than before. The advent of integrated circuit design has enabled us to construct in one piece smaller sized circuits that are many times more complicated and much more reliable than those circuits built with discrete elements. The technology that makes such *miniaturization* possible is called integrated circuit (microelectronic) technology. These integrated circuits (or chips) are abbreviated as IC's. The first such IC to be studied in this textbook is called an operational amplifier (or Op-Amp for short). Even though a typical Op-Amp is an active two-port device in which its internal circuit consists of various transistors, diodes, resistors and capacitors, we will consider this unit as a new network element. Thus, in this part, we explore some of its general characteristics as has been the case for transistors and/or other network components in the earlier experiments of this textbook. We bear in mind that each manufacturer or vendor of IC's has specific instructions about the general properties of its products. Therefore it is not essential for us to be too concerned about every little detail of these devices. Since these details may differ from one brand to another, we must be only concerned with the one we are going to use in our application. We will, however, give two typical manufacturers' specification data sheets for general purpose Op-Amps in Figs. 22 and 23, for the sake of some basic information on this device. The following comments, however, seem to be common among all types of Op-Amps.

Each Op-Amp has at least eight external pins that are numbered and, depending on the vendor, these pins may have different meanings. Each of these chips, however, has five important pins as shown in Fig. 1. The supply voltages needed for Op-Amps are usually in the range of ± 20 volts (or less). For input voltages that result in an output voltage in this range, the Op-Amps act as linear devices. If the input voltages yield an output voltage that theoretically increases beyond the $|V_{supply}|$, the Op-Amps will *saturate* and thus non-linearity will occur. Usually these supply voltage sources are understood and therefore are not shown in the figures. We should also mention that there are chips containing several of these Op-Amps in one package. (cf., Fig. VI-5-16).

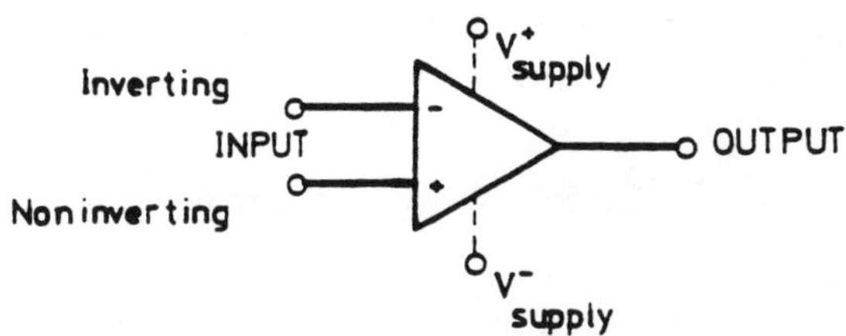

Fig. 1. A Schematic Symbol of An Op-Amp.

In the following we first define a mathematical (or an ideal) model for this device. Then we will study the actual or physical Op-Amp with its limitations and/or non-idealness problems. It is also to be noted that due to an extremely high gain factor, these devices are not generally used in an open-loop configuration. We will thus study the feedback configurations of this device with the associated Op-Amp models.

IV-3. MATHEMATICAL MODEL OF AN IDEAL OPEN-LOOP OP-AMP

Let us define a mathematical model for an Op-Amp (Fig. 2) that has the following ideal properties. This mathematical model should be considered as an element that does not physically exist, but it serves as a frame of reference to compare with the properties of a practical Op-Amp.

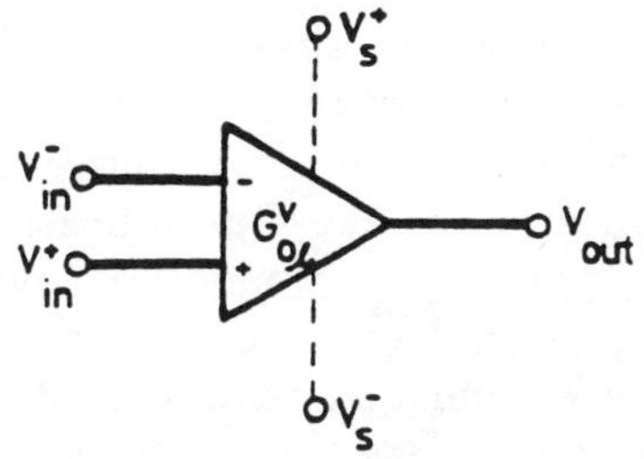

Fig. 2. An Ideal and Open-Loop Op-Amp.

IV-3A.   Infinite Open-Loop DC Gain $(G_{ol}^v = \infty)$

This gain is the ratio of the maximum output voltage swing with load to the change in the input voltage required to derive the output from zero to this value. For practical Op-Amps this gain is a finite, though large, number. A typical value for this gain is 200,000. This finite and in most cases frequency dependent open-loop gain is the source of many errors (i.e., the deviations from the idealness) in the Op-Amp. Also this gain is responsible for the range of the maximum output voltage that this device produces. For linear operation of the Op-Amp the input should not exceed $|V_{supply}|/G_{ol}^v$. In most applications we consider an Op-Amp a unit for which a small differential input of $(V_{in}^- - V_{in}^+)$ yields an output voltage which is

$$V_{out} = -G_{ol}^v(V_{in}^- - V_{in}^+).$$

This relationship is the fundamental definition of the operation of an Op-Amp. In the ideal case $G_{ol}^v \to \infty$ and $(V_{in}^- - V_{in}^+) \to 0$.

IV-3B.   Infinite Input Resistance ($R_{in}=\infty$)

The definition of an input resistance is clear (i.e., the ratio of input voltage over input current). In general this quantity is finite but very large - say on the order of megaohms. Equivalently, this is the same as saying no current can enter an ideal Op-Amp.

IV-3C.   Zero Output Resistance ($R_{out}=0$)

This definition is also clear. In most practical situations this quantity is on the order of only a few tens of ohms.

IV-3D.   Zero Input Offset Voltage ($V_{1o}=0$)

This means there is no offset voltage or simply that as $(V_{in}^- - V_{in}^+)\to 0$, then $V_{out}\to 0$.

IV-3E.   Infinite Bandwidth (BW$=\infty$)

Infinite bandwidth implies that all signals will be processed equally well, without regard to their frequencies. Real Op-Amps, on the other hand, are essentially DC devices, operating best at $f_{in}=0$.

Putting all the above definitions together, we will have the equivalent circuit of Fig. 3 for an ideal model of an Op-Amp. Although the assumptions that we have made will simplify our subsequent analysis drastically, in practice there is no such device as an ideal Op-Amp. The main source of discrepancy in the application of the Op-Amp as an ideal device is the fact that G_{ol}^v is large but not infinite. Also due to this large open-loop gain it is desirable to use this device in conjunction with a feedback configuration. Thus, we will briefly review the concept of a feedback amplifier. In closing this section we again remind our readers that the subsequent analysis is based on a comparison between the practical Op-Amp and the mathematical (or ideal) Op-Amp.

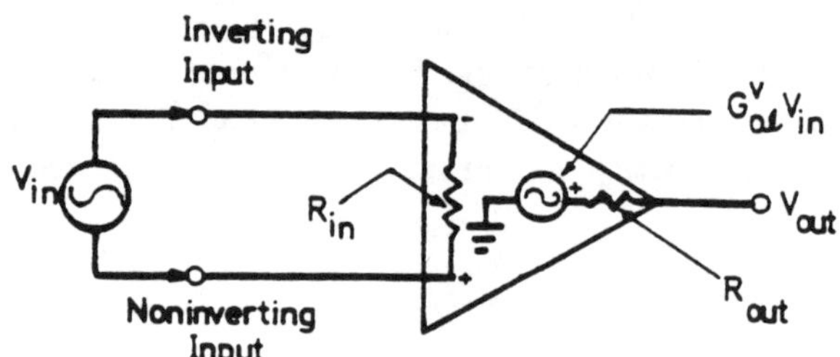

Fig. 3. Equivalent Circuit for an Ideal Op-Amp.

IV-3.1. MATHEMATICAL MODELS OF FEEDBACK CONFIGURATIONS WITH AN IDEAL OP-AMP

In the following we will study two commonly used feedback configurations with their mathematical models. In each of these two cases we choose the associated Op-Amp to be an ideal device in the sense that $(V_{in}^- - V_{in}^+)\to 0$. It is also assumed that no current enters this device.

Consider the *inverting* feedback amplifier circuit of Fig. 4. Here we have $V_{in}^+=0$. From the idealness of the Op-Amp we conclude that $V_{in}^-=0$. This defines a *virtual* ground. At node SP (summing point) we have $I_{in}=I_f$, since no current enters the Op-Amp.

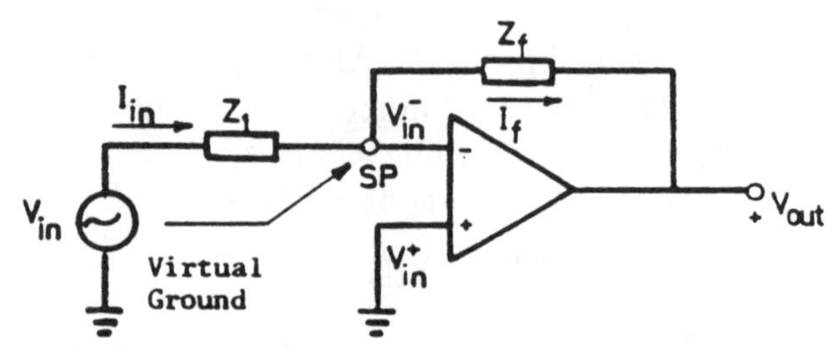

Fig. 4. An Inverting Feedback Amplifier.

A simple node analysis based on these assumptions yield the following fundamental relationship of an ideal Op-Amp in an inverting feedback configuration:

$$\frac{V_{out}(s)}{V_{in}(s)} = -\frac{Z_f(s)}{Z_1(s)} \; .$$

The ratio of $-Z_{(f)}/Z_1(s)$ is called the closed-loop gain. The term inverting refers to the minus sign in this equation.

Now, consider the *noninverting* circuit of Fig. 5. From the ideal properties of the Op-Amp we conclude that $V_{in} = V_{in}^- = V_{in}^+$, and $I_{in} = I_f$. A simple node analysis for this circuit yields

$$\frac{V_{out}(s)}{V_{in}(s)} = \frac{Z_1(s) + Z_f(s)}{Z_1(s)} \; .$$

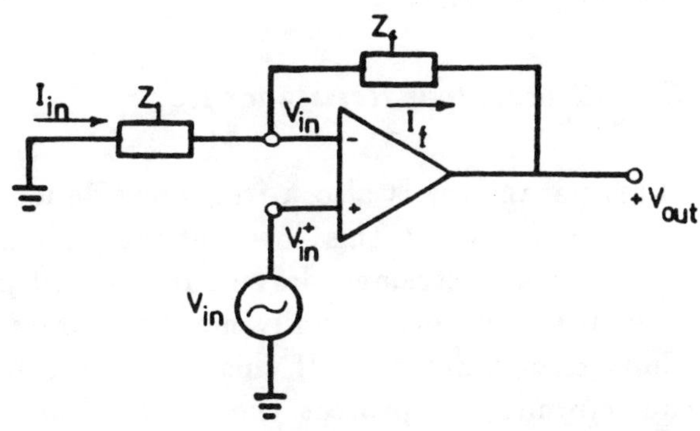

Fig. 5.   A Noninverting (Positive) Feedback Amplifier.

In this equation if we let $Z_f(s) \rightarrow 0$ (or $Z_1(s) \rightarrow \infty$ although not recommended), then we will have $V_{out} \equiv V_{in}$. This is called a unity gain feedback amplifier or a *voltage follower* circuit which has many applications such as buffering in a circuit design.

IV-4.   PARAMETERS OF A PRACTICAL OPEN-LOOP OP-AMP

The source of all discrepancies in the operation of an Op-Amp from being an ideal device is the fact that G_{ol}^v is not infinite. This gain or transfer function is indeed a frequency dependent quantity. For example, we can express a low-pass filter model for this device as follows:

$$G_{ol}^v(s) = \frac{G_0}{1 + \tau_0 s} = \frac{G_0}{1 + j\dfrac{\omega}{\omega_0}} \; .$$

Here τ_0 (the device time constant) is, of course, very small, but not zero. If τ_0 were zero we would have an ideal device. Correspondingly, to develop a relationship between the properties of an ideal amplifier and those of a practical amplifier, we must use the above expression for the gain of the Op-Amp, instead of the constant G_{ol}^v we used in Section IV-3. Based on this new expression we now reevaluate the Op-Amp parameters that we studied in Section IV-3.

IV-4A.   The Open-Loop Gain G_{ol}^v

As mentioned earlier, we now have a first-order low-pass transfer function to describe the open-loop gain of the Op-Amp. The *true* open-loop gain or transfer function of an Op-Amp, however, is more complicated than this first-order low-pass transfer function. Further elaboration on this gain is beyond the scope of this textbook. In any event, our main objective is to use the fundamental definition of the operation of an Op-Amp in a given circuit, in order to complete the analysis and/or design of this circuit. The only change that occurs in this fundamental relation is in the open-loop gain term that must now be replaced by that of the practical Op-Amp. This substitution is the only change we must make whenever we are using a practical Op-Amp. One may also interpret this situation by considering an ideal Op-Amp, inside of which we have a feedback amplifier consisting of an ideal Op-Amp; the overall feedback gain is now $G_{o\ell}^v(s)$.

IV-4B.   The Input Resistance R_{in}

Here, this parameter is generally a frequency dependent quantity whose amplitude is on the order of megaohms.

IV-4C.   The Output Resistance R_{out}

This parameter is also a frequency dependent quantity whose magnitude is on the order of tens of ohms. Because of this low output impedance, if by some mistake the output be shorted to ground, then an extremely large current will pass through this impedance that most likely will destroy the entire circuit. Usually manufacturers provide additional circuitry inside the IC chip to limit this short circuit current. If this protection has not been made for us, then we must use additional external circuitry to protect the device, but generally this addition will degrade the overall performance of the amplifier. This output impedance and the short-circuit current limiter must be consistent for proper functioning of the device. To show the importance of the proper choice of this output resistance, consider the unity gain circuit of Fig. 6a.

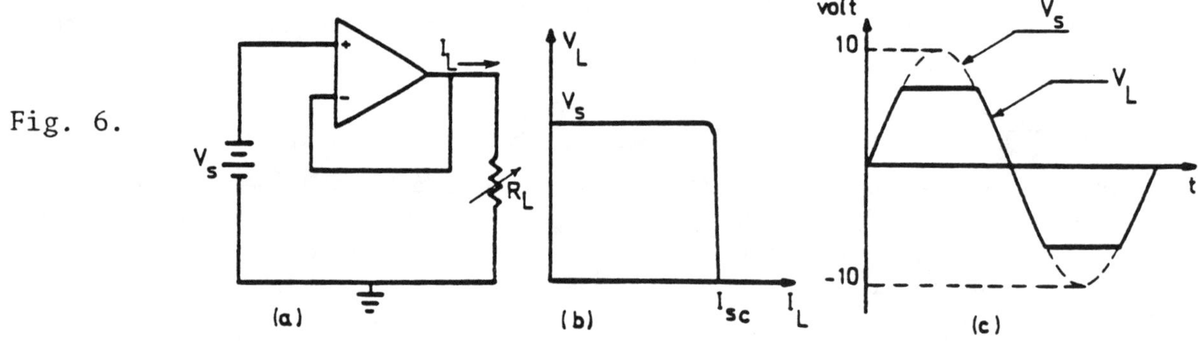

Fig. 6.

Here we can show that as R_L decreases the current I_L increases while the output voltage remains almost constant (cf., Fig. 6b). Gradually R_L drops to a value corresponding to the short-circuit current I_{sc} and further decrease will not change this short-circuit current. The current that is statically being limited actually saves the device. If instead of the constant DC voltage source we use $V_s = \sin\omega t$ and we choose a small loading resistor, then it is quite possible that the alternating V_L becomes that of Fig. 6c, which is clipped on both the positive and negative portions of the cycle [3].

IV-4D.   The Input Offset Voltage V_{io}

In a practical amplifier the offset voltages are not zero. The input offset V_{io}, is the voltage that must be applied between the input terminals through two equal resistors to obtain zero output voltage. This voltage may have different polarities. The output offset voltage V_{oo}, is the voltage which we have for an Op-Amp with $V_{in}^- = V_{in}^+$.

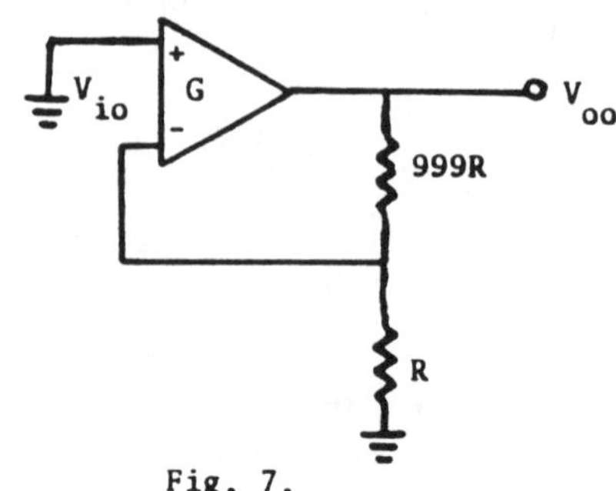

Fig. 7.

The circuit of Fig. 7 gives a simple procedure to find this offset voltage [2]. The resistor R is chosen such that $|V_{1o}| \gg |I^- R|$. Thus

$$V_{oo} = G\,(V_{1o} - 10^{-3}\,V_{oo})\,,$$

which can be written as

$$V_{oo} = GV_{1o}/(1 + 10^{-3}G) \simeq 10^3\,V_{1o}\,.$$

Using this expression and the fact that it is easy to read V_{oo} with a DMM, therefore V_{1o} is determined readily.

IV-4E.   *The Bandwidth* BW

We have noticed that for most practical purposes we must use the Op-Amp in a feedback configuration due to its very high open-loop gain. In general, even for an open-loop Op-Amp the BW is not infinite. To measure this parameter we must follow the general approach of drawing the amplitude of the transfer function versus frequency from which the BW becomes easily derived. For example for the choice of a first-order low-pass filter model of Op-Amp, we can draw its Bode diagram with the corresponding corner frequency to find the overall BW of the device. For a negative closed-loop case we have $G_{cl}^v(s) = G_{ol}^v(s)/[1 + G_{ol}^v(s)H(s)]$. If $|G_{ol}^v(j\omega)H(j\omega)| \gg 1$, then $G_{cl}^v(j\omega) \simeq 1/H(j\omega)$, and for $|G_{ol}^v(j\omega)H(\omega)| \ll 1$, we have $G_{cl}^v(j\omega) \simeq G_{ol}^v(j\omega)$. At the corner frequency of ω_c these two quantities become the same. Substituting for $|G_{ol}^v(j\omega_c)|$, and for $\omega_c > \omega_0$, this will result in

$$\frac{1}{|H(j\omega_c)|} = \frac{G_0\omega_0}{\omega_c} \quad \text{or} \quad G_0\omega_0 = \omega_c/|H(j\omega_c)|\,.$$

The above expression shows that the Gain-Bandwidth product (GBW) for an amplifier whose transfer function is given in the above is a constant and we call this $\omega_u.1$. Here ω_u stands for unity gain angular frequency.

There are several other parameters associated with a practical Op-Amp. Since we have not explicitly mentioned them for the case of an ideal Op-Amp, we will introduce them as follows.

IV-4F.   *The Input Bias Current* I_B

We must realize that in a practical Op-Amp some bias current must enter the device. The current that is required at the input of a single-input amplifier or the average of the two input currents for a *differential* input amplifier, when the output voltage is at zero, is called the input bias current. In the complete balance case these two input currents are zero. In general, $I_B = (I_B^- + I_B^+)/2$.

IV-4G.   *The Input Offset Current* I_{1o}

The difference in the currents into the two input terminals of a differential amplifier when the output voltage is at zero, is called the input offset current. But in general the dependence of both bias and offset current on the output voltage is minimal. Thus $I_{1o} = I_B^- - I_B^+$. We notice that I_B is of one polarity, while I_{1o} is of either polarity.

We may use the above relations to model I_B^- and I_B^+ as follows:

$$I_B^- = I_B + I_{1o}/2 \quad \text{and} \quad I_B^+ = I_B - I_{1o}/2\,.$$

To measure these quantities we look into the circuits of Fig. 8. In Fig. 8a, we have $V_{oo} = I_B^- R$, and in Fig. 8b, we have $V_{oo} = -I_B^+ R$. With a right choice of R we can measure V_{oo} and thus determine the I_B^- and I_B^+.

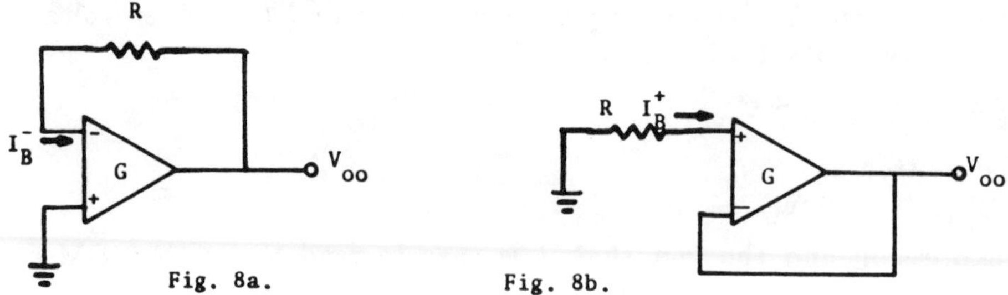

Fig. 8a.   Fig. 8b.

IV-4H. *The Common-Mode Rejection Ratio* CMRR

It is well known that not only do amplifiers respond to the differential or input signal, but they also respond to another signal that is called the common mode signal. To be more formal about this concept we define the following parameters:

$$V_{dm}(\text{differential—mode}) = V_{in}^- - V_{in}^+ , \quad \text{and} \quad V_{cm}(\text{common—mode}) = (V_{in}^- + V_{in}^+)/2.$$

Equivalently, we can state that:

$$V_{in}^- = V_{cm} + V_{dm}/2 \ , \quad \text{and} \quad V_{in}^+ = V_{cm} - V_{dm}/2 \ .$$

Since V_{out} is a function of both V_{dm} and V_{cm} , we can say that

$$V_{out} = G_{dm} V_{dm} + G_{cm} V_{cm}.$$

This equation means that we have two corresponding gain factors for each component of the output signal. Since V_{dm} is the signal that we have more control over than V_{cm} , it is natural to expect that G_{dm} should be much larger than G_{cm}. In other words, in order that the effect of V_{cm} be minimized, the ratio of G_{dm}/G_{cm} , must be maximized. This ratio is extremely large for any reasonable design, so much so that in most cases this ratio is expressed in terms of dB. This ratio, usually expressed in dB, is called the CMRR of the given amplifier. In future experiments we will look into measuring this parameter.

IV-4I. *The Slew Rate* (SR) *and Transient Response*

This parameter is another limiting factor in the application of an Op-Amp. Slew rate refers to "the maximum rate of change of the output voltage under the large signal conditions" [0]. This quantity is usually measured and specified for the closed-loop condition. This phenomenon is caused by current limiting and saturation occurring in some internal circuitry; in particular, the capacitance elements of the Op-Amp due to high frequency input signals [3]. Some typical outputs of an amplifier considering slew rate limiting is given by Fig. 9, [0].

We may now define the slew rate as $SR = \Delta V_o/\Delta t$. For a sinusoidal input signal $v(t) = V \sin \omega t$ the $SR = |dv(t)/dt|_{Max} = |V \omega \cos \omega t|_{Max} = V\omega$, or $f_{Max} \leq (SR)/2\pi V_{peak}$ to avoid distortion. This is called full power response. The physical dimension of $\Delta V_o/\Delta t$ reinforces the idea that slew rate is caused by changes in voltage of some internal or external capacitances. The capacitance that is responsible for limiting the SR is generally a compensation capacitance. At higher frequencies or higher rates of signal changes, the current available to charge and discharge this capacitance becomes exhausted and the slew rate limiting will occur [0].

In regard to the transient response of the Op-Amp, we should recall that a more realistic transfer function of an Op-Amp than a constant gain is the one with at least one pole. This transfer function must be then substituted for the open-loop gain of Op-Amp in order to find the overall network transfer function. From the inverse (time-domain) response corresponding to this transfer function we may find the relevant transient parameters of the network. In the next section we look

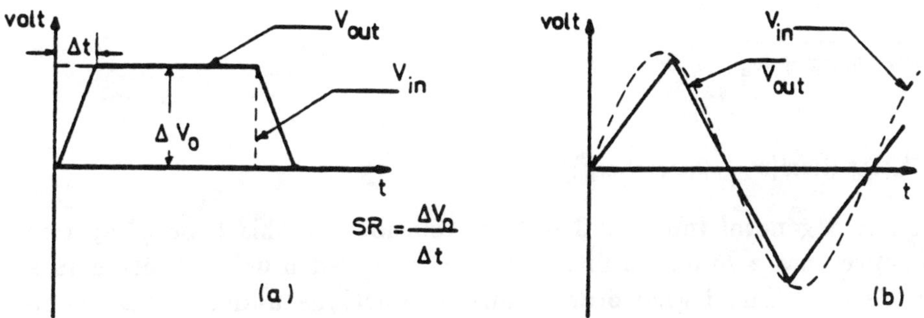

Fig. 9.　Example of Amplifier Output
Considering Slew Rate Limiting Effects.

into determination of such parameters in the case of a second-order system.

IV-4.1. FEEDBACK CONFIGURATIONS WITH PRACTICAL OP-AMP

To develop a comparable result for the circuit of Fig. 4, with a practical Op-Amp instead of an ideal Op-Amp, we use the following equations:

$$V_{out}(s) = -G_{ol}^v(s) [V_{in}^-(s) - V_{in}^+(s)] = -G_{ol}^v(s) V_{in}^-(s), \text{ and}$$

$$[V_{in}(s) - V_{in}^-(s)]/Z_1(s) = [V_{in}^-(s) - V_{out}(s)]/Z_f(s) .$$

Assuming $H(s) = Z_1(s)/[Z_1(s) + Z_f(s)]$ (the feedback factor), we have

$$\frac{V_{out}(s)}{V_{in}(s)} = \left(-\frac{Z_f(s)}{Z_1(s)}\right) \left(\underbrace{\frac{1}{1 + \dfrac{1}{H(s)\, G_{ol}^v(s)}}}\right) .$$

$$\underbrace{}_{\substack{\text{ideal} \\ \text{case}}} \underbrace{}_{\substack{\text{contribution of} \\ \text{nonidealness}}}$$

If for all s, $|G_{ol}^v(s)|$ is sufficiently large, we can approximate this equation to get

$$\frac{V_{out}(s)}{V_{in}(s)} = \left(-\frac{Z_f(s)}{Z_1(s)}\right) \left(1 - \frac{1}{H(s)\, G_{ol}^v(s)}\right) ,$$

where the term $(1 - 1/HG_{ol}^v)$ is called the gain error. The readers are invited to develop the parallel equation for the circuit in Fig. 5.

IV-5.　BASIC PROPERTIES OF A SECOND-ORDER SYSTEM

Consider the closed-loop second-order system in Fig. 10.

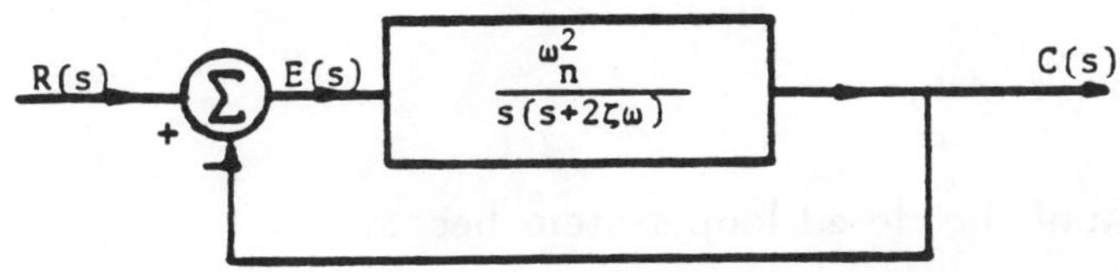

Fig. 10.　A Second-Order System.

For this system it is easy to show that

$$\frac{C(s)}{R(s)} = \frac{\omega_n^2}{s^2 + s\varsigma\omega_n s + \omega_n^2} \; .$$

Or equivalently

$$c(t) = \mathcal{L}^{-1}\left\{\omega_n^2\, R(s)/(s^2 + 2\varsigma\omega_n s + \omega_n^2)\right\} \; .$$

The above relation is a general input and output relation for this type of system. This particular transfer function represents a low-pass filter and has received much attention in system theory and in active network design. The higher-order transfer functions under certain conditions are decomposed into several blocks of second-order (as well as first-order) transfer functions of various types. Studying each of these blocks separately will enable us, in general, to gain the necessary information about the overall system transfer function. We will thus try to analyze various possible forms of the above transfer function thoroughly. The relevant properties of this type of transfer function in conjunction with both passive and active circuits are reviewed subsequently.

Consider $C(s)/R(s) = \omega_n^2/(s^2 + 2\varsigma\omega_n s + \omega_n^2)$. In this transfer function ω_n is called the natural frequency and ς is called the damping ratio. Knowing these two parameters is sufficient to know the complete behavior of this transfer function. Depending on what values ς assumes, this behavior changes.

IV-5A.   Underdamped Case $(0 < \varsigma < 1)$

In this case we have

$$\frac{C(s)}{R(s)} = \frac{\omega_n^2}{(s + \varsigma\omega_n + j\omega_d)\,(s + \varsigma\omega_n - j\omega_d)} \; ,$$

where $\omega_d = \omega_n\sqrt{1 - \varsigma^2}$ is the damped natural frequency. The step response of this system becomes $C(s) = \omega_n^2/s(s + \varsigma\omega_n + j\omega_d)(s + \varsigma\omega_n - j\omega_d)$. From which we have

$$c(t) = 1 - [\exp(-\varsigma\omega_n t)]\left(\cos \omega_d t + \frac{\varsigma}{\sqrt{1 - \varsigma^2}} \sin \omega_d t\right),$$

$$= 1 - \frac{[\exp(-\varsigma\omega_n t)]}{\sqrt{1 - \varsigma^2}} \sin\left(\omega_d t + \tan^{-1} \frac{\sqrt{1 - \varsigma^2}}{\varsigma}\right), \; t \geq 0 \; .$$

It is easy to show that for this system $e(t) = [r(t) - c(t)] \to 0$ as $t \to \infty$. If $\varsigma = 0$, then we have a pure sinusoidal oscillation as follows: $c(t) = 1 - \cos \omega_n t$, $t \geq 0$.

IV-5B.   Critically Damped Case $(\varsigma = 1)$

If the two poles of the transfer function are close to each other, which is the same as $\varsigma \to 1$, then we have $C(s) = \omega_n^2 R(s)/(s + \omega_n)^2$. The unit step response of this system becomes $c(t) = 1 - [\exp(-\omega_n t)](1 + \omega_n t)$, $t \geq 0$.

IV-5C.   Overdamped Case $(\varsigma > 1)$

The unit step response of the closed-loop system becomes

$$C(s) = \omega_n/s(s + \varsigma\omega_n + \omega_n\sqrt{\varsigma^2 - 1})\,(s + \varsigma\omega_n - \omega_n\sqrt{\varsigma^2 - 1}).$$

From which we have

$$c(t) = 1 + \frac{1}{2\sqrt{\varsigma^2-1}\,(\varsigma+\sqrt{\varsigma^2-1})}\exp[-(\varsigma+\sqrt{\varsigma^2-1})\,\omega_n t]$$

$$- \frac{1}{2\sqrt{\varsigma^2-1}\,(\varsigma-\sqrt{\varsigma^2-1})}\exp[-(\varsigma-\sqrt{\varsigma^2-1})\omega_n t]\ ,\ t \geq 0.$$

Clearly $c(t)$ consists of two decaying exponential terms. When $\varsigma \gg 1$ one of these two exponential terms goes to zero much faster than the other one. Thus $c(t) = 1 - \alpha\exp(-p_1 t)$, where p_1 is the *dominant* (closer to the imaginary-axis) pole of the system and α is a constant. More detailed analysis of this problem can be found in most standard textbooks of control system such as [1]. We may summarize the above analysis in the family of curves that we draw for various values of ς, where the abscissa is the dimensionless variable $\omega_n t$.

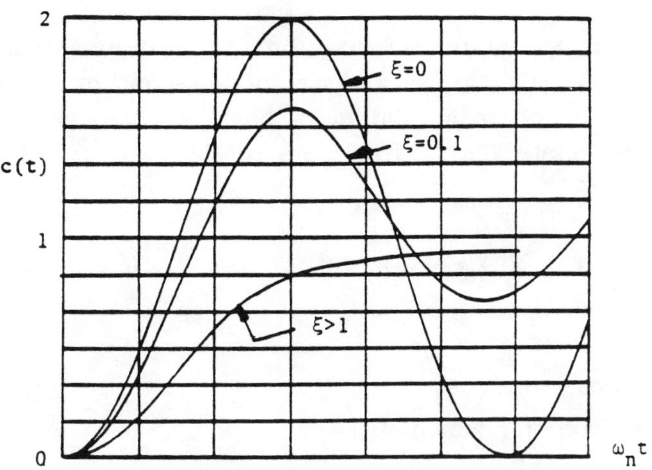

Fig. 11. Unit-Step Response Curves of the
System in Fig. 10.

IV-5.1. DEFINITIONS OF TRANSIENT RESPONSE PARAMETERS

Consider the unit-response of an underdamped second-order system, as shown in Fig. 12. From this figure we have following important parameters which are essential data for the behavior of the second-order system under study.

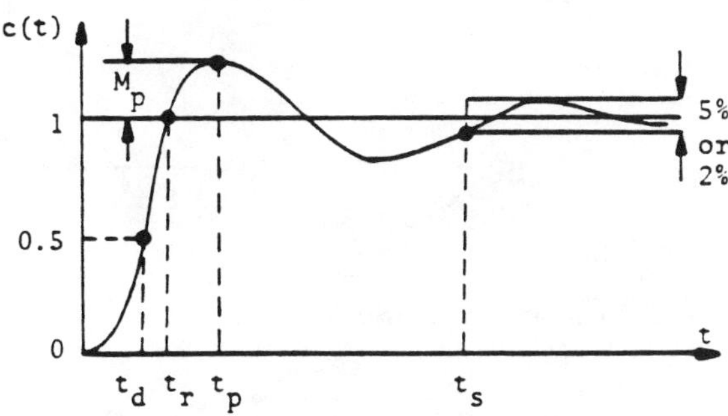

Fig. 12. Unit-Step Response.

IV-5.1A. Delay Time t_d

The time required for the response to reach half its final value the first time.

IV-5.1B. Rise Time t_r

The time required for the response to rise from 0 to 100% (underdamped); 5 to 95% (critically damped); and 10 to 90% (overdamped) of its final value.

IV-5.1C. Peak Time t_p

The time required for the response to reach its first peak of the overshoot.

IV-5.1D. Settling Time t_s

The time required for the response to reach and to stay within a range of 2 to 5% of its final value.

IV-5.1E. Maximum Overshoot M_p

The maximum peak of response curve measured from unity. If the final value is different from unity then the maximum percent overshoot is defined as $M_p = [c(t_p) - c(\infty)] \times 100\%/c(\infty)$.

Our objective is to use the above information in the design of a second-order system and/or from this information to identify the parameters of a second-order system. We will study some relevant properties of the second-order system in both of these two directions. Before proceeding further, however, we must relate these parameters to the two system parameters ς and ω_n that we originally introduced.

IV-5.2. DERIVATIONS OF TRANSIENT RESPONSE PARAMETERS

IV-5.2A. Rise Time t_r

By definition t_r is the time for which $c(t_r) = 1$.
Therefore we have

$$1 = 1 - [\exp(-\omega_n t_r)](\cos\omega_d t_r + \frac{\varsigma}{\sqrt{1-\varsigma^2}}\sin\omega_d t_r).$$

From which and referring to Fig. 13, we have

$$\tan(\omega_d t_r) = -\frac{\sqrt{1-\varsigma^2}}{\varsigma} = -\frac{\omega_d}{\sigma}.$$

Thus

$$t_r = \frac{1}{\omega_d}\tan^{-1}\left(-\frac{\omega_d}{\sigma}\right) = \frac{\pi - \beta}{\omega_d} = \frac{\pi - \beta}{\omega\sqrt{1-\varsigma^2}}.$$

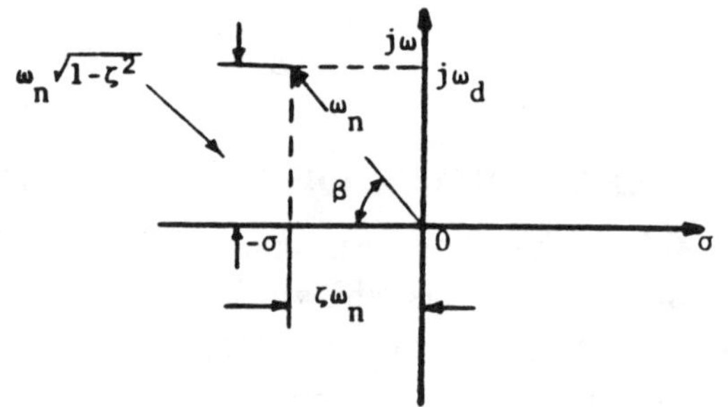

Fig. 13.   Definition of the Angle β.

IV-5.2B. Peak Time t_p

By definition t_p is such that $|\frac{dc(t)}{dt}|_{t=t_p} = 0$, which will result in $\sin\omega_d t_p = 0$, or $t_p = \frac{\pi}{\omega_d}$.

IV-5.2C. Maximum Overshoot M_p

This quantity corresponds to $c(t_p)$. Thus

$$M_p \triangleq c(t_p) - 1 = \exp(-\varsigma\pi/\sqrt{1-\varsigma^2}).$$

This quantity depends solely on ς and not on ω_n. Also it is evident that if we make t_r small (i.e., fast response), then M_p becomes large, which, in general, is not desirable. The maximum per cent overshoot (which is defined for cases of non-unity input) is $[\exp(-\sigma\pi/\omega_d)] \times 100\%$.

IV-5.2D. Settling Time t_s

This time corresponds to the time that $c(t_s)$ becomes $\pm 2\%$ or $\pm 5\%$ of its final value. We can show that [1]

$$t_s = \begin{cases} \dfrac{4}{\varsigma \omega_n} = \dfrac{4}{\sigma} & \text{for 2\% criterion,} \\[2ex] \dfrac{3}{\varsigma \omega_n} = \dfrac{3}{\sigma} & \text{for 5\% criterion.} \end{cases}$$

We notice here that t_s is inversely dependent on $\varsigma \omega_n$, because ς is chosen upon the requirements on the maximum overshoot and therefore t_s is determined primarily from the undamped natural frequency ω_n.

IV-5.3. PASSIVE AND ACTIVE SECOND-ORDER NETWORKS

IV-5.3A. General Comments on Passive Second-Order Networks

In the following we choose a passive and an active second-order network to study their behavior in relation to a general second-order system. First, consider the simple RLC network in Fig. 14, where r is a resistance representing total resistances of all elements.

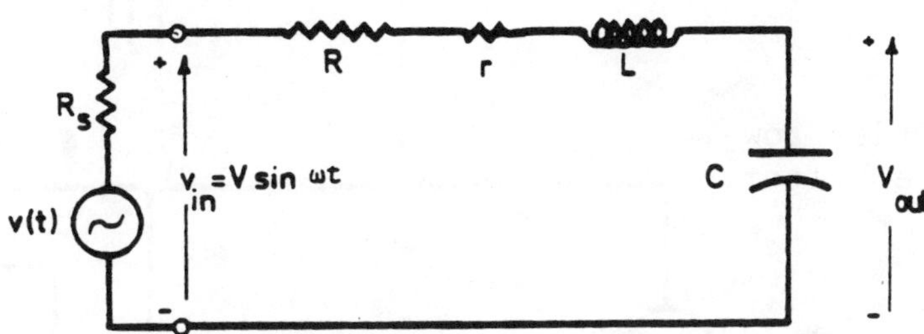

Fig. 14.   A Simple RLC Network.

It is easy to show that

$$\frac{V_{out}(s)}{V_{in}(s)} = \frac{1/sC}{R + r + sL + 1/sC}$$

$$= \frac{1/LC}{[s^2 + (R + r)s/L + 1/LC]} \; .$$

Comparing this transfer function with that of a general second-order system yields the natural frequency $\omega_n = 1/\sqrt{LC}$, and the damping ratio $\varsigma = (R+r)\sqrt{C/L}/2$.

Thus by changing these parameters -- say, changing R and/or L, we can study $v_{out}(t)$ and observe all the cases that we have seen with regard to the system being underdamped, critically damped, or overdamped. On the other hand, if we have no idea about the values of, at most, two parameters -- say, R and L, then we can measure and/or record $v_{out}(t)$ and by studying its transient response behavior we can identify these two parameters. In fact this type of system identification is quite popular for many other second-order systems such as an electromechanical system (mass, spring and friction problems). Actually, in the early stage of the servomechanism, literally, every system was assumed to be, at most, a second-order system. By studying this kind of transient response analysis a set of parameters could be associated with the system. Nowadays, for a given system we *may* start identifying its parameters by assuming that the system is a second-order system. If this model is not satisfactory, then we will increase the dimension of the system model to complete the identification process and thus obtain a model that matches our system's actual response more accurately. As we have also mentioned before, for a general case of a second-order system, we can choose any design criterion. For example we may wish to have a certain maximum overshoot; or to have a certain speed of the system response, *etc.* Correspondingly, we may adjust any of these parameters R, L or C to get the desired performance. It is also interesting to note that for the circuit in Fig. 14, at $s = j\omega_n$, we have

$$|V_{out}(j\omega_n)/V_{in}(j\omega_n)| = 1/2\varsigma\,.$$

In closing this matter we remind our readers that as we learned in EXP. II-6, the synthesis of passive networks (with RLC and transformers) is an important subject and that electrical engineering students must be familiar with its fundamental concepts. In the following we look into two typical second-order active RC networks. Discussions on the synthesis of such networks are deferred to EXP. VI-5.

IV-5.3B. General Comments on Active Second-Order Networks

Consider the second-order low-pass active filter in Fig. 15.

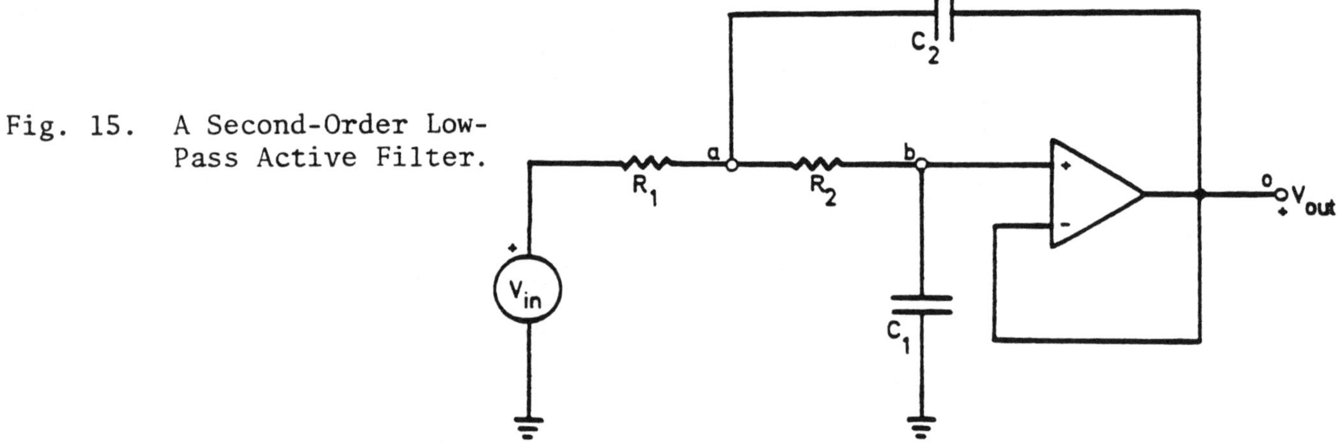

Fig. 15. A Second-Order Low-Pass Active Filter.

To show that the overall transfer function is a second-order low-pass active filter we must analyze this circuit. To analyze this type of circuit, customarily we use node equations that make the analysis very convenient. To perform such analysis we must first identify all of the existing nodes. In our circuit, if we know the node voltages, say, at (a), (b) and (o), then we completely know this circuit. In general, we must write as many node equations as we have unknown node voltages to analyze a given network. In this type of network, however, one equation must always be written for each Op-Amp relating its input and output voltages. Therefore one less node equation is needed to analyze this network. To solve for $V_{out}(s)$, we have the following set of equations.

For node (a)

$$G_1 V_{in}(s)\ (\triangleq \text{total current coming to node (a)}) =$$

$$+ (G_1 + G_2 + sC_2)\ (\triangleq \text{total admittance of node (a)}) \times V_a(s)$$

$$- (G_2)\ (\triangleq \text{total admittance between nodes (a) and (b)}) \times V_b(s)$$

$$- (sC_2)\ (\triangleq \text{total admittance between nodes (a) and (o)}) \times V_o(s)\,.$$

For node (b)

$$0 = + (G_2 + sC_1) \, (\triangleq \text{ total admittance of node (b)}) \times V_b(s)$$

$$- (G_2) \, (\triangleq \text{ total admittance between nodes (a) and (b)}) \times V_a(s)$$

$$- (\text{zero}) \, (\triangleq \text{ total admittance between nodes (b) and (o)}) \times V_o(s) \, .$$

For Op-Amp

$V_b = V_o$ (from the idealness property of the Op-Amp).

Eliminating $V_a(s)$ and $V_b(s)$ from these three equations will result in

$$\frac{V_o(s)}{V_{in}(s)} = \frac{1}{R_1 R_2 C_1 C_2 s^2 + (R_1 + R_2)C_1 s + 1} \, .$$

This transfer function is in a standard low-pass second-order form. We may write

$$\omega_n = 1/\sqrt{R_1 R_2 C_1 C_2} \, ,$$

and

$$\varsigma = \frac{R_1 + R_2}{2\sqrt{R_1 R_2}} \sqrt{\frac{C_1}{C_2}} \, .$$

If we let $R_1 = R_2 = R$, then we have

$$\omega_n = 1/R\sqrt{C_1 C_2} \, , \text{ and } \varsigma = \sqrt{C_1/C_2} \, .$$

Similarly, if we consider the circuit in Fig. 16,

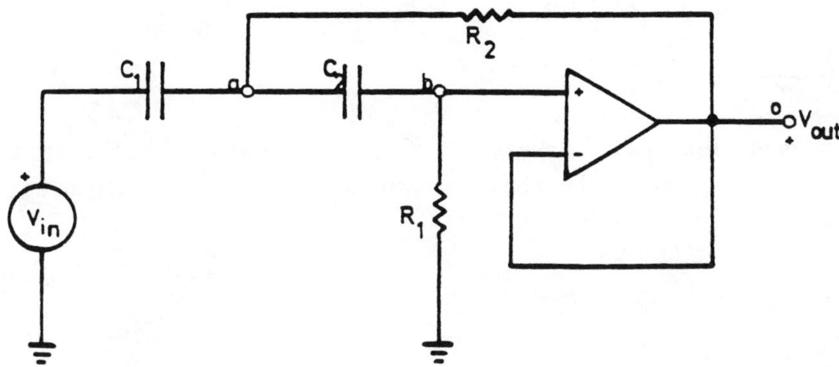

Fig. 16.   A Second-Order High-Pass Active Filter.

then we can show that (try this example),

$$\frac{V_o(s)}{V_{in}(s)} = \frac{R_1 R_2 C_1 C_2 s^2}{R_1 R_2 C_1 C_2 s^2 + R_2(C_1 + C_2)s + 1} \, .$$

This transfer function is in a standard high-pass second-order form. If we let $C_1 = C_2 = C$, then we get

$$\frac{V_o(s)}{V_{in}(s)} = \frac{(s/\omega_n)^2}{(s/\omega_n)^2 + 2\varsigma(s/\omega_n) + 1} \, ,$$

where

$$\omega_n = 1/C \sqrt{R_1 R_2} \ \text{and} \ \varsigma = \sqrt{R_2/R_1} \ .$$

The above two transfer functions, in particular the low-pass filter, can exhibit the properties that we have stated for a general second-order system.

We may briefly comment that there are, in general, several ways of designing a circuit with both active and passive elements. The passive network synthesis is more systematic than the active network synthesis. We have already seen in EXP. II-6 the passive network synthesis. We will look into active network synthesis in EXP. IV-5 and EXPs. VI-5 to 7.

IV-6.   FUNDAMENTAL APPLICATIONS OF OP-AMPS

The Op-Amp allows someone with limited knowledge of electronics to perform subtle manipulations with electrical signals. For example, consider the inverting feedback amplifier circuit of Fig. 17. We have seen that for this circuit and for sufficiently large G we have:

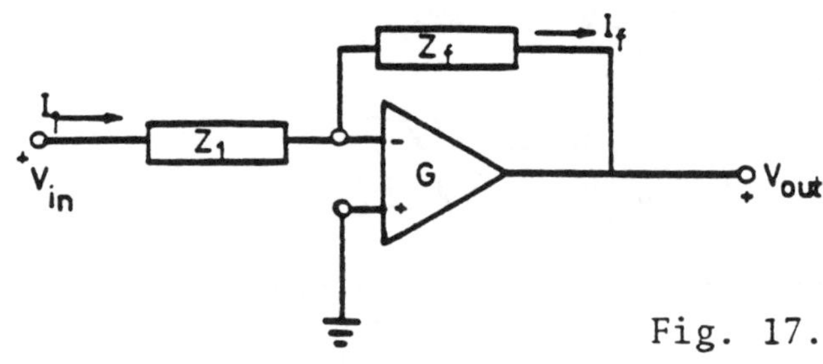

$$\frac{V_{out}(s)}{V_{in}(s)} = -\frac{Z_f(s)}{Z_1(s)} \ .$$

IV-6A.   An Integrator

Suppose $Z_1(s) = R$ and $Z_f(s) = 1/sC$, then we have

$$V_{out}(s) = (-\frac{1}{RC})(V_{in}(s)/s) \ .$$

We recall that $1/s$ is an integrator operator. Thus this relation implies that the simple inverting feedback amplifier of Fig. 17 with the above arrangement can act as an integrator of the input signal.

IV-6B.   A Differentiator

Suppose $Z_1(s) = 1/sC$ and $Z_f(s) = R$, then we have

$$V_{out}(s) = (-RC)(sV_{in}(s)) \ .$$

We recall that s is the differentiator operator. Thus this relation implies that the inverting feedback amplifier of Fig. 17, with the above arrangement can act as a differentiator of the input signal. Building a differentiator is usually undesirable in practice due to noise amplification. Maybe we should point out how this amplification takes place. Associated with each signal of frequency f, there may be a noise signal which is of much higher frequency than f, say, 1000 f. This high frequency signal will become extremely large in amplitude when it is being differentiated, while the integration of the same signal will result in a signal with very small amplitude. For example, let x(t) = sin 1000t, (amplitude is one), then dx/dt = 1000 cos 1000t and $\int$xdt = –cos 1000t/1000. Clearly one can see how small the amplitude of the integrated signal is and how large the amplitude of the differentiated signal is. This is why integration is considered to be a stable operation and differentiation is always subject to possible difficulty and/or instability.

IV-6C.   A Constant Multiplier

In the circuit of Fig. 17, if we let $Z_1(s) = R_1$ and $Z_f(s) = R_f$, then we have

$$V_{out}(s) = (-R_f / R_1)V_{in}(s) .$$

This relation shows that the output signal is a constant multiple of the input signal with 180° phase shift.

IV-6D.   An Adder or Summer

If an Op-Amp is used in its linear operation range then the superposition rule can be applied. Using this concept we can add several signals with the corresponding gain factor as depicted in the circuit of Fig. 18. The result is

$$V_{out}(s) = \sum_{j=1}^{n} -(\frac{R_f}{R_j})V_{in}^j(s) .$$

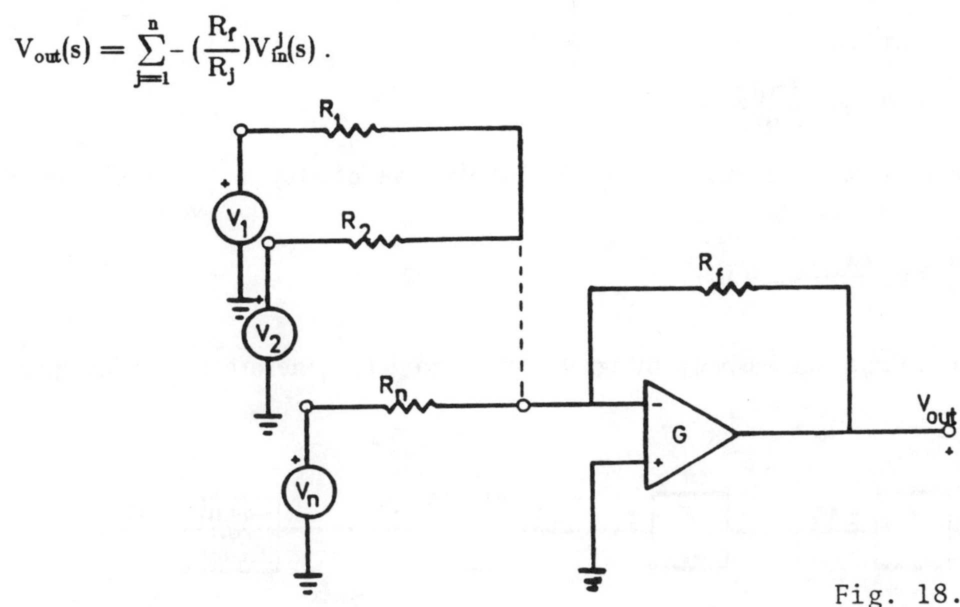

Fig. 18.

IV-6.1. ANALOG COMPUTATIONS

One of the most important applications of the Op-Amp is in the area of analog computers. Still there are many cases where we must use an analog computer to simulate a dynamic trajectory. We may do this quite easily thanks to the above four operations that can be performed with a basic inverting feedback amplifier using Op-Amps and simple R's and C's. Thus in this section we propose a simple technique to solve an nth-order linear differential equation with constant coefficients.

Suppose we want to solve the following differential equation in terms of x(t).

$$\frac{d^n x}{dt^n} + a_1 \frac{d^{n-1}x}{dt^{n-1}} + ... + a_j \frac{d^{n-j}x}{dt^{n-j}} + \cdots + a_n x = f(t).$$

We also assume that $\frac{d^{n-j}x(0)}{dt^{n-j}} = 0$, j = 1 to n, for the time being. Rearranging the terms will result in

$$\frac{d^n x(t)}{dt^n} = -\sum_{j=1}^{n} a_j \frac{d^{n-j}x}{dt^{n-j}} + f(t) .$$

In this equation we have expressed the highest-order derivative of the response (or unknown) signal $x(t)$ in terms of the lower-order derivatives of $x(t)$ and the forcing (or input) function $f(t)$. Using integrator, constant multiplier, and adder circuits that we have studied earlier, we can simulate the above equation to find $x(t)$. This idea can be pursued as follows. Suppose we have the $\frac{d^n x(t)}{dt^n}$, i.e., the highest-order derivative of unknown signal in the differential equation. Then by sucessive integration of this signal we can generate $x(t)$. This concept is the sole idea behind the analog computer programming. Having at first superficially generated $x(t)$, $\frac{dx(t)}{dt}$,..., $\frac{d^{n-1} x(t)}{dt^{n-1}}$, we can complete the simulation process by inserting the appropriate adder and/or multiplier circuits to finish the integration process.

As an example, consider the following differential equation:

$$\frac{d^3 x}{dt^3} + a_1 \frac{d^2 x}{dt^2} + a_2 \frac{dx}{dt} + a_3 x = f(t) \,,$$

where the initial conditions are

$$x(0) = 0, \ \frac{dx(0)}{dt} = 0, \text{ and } \frac{d^2 x(0)}{dt^2} = 0 \,.$$

Using the above procedure we can express the third-derivative of $x(t)$ in terms of the remaining terms of this differential equation as

$$\frac{d^3 x}{dt^3} = -a_1 \frac{d^2 x}{dt^2} - a_2 \frac{dx}{dt} - a_3 x + f(t) \,.$$

If we have $\frac{d^3 x}{dt^3}$, then we can successively integrate this signal to generate $x(t)$. The procedure is shown in Fig. 19.

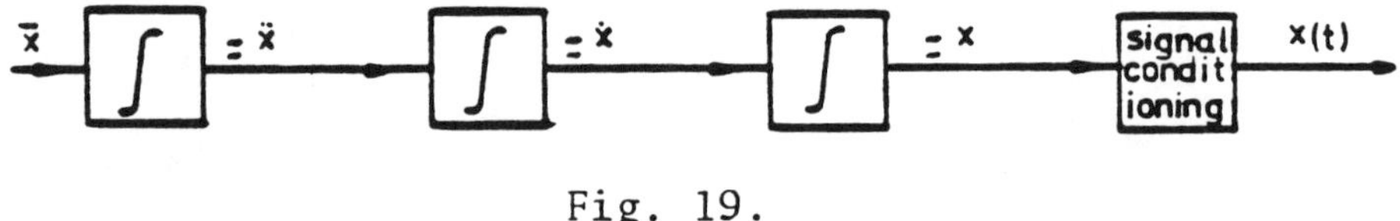

Fig. 19.

As illustrated in Fig. 19, the output of each integrator, in general, is *proportional* to the integral of the input signal. Therefore, we must do the necessary signal conditionings in order to make this simulation meaningful. This simulation is carried out next, as shown in Fig. 20.

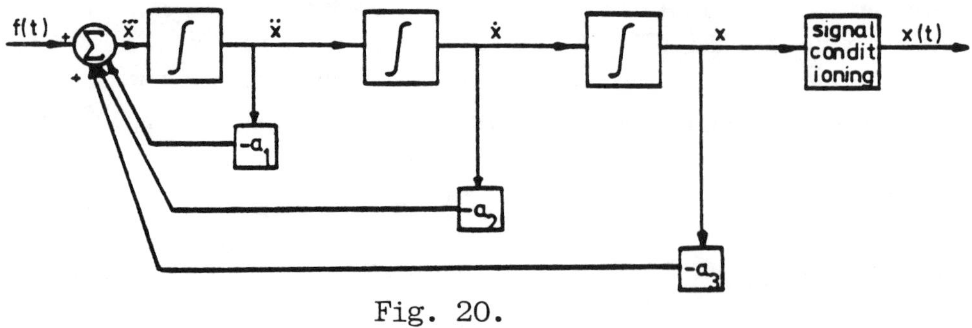

Fig. 20.

We must point out that the block diagram of Fig. 20, is not intended to serve as a simulator diagram in a real analog computer case. It is only for the purpose of illustrating the concept. In real situations the output of each integrator has a minus sign plus a specific gain that must be accounted for.

This sort of integration of a differential equation can be done by an analog computer. Each analog computer has several operational amplifiers that can be used with the help of capacitors and resistors already in the computer to build integrators, constant multipliers, summers or adders. Therefore, programming on an analog computer is simply an easy way to simulate a differential equation through the use of these operational blocks. Provisions have been made in these computers to have certain signal multipliers in order to simulate the non-linear differential equations. Also we can incorporate the initial conditions of the differential equation in the integrating process by inserting these signals into the appropriate integrators.

We must mention that in using analog computers the independent variable is always time and the dependent variable is voltage. Thus if we have a differential equation in terms of other physical quantities we must rescale our vertical and horizontal readings from any output devices such as an oscilloscope, or X-Y plotter, *etc.*, into the corresponding physical quantities of our interest. Also we must point out that in certain cases the amplitude of the signal may become very large which may saturate the Op-Amps. There is a technique called amplitude scaling which rectifies this problem by scaling down the $x(t)$ to below the saturation limit. Finally, if the speed of response is not convenient, and it may cause us difficulty, we have the option of frequency scaling to modify the differential equation to avoid this difficulty. The scaling concept is described in EXP. VI-5.

IV-6.2. A GENERAL SYNTHESIS PROCEDURE

To show another important application of the Op-Amp we present the following general synthesis procedure. There are many other ways, from system theoretical point of view, reported in the literature that result in just the same final product. But we believe the following approach is a straightforward method that serves to indicate this synthesis concept. Consider the following transfer function,

$$\text{TF} \triangleq \frac{N(s)}{D(s)} = K \frac{s^m + b_1 s^{m-1} + \dots + b_j s^{m-j} + \dots + b_m}{s^n + a_1 s^{n-1} + \dots + a_j s^{n-j} + \dots + a_n},$$

where K is a constant and $m \leq n$. The $N(s)$ and the $D(s)$ are two arbitrary signals. If we introduce an auxiliary signal $X(s)$ and multiply both numerator and denominator of the right-hand side of the above transfer function with $X(s)$, we will have:

$$N(s) = K \sum_{i=0}^{m} b_i s^{m-i} X(s), \quad b_0 = 1,$$

and

$$D(s) = \sum_{j=0}^{n} a_j s^{n-j} X(s), \quad a_0 = 1.$$

Each of the above two equations is a linear differential equation. We have studied their simulation in the previous section. Because $X(s)$ is an auxiliary signal and plays no role in our final synthesis we may assume all of its initial conditions are zero. Since in general we assume $m \leq n$, we therefore make the following rearrangement for an input signal, $D(s)$.

$$s^n X(s) = D(s) - \sum_{j=1}^{n} a_j s^{n-j} X(s).$$

Having made the above simplifications, we can present the block diagram of Fig. 21, as a synthesis approach to simulate this transfer function. We notice that $N(s)$ has terms that are generated in the above differential equation. One may consider that the detailed circuitry of the block diagram in Fig. 21 is the complete synthesis of this transfer function.

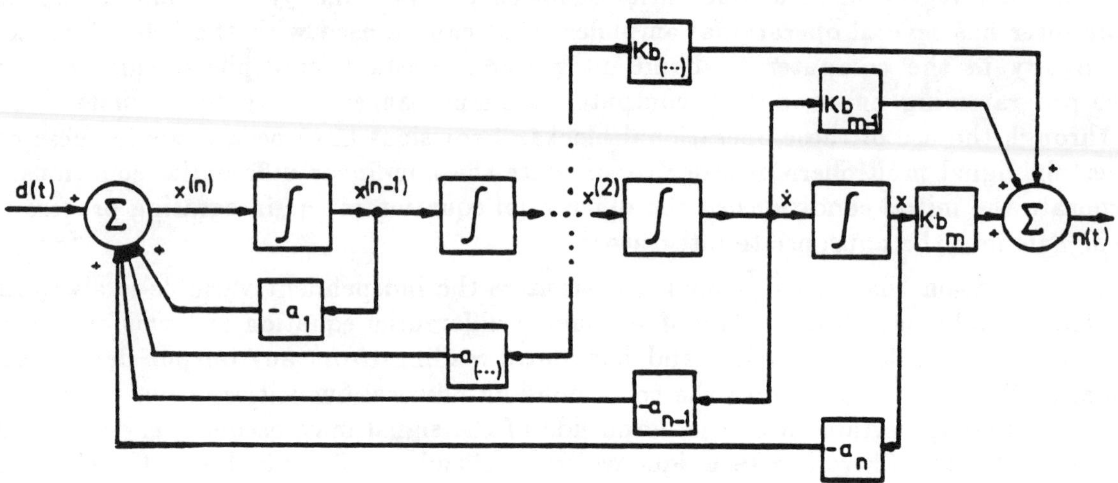

Fig. 21.   A General Simulation Diagram of a Transfer Function.

In conclusion, there is a wide range of other applications of Op-Amps in electronic circuits design that we have not mentioned. The active network theory plays an extremely important role in the circuit design and it depends heavily on the applications of Op-Amps. The few examples that we have described in the previous sections and the materials that are covered in the references of this part serve as a good starting point for someone who wishes to learn more about this device and its applications.

IV-7.   MANUFACTURERS' SPECIFICATION DATA SHEETS

On the following pages (cf., Figs. 22 and 23), we present the specification data sheets of two different general purpose operational amplifiers for the sake of initial review and familiarization with the basic performance of this device [6] and [7].

National Semiconductor

Operational Amplifiers/Buffers

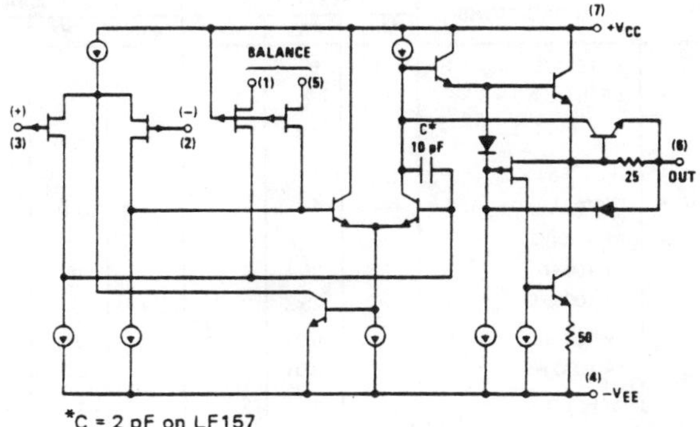

LF155/LF156/LF157 Series Monolithic JFET Input Operational Amplifiers

LF155, LF155A, LF255, LF355, LF355A, LF355B low supply current
LF156, LF156A, LF256, LF356, LF356A, LF356B wide band
LF157, LF157A, LF257, LF357, LF357A, LF357B wide band decompensated (Av_{MIN} = 5)

General Description

These are the first monolithic JFET input operational amplifiers to incorporate well matched, high voltage JFETs on the same chip with standard bipolar transistors (BI-FET Technology). These amplifiers feature low input bias and offset currents, low offset voltage and offset voltage drift, coupled with offset adjust which does not degrade drift or common-mode rejection. The devices are also designed for high slew rate, wide bandwidth, extremely fast settling time, low voltage and current noise and a low 1/f noise corner.

Advantages

- Replace expensive hybrid and module FET op amps
- Rugged JFETs allow blow-out free handling compared with MOSFET input devices
- Excellent for low noise applications using either high or low source impedance—very low 1/f corner
- Offset adjust does not degrade drift or common-mode rejection as in most monolithic amplifiers
- New output stage allows use of large capacitive loads (10,000 pF) without stability problems
- Internal compensation and large differential input voltage capability

Applications

- Precision high speed integrators
- Fast D/A and A/D converters
- High impedance buffers
- Wideband, low noise, low drift amplifiers
- Logarithmic amplifiers
- Photocell amplifiers
- Sample and Hold circuits

Common Features

(LF155A, LF156A, LF157A)

- Low input bias current — 30 pA
- Low Input Offset Current — 3 pA
- High input impedance — $10^{12}\Omega$
- Low input offset voltage — 1 mV
- Low input offset voltage temperature drift — $3\mu V/°C$
- Low input noise current — 0.01 pA/$\sqrt{Hz}$
- High common-mode rejection ratio — 100 dB
- Large dc voltage gain — 106 dB

Uncommon Features

	LF155A	LF156A	LF157A (A_V = 5)*	UNITS
Extremely fast settling time to 0.01%	4	1.5	1.5	µs
Fast slew rate	5	12	50	V/µs
Wide gain bandwidth	2.5	5	20	MHz
Low input noise voltage	20	12	12	nV/$\sqrt{Hz}$

Simplified Schematic

Fig. 22. General Purpose Op-Amp (Courtesy of National [6]).

Absolute Maximum Ratings

	LF155A/6A/7A	LF155/6/7	LF355B/6B/7B LF255/6/7 LF355B/6B/7B	LF355A/6A/7A LF355/6/7
Supply Voltage	±22V	±22V	±22V	±18V
Power Dissipation (P_d at 25°C) and Thermal Resistance (θ_{jA}) (Note 1)				
T_jMAX				
(H and J Package)	150°C	150°C	115°C	115°C
(N Package)			100°C	100°C
(H Package) P_d	670 mW	670 mW	570 mW	570 mW
θ_{jA}	150°C/W	150°C/W	150°C/W	150°C/W
(J Package) P_d	670 mW	670 mW	570 mW	570 mW
θ_{jA}	140°C/W	140°C/W	140°C/W	140°C/W
(N Package) P_d			500 mW	500 mW
θ_{jA}			155°C/W	155°C/W
Differential Input Voltage	±40V	±40V	±40V	±30V
Input Voltage Range (Note 2)	±20V	±20V	±20V	±16V
Output Short Circuit Duration	Continuous	Continuous	Continuous	Continuous
Storage Temperature Range	−65°C to +150°C	−65°C to +150°C	−65°C to +150°C	−65°C to +150°C
Lead Temperature (Soldering, 10 seconds)	300°C	300°C	300°C	300°C

DC Electrical Characteristics (Note 3)

SYMBOL	PARAMETER	CONDITIONS	LF155A/6A/7A			LF355A/6A/7A			UNITS
			MIN	TYP	MAX	MIN	TYP	MAX	
V_{OS}	Input Offset Voltage	$R_S = 50\Omega$, $T_A = 25°C$		1	2		1	2	mV
		Over Temperature			2.5			2.3	mV
$\Delta V_{OS}/\Delta T$	Average TC of Input Offset Voltage	$R_S = 50\Omega$		3	5		3	5	μV/°C
$\Delta TC/\Delta V_{OS}$	Change in Average TC with V_{OS} Adjust	$R_S = 50\Omega$, (Note 4)		0.5			0.5		μV/°C per mV
I_{OS}	Input Offset Current	$T_j = 25°C$, (Notes 3, 5)		3	10		3	10	pA
		$T_j \leq T_{HIGH}$			10			1	nA
I_B	Input Bias Current	$T_J = 25°C$, (Notes 3, 5)		30	50		30	50	pA
		$T_J \leq T_{HIGH}$			25			5	nA
R_{IN}	Input Resistance	$T_J = 25°C$		10^{12}			10^{12}		Ω
A_{VOL}	Large Signal Voltage Gain	$V_S = \pm15V$, $T_A = 25°C$ $V_O = \pm10V$, $R_L = 2k$	50	200		50	200		V/mV
		Over Temperature	25			25			V/mV
V_O	Output Voltage Swing	$V_S = \pm15V$, $R_L = 10k$	±12	±13		±12	±13		V
		$V_S = \pm15V$, $R_L = 2k$	±10	±12		±10	±12		V
V_{CM}	Input Common-Mode Voltage Range	$V_S = \pm15V$	±11	+15.1		±11	+15.1		V
				−12			−12		V
CMRR	Common-Mode Rejection Ratio		85	100		85	100		dB
PSRR	Supply Voltage Rejection Ratio	(Note 6)	85	100		85	100		dB

AC Electrical Characteristics $T_A = 25°C$, $V_S = \pm15V$

SYMBOL	PARAMETER	CONDITIONS	LF155A/355A			LF156A/356A			LF157A/357A			UNITS
			MIN	TYP	MAX	MIN	TYP	MAX	MIN	TYP	MAX	
SR	Slew Rate	LF155A/6A; $A_V = 1$,	3	5		10	12					V/μs
		LF157A; $A_V = 5$							40	50		V/μs
GBW	Gain Bandwidth Product			2.5		4	4.5		15	20		MHz
t_s	Settling Time to 0.01%	(Note 7)		4			1.5			1.5		μs
e_n	Equivalent Input Noise Voltage	$R_S = 100\Omega$										
		f = 100 Hz		25			15			15		$nV/\sqrt{Hz}$
		f = 1000 Hz		25			12			12		$nV/\sqrt{Hz}$
i_n	Equivalent Input Noise Current	f = 100 Hz		0.01			0.01			0.01		$pA/\sqrt{Hz}$
		f = 1000 Hz		0.01			0.01			0.01		$pA/\sqrt{Hz}$
C_{IN}	Input Capacitance			3			3			3		pF

Fig. 22. Continued.

DC Electrical Characteristics (Note 3)

SYMBOL	PARAMETER	CONDITIONS	LF155/6/7			LF255/6/7 LF355B/6B/7B			LF355/6/7			UNITS
			MIN	TYP	MAX	MIN	TYP	MAX	MIN	TYP	MAX	
V_{OS}	Input Offset Voltage	$R_S = 50\Omega$, $T_A = 25°C$		3	5		3	5		3	10	mV
		Over Temperature			7			6.5			13	mV
$\Delta V_{OS}/\Delta T$	Average TC of Input Offset Voltage	$R_S = 50\Omega$		5			5			5		μV/°C
$\Delta TC/\Delta V_{OS}$	Change in Average TC with V_{OS} Adjust	$R_S = 50\Omega$, (Note 4)		0.5			0.5			0.5		μV/°C per mV
I_{OS}	Input Offset Current	$T_j = 25°C$, (Notes 3, 5)		3	20		3	20		3	50	pA
		$T_j \leq T_{HIGH}$			20			1			2	nA
I_B	Input Bias Current	$T_J = 25°C$, (Notes 3, 5)		30	100		30	100		30	200	pA
		$T_J \leq T_{HIGH}$			50			5			8	nA
R_{IN}	Input Resistance	$T_J = 25°C$		10^{12}			10^{12}			10^{12}		Ω
A_{VOL}	Large Signal Voltage Gain	$V_S = \pm15V$, $T_A = 25°C$ $V_O = \pm10V$, $R_L = 2k$	50	200		50	200		25	200		V/mV
		Over Temperature	25			25			15			V/mV
V_O	Output Voltage Swing	$V_S = \pm15V$, $R_L = 10k$	±12	±13		±12	±13		±12	±13		V
		$V_S = \pm15V$, $R_L = 2k$	±10	±12		±10	±12		±10	±12		V
V_{CM}	Input Common-Mode Voltage Range	$V_S = \pm15V$	±11	+15.1 −12		±11	+15.1 −12		±10	+15.1 −12		V V
CMRR	Common-Mode Rejection Ratio		85	100		85	100		80	100		dB
PSRR	Supply Voltage Rejec- Ratio	(Note 6)	85	100		85	100		80	100		dB

DC Electrical Characteristics $T_A = 25°C$, $V_S = \pm15V$

PARAMETER	LF155A/155, LF255, LF355A/355B		LF355		LF156A/156, LF256/356B		LF356A/356		LF157A/157 LF257/357B		LF357A/357		UNITS
	TYP	MAX	TYP	MAX	TYP	MAX	TYP	MAX	TYP	MAX	TYP	MAX	
Supply Current	2	4	2	4	5	7	5	10	5	7	5	10	mA

AC Electrical Characteristics $T_A = 25°C$, $V_S = \pm15V$

SYMBOL	PARAMETER	CONDITIONS	LF155/255/ 355/355B	LF156/256, LF356B	LF156/256/ 356/356B	LF157/257/ LF357B	LF157/257/ 357/357B	UNITS
			TYP	MIN	TYP	MIN	TYP	
SR	Slew Rate	LF155/6: $A_V = 1$, LF157: $A_V = 5$	5	7.5	12	30	50	V/μs V/μs
GBW	Gain Bandwidth Product		2.5		5		20	MHz
t_s	Settling Time to 0.01%	(Note 7)	4		1.5		1.5	μs
e_n	Equivalent Input Noise Voltage	$R_S = 100\Omega$						
		$f = 100$ Hz	25		15		15	nV/$\sqrt{Hz}$
		$f = 1000$ Hz	20		12		12	nV/$\sqrt{Hz}$
i_n	Equivalent Input Current Noise	$f = 100$ Hz	0.01		0.01		0.01	pA/$\sqrt{Hz}$
		$f = 1000$ Hz	0.01		0.01		0.01	pA/$\sqrt{Hz}$
C_{IN}	Input Capacitance		3		3		3	pF

Fig. 22. Continued.

LF155/LF156/LF157 Series

Notes for Electrical Characteristics

Note 1: The maximum power dissipation for these devices must be derated at elevated temperatures and is dictated by T_{jMAX}, θ_{jA}, and the ambient temperature, T_A. The maximum available power dissipation at any temperature is $P_d = (T_{jMAX} - T_A)/\theta_{jA}$ or the $25°C$ P_{dMAX}, whichever is less.

Note 2: Unless otherwise specified the absolute maximum negative input voltage is equal to the negative power supply voltage.

Note 3: Unless otherwise stated, these test conditions apply:

	LF155A/6A/7A LF155/6/7	LF255/6/7	LF355A/6A/7A	LF355B/6B/7B	LF355/6/7
Supply Voltage, V_S	$\pm15V \leq V_S \leq \pm20V$	$\pm15V \leq V_S \leq \pm20V$	$\pm15V \leq V_S \leq \pm18V$	$\pm15V \leq V_S \leq \pm20V$	$V_S = \pm15V$
T_A	$-55°C \leq T_A \leq +125°C$	$-25°C \leq T_A \leq +85°C$	$0°C \leq T_A \leq +70°C$	$0°C \leq T_A \leq +70°C$	$0°C \leq T_A \leq +70°C$
T_{HIGH}	$+125°C$	$+85°C$	$+70°C$	$+70°C$	$+70°C$

and V_{OS}, I_B and I_{OS} are measured at $V_{CM} = 0$.

Note 4: The Temperature Coefficient of the adjusted input offset voltage changes only a small amount ($0.5\mu V/°C$ typically) for each mV of adjustment from its original unadjusted value. Common-mode rejection and open loop voltage gain are also unaffected by offset adjustment.

Note 5: The input bias currents are junction leakage currents which approximately double for every $10°C$ increase in the junction temperature, T_j. Due to limited production test time, the input bias currents measured are correlated to junction temperature. In normal operation the junction temperature rises above the ambient temperature as a result of internal power dissipation, Pd. $T_j = T_A + \theta_{jA}$ Pd where θ_{jA} is the thermal resistance from junction to ambient. Use of a heat sink is recommended if input bias current is to be kept to a minimum.

Note 6: Supply Voltage Rejection is measured for both supply magnitudes increasing or decreasing simultaneously, in accordance with common practice.

Note 7: Settling time is defined here, for a unity gain inverter connection using 2 kΩ resistors for the LF155/6. It is the time required for the error voltage (the voltage at the inverting input pin on the amplifier) to settle to within 0.01% of its final value from the time a 10V step input is applied to the inverter. For the LF157, $A_V = -5$, the feedback resistor from output to input is 2 kΩ and the output step is 10V (See Settling Time Test Circuit, page 3-9).

Typical DC Performance Characteristics

Curves are for LF155, LF156 and LF157 unless otherwise specified.

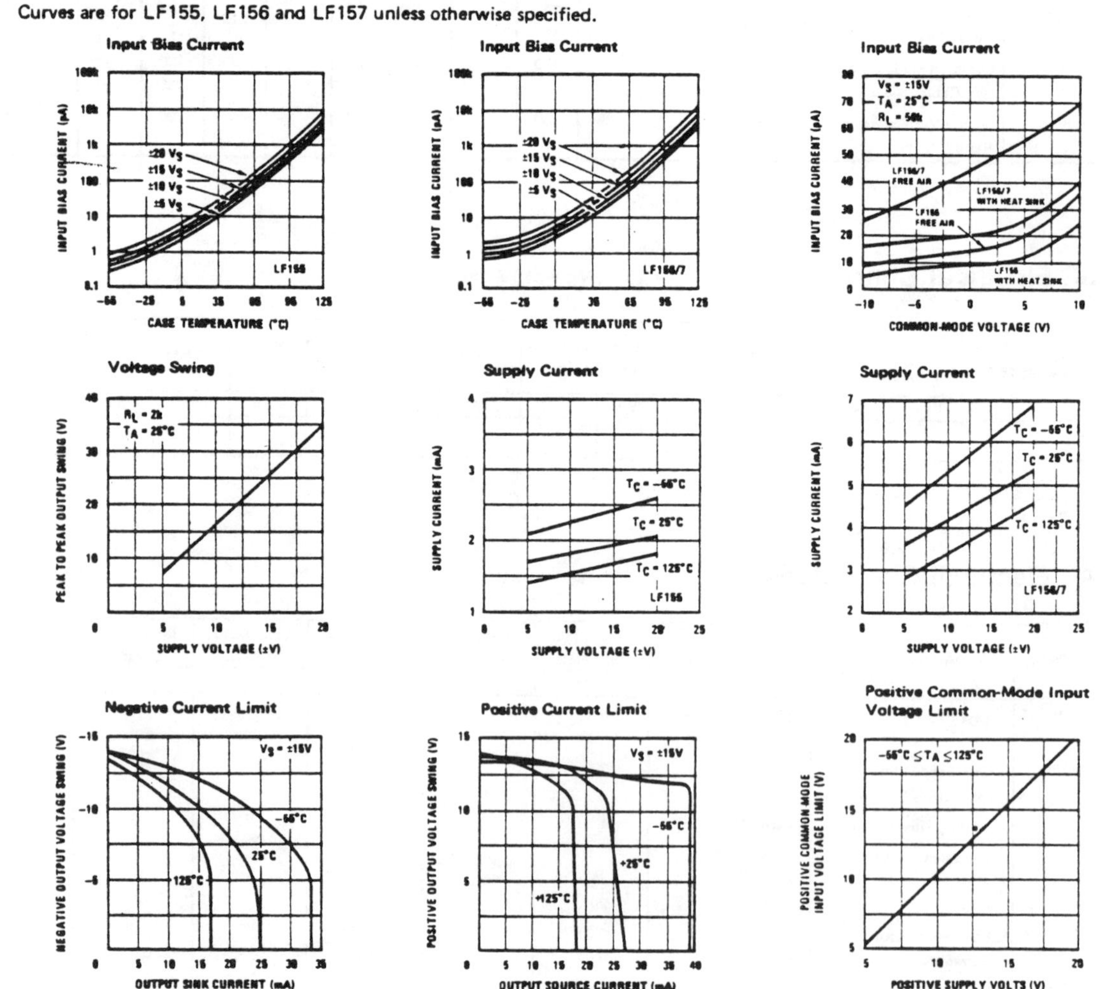

Fig. 22. Continued.

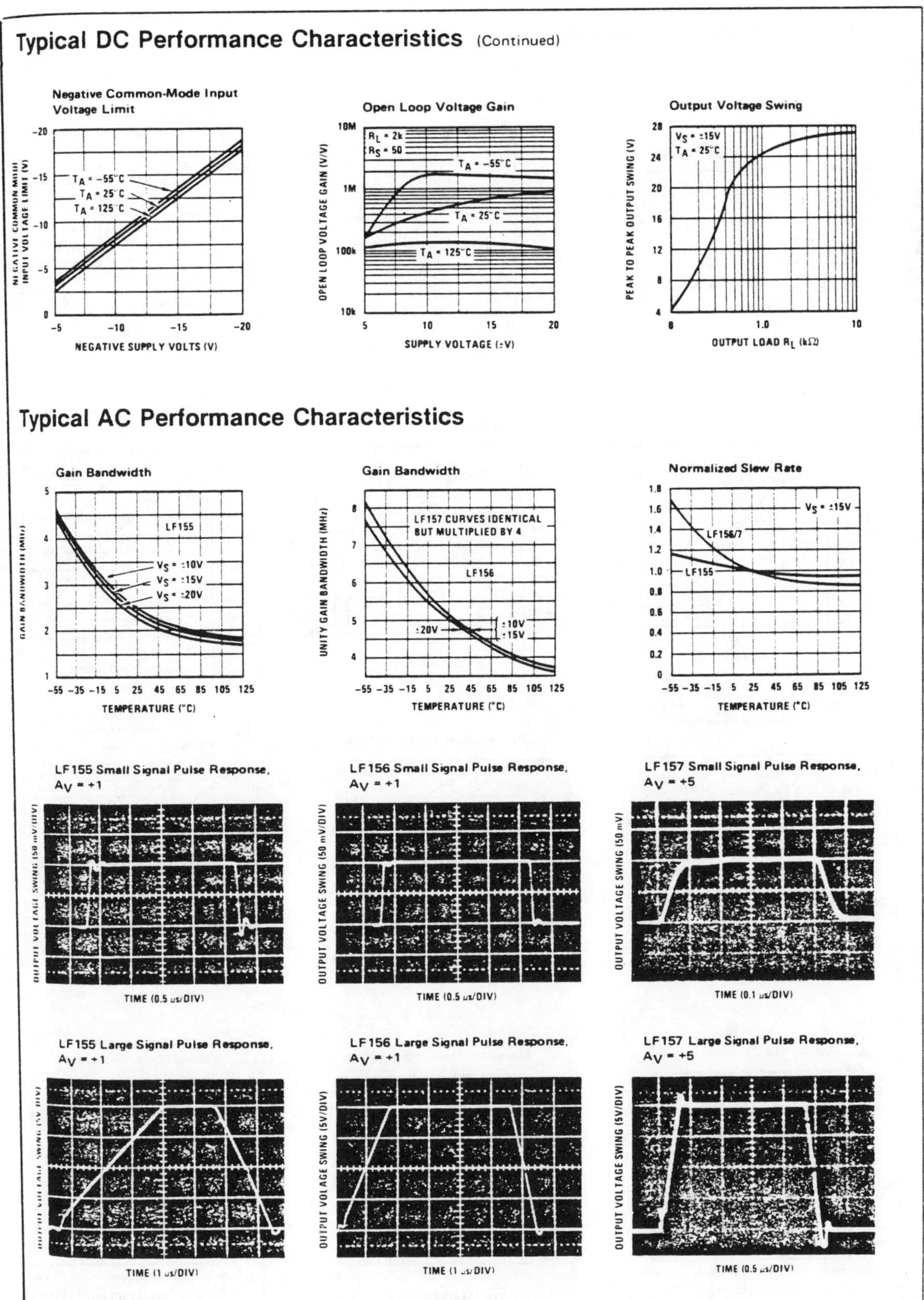

Fig. 22. Continued.

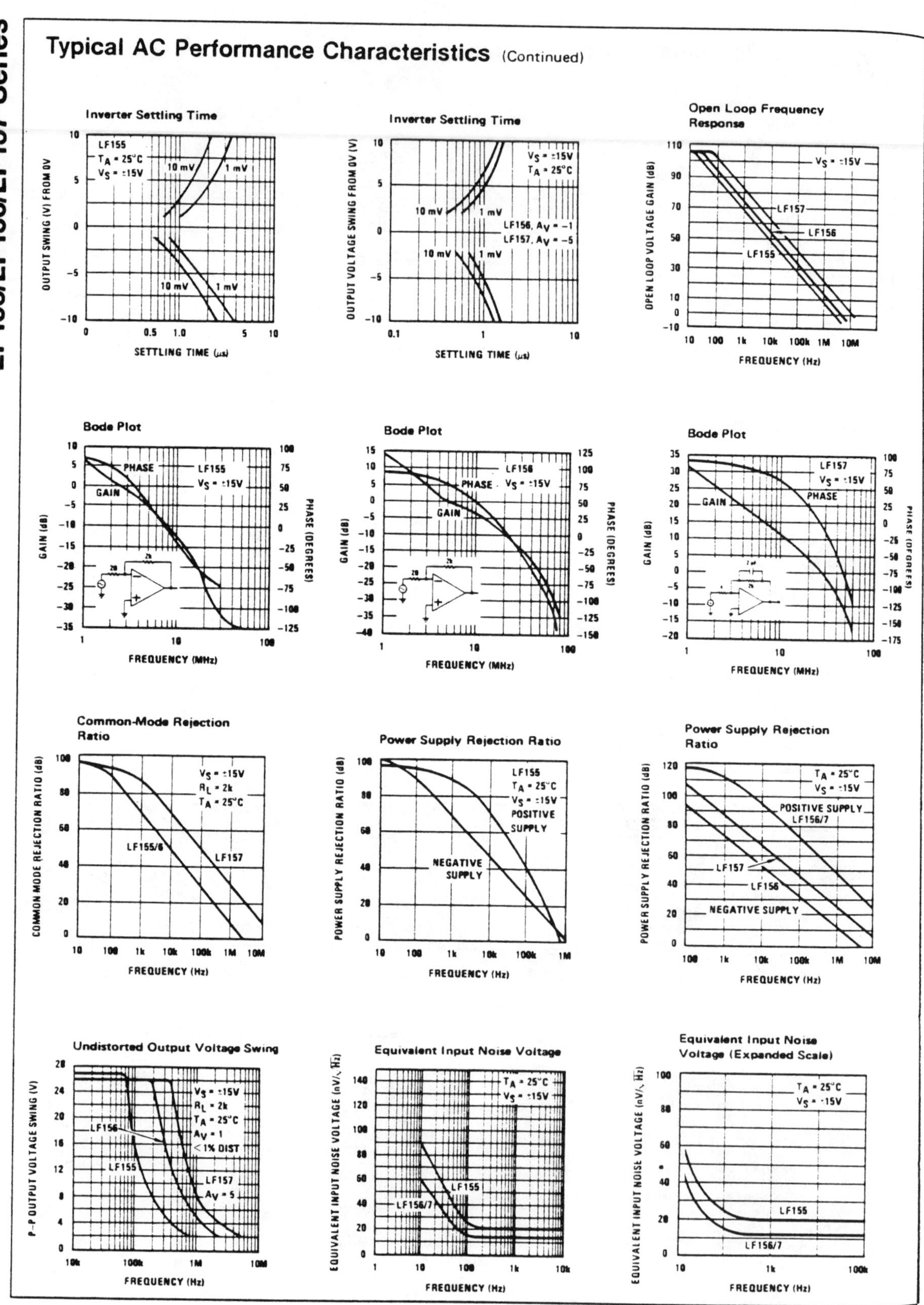

Fig. 22. Continued.

LF155/LF156/LF157 Series

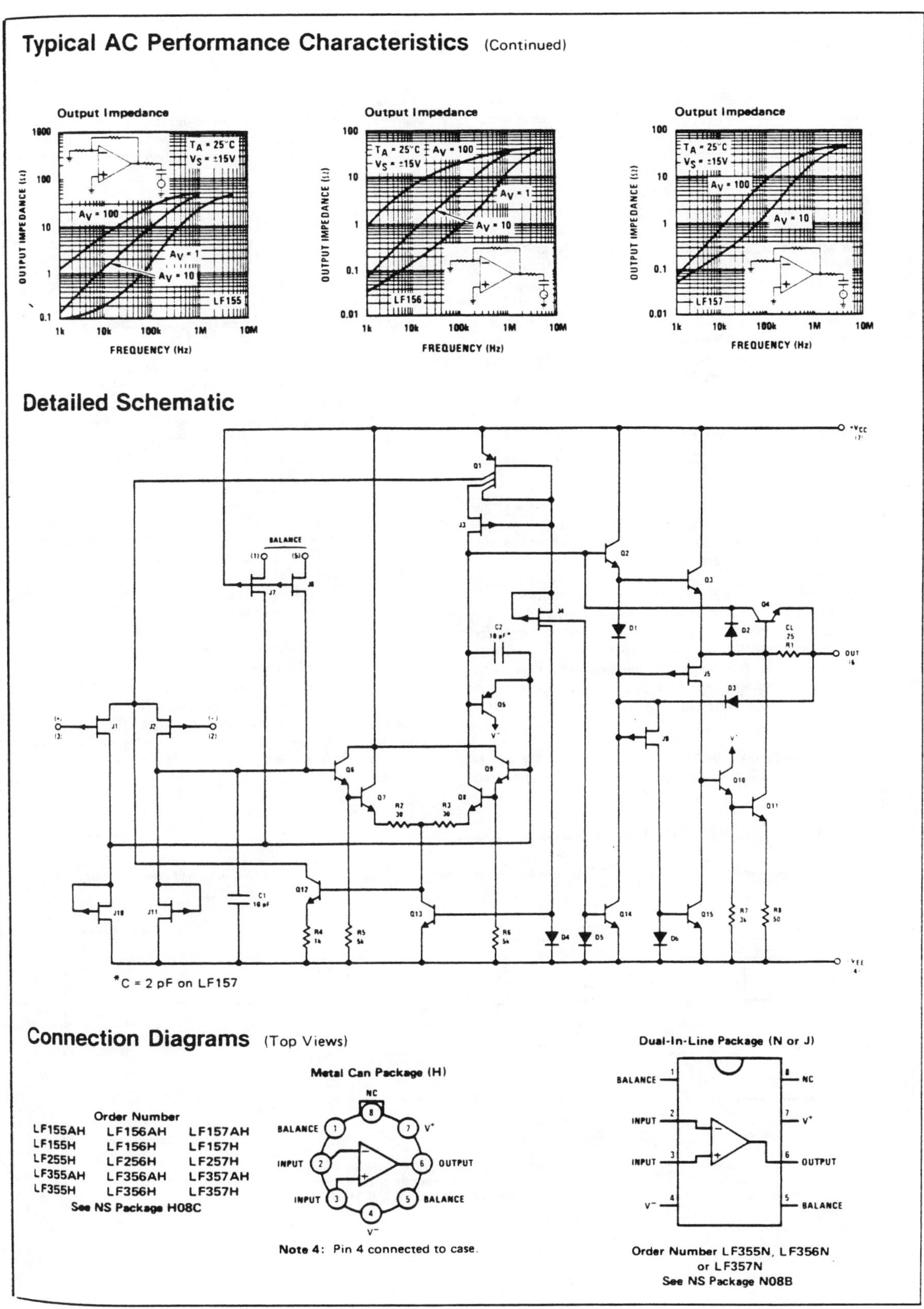

Fig. 22. Continued.

LF155/LF156/LF157 Series

Application Hints

The LF155/6/7 series are op amps with JFET input devices. These JFETs have large reverse breakdown voltages from gate to source and drain eliminating the need for clamps across the inputs. Therefore large differential input voltages can easily be accomodated without a large increase in input current. The maximum differential input voltage is independent of the supply voltages. However, neither of the input voltages should be allowed to exceed the negative supply as this will cause large currents to flow which can result in a destroyed unit.

Exceeding the negative common-mode limit on either input will cause a reversal of the phase to the output and force the amplifier output to the corresponding high or low state. Exceeding the negative common-mode limit on both inputs will force the amplifier output to a high state. In neither case does a latch occur since raising the input back within the common-mode range again puts the input stage and thus the amplifier in a normal operating mode.

Exceeding the positive common-mode limit on a single input will not change the phase of the output however, if both inputs exceed the limit, the output of the amplifier will be forced to a high state.

These amplifiers will operate with the common-mode input voltage equal to the positive supply. In fact, the common-mode voltage can exceed the positive supply by approximately 100 mV independent of supply voltage and over the full operating temperature range. The positive supply can therefore be used as a reference on an input as, for example, in a supply current monitor and/or limiter.

Precautions should be taken to ensure that the power supply for the integrated circuit never becomes reversed in polarity or that the unit is not inadvertently installed backwards in a socket as an unlimited current surge through the resulting forward diode within the IC could cause fusing of the internal conductors and result in a destroyed unit.

Because these amplifiers are JFET rather than MOSFET input op amps they do not require special handling.

All of the bias currents in these amplifiers are set by FET current sources. The drain currents for the amplifiers are therefore essentially independent of supply voltage.

As with most amplifiers, care should be taken with lead dress, component placement and supply decoupling in order to ensure stability. For example, resistors from the output to an input should be placed with the body close to the input to minimize "pickup" and maximize the frequency of the feedback pole by minimizing the capacitance from the input to ground.

A feedback pole is created when the feedback around any amplifier is resistive. The parallel resistance and capacitance from the input of the device (usually the inverting input) to ac ground set the frequency of the pole. In many instances the frequency of this pole is much greater than the expected 3 dB frequency of the closed loop gain and consequently there is negligible effect on stability margin. However, if the feedback pole is less than approximately six times the expected 3 dB frequency a lead capacitor should be placed from the output to the input of the op amp. The value of the added capacitor should be such that the RC time constant of this capacitor and the resistance it parallels is greater than or equal to the original feedback pole time constant.

Typical Circuit Connections

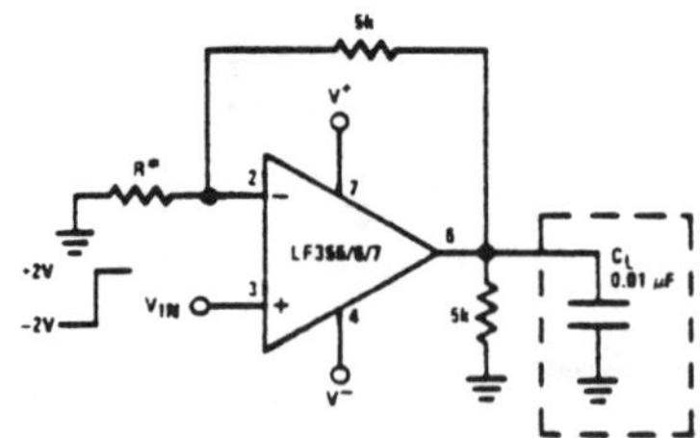

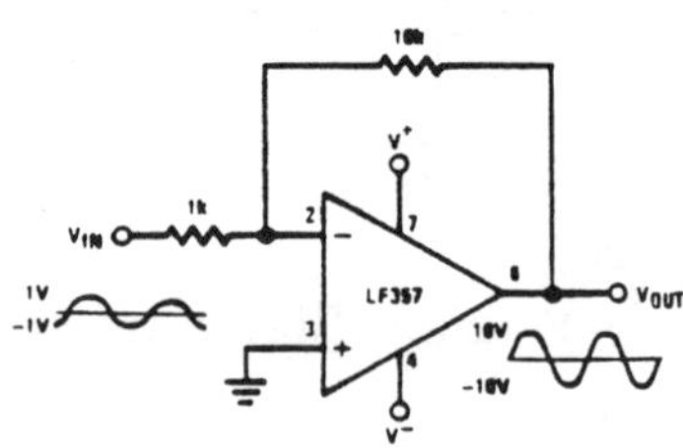

- V_OS is adjusted with a 25k potentiometer
- The potentiometer wiper is connected to V+
- For potentiometers with temperature coefficient of 100 ppm/°C or less the additional drift with adjust is ≈ 0.5 μV/°C/mV of adjustment
- Typical overall drift: 5 μV/°C ±(0.5 μV/°C/mV of adj.)

*LF155/6 R = 5k
LF157   R = 1.25k

Due to a unique output stage design, these amplifiers have the ability to drive large capacitive loads and still maintain stability. $C_{L(MAX)} \cong 0.01$ μF.

Overshoot ≤ 20%

Settling time (t_s) ≅ 5 μs

For distortion ≤ 1% and a 20 Vp-p V_OUT swing, power bandwidth is: 500 kHz.

Fig. 22. Continued.

GENERAL PURPOSE OPERATIONAL AMPLIFIER μA741/ μA741C/SA741C

DESCRIPTION

The μA741 is a high performance operational amplifier with high open loop gain, internal compensation, high common mode range and exceptional temperature stability. The μA741 is short-circuit protected and allows for nulling of offset voltage.

FEATURES

- Internal frequency compensation
- Short circuit protection
- Excellent temperature stability
- High input voltage range

PIN CONFIGURATION

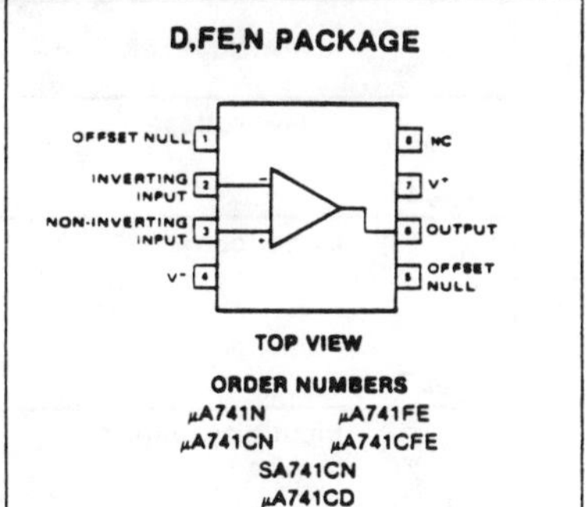

ABSOLUTE MAXIMUM RATINGS

PARAMETER	RATING	UNIT
Supply voltage		
μA741C	±18	V
μA741	±22	V
Internal power dissipation		
N package	500	mW
FE package	1000	mW
Differential input voltage	±30	V
Input voltage[1]	±15	V
Output short-circuit duration	Continuous	
Operating temperature range		
μA741C	0 to +70	°C
SA741C	−40 to +85	°C
μA741	−55 to +125	°C
Storage temperature range	−65 to +150	°C
Lead temperature (soldering 60sec)	300	°C

NOTE

1. For supply voltages less than ±15V, the absolute maximum input voltage is equal to the supply voltage.

EQUIVALENT SCHEMATIC

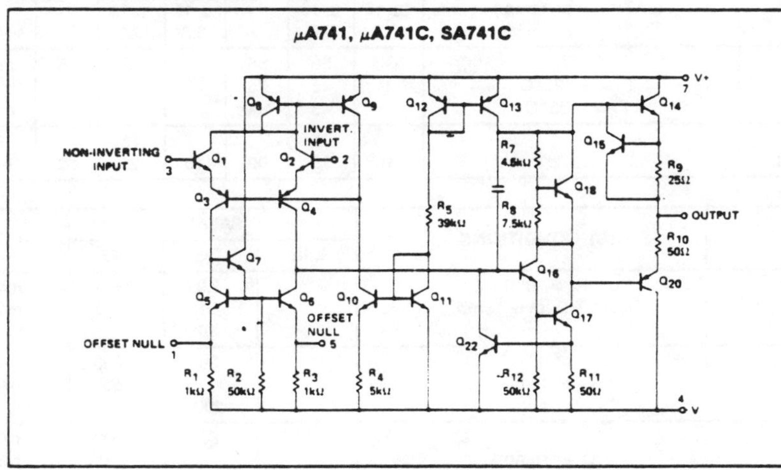

Signetics

Fig. 23. General Purpose Op-Amp (Courtesy of Signetics [7]).

GENERAL PURPOSE OPERATIONAL AMPLIFIER μA741/ μA741C/SA741C

DC ELECTRICAL CHARACTERISTICS $T_A = 25°C$, $V_S = \pm 15V$, unless otherwise specified.

PARAMETER		TEST CONDITIONS	μA741			μA741C			UNIT
			Min	Typ	Max	Min	Typ	Max	
V_{OS}	Offset voltage	$R_S = 10k\Omega$		1.0	5.0		2.0	6.0	mV
		$R_S = 10k\Omega$, over temp.		1.0	6.0			7.5	mV
$\Delta V_{OS}/\Delta T$				10			10		μV/°C
I_{OS}	Offset current			20	200		20	200	nA
		Over temp.						300	nA
		$T_A = +125°C$		7.0	200				nA
		$T_A = -55°C$		20	500				nA
$\Delta I_{OS}/\Delta T$				200			200		pA/°C
I_{BIAS}	Input bias current			80	500		80	500	nA
		Over temp.						800	nA
		$T_A = +125°C$		30	500				nA
		$T_A = -55°C$		300	1500				nA
$\Delta I_B/\Delta T$				1			1		nA/°C
V_{OUT}	Output voltage swing	$R_L = 10k\Omega$	± 12	± 14		± 12	± 14		V
		$R_L = 2k\Omega$, over temp.	± 10	± 13		± 10	± 13		V
A_{VOL}	Large signal voltage gain	$R_L = 2k\Omega$, $V_O = \pm 10V$	50	200		20	200		V/mV
		$R_L = 2k\Omega$, $V_O = \pm 10V$, over temp.	25			15			V/mV
	Offset voltage adjustment range			± 30			± 30		mV
PSRR	Supply voltage rejection ratio	$R_S \leq 10k\Omega$					10	150	μV/V
		$R_S \leq 10k$, over temp.		10	150				μV/V
CMRR	Common mode rejection ratio								dB
		Over temp.	70	90					dB
I_{CC}	Supply current			1.4	2.8		1.4	2.8	mA
		$T_A = +125°C$		1.5	2.5				mA
		$T_A = -55°C$		2.0	3.3				mA
V_{IN}	Input voltage range	(μA741, over temp.)	± 12	± 13		± 12	± 13		V
R_{IN}	Input resistance		0.3	2.0		0.3	2.0		MΩ
P_d	Power consumption			50	85		50	85	mW
		$T_A = +125°C$		45	75				mW
		$T_A = -55°C$		45	100				mW
R_{OUT}	Output resistance			75			75		Ω
I_{SC}	Output short-circuit current		10	25	60	10	25	60	mA

PARAMETER		TEST CONDITIONS	SA741C			UNIT
			Min	Typ	Max	
V_{OS}	Offset voltage	$R_S = 10k\Omega$		2.0	6.0	mV
		$R_S = 10k\Omega$, over temp.			7.5	mV
$\Delta V_{OS}/\Delta T$				10		μV/°C
I_{OS}	Offset current			20	200	nA
		Over temp.			500	nA
$\Delta I_{OS}/\Delta T$				200		pA/°C
I_{BIAS}	Input bias current			80	500	nA
		Over temp.			1500	nA
$\Delta I_B/\Delta T$				1		nA/°C
V_{OUT}	Output voltage swing	$R_L = 10k\Omega$	± 12	± 14		V
		$R_L = 2k\Omega$, over temp.	± 10	± 13		V
A_{VOL}	Large signal voltage gain	$R_L = 2k\Omega$, $V_O = \pm 10V$	20	200		V/mV
		$R_L = 2k\Omega$, $V_O = \pm 10V$, over temp.	15			V/mV
	Offset voltage adjustment range			± 30		mV

Signetics

Fig. 23. Continued.

GENERAL PURPOSE OPERATIONAL AMPLIFIER μA741/μA741C/SA741C

DC ELECTRICAL CHARACTERISTICS (Cont'd) $T_A = 25\,°C$, $V_S = \pm 15V$, unless otherwise specified.

PARAMETER		TEST CONDITIONS	SA741C			UNIT
			Min	Typ	Max	
PSRR	Supply voltage rejection ratio	$R_S \leq 10k\Omega$		10	150	μV/V
CMRR	Common mode rejection ratio					dB
V_{IN}	Input voltage range Input resistance	(μA741, over temp.)	± 12 0.3	± 13 2.0		V MΩ
P_d	Power consumption			50	85	mW
R_{OUT}	Output resistance			75		Ω
I_{SC}	Output short-circuit current			25		mA

AC ELECTRICAL CHARACTERISTICS $T_A = 25°C$, $V_S = \pm15V$, unless otherwise specified.

PARAMETER	TEST CONDITIONS	μA741, μA741C			UNIT
		Min	Typ	Max	
Parallel input resistance	Open loop, f = 20Hz				MΩ
Parallel input capacitance	Open loop, f = 20Hz		1.4		pF
Unity gain crossover frequency	Open loop		1.0		MHz
Transient response unity gain	$V_{IN} = 20mV$, $R_L = 2k\Omega$, $C_L \leq 100pf$				
Rise time			0.3		μs
Overshoot			5.0		%
Slew rate	$C \leq 100pf$, $R_L \geq 2k$, $V_{IN} = \pm10V$		0.5		V/μs

TYPICAL PERFORMANCE CHARACTERISTICS

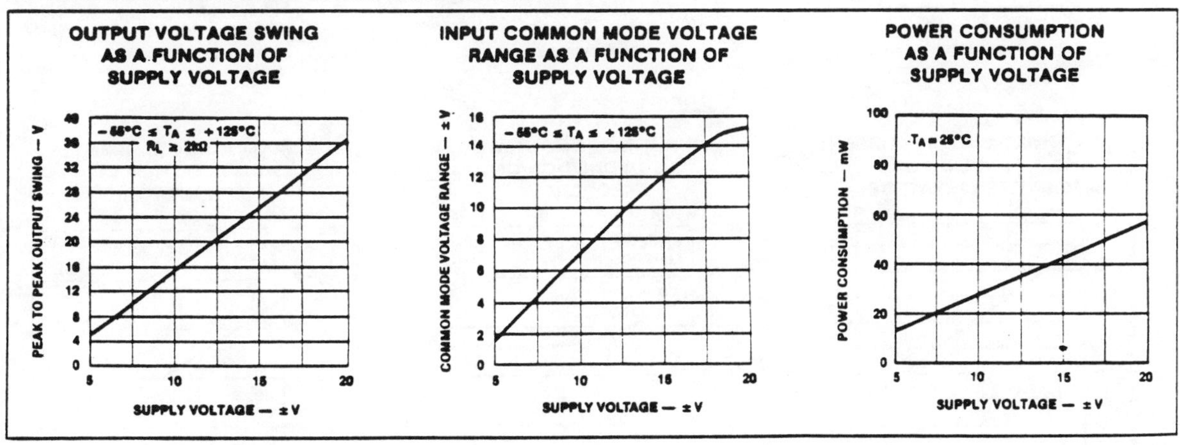

Signetics

Fig. 23. Continued.

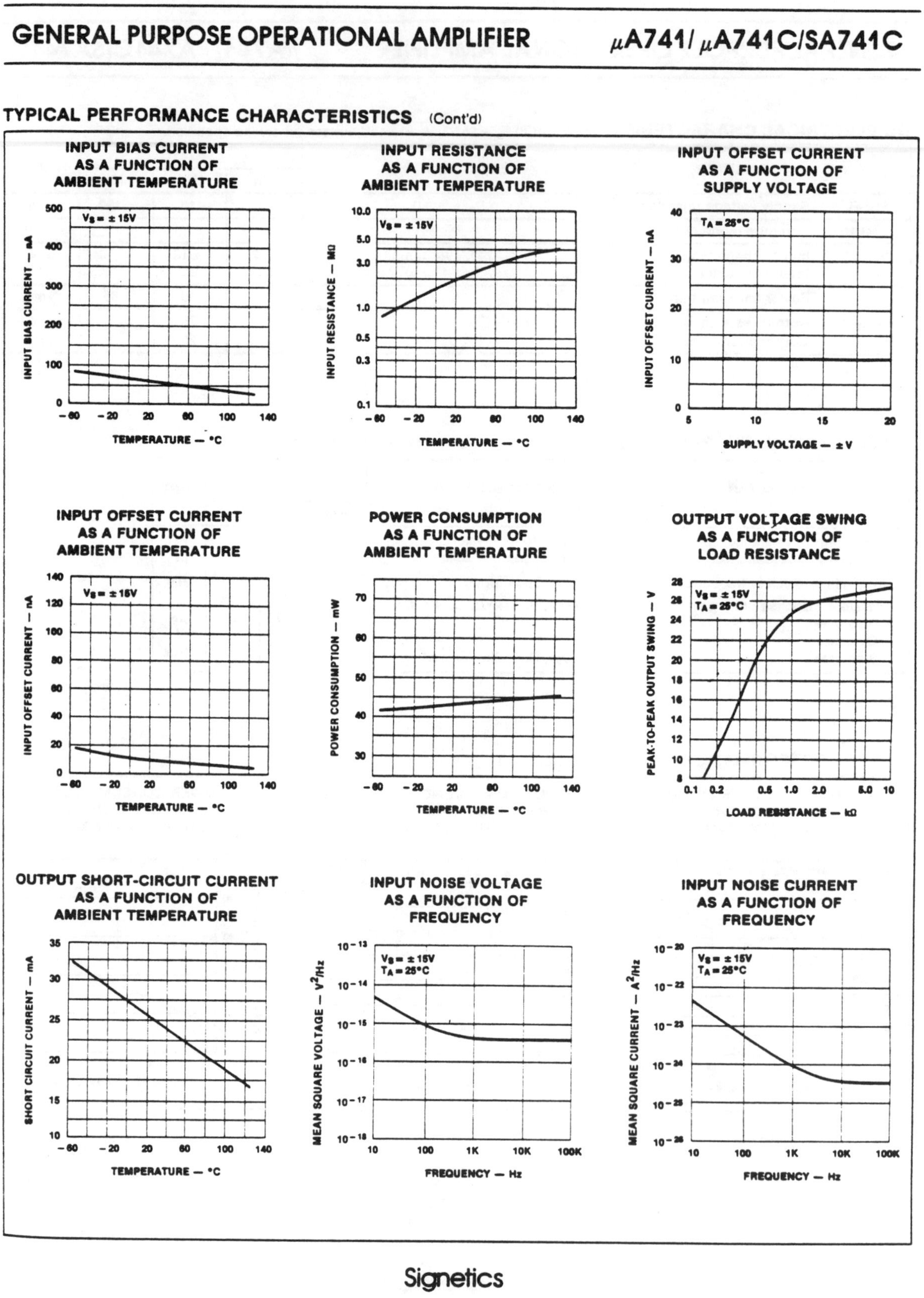

GENERAL PURPOSE OPERATIONAL AMPLIFIER   μA741/ μA741C/SA741C

TYPICAL PERFORMANCE CHARACTERISTICS (Cont'd)

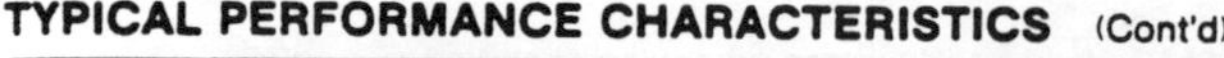

Signetics

Fig. 23. Continued.

GENERAL PURPOSE OPERATIONAL AMPLIFIER $\quad \mu A741/\mu A741C/SA741C$

TYPICAL PERFORMANCE CHARACTERISTICS (Cont'd)

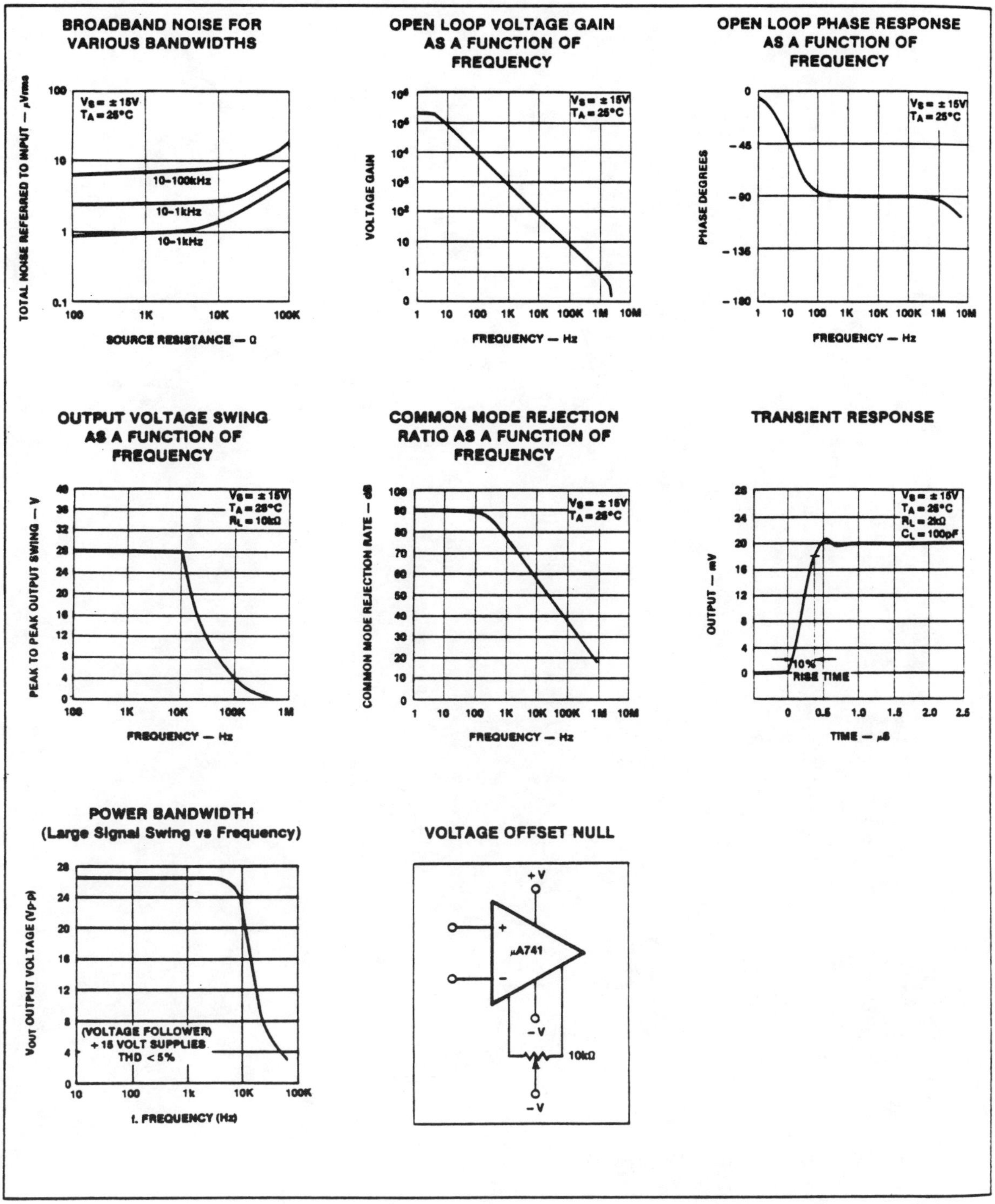

Signetics

Fig. 23. Continued.

WORKSHEET

WORKSHEET

EXPERIMENT IV-1

PRACTICAL OP-AMP PARAMETERS

EQUIPMENT:

EEP
OP-AMPs (3)
CAPACITORS (0.01 μF/2, 470 pF)
RESISTORS (100 Ω, 1 kΩ/2, 2 kΩ, 10 kΩ, 100 kΩ/2, 1 MΩ, 10 MΩ/2)

Table of Contents   Page No.

1. PURPOSE

The purpose of this experiment is to acquaint the student with a few techniques of measuring the practical parameters of an Op-Amp and to compare these values with those given by the manufacturers.

2. INTRODUCTION

All the integrated circuits in today's market are different from the ideal device introduced in Section IV-3. Nevertheless at room temperature the typical characteristics of these IC's made by different vendors are almost the same. In the following, we directly use various circuits to measure these parameters. We also recall that these active devices need dual DC supply voltage. Extreme care must be exercised to avoid connecting this supply voltage to this device with the wrong or opposite polarity. We also recall that each Op-Amp has at least eight pins that are numbered and each number is described by the manufacturer. As shown in Fig. 1, for a typical Op-Amp, we must put this device in the center of the bread board and push it gently downward to fit in the holes. Note that, in this figure the distance between the pins are exaggerated for the sake of illustration.

To avoid wasting too much time always check your Op-Amp separately to make sure it is not defective before using it in a circuit. This is easily done by applying a low level signal and monitoring the output.

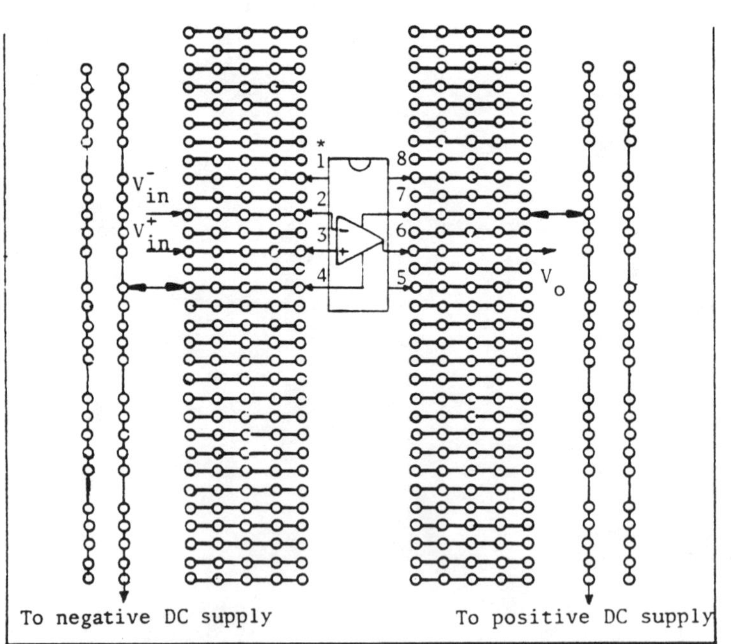

*: The distance between the pins are exaggerated for the sake of illustration.

Fig. 1. Schematic Diagram of A Typical Op-Amp.

3. LABORATORY PROCEDURE

3.1. PRE-LABORATORY WORK

Carefully review the materials given earlier in this part as well as the specific data sheets regarding your own Op-Amp.

3.2. LABORATORY WORK

3.2.1. *DC (Low Frequency) Open-Loop Gain*

Construct the circuit in Fig. 2. For this circuit *use* the DC supply voltage that is less than 15 volts in amplitude. We can easily show that

$$G_{ol}^{v} = -\frac{V_o}{V}\,\frac{R_1 + R_2}{R_2}$$

if the input frequency is low - say about 20 Hz. (Why?). *Determine* the value of this parameter for your Op-Amp and *save* this circuit for the next experiment.

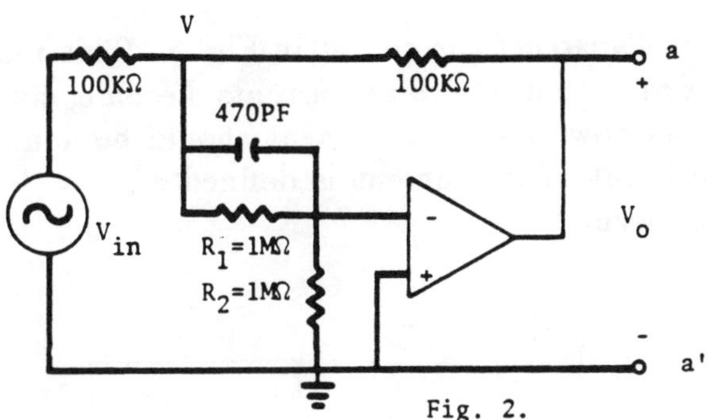

Fig. 2.

3.2.2. *Output Impedance*

Turn off the power supply and function generator before making the following circuit changes. *Put* an $R_L = 200\,\Omega$ between nodes a and a' of the circuit of Fig. 2. *Increase* the input signal frequency to about 25 kHz and *measure* $V_{aa'}$ (with R_L). The output impedance is

$$R_o = \left[\,\frac{V_o(\text{without } R_L)}{V_{aa'}\,(\text{with } R_L)} - 1\,\right] R_L\,. \quad (\text{Why?})$$

3.2.3. *The Input Offset Voltage*

Construct the circuit of Fig. 3. Here we can show that $V_{io} = V_o/1001$. *Measure* V_o first, then compute V_{io}.

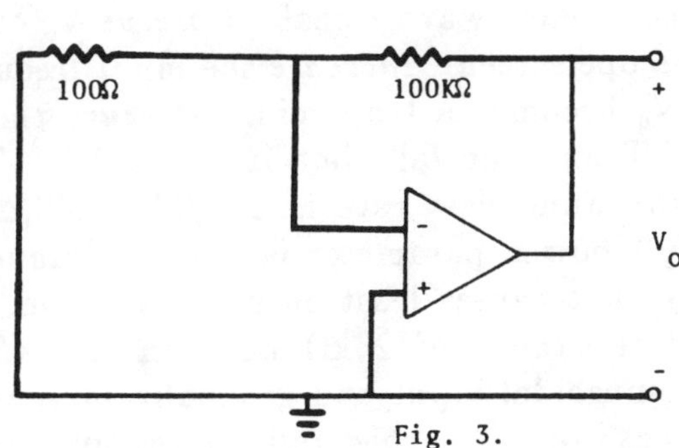

Fig. 3.

3.2.4. The Output Offset Voltage

Construct the circuit in Fig. 4. *Measure* V_{oo} . This is the output offset voltage. In most Op-Amps this parameter can be reduced to zero by utilizing an external potentiometer to compensate for the undesirable differential imbalance.

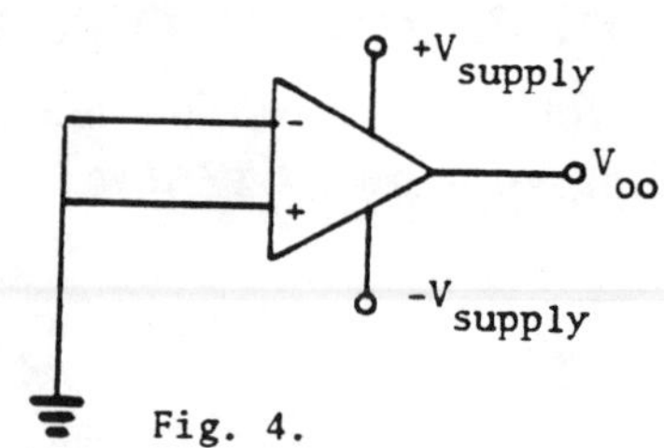

Fig. 4.

3.2.5. Input Bias Currents and Input Offset Current

Construct the circuit in Fig. 5. This measurement depends on R_1 and R_2 to be sufficiently large in order that the bias currents become much larger than the offset current. The table in Fig. 5 shows how this measurement should be done. The bias current I_B is defined by $I_B = (I_{B1} + I_{B2})/2$ and the input offset current is defined as $I_{1o} = |I_{B1} - I_{B2}|$. What is the purpose of the two capacitors in this circuit?

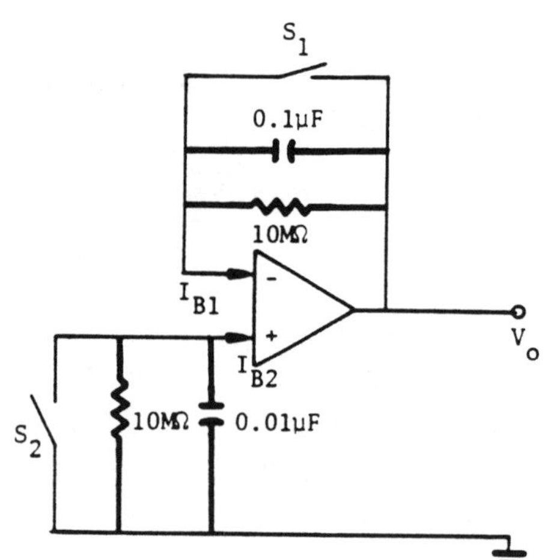

S_1	S_2	V_o
open	open	$10M \times I_{io}$
open	close	$10M \times I_{B1}$
close	open	$10M \times I_{B2}$

Fig. 5.

3.2.6. Slew Rate Measurement

Construct the circuit in Fig. 6. *Let* V_{in} be a fast rise square wave signal. *Observe* V_o on the oscilloscope screen. *Increase* the input frequency until V_o becomes a trapezoid. *Measure* the rise time ΔT and the fall time $\Delta T'$ for V_o . Then the maximum slew rate is $\Delta V / Max(\Delta T, \Delta T')$. The full power parameter is the maximum frequency that causes least output distortion. *Put* a load resistance of 2 kΩ between aa' . Then *use* a sinusoidal input and gradually *increase* its frequency to find this full power parameter.

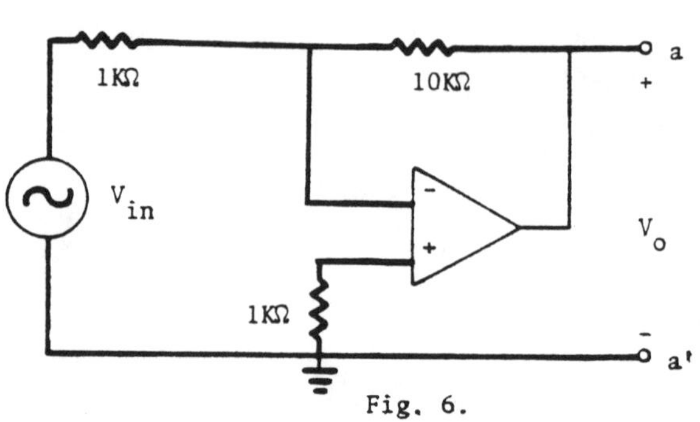

Fig. 6.

3.2.7.   Gain-Bandwidth Product

Construct the circuit of Fig. 7. The frequency at which $|V_o/V_{in}| = 1$ can be taken to be the Op-Amp bandwidth. *Measure* this parameter for your Op-Amp.

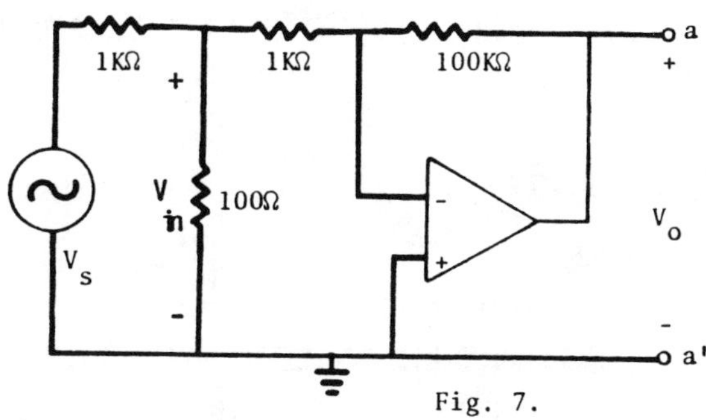

Fig. 7.

4. PROBLEM SET

1. The interesting circuit in Fig. 8 is called a Gyrator [2]. Find its input impedance in terms of the element Z. Describe your conclusion about the function of this network.

2. Find the transfer function of the circuit in Fig. 9, using the theory that has been developed for interconnecting two networks in Section III-2.2. This circuit will be revisited in EXP. IV-4.

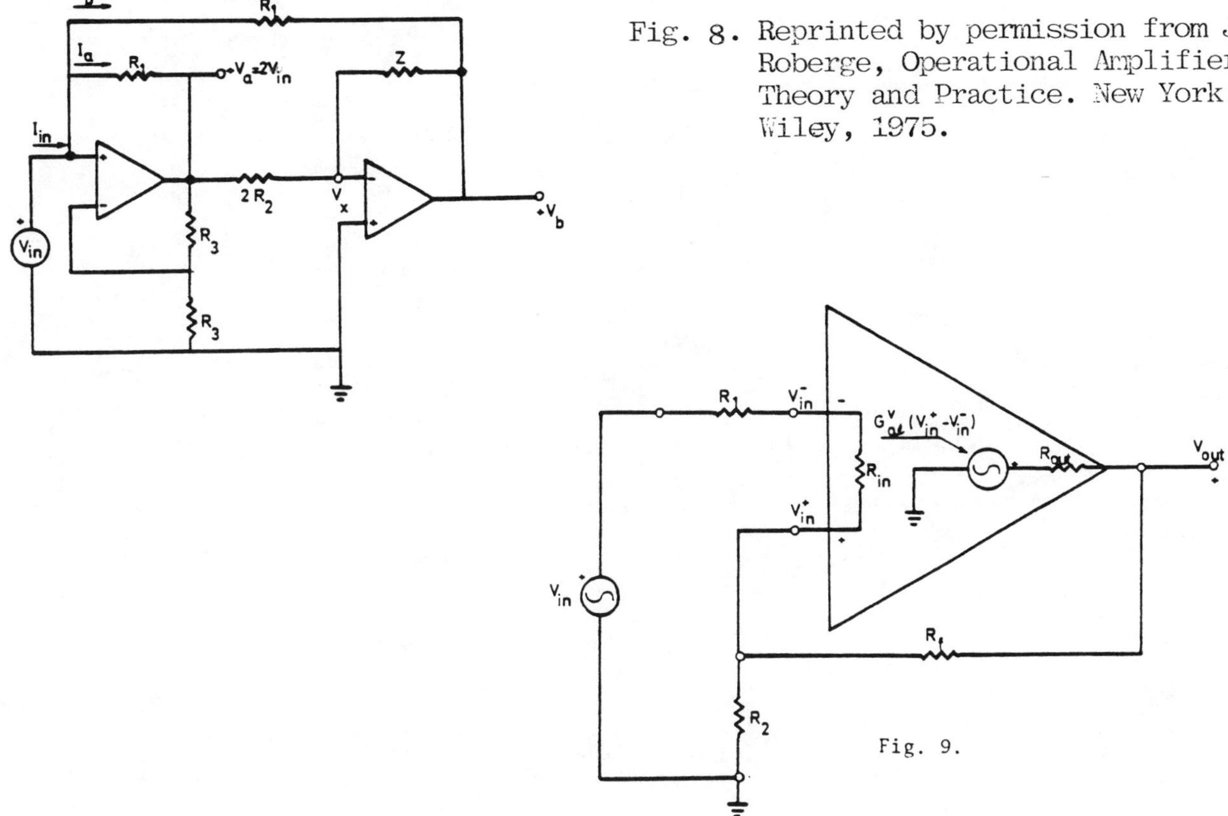

Fig. 8. Reprinted by permission from J.K. Roberge, Operational Amplifiers: Theory and Practice. New York: Wiley, 1975.

Fig. 9.

5. FOR YOUR REPORT

Please submit your data with a comparison with those given by the manufacturer. Also submit the solutions of the above two problems. The report must be in our standard form.

EXPERIMENT IV-2

INSTRUMENTATION AMPLIFIERS AND TWO BASIC
NON-LINEAR OP-AMP CIRCUITS

EQUIPMENT:

>EEP
>OP-AMPs (3)
>POTS (20 kΩ, 100 kΩ)
>RESISTORS (100 Ω/2, 10 kΩ/2, 100 kΩ/2)

Table of Contents   Page No.

1. PURPOSE

The purpose of this experiment is to acquaint the student with a way of measuring the common-mode rejection ratio (CMRR) of an Op-Amp and a more general amplifier configuration called an instrumentation amplifier. We will also look into some nonlinear behaviors of the Op-Amp.

2. INTRODUCTION

In this section we first review some basic applications of Op-Amps. We have studied the meaning and definition of CMRR in Section IV-4. This parameter is a figure of merit of a differential amplifier. It gives an indication of how to classify an amplifier. Consider the amplifier circuits of Figs. 1 and 2. Both are intended to respond to the differential input voltages. We know that in each of these cases we have a differential input shown by V_{dm} and the output voltage is

$$V_o = G_{dm} V_{dm} + G_{cm} V_{cm} ,$$

$$CMRR = G_{dm} / G_{cm} .$$

This relation is well described by Fig. 3.

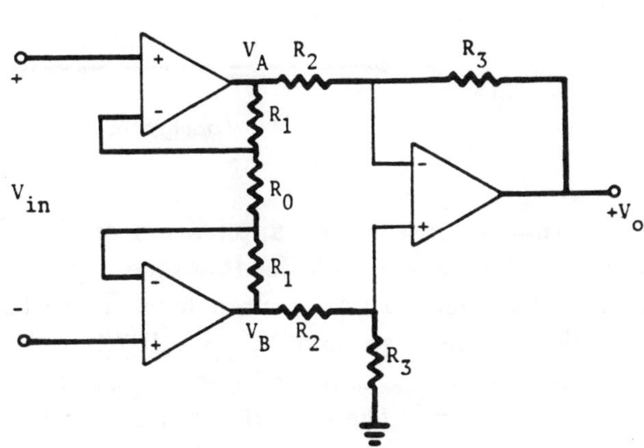

Fig. 1. Instrumentation Amplifier.

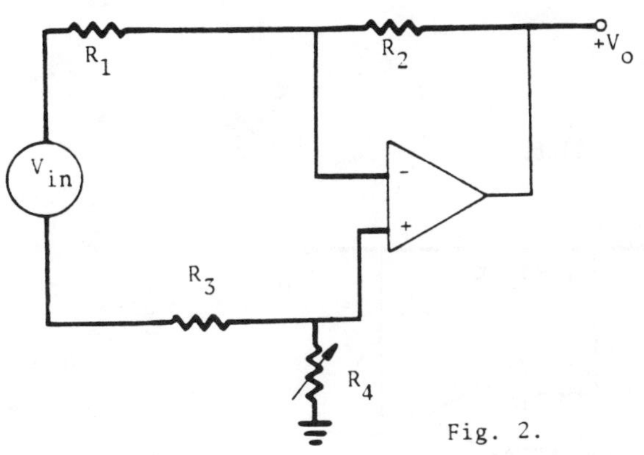

Fig. 2.

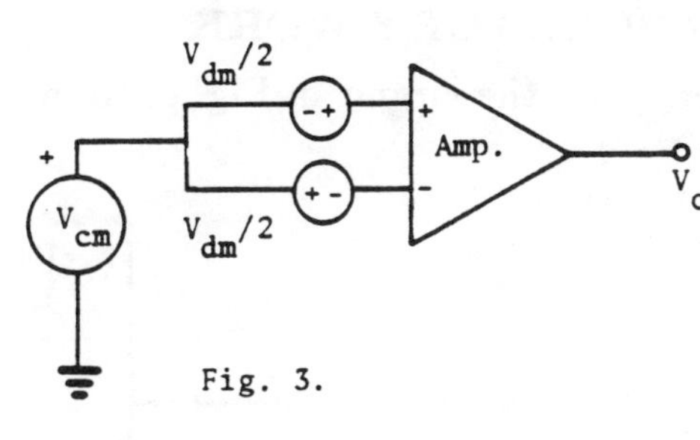

Fig. 3.

There are several useful applications of Op-Amps in which these are used in their non-linear modes. One such application is called *comparator*, wherein the Op-Amp is acted as a signal comparator. The circuit of Fig. 4 shows the principle of this mechanism. In this circuit if $|G_{ol}^v (V_{in}^- - V_{in}^+)|$ exceeds the supply voltage then V_o saturates, without any feedback, at the plus or minus DC supply voltage level. This is the basis of the comparator that gives us a way of saying which of the two voltage levels in a given application is more positive than the other. The input difference at which the output changes from plus to minus supply voltage is called *switching point*. In a comparator this switching point is on the order of millivolts or less. We will further study comparator circuits in EXP. VI-2.

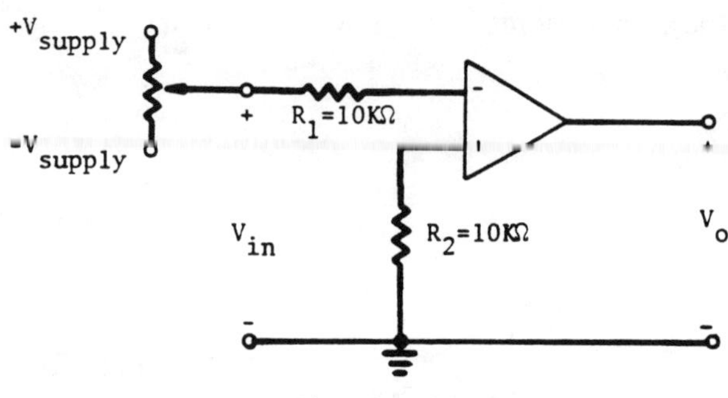

Fig. 4. Comparator.

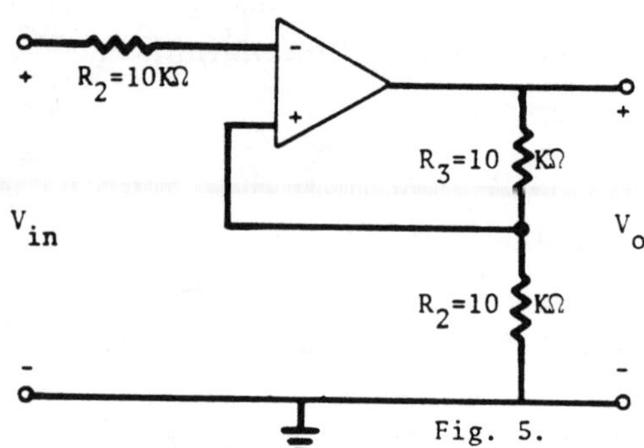

Fig. 5.

Another interesting application of this device is in a scheme called a *Schmitt-Trigger* shown in Fig. 5. This is a simple feedback configuration. In this case the distance between switching points can be widened to any desired level up to the supplied voltage. The way that this circuit works is as follows. Here we have $V_{in}^+ = V_o R_2/(R_2 + R_3)$. Let $V_o = V_{supply}$, then if $V_{in} > V_{supply} R_2/(R_2 + R_3)$, V_o switches to $-V_{supply}$ and for $V_{in} < -V_{supply} R_2/(R_2 + R_3)$, the V_o switches to $+V_{supply}$. When V_{in} is between these two bounds the output remains in its previous state which is either V_{supply} or $-V_{supply}$.

3. LABORATORY PROCEDURE

3.1. PRE-LABORATORY WORK

Derive all the necessary equations that are used in this experiment.

3.2. LABORATORY WORK

Construct the circuit in Fig. 6, to measure the CMRR.

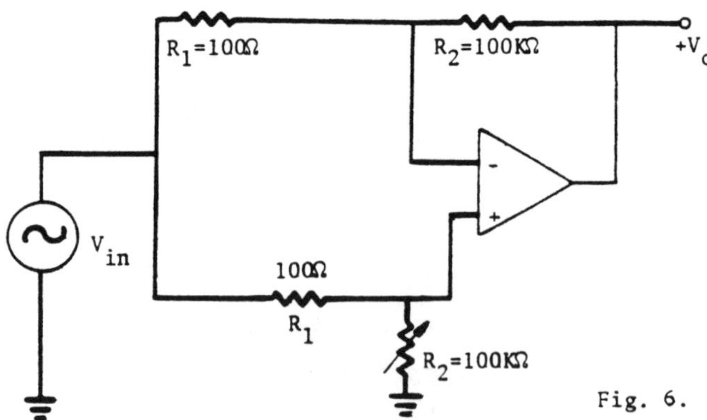

Fig. 6.

Here we can show that $\quad \text{CMRR} = \dfrac{R_2}{R_1} \dfrac{V_{in}}{V_o}$. (How?)

What is the expression for the CMRR of the circuit in Fig. 1?

Construct the circuit in Fig. 1, and measure its CMRR. (Is it the same as expected theoretically?) Then *compare* these two designs in terms of this particular parameter. Which design is superior? *Use* for R_2 in the positive input a 100 kΩ pot to compensate for various tolerances of the resistors that are being used in these circuits.

Construct the circuit in Fig. 4, with 20 kΩ potentiometer. *Use* the DMM to monitor V_{in} and an oscilloscope to monitor V_o or *put* V_{in} in CH1 and V_o in CH2 of the oscilloscope. *Find* the value of v_{in} which results in V_o that reaches to plus or minus supply voltage. Instead of the potentiometer we can use a signal generator without DC offset for $v_{in}(t)$. Then *increase* the DC offset of this V_{in} and *observe* its effects on $v_o(t)$.

Construct the circuit in Fig. 7. First *find* the switching point, i.e., values of V_{in} for which V_o is plus or minus supplied voltage. Then *apply* a variable amplitude sinusoidal at point A and *study* its effect at the output. *Start* with zero DC offset. Then *increase* this DC offset and *observe* its effects on V_o.

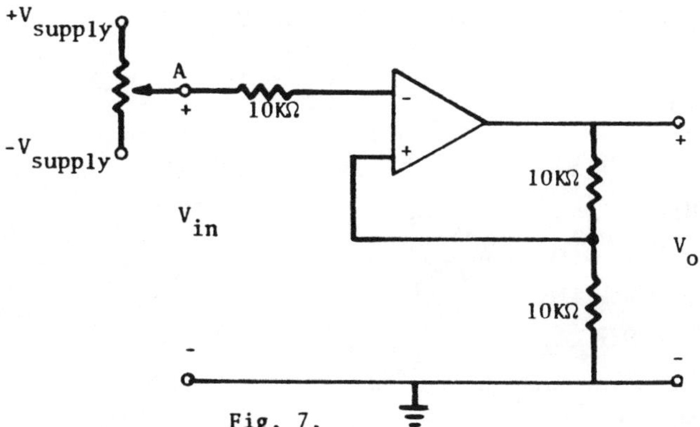

Fig. 7.

4. PROBLEM SET

Find the CMRR for the circuit in Fig. 8, [2].

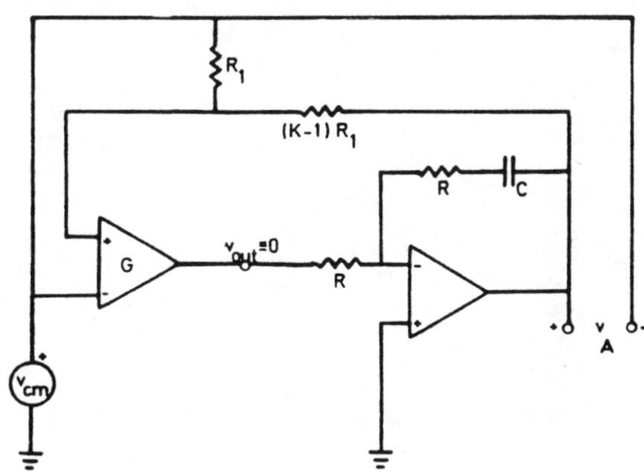

Fig. 8. Reprinted by permission from J.K. Roberge, Operational Amplifiers: Theory and Practice. New York: Wiley, 1975.

5. FOR YOUR REPORT

Submit your data and the solution of the above problem set in our standard report form. Discuss any conclusions you have made regarding the various amplifier circuits you have studied in this experiment.

EXPERIMENT IV-3

NOISE MEASUREMENT

EQUIPMENT:

EEP
X-Y RECORDER
OP-AMPs (3)
COMPARATOR (CF., FIG. VI-2-9)
CAPACITORS (0.1 μF/2, 470 pF, 1 μF/2)
POT (10 kΩ)
GR-DECADE RESISTANCE BOX
RESISTORS (100 Ω, 4.7 kΩ, 6.8 kΩ/2, 10 kΩ, 100 kΩ/2, 1 MΩ/2, 4.7 MΩ)

Table of Contents   Page No.

1. PURPOSE

The purpose of this experiment is to measure the internal noise (voltage or current) source of an Op-Amp.

2. INTRODUCTION

This experiment is developed in collaboration with a former colleague Dr. J.A. Svoboda.

Using the Op-Amp to amplify small signals may result in inaccurate measurements due to amplification of the internal noise of the Op-Amp. Fig. 1 shows an internal model of the Op-Amp that will be used for modeling this noise source. If this model is used in the circuit of Fig. 2, then we can show that the RMS of the noise source is

$$V_{no} = \frac{R_1 + R_3}{R_1} \sqrt{V_n^2 + (R_2 I_{n2})^2 + \left[\frac{R_1 R_3}{R_1 + R_3} I_{n1}\right]^2} \ .$$

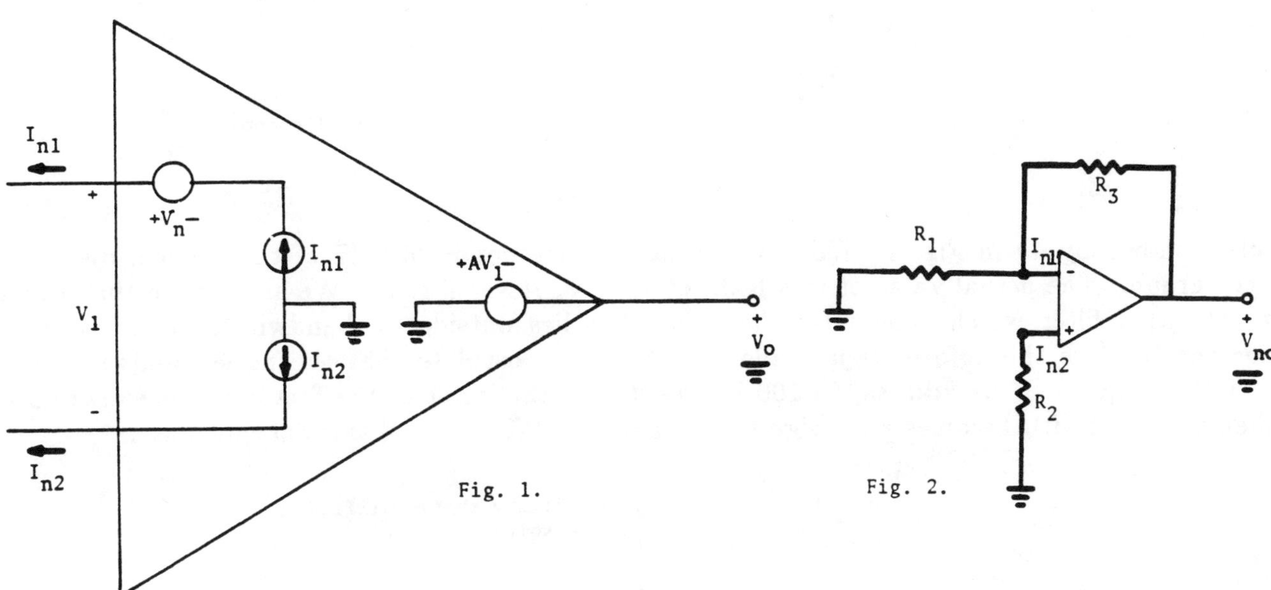

In this experiment we must first find the parameters V_n and I_{ni} for i=1,2, then generate a cumulative distribution function for the noise source. From this cdf plot we can find the mean and the standard deviation of the noise source. These two parameters can be related to the actual parameters of this noise source as shown later. The reason for generation of this plot is as follows. Consider the circuit in Fig. 3. This circuit works as follows. The noise source that is generated in the first Op-Amp is amplified by a large gain in the first two-stage. This amplified signal then enters into the comparator that is being compared with a known DC level. The output of the comparator is then integrated as a function of input random noise. The result is a cumulative distribution function cdf that expresses the cumulative probability of the random noise be the DC level that is set by the 10 kΩ pot at each point. A typical of this cdf is shown in Fig. 4. It is well known that the RMS (times gain) of the noise source is

$$V_{no} = \sqrt{m^2 + \sigma^2} \ .$$

Here m is the mean (horizontal distance of the center from cdf plot), and σ is the standard deviation

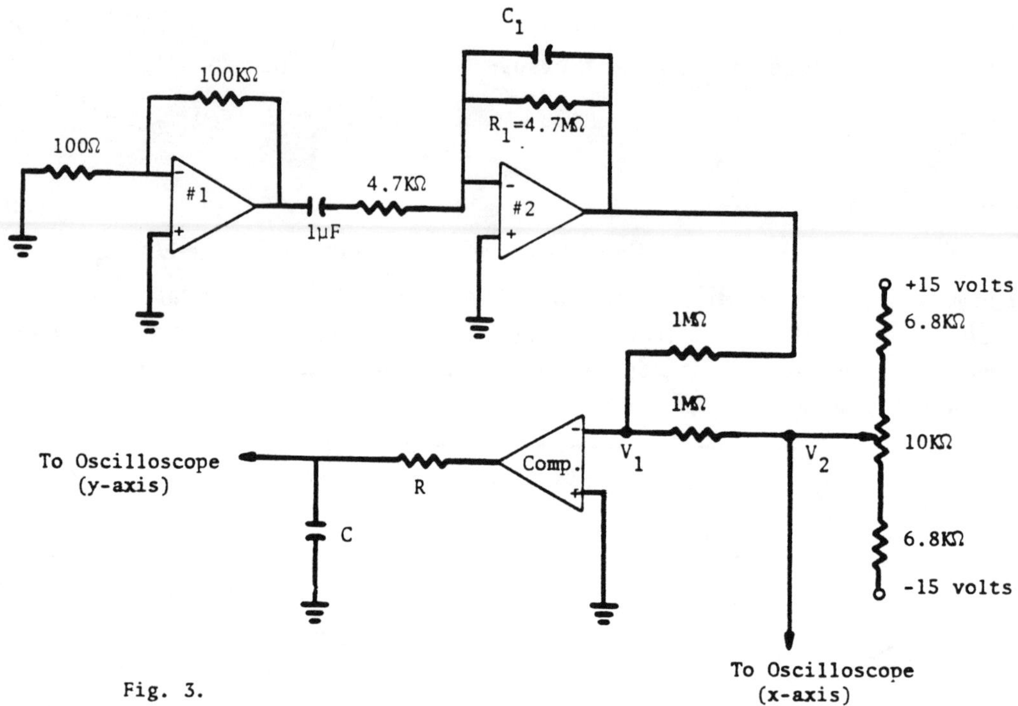

Fig. 3.

which is shown by 4σ in Fig. 4. This 4σ parameter corresponds to 95% of the vertical distance on the cdf graph. The actual value of this RMS is $V_n = V_{no}/$(overall gain). We also notice that this circuit acts as a filter which attenuates the noise that lies outside its bandwidth. It is, of course, important to have this information available. We may postulate that the noise bandwidth is the min $(1/RC, 1/R_1C_1)$. If we add, say, a 100 kΩ resistor at the input of the first stage as shown in Fig. 5, then the effect of I_n becomes more significant than V_n. We may call this current noise.

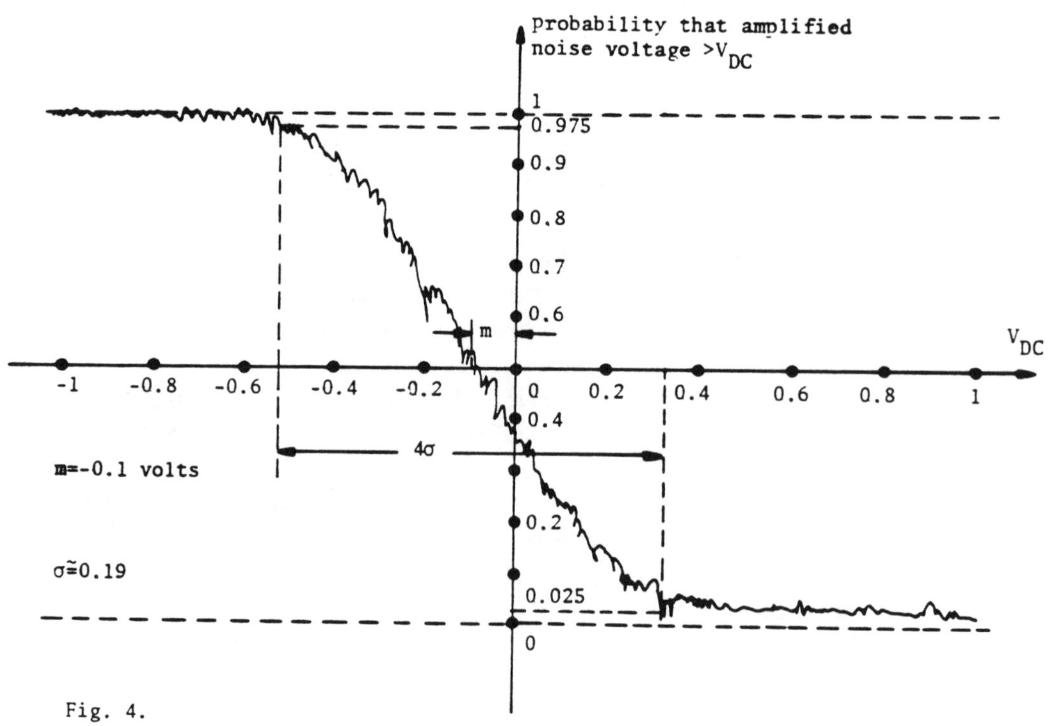

Fig. 4.

3. LABORATORY PROCEDURE

3.1. PRE-LABORATORY WORK

Carefully study the Section 0-9.3 on basics of probability theory.

3.2. LABORATORY WORK

Construct the circuit in Fig. 5. *Choose* suitable values of C_1, C and R. *Investigate* the noise source in different bandwidths. *Generate* the cdf curve of the noise source. From this result *obtain* the values of m and σ. *Compute* the overall gain of this circuit. Finally *find* the RMS value of V_n or I_n as described in Section 2.

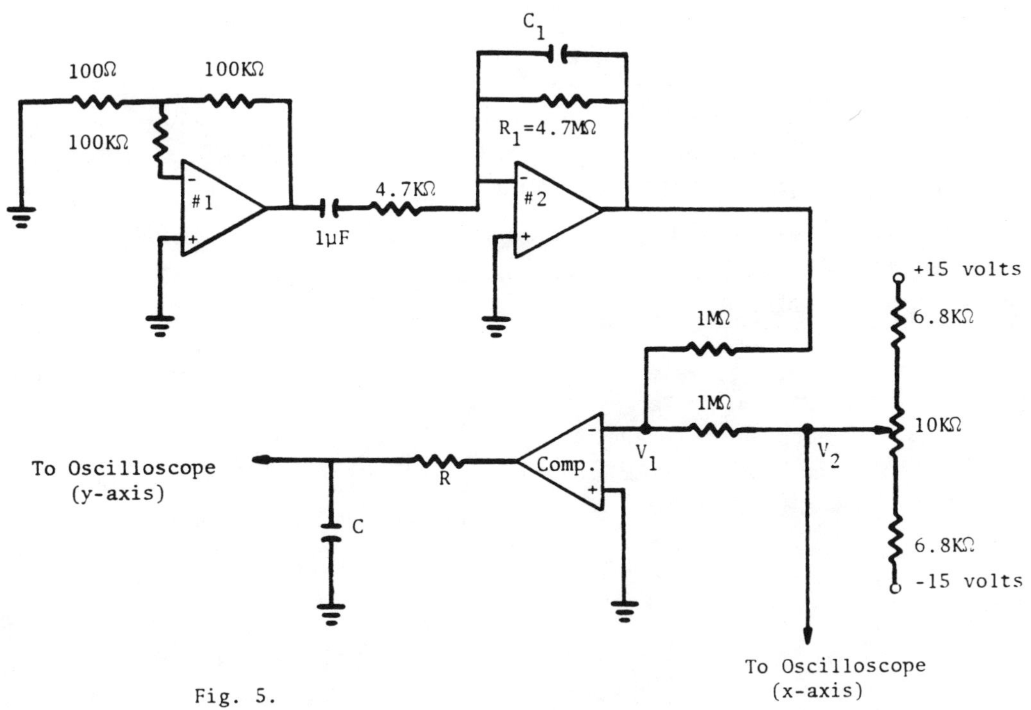

Fig. 5.

4. PROBLEM SET

1. What is the effect of output offset voltage of Op-Amp #2, on the mean value computed in the above experiment?

2. What is the effect of nonsymmetric output voltage swing in the comparator?

3. What is the theoretical value of m for this circuit configuration?

4. (Optional) What is the mathematical nature of the actual integration process that is used in the last stage of this experiment?

5. FOR YOUR REPORT

Submit your data in our standard report form with answers to the above questions.

EXPERIMENT IV-4

BASIC LINEAR APPLICATIONS OF FEEDBACK OP-AMP AND ANALOG COMPUTATIONS

EQUIPMENT:

EEP
OP-AMPs (3)
CAPACITORS (0.01 μF/2, 1 μF)
RESISTORS (1 kΩ/4, 10 kΩ/2, 100 kΩ, 4.7 MΩ)

Table of Contents Page No.

1. PURPOSE

The purpose of this experiment is to acquaint the student with basic feedback configurations using Op-Amps. Applications of some of these schemes in analog computations are also included.

2. INTRODUCTION

The theory behind this experiment is already described in Section III-2.2 and Section IV-6 of this textbook. Thus we urge our reader to refer to these two sections. Also the second problem of EXP. IV-1, should be looked into for additional preparation for this experiment.

3. LABORATORY PROCEDURE

3.1. PRE-LABORATORY PROCEDURE

Study the relevant theoretical materials needed for this experiment from Section III-2.2 and Section IV-6.

> *Warning: To avoid damaging the Op-Amp and to save time in each of the following circuit first turn off the power supply then only change R's and C's as being described.*

3.2. LABORATORY WORK

The next four tests are based on the circuit of Fig. 1. This is a simple inverting feedback amplifier.

3.2.1.   An Amplifier

Construct the circuit in Fig. 1, with $Z_1 = R_1 = 10\,k\Omega$ and $Z_f = R_2 = 100\,k\Omega$. *Choose* the DC supply voltage levels at ± 15 volts. *Apply* an input signal $v_{in}(t)$, such that its maximum is less than $|V_{supply}|/(Gain)$ to avoid saturation, on CH1 of the oscilloscope. *Observe* the output waveform on the CH2 of the oscilloscope. What is the gain of this feedback configuration circuit? *Keep* the amplitude of input signal low and *increase* its frequency to see the slew-rate effect on the output. Then *decrease* the frequency and *increase* the amplitude of the input signal to see its effect on the output.

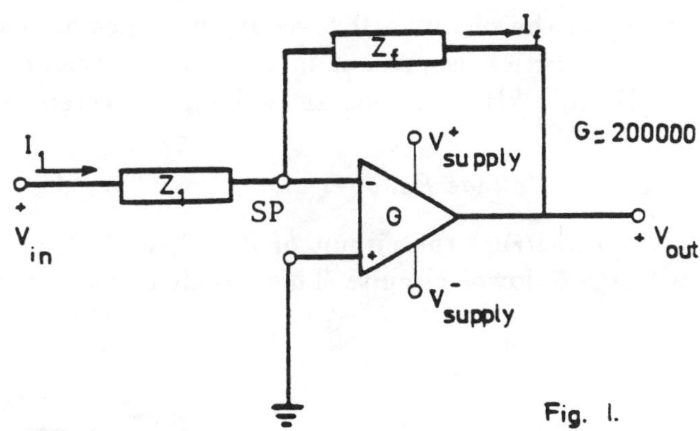

3.2.2.   An Adder

Connect an additional 10 kΩ resistor (there are now two) to the summing point of the Op-Amp to make an adder. Now, you can apply two input signals to the Op-Amp. *Use* your FG as one input source and *generate* another input signal from the PS or an auxiliary FG if available. *Choose* several different input signals and *observe* V_{out} on the oscilloscope.

3.2.3. An Integrator

Construct the circuit in Fig. 1, with $Z_1 = R_1 = 100$ kΩ and $Z_f = 1/sC$, C$=1.0$ μF. An integrator is initialized by shorting the capacitor to remove stored charge. *Use* a 1 kΩ resistor for this purpose to limit the current surge. When removing this resistor, you will probably observe the output gradually drifting away from zero volts, even though the input voltage is zero, *short* R_1 to ground to guarantee this point. Assuming zero capacitor leakage, the drift is due to non-ideal offset voltage and current in the Op-Amp. The output voltage will tend to drift to offset current I at a rate of I/C, so that even if I is only 10^{-8}A and C$=1$ μF, the output will drift to 1 volt in only 100 sec. Offset current is thus a prime consideration in choosing an Op-Amp as an integrator. Offset voltage can also produce drift, driving a current through the input resistor which has to be matched by a current through the feedback capacitor, creating output drift. *Observe* the rate of drift in the integrator. The drift is primarily due to offset current if the voltage offset is earlier annihilated carefully and if it has not drifted since. To improve the integrator circuit *put* a 4.7 MΩ resistor in parallel with C to reduce the low frequency gain of the integrator. *Observe* the integrator action for a variety of waveforms. *Try* a small constant voltage input and *verify* that the output rate is governed by the RC. *Observe* the output amplitude with a sinusoidal input as the frequency is varied. What is the relationship between input and output amplitude as a function of input frequency?

3.2.4. A Differentiator

Construct the circuit of Fig. 1, with $Z_f = R = 10$ kΩ in parallel with a capacitor of 100 pF and $Z_1 = 1/sC_1$, $C_1 = 0.1$ μF. *Apply* a small sinusoidal signal of about 100 Hz and insure that the amplifier is not saturating. Using this input, *verify* that the circuit has the frequency-dependent gain of a true differentiator. If the same RC components were used as a passive differentiator, at what frequency would it no longer differentiate? Now, *use* a square-wave input; *observe* and *describe* the output. If the output spikes appear to be non-ideal, do not neglect the possibility that this distortion is present in the input waveform. *Check* whether the circuit differentiates well from 10 Hz to 1 kHz. A constant spike is expected: is the width related to the input waveform?

3.2.5. Voltage Follower

Construct the circuit of Fig. 2, with $Z_1 = 1$ kΩ and $Z_f = 0$. *Verify* the property of a unity gain voltage follower circuit. This circuit is used as a buffer.

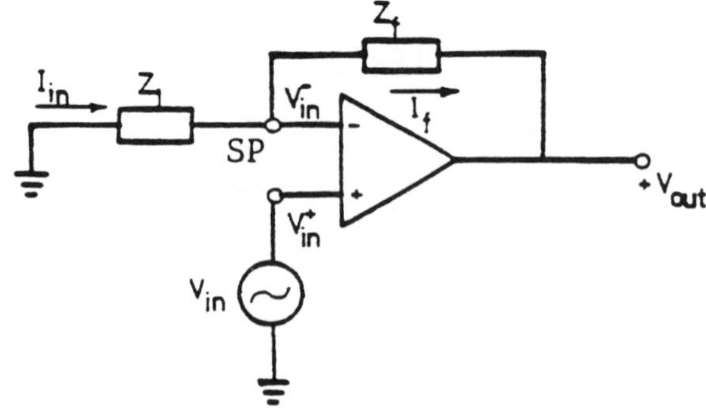

Fig. 2.

3.2.6. Analog Computation

Using the knowledge gained in the preceeding experiments *draw* a detailed simulation diagram to solve the following differential equations. Set all the initial conditions to zero. *Draw* this output x(t) of your circuit on the chart of Fig. 3.

$$\frac{d^2x}{dt^2} + 0.4\frac{dx}{dt} + x(t) = \sin 62.8t \ .$$

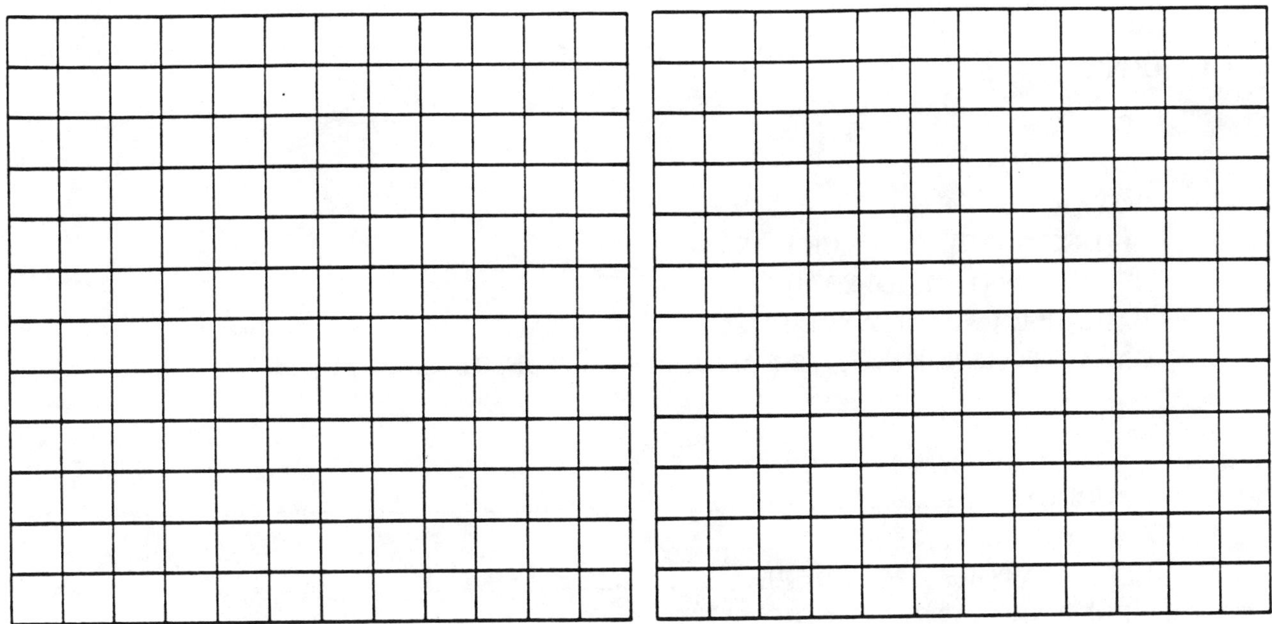

Fig. 3.

3.2.7.   Feedback Amplifier

Construct the circuit of Fig. 4, *to verify* and *to compare* both theoretically as well as experimentally the behavior of the $|V_o(j\omega)/V_{in}(j\omega)|$ for a reasonable range of frequency as determined from the slew rate figure of the Op-Amp. This circuit is similar to the second problem of EXP. IV-1.

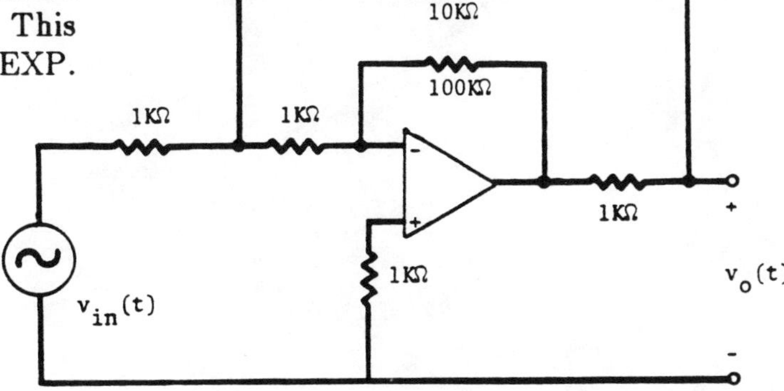

Fig. 4.

4. FOR YOUR REPORT

In your report give the necessary steps that must be taken to find $|V_o(j\omega)/V_{in}(j\omega)|$ for the circuit in Fig. 4, according to theory that is developed in Section III-2.2. Compare theoretical as well as experimental data for the circuit in Fig. 4. Then submit your report including these discussions in our standard report form.

EXPERIMENT IV-5

ACTIVE FIRST- AND SECOND-ORDER NETWORKS

EQUIPMENT:

    EEP
    OP-AMPs (3)
    POT (10 kΩ)
    GR-DECADE RESISTANCE BOX
    RESISTORS (1 kΩ, 100 kΩ/3)
    VARIABLE CAPACITANCE BOX
    CAPACITORS (0.001 μF/3, 0.1 μF, 0.22 μF, 0.47 μF, 0.68 μF, 1 μF)

Table of Contents   Page No.

1. PURPOSE

The purpose of this experiment is to acquaint the student with the basic properties of first-and second-order active networks.

2. INTRODUCTION

We have studied basic filter theory in EXP. II-4, using only passive elements. We have also studied some general theory behind active second-order networks in Section IV-5.3. In general we can design any type of filter using active elements. But the most important ones are the first-order and second-order active networks that play a fundamental role in the overall network synthesis with active elements. In this experiment we build one circuit for each of these two types of filters. For more information on this subject please refer to EXP. VI-5.

3. LABORATORY PROCEDURE

3.1. PRE-LABORATORY WORK

For circuits that are to be constructed in this experiment derive the transfer functions between input and output quantities as shown in each circuit. Also in the case of second-order networks find the relation between ς (and ω_n) and the network parameters. Decide which parameter and how much that must be changed in order to see a complete graph of second-order response as ς changes from zero to one.

Warning: Turn off the power supply before making any changes in your circuits.

3.2. LABORATORY WORK

In each of these experiments apply the input on CH1 and output on CH2 of your oscilloscope.

3.2.1.   Active First-Order Networks

Construct the active first-order circuits of Figs. 1 and 2, with ± 15 volts DC supply voltage. *Find* the Bode diagram of $|V_o(j\omega)/V_{in}(j\omega)|$ as a function of frequency. *Plot* this diagram on the chart of Fig. 3.

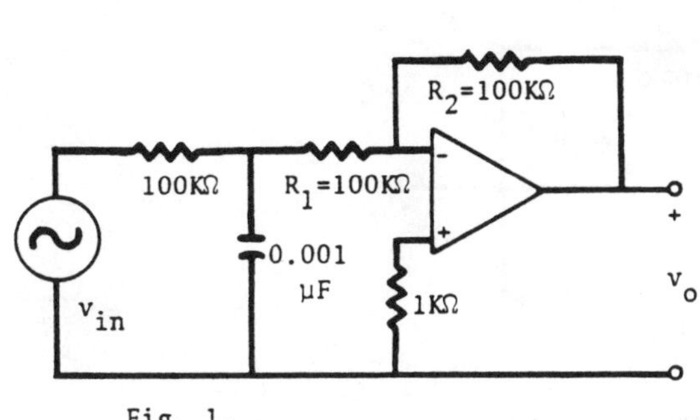

Fig. 1.

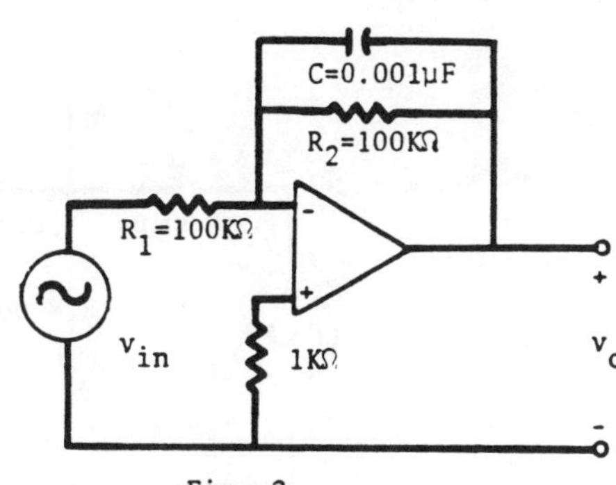

Fig. 2.

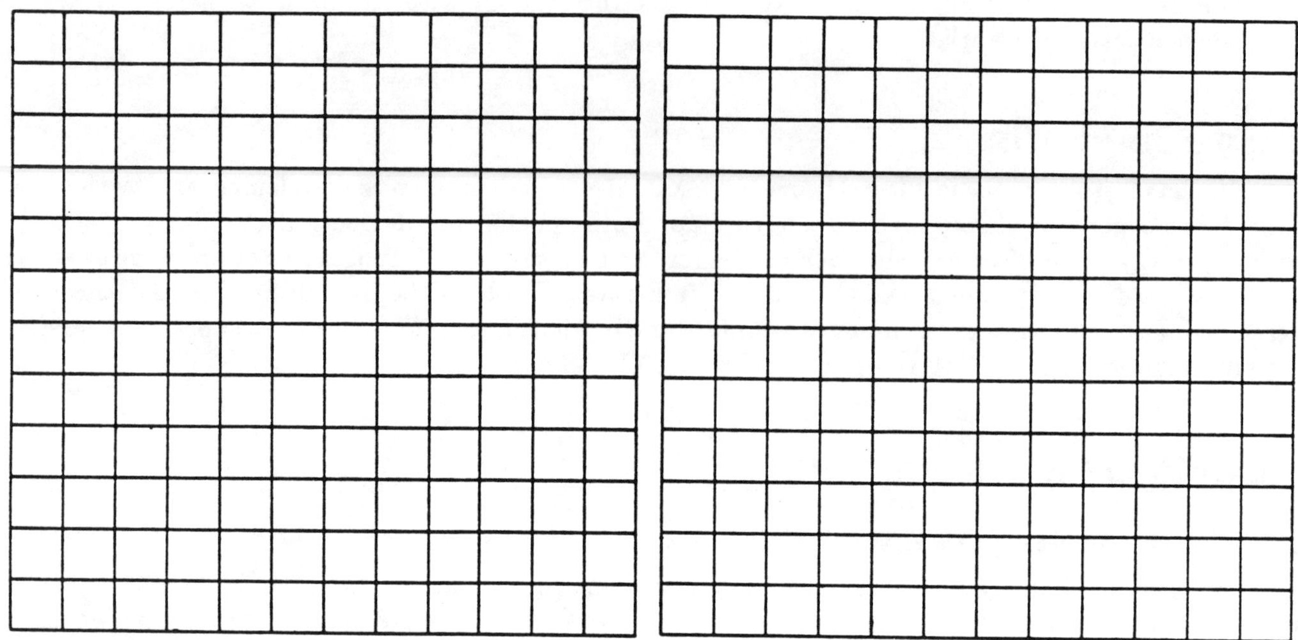

Fig. 3.

3.2.2. *Active Second-Order Networks*

Construct the standard low-pass circuit of Fig. 4, with ±15 volts DC supply voltage. *Change* C_1 or C_2, depending on your pre-laboratory computations, such that you can see several outputs of this second-order system for various input signals including unit step input. Here you may apply a low-frequency rectangular input instead of the unit-step input. Also *apply* several other input functions, as your FG permits, to this circuit and review the output waveforms. *Draw* these outputs on the chart of Fig. 6. *Compare* these with the outputs you expect theoretically.

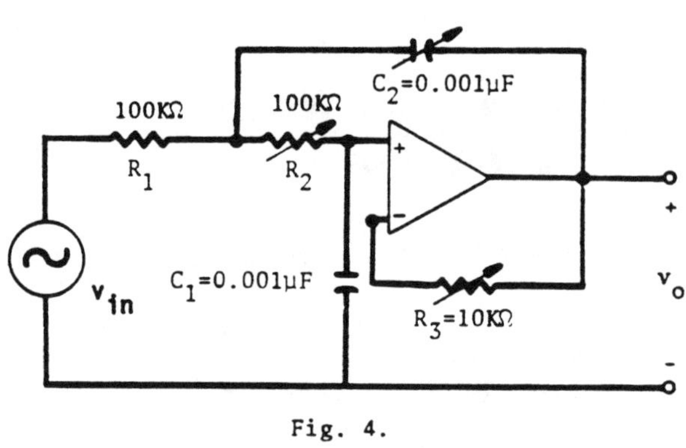

Fig. 4.

Repeat the above test for the high-pass filter of Fig. 5. *Use* the chart of Fig. 6 for your drawing.

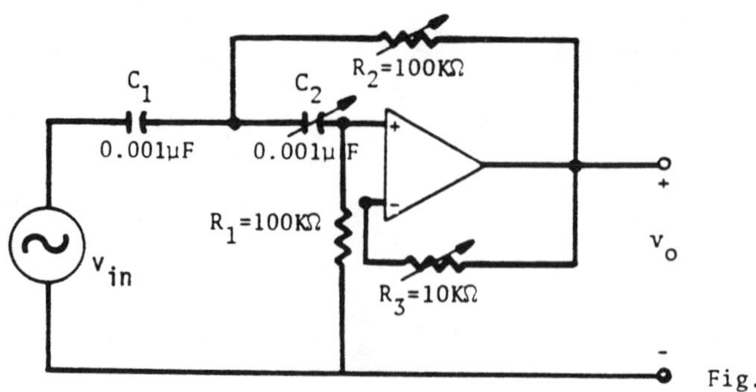

Fig. 5.

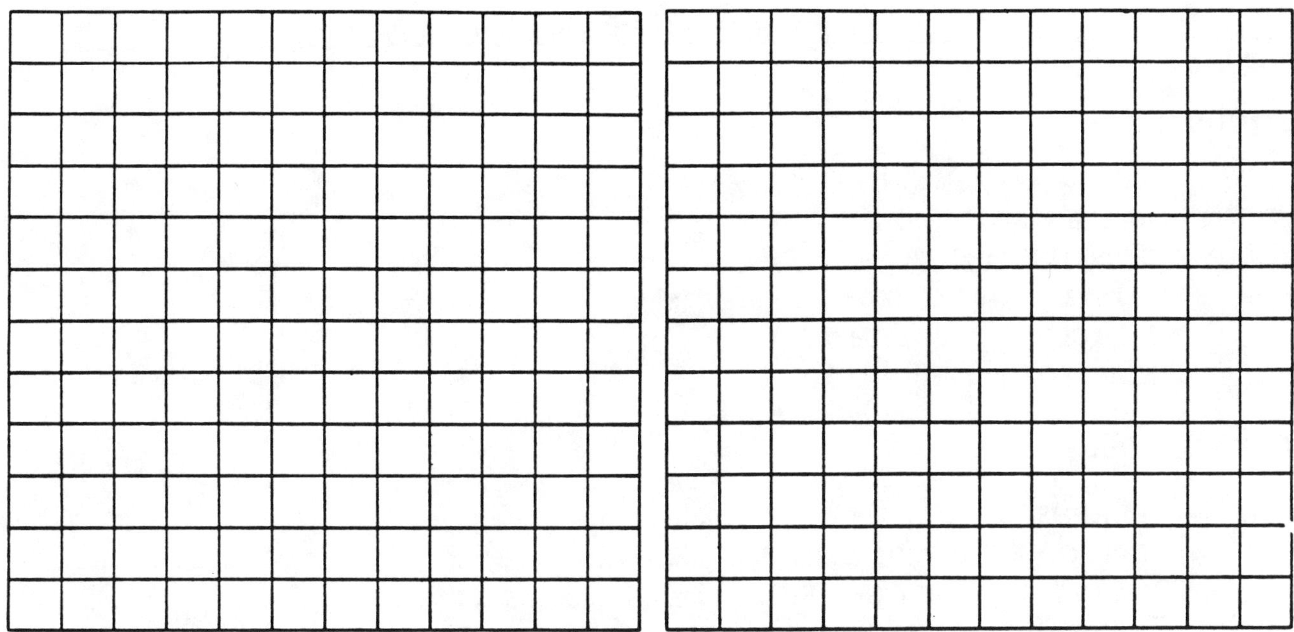

Fig. 6.

4. FOR YOUR REPORT

Please submit your data in our standard report form.

EXPERIMENT IV-6

NEGATIVE RESISTANCE CIRCUITS

EQUIPMENT:

EEP
OP-AMPs (3)
RESISTORS (1 kΩ/2)
POTS (1 kΩ, 10 TO 5000 Ω)
CAPACITORS (0.01 μF, 1.0 μF)
INDUCTOR (0.15 H)

Table of Contents

1. PURPOSE

The purpose of this experiment is to demonstrate how an Op-Amp can be used to generate a negative resistance in a dynamic sense.

2. INTRODUCTION

This discussion is originally due to Emeritus Professor J.B. Miller of the University of Wisconsin-Madison.

Consider the circuit in Fig. 1. It is easy to show that the input resistance of this circuit is

$$v/i = -R(R_1/R_2) \triangleq R_{in\,(active)} \cdot$$

For a large enough input voltage that saturates this circuit we get a characteristic curve similar to that of Fig. 2. Since the current uniquely determines the voltage, this is called a current- controlled negative resistance or CNR.

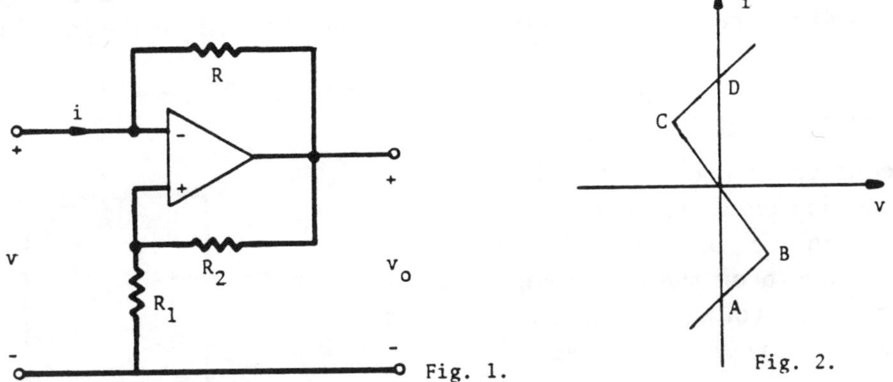

Fig. 1.

Fig. 2.

The reason for the plot of C through D or of B through A is because at B and C, the Op-Amp voltage saturates, the output voltage is now determined by the bias supply. If the current increases beyond its value at C the voltage will increase toward D and beyond.

If a capacitor is connected to the input of the circuit in Fig. 1, it will result in a negative time constant and thus instability. The voltage will tend toward C or B but neither of these cases are stable points because the current will not be zero. Thus in effect the current is jumping from one positive slope segment to another one at a constant voltage. The voltage will decay exponentially to B or C and the process continues. The circuit is that of an astable multivibrator.

If an inductor is used instead of the capacitor, then the current will be that of A or D and the voltage across the inductor will be zero. This is a bistable condition since either A or D currents satisfy the necessary conditions.

If a series combination of inductance and capacitance is connected at the input terminals, a steady current cannot exist because of the capacitor and again astable operation takes place. However, the waveform of the current of the series inductor-capacitor combination must in the limit be sinusoidal if no losses occur. This actually can be done by adjusting the negative resistance such that it cancels the resistance of the series inductor.

If the circuit of Fig. 1 is slightly changed to that of the circuit in Fig. 3, then the characteristic curve becomes that of Fig. 4. This is called a voltage-controlled negative resistance or VNR. Adding an inductor to the input terminals of circuit in Fig. 3 will result in astable conditions (oscillations), but a capacitor simply results in bistable operation.

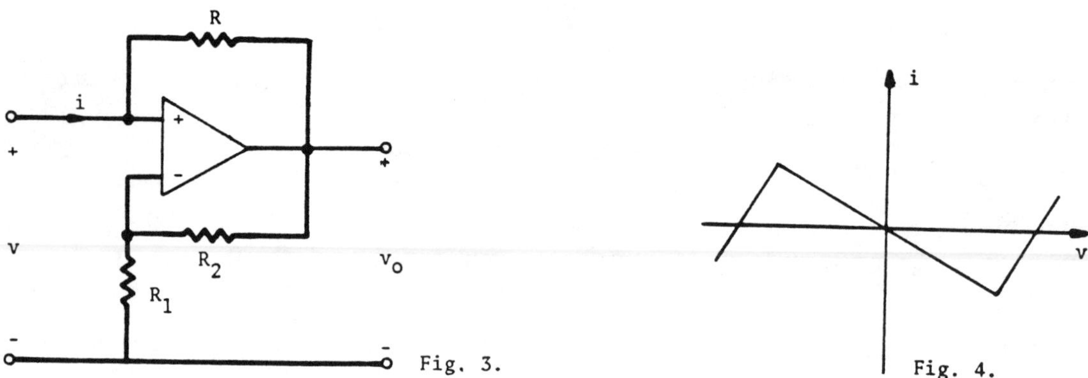

Fig. 3.

Fig. 4.

3. LABORATORY PROCEDURE

3.1. PRE-LABORATORY WORK

Design an alternative negative resistance circuit. Also analyze the circuits of Fig. 1 and Fig. 2, when both inductor and capacitor are connected to the input terminals as described in Section 2.

3.2. LABORATORY WORK

Construct the circuit of Fig. 5, with ±3 volts supply voltages. *Use* a sinusoidal input of 1 kHz. *Make* the necessary change in Fig. 5, so that the input V-I characteristic can be shown on the oscilloscope. *Measure* the slopes and the voltage boundaries of this characteristic curve *compare* with that expected from your pre-laboratory analysis.

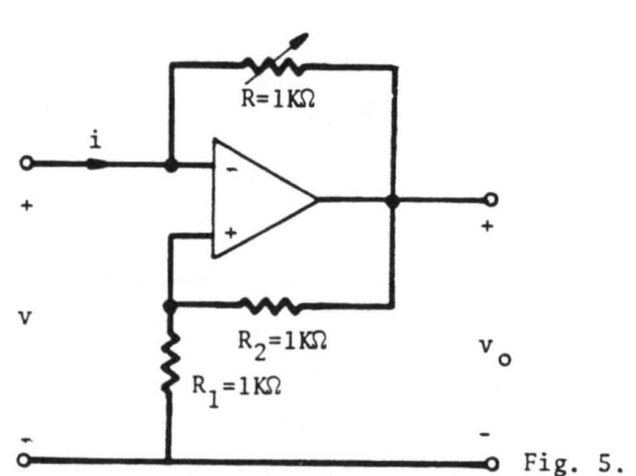

Fig. 5.

Connect the 1.0 μF capacitor across the input terminals. *Remove* the sinusoidal input. Does this circuit now act as an astable multivibrator? If not how can it be changed to make it work as on astable circuit? *Check* to see if you have an oscillation or not? You may have to change R to have oscillation. (To recall definitions of various multivibrator circuits please refer to EXP. VII-1).

Connect a series combination of L=0.15 H and C=0.01 μF at the input terminals. *Change* R with R_1 while the power supply is off. *Adjust* R_1 (which is now 1 kΩ pot) to get oscillation. If this oscillation is not sinusoidal *adjust* the negative resistance to cancel the resistance of the inductor. *Measure* the oscillation frequency and the Q of the inductor at this frequency.

Construct the circuit of Fig. 3, instead of Fig. 5, i.e., *change* the input terminals of the Op-Amp in Fig. 5. *Repeat* the above three steps for this circuit. *Discuss* your observations in regard to the differences between the performance of these two circuits.

4. FOR YOUR REPORT

Please submit your data and analysis in our standard report form.

WORKSHEET

WORKSHEET

INDEX

A

B

C

D

G

H

flyback time, 59
    retrace time, 59
    sawtooth signal, 58, 59
Hoyt, W.H., 41
Huelsman, L.P., 298, 524, 577

I

Immittance, 82
    matrix, 553
Inductor, 7
    quality factor (Q), 82, 83, 354, 554
Instrumentation Amplifier, 337
Integral Operator, 6
Integrated, 216
Integrator,
    Op-Amp, 312, 346
    RC circuits, 144

J

Jaeger, R.C., 657
Jung, W.G., 298

K

Karnaugh, 369
Kemmerly, J.E., 41
Kirchhoff's Laws, 5,6
    mesh and node equations, 6
Kohonen, T., 360
Korn, G.A., 298
Kouwenhoven, W.B., 4

L

Laboratory
    common procedures, 41
    conduct, 21
    general guidelines, 21
Laplace
    transform, 6
    variable, 6
Large-Scale Integration (LSI), 368
Leakage Current
    collector, 251
    emitter, 251
LED, 396
Lee, W.R., 4
Lim, K.Y., xiii
Linearity of Measurement, 633, 634
Linear Variable Differential Transformer, 636,
    637, 734
    moving core, 637
    ring demodulator, 638
    signal conditioner, 638 to 640
    translational, 637
Lissajous Figures, 64, 65
Logic Design, 359
Logic Design Accessories, 396
    pulser, 397
    pushbutton, 397
Logic Diagram Simplification, 368
Lowman, C.E., 103

M

Magnetic Material
    Coercive force, 165
    Coercivity, 165, 174, 176
    core excitation, 166
    degaussed, 173
    Deltamax, 168, 170
    Eddy-current, 166
    equalization, 171
    exciting current, 175
    Faraday's law, 166, 168

O

Octal, 362, 363
Ogata, K., 298
Ohmmeter, 46
    loading, 50
    mid-scale, 46
    range, 46
    series, 46
    use in circuits, 49
    voltage range extension, 48
Ohm's Law, 5
Operational Amplifier, 183, 297, 298, 317, 325
    CMRR, 304
    equivalent circuit, 300
    gain-bandwidth product, 303, 335
    ideal feedback model, 300
    ideal open-loop model, 299
    inverting, 300
    noise measurement, 340
    noninverting, 301
    practical feedback model, 305
    practical open-loop model, 301
    practical parameters, 331
    slew-rate, 304
    summing point, 300
    supply voltage, 299
    transient response, 304,305
    virtual ground, 300
Op-Amp Applications, 312
    adder, 313, 345
    amplifier, 345
    analog computation, 313, 346
    constant multiplier, 313
    differentiator, 312, 346
    feedback amplifier, 347
    integrator, 312, 346
    negative resistance, 353, 354
    summer, 313
    synthesis, 315
    voltage follower, 346
OR gate, 365
Oscillation, 190, 552
    Barkhausen criterion, 552
    Colpitts, 554
    frequency, 354, 553
    Hartley, 554
    oscillator circuits, 551, 553 to 555
    phase shift, 555
    Pierce crystal, 554
Oscilloscope, 44
    AC, Ground, DC switch, 56
    Cathode Ray Tube, 55
    control grid, 55
    coupling switch, 56
    display, 55
    dual trace, 56
    electron gun, 55
    general purpose, 55
    horizontal amplifier, 57
    horizontal deflection plates, 55
    intensity control switch, 55
    position switch, 56
    probe, 56
    rise-time measurement,
    sensitivity control switch, 56
    time-base amplifier, 57
    triggering, 60
    vertical amplifier, 56
    vertical deflection plates, 55
    x-y mode, 56
Output Meter, 73

P

Passive Network
    Brune, Bott and Duffin, 161
    Cauer's design, 161
    Foster's design, 159
    minimum function, 161
    multi-port, 157
    one-port, 157
    positive-real, 158
    second-order, 309
    synthesis, 156, 157
    Wattmeter connection, 85
Phase Shift Meter, 675, 676
p-n Junctions, 196, 197
    abrupt, 196
    built-in-electric field, 196
    electrostatic potential, 196
    energy diagram, 196
    Gauss's equation, 196
    impurity density profile, 196

Q

R

W

X

Y

Z